T0190362

Lecture Notes in Computer Science 13247

More information about this series at https://link.springer.com/bookseries/558

Lecture Notes in Computer Science 13247

More information about this series at https://link.springer.com/bookseries/558

Arnab Bhattacharya · Janice Lee Mong Li ·
Divyakant Agrawal · P. Krishna Reddy ·
Mukesh Mohania · Anirban Mondal ·
Vikram Goyal · Rage Uday Kiran (Eds.)

Database Systems for Advanced Applications

27th International Conference, DASFAA 2022
Virtual Event, April 11–14, 2022
Proceedings, Part III

Editors
Arnab Bhattacharya
Indian Institute of Technology Kanpur
Kanpur, India

Divyakant Agrawal
University of California, Santa Barbara
Santa Barbara, CA, USA

Mukesh Mohania
Indraprastha Institute of Information
Technology Delhi
New Delhi, India

Vikram Goyal
Indraprastha Institute of Information
Technology Delhi
New Delhi, India

Janice Lee Mong Li
National University of Singapore
Singapore, Singapore

P. Krishna Reddy 🆔
IIIT Hyderabad
Hyderabad, India

Anirban Mondal
Ashoka University
Sonepat, Haryana, India

Rage Uday Kiran
University of Aizu
Aizu, Japan

ISSN 0302-9743   ISSN 1611-3349 (electronic)
Lecture Notes in Computer Science
ISBN 978-3-031-00128-4   ISBN 978-3-031-00129-1 (eBook)
https://doi.org/10.1007/978-3-031-00129-1

This Springer imprint is published by the registered company Springer Nature Switzerland AG
The registered company address is: Gewerbestrasse 11, 6330 Cham, Switzerland

General Chairs' Preface

On behalf of the Organizing Committee, it is our great pleasure to welcome you to the proceedings of the 27th International Conference on Database Systems for Advanced Applications (DASFAA 2022), which was held during April 11–14, 2022, in Hyderabad, India. The conference has returned to India for the second time after a gap of 14 years, moving from New Delhi in 2008 to Hyderabad in 2022. DASFAA has long established itself as one of the leading international conferences in database systems. We were expecting to welcome you in person and give you a feel of our renowned Indian hospitality. However, unfortunately, given the Omicron wave of COVID-19 and the pandemic circumstances, we had to move the conference to a fully online mode.

Our gratitude goes first and foremost to the researchers, who submitted their work to the DASFAA 2022 main conference, workshops, and the data mining contest. We thank them for their efforts in submitting the papers, as well as in preparing high-quality online presentation videos. It is our distinct honor that five eminent keynote speakers graced the conference: Sunita Sarawagi of IIT Bombay, India, Guoliang Li of Tsinghua University, China, Gautam Das of the University of Texas at Arlington, Ioana Manolescu of Inria and Institut Polytechnique de Paris, and Tirthankar Lahiri of the Oracle Corporation. Each of them is a leader of international renown in their respective areas, and their participation significantly enhanced the conference. The conference program was further enriched with a panel, five high–quality tutorials, and six workshops on cutting-edge topics.

We would like to express our sincere gratitude to the contributions of the Senior Program Committee (SPC) members, Program Committee (PC) members, and anonymous reviewers, led by the PC chairs, Arnab Bhattacharya (IIT Kanpur), Lee Mong Li Janice (National University of Singapore), and Divyakant Agrawal (University of California, Santa Barbara). It is through their untiring efforts that the conference had an excellent technical program. We are also thankful to the other chairs and Organizing Committee members: industry track chairs, Prasad M. Deshpande (Google), Daxin Jiang (Microsoft), and Rajasekar Krishnamurthy (Adobe); demo track chairs, Rajeev Gupta (Microsoft), Koichi Takeda (Nagoya University), and Ladjel Bellatreche (ENSMA); workshop chairs, Maya Ramanath (IIT Delhi), Wookey Lee (Inha University), and Sanjay Kumar Madria (Missouri Institute of Technology); tutorial chairs, P. Sreenivasa Kumar (IIT Madras), Jixue Liu (University of South Australia), and Takahiro Hara (Osaka university); panel chairs, Jayant Haritsa (Indian Institute of Science), Reynold Cheng (University of Hong Kong), and Georgia Koutrika (Athena Research Center); Ph.D. consortium chairs, Vikram Pudi (IIIT Hyderabad), Srinath Srinivasa (IIIT Bangalore), and Philippe Fournier-Viger (Harbin Institute of Technology); publicity chairs, Raj Sharma (Goldman Sachs), Jamshid Bagherzadeh Mohasefi (Urmia University), and Nazha Selmaoui-Folcher (University of New Caledonia); publication chairs, Vikram Goyal (IIIT Delhi), and R. Uday Kiran (University of Aizu); and registration/local arrangement chairs, Lini Thomas (IIIT Hyderabad), Satish Narayana Srirama (University of Hyderabad), Manish Singh (IIT Hyderabad), P. Radha Krishna (NIT Warangal), Sonali Agrawal (IIIT Allahabad), and V. Ravi (IDRBT).

We appreciate the hosting organization IIIT Hyderabad, which is celebrating its silver jubilee in 2022. We thank the researchers at the Data Sciences and Analytics Center (DSAC) and the Kohli Center on Intelligent Systems (KCIS) at IIIT Hyderabad for their support. We also thank the administration and staff of IIIT Hyderabad for their help. We thank Google for the sponsorship. We feel indebted to the DASFAA Steering Committee for its continuing guidance.

Finally, our sincere thanks go to all the participants and volunteers. There would be no conference without them. We hope all of you enjoy these DASFAA 2022 proceedings.

February 2022   P. Krishna Reddy
                                                                 Mukesh Mohania
                                                                 Anirban Mondal

Program Chairs' Preface

It is our great pleasure to present the proceedings of the 27th International Conference on Database Systems for Advanced Applications (DASFAA 2022). DASFAA is a premier international forum for exchanging original research results and practical developments in the field of databases.

For the research track, we received 488 research submissions from across the world. We performed an initial screening of all submissions, leading to the desk rejection of 88 submissions due to violations of double-blind and page limit guidelines. For submissions entering the double-blind review process, each paper received at least three reviews from Program Committee (PC) members. Further, an assigned Senior Program Committee (SPC) member also led a discussion of the paper and reviews with the PC members. The PC co-chairs then considered the recommendations and meta-reviews from SPC members in making the final decisions. As a result, 72 submissions were accepted as full papers (acceptance ratio of 18%), and 76 submissions were accepted as short papers (acceptance ratio of 19%). For the industry track, 13 papers were accepted out of 36 submissions. Nine papers were accepted out of 16 submissions for the demo track. For the Ph.D. consortium, two papers were accepted out of three submissions. Four short research papers and one industry paper were withdrawn. The review process was supported by Microsoft's Conference Management Toolkit (CMT).

The conference was conducted in an online environment, with accepted papers presented via a pre-recorded video presentation with a live Q&A session. The conference program also featured five keynotes from distinguished researchers in the community, a panel, five high–quality tutorials, and six workshops on cutting-edge topics.

We wish to extend our sincere thanks to all SPC members, PC members, and external reviewers for their hard work in providing us with thoughtful and comprehensive reviews and recommendations. We especially thank the authors who submitted their papers to the conference. We hope that the readers of the proceedings find the content interesting, rewarding, and beneficial to their research.

March 2022

Arnab Bhattacharya
Janice Lee Mong Li
Divyakant Agrawal
Prasad M. Deshpande
Daxin Jiang
Rajasekar Krishnamurthy
Rajeev Gupta
Koichi Takeda
Ladjel Bellatreche
Vikram Pudi
Srinath Srinivasa
Philippe Fournier-Viger

Organization

DASFAA 2022 was organized by IIIT Hyderabad, Hyderabad, Telangana, India.

Steering Committee Chair

Lei Chen   Hong Kong University of Science and
              Technology, Hong Kong

Honorary Chairs

P. J. Narayanan   IIIT Hyderabad, India
S. Sudarshan   IIT Bombay, India
Masaru Kitsuregawa   University of Tokyo, Japan

Steering Committee Vice Chair

Stephane Bressan   National University of Singapore, Singapore

Steering Committee Treasurer

Yasushi Sakurai   Osaka University, Japan

Steering Committee Secretary

Kyuseok Shim   Seoul National University, South Korea

General Chairs

P. Krishna Reddy   IIIT Hyderabad, India
Mukesh Mohania   IIIT Delhi, India
Anirban Mondal   Ashoka University, India

Program Committee Chairs

Arnab Bhattacharya   IIT Kanpur, India
Lee Mong Li Janice   National University of Singapore, Singapore
Divyakant Agrawal   University of California, Santa Barbara, USA

Steering Committee

Zhiyong Peng	Wuhan University, China
Zhanhuai Li	Northwestern Polytechnical University, China
Krishna Reddy	IIIT Hyderabad, India
Yunmook Nah	Dankook University, South Korea
Wenjia Zhang	University of New South Wales, Australia
Zi Huang	University of Queensland, Australia
Guoliang Li	Tsinghua University, China
Sourav Bhowmick	Nanyang Technological University, Singapore
Atsuyuki Morishima	University of Tsukaba, Japan
Sang-Won Lee	Sungkyunkwan University, South Korea
Yang-Sae Moon	Kangwon National University, South Korea

Industry Track Chairs

Prasad M. Deshpande	Google, India
Daxin Jiang	Microsoft, China
Rajasekar Krishnamurthy	Adobe, USA

Demo Track Chairs

Rajeev Gupta	Microsoft, India
Koichi Takeda	Nagoya University, Japan
Ladjel Bellatreche	ENSMA, France

PhD Consortium Chairs

Vikram Pudi	IIIT Hyderabad, India
Srinath Srinivasa	IIIT Bangalore, India
Philippe Fournier-Viger	Harbin Institute of Technology, China

Panel Chairs

Jayant Haritsa	Indian Institute of Science, India
Reynold Cheng	University of Hong Kong, China
Georgia Koutrika	Athena Research Center, Greece

Sponsorship Chair

P. Krishna Reddy	IIIT Hyderabad, India

Publication Chairs

Vikram Goel   IIIT Delhi, India
R. Uday Kiran   University of Aizu, Japan

Workshop Chairs

Maya Ramanath   IIT Delhi, India
Wookey Lee   Inha University, South Korea
Sanjay Kumar Madria   Missouri Institute of Technology, USA

Tutorial Chairs

P. Sreenivasa Kumar   IIT Madras, India
Jixue Liu   University of South Australia, Australia
Takahiro Hara   Osaka University, Japan

Publicity Chairs

Raj Sharma   Goldman Sachs, India
Jamshid Bagherzadeh Mohasefi   Urmia University, Iran
Nazha Selmaoui-Folcher   University of New Caledonia, New Caledonia

Organizing Committee

Lini Thomas   IIIT Hyderabad, India
Satish Narayana Srirama   University of Hyderabad, India
Manish Singh   IIT Hyderabad, India
P. Radha Krishna   NIT Warangal, India
Sonali Agrawal   IIIT Allahabad, India
V. Ravi   IDRBT, India

Senior Program Committee

Avigdor Gal   Technion - Israel Institute of Technology, Israel
Baihua Zheng   Singapore Management University, Singapore
Bin Cui   Peking University, China
Bin Yang   Aalborg University, Denmark
Bingsheng He   National University of Singapore, Singapore
Chang-Tien Lu   Virginia Tech, USA
Chee-Yong Chan   National University of Singapore, Singapore
Gautam Shroff   Tata Consultancy Services Ltd., India
Hong Gao   Harbin Institute of Technology, China

Jeffrey Xu Yu	Chinese University of Hong Kong, China
Jianliang Xu	Hong Kong Baptist University, China
Jianyong Wang	Tsinghua University, China
Kamalakar Karlapalem	IIIT Hyderabad, India
Kian-Lee Tan	National University of Singapore, Singapore
Kyuseok Shim	Seoul National University, South Korea
Ling Liu	Georgia Institute of Technology, USA
Lipika Dey	Tata Consultancy Services Ltd., India
Mario Nascimento	University of Alberta, Canada
Maya Ramanath	IIT Delhi, India
Mohamed Mokbel	University of Minnesota, Twin Cities, USA
Niloy Ganguly	IIT Kharagpur, India
Sayan Ranu	IIT Delhi, India
Sourav S. Bhowmick	Nanyang Technological University, Singapore
Srikanta Bedathur	IIT Delhi, India
Srinath Srinivasa	IIIT Bangalore, India
Stephane Bressan	National University of Singapore, Singapore
Tok W. Ling	National University of Singapore, Singapore
Vana Kalogeraki	Athens University of Economics and Business, Greece
Vassilis J. Tsotras	University of California, Riverside, USA
Vikram Pudi	IIIT Hyderabad, India
Vincent Tseng	National Yang Ming Chiao Tung University, Taiwan
Wang-Chien Lee	Pennsylvania State University, USA
Wei-Shinn Ku	Auburn University, USA
Wenjie Zhang	University of New South Wales, Australia
Wynne Hsu	National University of Singapore, Singapore
Xiaofang Zhou	Hong Kong University of Science and Technology, China
Xiaokui Xiao	National University of Singapore, Singapore
Xiaoyong Du	Renmin University of China, China
Yoshiharu Ishikawa	Nagoya University, Japan
Yufei Tao	Chinese University of Hong Kong, China

Program Committee

Abhijnan Chakraborty	IIT Delhi, India
Ahmed Eldawy	University of California, Riverside, USA
Akshar Kaul	IBM Research, India
Alberto Abell	Universitat Politecnica de Catalunya, Spain
An Liu	Soochow University, China
Andrea Cali	Birkbeck, University of London, UK

Andreas Züfle	George Mason University, USA
Antonio Corral	University of Almeria, Spain
Atsuhiro Takasu	National Institute of Informatics, Japan
Bin Wang	Northeastern University, China
Bin Yao	Shanghai Jiao Tong University, China
Bo Jin	Dalian University of Technology, China
Bolong Zheng	Huazhong University of Science and Technology, China
Chandramani Chaudhary	National Institute of Technology, Trichy, India
Changdong Wang	Sun Yat-sen University, China
Chaokun Wang	Tsinghua University, China
Cheng Long	Nanyang Technological University, Singapore
Chenjuan Guo	Aalborg University, Denmark
Cheqing Jin	East China Normal University, China
Chih-Ya Shen	National Tsing Hua University, Taiwan
Chittaranjan Hota	BITS Pilani, India
Chi-Yin Chow	Social Mind Analytics (Research and Technology) Limited, Hong Kong
Chowdhury Farhan Ahmed	University of Dhaka, Bangladesh
Christos Doulkeridis	University of Pireaus, Greece
Chuan Xiao	Osaka University and Nagoya University, Japan
Cindy Chen	University of Massachusetts Lowell, USA
Cuiping Li	Renmin University of China, China
Dan He	University of Queensland, Australia
Demetrios Zeinalipour-Yazti	University of Cyprus, Cyprus
De-Nian Yang	Academia Sinica, Taiwan
Dhaval Patel	IBM TJ Watson Research Center, USA
Dieter Pfoser	George Mason University, USA
Dimitrios Kotzinos	University of Cergy-Pontoise, France
Fan Zhang	Guangzhou University, China
Ge Yu	Northeast University, China
Goce Trajcevski	Iowa State University, USA
Guoren Wang	Beijing Institute of Technology, China
Haibo Hu	Hong Kong Polytechnic University, China
Haruo Yokota	Tokyo Institute of Technology, Japan
Hiroaki Shiokawa	University of Tsukuba, Japan
Hongzhi Wang	Harbin Institute of Technology, China
Hongzhi Yin	University of Queensland, Australia
Hrishikesh R. Terdalkar	IIT Kanpur, India
Hua Lu	Roskilde University, Denmark
Hui Li	Xidian University, China
Ioannis Konstantinou	University of Thessaly, Greece

Iouliana Litou	Athens University of Economics and Business, Greece
Jagat Sesh Challa	BITS Pilani, India
Ja-Hwung Su	Cheng Shiu University, Taiwan
Jiali Mao	East China Normal University, China,
Jia-Ling Koh	National Taiwan Normal University, Taiwan
Jian Dai	Alibaba Group, China
Jianqiu Xu	Nanjing University of Aeronautics and Astronautics, China
Jianxin Li	Deakin University, Australia
Jiawei Jiang	ETH Zurich, Switzerland
Jilian Zhang	Jinan University, China
Jin Wang	Megagon Labs, USA
Jinfei Liu	Zhejiang University, China
Jing Tang	Hong Kong University of Science and Technology, China
Jinho Kim	Kangwon National University, South Korea
Jithin Vachery	National University of Singapore, Singapore
Ju Fan	Renmin University of China, China
Jun Miyazaki	Tokyo Institute of Technology, Japan
Junjie Yao	East China Normal University, China
Jun-Ki Min	Korea University of Technology and Education, South Korea
Kai Zeng	Alibaba Group, China
Karthik Ramachandra	Microsoft Azure SQL, India
Kento Sugiura	Nagoya University, Japan
Kesheng Wu	Lawrence Berkeley National Laboratory, USA
Kjetil Nørvåg	Norwegian University of Science and Technology, Norway
Kostas Stefanidis	Tempere University, Finland
Kripabandhu Ghosh	Indian Institute of Science Education and Research Kolkata, India
Kristian Torp	Aalborg University, Denmark
Kyoung-Sook Kim	Artificial Intelligence Research Center, Japan
Ladjel Bellatreche	ENSMA, France
Lars Dannecker	SAP, Germany
Lee Roy Ka Wei	Singapore University of Technology and Design, Singapore
Lei Cao	Massachusetts Institute of Technology, USA
Leong Hou U.	University of Macau, China
Lijun Chang	University of Sydney, Australia
Lina Yao	University of New South Wales Australia
Lini Thomas	IIIT Hyderabad, India

Liping Wang	East China Normal University, China
Long Yuan	Nanjing University of Science and Technology, China
Lu-An Tang	NEC Labs America, USA
Makoto Onizuka	Osaka University, Japan
Manish Kesarwani	IBM Research, India
Manish Singh	IIT Hyderabad, India
Manolis Koubarakis	University of Athens, Greece
Marco Mesiti	University of Milan, Italy
Markus Schneider	University of Florida, USA
Meihui Zhang	Beijing Institute of Technology, China
Meng-Fen Chiang	University of Auckland, New Zealand
Mirella M. Moro	Universidade Federal de Minas Gerais, Brazil
Mizuho Iwaihara	Waseda University, Japan
Navneet Goyal	BITS Pilani, India
Neil Zhenqiang Gong	Iowa State University, USA
Nikos Ntarmos	Huawei Technologies R&D (UK) Ltd., UK
Nobutaka Suzuki	University of Tsukuba, Japan
Norio Katayama	National Institute of Informatics, Japan
Noseong Park	George Mason University, USA
Olivier Ruas	Inria, France
Oscar Romero	Universitat Politècnica de Catalunya, Spain
Oswald C.	IIT Kanpur, India
Panagiotis Bouros	Johannes Gutenberg University Mainz, Germany
Parth Nagarkar	New Mexico State University, USA
Peer Kroger	Christian-Albrecht University of Kiel, Germany
Peifeng Yin	Pinterest, USA
Peng Wang	Fudan University, China
Pengpeng Zhao	Soochow University, China
Ping Lu	Beihang University, China
Pinghui Wang	Xi'an Jiaotong University, China
Poonam Goyal	BITS Pilani, India
Qiang Yin	Shanghai Jiao Tong University, China
Qiang Zhu	University of Michigan – Dearborn, USA
Qingqing Ye	Hong Kong Polytechnic University, China
Rafael Berlanga Llavori	Universitat Jaume I, Spain
Rage Uday Kiran	University of Aizu, Japan
Raghava Mutharaju	IIIT Delhi, India
Ravindranath C. Jampani	Oracle Labs, India
Rui Chen	Samsung Research America, USA
Rui Zhou	Swinburne University of Technology, Australia
Ruiyuan Li	Xidian University, China

Sabrina De Capitani di Vimercati	Università degli Studi di Milano, Italy
Saiful Islam	Griffith University, Australia
Sanghyun Park	Yonsei University, South Korea
Sanjay Kumar Madria	Missouri University of Science and Technology, USA
Saptarshi Ghosh	IIT Kharagpur, India
Sebastian Link	University of Auckland, New Zealand
Shaoxu Song	Tsinghua University, China
Sharma Chakravarthy	University of Texas at Arlington, USA
Shiyu Yang	Guangzhou University, China
Shubhadip Mitra	Tata Consultancy Services Ltd., India
Shubhangi Agarwal	IIT Kanpur, India
Shuhao Zhang	Singapore University of Technology and Design, Singapore
Sibo Wang	Chinese University of Hong Kong, China
Silviu Maniu	Université Paris-Saclay, France
Sivaselvan B.	IIIT Kancheepuram, India
Stephane Bressan	National University of Singapore, Singapore
Subhajit Sidhanta	IIT Bhilai, India
Sungwon Jung	Sogang University, South Korea
Tanmoy Chakraborty	Indraprastha Institute of Information Technology Delhi, India
Theodoros Chondrogiannis	University of Konstanz, Germany
Tien Tuan Anh Dinh	Singapore University of Technology and Design, Singapore
Ting Deng	Beihang University, China
Tirtharaj Dash	BITS Pilani, India
Toshiyuki Amagasa	University of Tsukuba, Japan
Tsz Nam (Edison) Chan	Hong Kong Baptist University, China
Venkata M. Viswanath Gunturi	IIT Ropar, India
Verena Kantere	National Technical University of Athens, Greece
Vijaya Saradhi V.	IIT Guwahati, India
Vikram Goyal	IIIT Delhi, India
Wei Wang	Hong Kong University of Science and Technology (Guangzhou), China
Weiwei Sun	Fudan University, China
Weixiong Rao	Tongji University, China
Wen Hua	University of Queensland, Australia
Wenchao Zhou	Georgetown University, USA
Wentao Zhang	Peking University, China
Werner Nutt	Free University of Bozen-Bolzano, Italy
Wolf-Tilo Balke	TU Braunschweig, Germany

Wookey Lee	Inha University, South Korea
Woong-Kee Loh	Gacheon University, South Korea
Xiang Lian	Kent State University, USA
Xiang Zhao	National University of Defence Technology, China
Xiangmin Zhou	RMIT University, Australia
Xiao Pan	Shijiazhuang Tiedao University, China
Xiao Qin	Amazon Web Services, USA
Xiaochun Yang	Northeastern University, China
Xiaofei Zhang	University of Memphis, USA
Xiaofeng Gao	Shanghai Jiao Tong University, China
Xiaowang Zhang	Tianjin University, China
Xiaoyang Wang	Zhejiang Gongshang University, China
Xin Cao	University of New South Wales, Australia
Xin Huang	Hong Kong Baptist University, China
Xin Wang	Tianjin University, China
Xu Xie	Peking University, China
Xuequn Shang	Northwestern Polytechnical University, China
Xupeng Miao	Peking University, China
Yan Shi	Shanghai Jiao Tong University, China
Yan Zhang	Peking University, China
Yang Cao	Kyoto University, Japan
Yang Chen	Fudan University, China
Yanghua Xiao	Fudan University, China
Yang-Sae Moon	Kangwon National University, South Korea
Yannis Manolopoulos	Aristotle University of Thessaloniki, Greece
Yi Yu	National Institute of Informatics, Japan
Yingxia Shao	Beijing University of Posts and Telecommunication, China
Yixiang Fang	Chinese University of Hong Kong, China
Yong Tang	South China Normal University, China
Yongxin Tong	Beihang University, China
Yoshiharu Ishikawa	Nagoya University, Japan
Yu Huang	National Yang Ming Chiao Tung University, Taiwan
Yu Suzuki	Gifu University, Japan
Yu Yang	City University of Hong Kong, China
Yuanchun Zhou	Computer Network Information Center, China
Yuanyuan Zhu	Wuhan University, China
Yun Peng	Hong Kong Baptist University, China
Yuqing Zhu	California State University, Los Angeles, USA
Zeke Wang	Zhejiang University, China

Zhaojing Luo	National University of Singapore, Singapore
Zhenying He	Fudan University, China
Zhi Yang	Peking University, China
Zhixu Li	Soochow University, China
Zhiyong Peng	Wuhan University, China
Zhongnan Zhang	Xiamen University, China

Industry Track Program Committee

Karthik Ramachandra	Microsoft, India
Akshar Kaul	IBM Research, India
Sriram Lakshminarasimhan	Google Research, India
Rajat Venkatesh	LinkedIn, India
Prasan Roy	Sclera, India
Zhicheng Dou	Renmin University of China, China
Huang Hu	Microsoft, China
Shan Li	LinkedIn, USA
Bin Gao	Facebook, USA
Haocheng Wu	Facebook, USA
Shivakumar Vaithyanathan	Adobe, USA
Abdul Quamar	IBM Research, USA
Pedro Bizarro	Feedzai, Portugal
Xi Yin	International Digital Economy Academy, China
Xiangyu Niu	Facebook

Demo Track Program Committee

Ahmed Awad	University of Tartu, Estonia
Beethika Tripathi	Microsoft, India
Carlos Ordonez	University of Houston, USA
Djamal Benslimane	Université Claude Bernard Lyon 1, France
Nabila Berkani	Ecole Nationale Supérieure d'Informatique, Algeria
Philippe Fournier-Viger	Shenzhen University, China
Ranganath Kondapally	Microsoft, India
Soumia Benkrid	Ecole Nationale Supérieure d'Informatique, Algeria

Sponsoring Institutions

Google, India

INTERNATIONAL INSTITUTE OF
INFORMATION TECHNOLOGY
H Y D E R A B A D

IIIT Hyderabad, India

Contents – Part III

Text and Image Processing

Emotion-Aware Multimodal Pre-training for Image-Grounded Emotional
Response Generation 3
    Zhiliang Tian, Zhihua Wen, Zhenghao Wu, Yiping Song, Jintao Tang,
    Dongsheng Li, and Nevin L. Zhang

Information Networks Based Multi-semantic Data Embedding for Entity
Resolution 20
    Chenchen Sun, Derong Shen, and Tiezheng Nie

Semantic-Based Data Augmentation for Math Word Problems 36
    Ailisi Li, Yanghua Xiao, Jiaqing Liang, and Yunwen Chen

Empowering Transformer with Hybrid Matching Knowledge for Entity
Matching 52
    Wenzhou Dou, Derong Shen, Tiezheng Nie, Yue Kou, Chenchen Sun,
    Hang Cui, and Ge Yu

Tracking the Evolution: Discovering and Visualizing the Evolution
of Literature 68
    Siyuan Wu and Leong Hou U

Incorporating Commonsense Knowledge into Story Ending Generation
via Heterogeneous Graph Networks 85
    Jiaan Wang, Beiqi Zou, Zhixu Li, Jianfeng Qu, Pengpeng Zhao, An Liu,
    and Lei Zhao

Open-Domain Dialogue Generation Grounded with Dynamic Multi-form
Knowledge Fusion 101
    Feifei Xu, Shanlin Zhou, Yunpu Ma, Xinpeng Wang, Wenkai Zhang,
    and Zhisong Li

KdTNet: Medical Image Report Generation via Knowledge-Driven
Transformer 117
    Yiming Cao, Lizhen Cui, Fuqiang Yu, Lei Zhang, Zhen Li, Ning Liu,
    and Yonghui Xu

Fake Restaurant Review Detection Using Deep Neural Networks
with Hybrid Feature Fusion Method 133
    Yifei Jian, Xingshu Chen, and Haizhou Wang

Aligning Internal Regularity and External Influence of Multi-granularity
for Temporal Knowledge Graph Embedding 149
  Tingyi Zhang, Zhixu Li, Jiaan Wang, Jianfeng Qu, Lin Yuan, An Liu,
  Lei Zhao, and Zhigang Chen

AdCSE: An Adversarial Method for Contrastive Learning of Sentence
Embeddings 165
  Renhao Li, Lei Duan, Guicai Xie, Shan Xiao, and Weipeng Jiang

HRG: A Hybrid Retrieval and Generation Model in Multi-turn Dialogue 181
  Deji Zhao, Xinyi Liu, Bo Ning, and Chengfei Liu

FALCON: A Faithful Contrastive Framework for Response Generation
in TableQA Systems 197
  Shineng Fang, Jiangjie Chen, Xinyao Shen, Yunwen Chen,
  and Yanghua Xiao

Tipster: A Topic-Guided Language Model for Topic-Aware Text
Segmentation 213
  Zheng Gong, Shiwei Tong, Han Wu, Qi Liu, Hanqing Tao, Wei Huang,
  and Runlong Yu

SimEmotion: A Simple Knowledgeable Prompt Tuning Method for Image
Emotion Classification 222
  Sinuo Deng, Ge Shi, Lifang Wu, Lehao Xing, Wenjin Hu, Heng Zhang,
  and Ye Xiang

Predicting Rumor Veracity on Social Media with Graph Structured
Multi-task Learning 230
  Yudong Liu, Xiaoyu Yang, Xi Zhang, Zhihao Tang, Zongyi Chen,
  and Zheng Liwen

Knowing What I Don't Know: A Generation Assisted Rejection
Framework in Knowledge Base Question Answering 238
  Junyang Huang, Xuantao Lu, Jiaqing Liang, Qiaoben Bao,
  Chen Huang, Yanghua Xiao, Bang Liu, and Yunwen Chen

Medical Image Fusion Based on Pixel-Level Nonlocal Self-similarity
Prior and Optimization 247
  Rui Zhu, Xiongfei Li, Yu Wang, and Xiaoli Zhang

Knowledge-Enhanced Interactive Matching Network for Multi-turn
Response Selection in Medical Dialogue Systems 255
  Ying Zhu, Shi Feng, Daling Wang, Yifei Zhang, and Donghong Han

KAAS: A Keyword-Aware Attention Abstractive Summarization Model
for Scientific Articles 263
  Shuaimin Li and Jungang Xu

E-Commerce Knowledge Extraction via Multi-modal Machine Reading
Comprehension 272
  Chaoyu Bai

PERM: Pre-training Question Embeddings via Relation Map for Improving
Knowledge Tracing 281
  Wentao Wang, Huifang Ma, Yan Zhao, Fanyi Yang, and Liang Chang

A Three-Stage Curriculum Learning Framework with Hierarchical Label
Smoothing for Fine-Grained Entity Typing 289
  *Bo Xu, Zhengqi Zhang, Chaofeng Sha, Ming Du, Hui Song,
  and Hongya Wang*

PromptMNER: Prompt-Based Entity-Related Visual Clue Extraction
and Integration for Multimodal Named Entity Recognition 297
  *Xuwu Wang, Junfeng Tian, Min Gui, Zhixu Li, Jiabo Ye, Ming Yan,
  and Yanghua Xiao*

TaskSum: Task-Driven Extractive Text Summarization for Long News
Documents Based on Reinforcement Learning 306
  *Moming Tang, Dawei Cheng, Cen Chen, Yuqi Liang, Yifeng Luo,
  and Weining Qian*

Concurrent Transformer for Spatial-Temporal Graph Modeling 314
  *Yi Xie, Yun Xiong, Yangyong Zhu, Philip S. Yu, Cheng Jin, Qiang Wang,
  and Haihong Li*

Towards Personalized Review Generation with Gated Multi-source Fusion
Network 322
  *Hongtao Liu, Wenjun Wang, Hongyan Xu, Qiyao Peng, Pengfei Jiao,
  and Yueheng Sun*

Definition-Augmented Jointly Training Framework for Intention Phrase
Mining 331
  *Denghao Ma, Yueguo Chen, Changyu Wang, Hongbin Pei, Yitao Zhai,
  Gang Zheng, and Qi Chen*

Modeling Uncertainty in Neural Relation Extraction 340
  Yu Hong, Yanghua Xiao, Wei Wang, and Yunwen Chen

Industry Papers

A Joint Framework for Explainable Recommendation with Knowledge
Reasoning and Graph Representation 351
  Luhao Zhang, Ruiyu Fang, Tianchi Yang, Maodi Hu, Tao Li, Chuan Shi,
  and Dong Wang

XDM: Improving Sequential Deep Matching with Unclicked User
Behaviors for Recommender System 364
  Fuyu Lv, Mengxue Li, Tonglei Guo, Changlong Yu, Fei Sun, Taiwei Jin,
  and Wilfred Ng

Mitigating Popularity Bias in Recommendation via Counterfactual
Inference 377
  Ming He, Changshu Li, Xinlei Hu, Xin Chen, and Jiwen Wang

Efficient Dual-Process Cognitive Recommender Balancing Accuracy
and Diversity 389
  Yixu Gao, Kun Shao, Zhijian Duan, Zhongyu Wei, Dong Li, Bin Wang,
  Mengchen Zhao, and Jianye Hao

Learning and Fusing Multiple User Interest Representations for Sequential
Recommendation 401
  Ming He, Tianshuo Han, and Tianyu Ding

Query-Document Topic Mismatch Detection 413
  Sahil Chelaramani, Ankush Chatterjee, Sonam Damani,
  Kedhar Nath Narahari, Meghana Joshi, Manish Gupta,
  and Puneet Agrawal

Beyond QA: 'Heuristic QA' Strategies in JIMI 425
  Shuangyong Song, Bo Zou, Jianghua Lin, Xiaoguang Yu,
  and Xiaodong He

SQLG+: Efficient *k*-hop Query Processing on RDBMS 430
  Li Zeng, Jinhua Zhou, Shijun Qin, Haoran Cai, Rongqian Zhao,
  and Xin Chen

Modeling Long-Range Travelling Times with Big Railway Data 443
  Wenya Sun, Tobias Grubenmann, Reynold Cheng, Ben Kao,
  and Waiki Ching

Multi-scale Time Based Stock Appreciation Ranking Prediction via Price
Co-movement Discrimination 455
  Ruyao Xu, Dawei Cheng, Cen Chen, Siqiang Luo, Yifeng Luo,
  and Weining Qian

RShield: A Refined Shield for Complex Multi-step Attack Detection
Based on Temporal Graph Network 468
Weiyong Yang, Peng Gao, Hao Huang, Xingshen Wei, Wei Liu,
Shishun Zhu, and Wang Luo

Inter-and-Intra Domain Attention Relational Inference for Rack
Temperature Prediction in Data Center 481
Fang Shen, Zhan Li, Bing Pan, Ziwei Zhang, Jialong Wang,
Wendy Zhao, Xin Wang, and Wenwu Zhu

DEMO Papers

An Interactive Data Imputation System 495
Yangyang Wu, Xiaoye Miao, Yuchen Peng, Lu Chen, Yunjun Gao,
and Jianwei Yin

FoodChain: A Food Delivery Platform Based on Blockchain for Keeping
Data Privacy 500
Rodrigo Folha, Valéria Times, Arthur Carvalho, André Araújo,
Henrique Couto, and Flaviano Viana

A Scalable Lightweight RDF Knowledge Retrieval System 505
Yuming Lin, Chuangxin Fang, Youjia Jiang, and You Li

CO-AutoML: An Optimizable Automated Machine Learning System 509
Chunnan Wang, Hongzhi Wang, Bo Xu, Xintong Song, Xiangyu Shi,
Yuhao Bao, and Bo Zheng

OIIKM: A System for Discovering Implied Knowledge from Spatial
Datasets Using Ontology 514
Liang Chang, Long Wang, Xuguang Bao, and Tianlong Gu

IDMBS: An Interactive System to Find Interesting Co-location Patterns
Using SVM 518
Liang Chang, Yuxiang Zhang, Xuguang Bao, and Tianlong Gu

SeTS^3: A Secure Trajectory Similarity Search System 522
Yiping Teng, Fanyou Zhao, Jiayv Liu, Mengfan Zhang, Jihang Duan,
and Zhan Shi

Data-Based Insights for the Masses: Scaling Natural Language Querying
to Middleware Data 527
Kausik Lakkaraju, Vinamra Palaiya, Sai Teja Paladi,
Chinmayi Appajigowda, Biplav Srivastava, and Lokesh Johri

Identifying Relevant Sentences for Travel Blogs from Wikipedia Articles 532
   Arnav Kapoor and Manish Gupta

PhD Constorium

Neuro-Symbolic XAI: Application to Drug Repurposing for Rare Diseases 539
   Martin Drancé

Leveraging Non-negative Matrix Factorization for Document
Summarization 544
   Alka Khurana

Author Index 549

Text and Image Processing

Emotion-Aware Multimodal Pre-training for Image-Grounded Emotional Response Generation

Zhiliang Tian[1], Zhihua Wen[2], Zhenghao Wu[1], Yiping Song[3], Jintao Tang[3], Dongsheng Li[2(✉)], and Nevin L. Zhang[1]

[1] The Hong Kong University of Science and Technology, Sai Kung, Hong Kong SAR, China
{ztianac,lzhang}@cse.ust.hk, zwubq@connect.ust.hk
[2] Science and Technology on Parallel and Distributed Laboratory, National University of Defense Technology, Changsha, Hunan, China
{zhwen,dsli}@nudt.edu.cn
[3] National University of Defense Technology, Changsha, Hunan, China
{songyiping,tangjintao}@nudt.edu.cn

Abstract. Face-to-face communication leads to better interactions between speakers than text-to-text conversations since the speakers can capture both textual and visual signals. Image-grounded emotional response generation (IgERG) tasks requires chatbots to generate a response with the understanding of both textual contexts and speakers' emotions in visual signals. Pre-training models enhance many NLP and CV tasks and image-text pre-training also helps multimodal tasks. However, existing image-text pre-training methods typically pre-train on images by recognizing or modeling objects, but ignore the emotions expressed in the images. In this paper, we propose several pre-training tasks in a unified framework that not only captures emotions from images but also learns to incorporate the emotion into text generation. The pre-training involves single-modal learning to strengthen the ability to understand images and generate texts. It also involves cross-modal learning to enhance interactions between images and texts. The experiments verify our method in appropriateness, informativeness, and emotion consistency.

Keywords: Multimodal · Conversation · Emotion · Pre-training · Generation

1 Introduction

Most conversation systems [21,45,59] lead a text-to-text dialog between users and chatbots. However, most people prefer face-to-face communication due to the accessibility of the speaker's visual signals, like facial expressions and body language. After analysing those signals, chatbots can garner speakers' emotional

---

Z. Tian and Z. Wen–The two authors contributed equally to this work.

A. Bhattacharya et al. (Eds.): DASFAA 2022, LNCS 13247, pp. 3–19, 2022.
https://doi.org/10.1007/978-3-031-00129-1_1

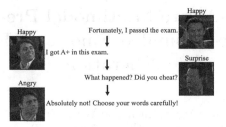

Fig. 1. An example of IgERG. Images hint speakers' emotions of in multi-turn dialogs.

states and make empathetic responses. Image-grounded emotional response generation (IgERG) [17] generates a response to context utterances while understanding both the users' textual contexts and emotional states reflected in images, where images contain users' facial expressions and gestures. IgERG has several practical applications, including AI therapist capturing a counselee's emotions via facial expressions, and AI baby-sitter conversing with kids who are moody and unstable (Fig. 1).

Deep learning achieves impressive performances on conversation systems but relies highly on large-scale corpora [44,56]. IgERG's training samples are hard to collect and limited since each sample contains conversations and the aligned speakers' visual signals. Pre-training models [9] help many NLP tasks without large-scale corpus by pre-training on large-scale unsupervised corpora.

Researchers propose multimodal pre-training for tasks involving both textual and visual data. To facilitate image understanding, they apply pre-training tasks of object classification [20] and object region modeling [5]. To ensure image-text alignment, they pre-train them with image captioning [16], text-conditioned object classification [23,51], and image-text entity matching [22]. Those tasks typically pre-train by recognizing objects in images, modeling objects, and interacting with other modalities (i.e. text) about object-related information. Those tasks ignore the styles or emotions reflected in the images. Such pre-training fits downstream tasks that require an understanding of the objects in images (e.g. image-text retrieval [3] and visual question answering [2]).

However, images imply both object-related and object-independent (e.g. styles, sentiments, or emotions) information. Object-independent information is more helpful to some applications, including multimodal sentiment analysis [62], multimodal style transfer [18,65], and IgERG [17]. Thus, pre-training of capturing object-independent signals enhances those tasks but faces two challenges: 1. object-independent information is harder to define and describe than objects. Intuitively, recognizing a human face is easier than describing the emotion seen in the facial expressions; 2. if the application involves multiple images (e.g. speakers' images in multi-turn conversations of IgERG)[1], the development of emotions reflected from a sequence of images is hard to model and utilize.

---

[1] 81.1% samples contain multiple emotions in IgERG.

In this paper, we propose a multimodal pre-training method that enhances the ability to capture emotions from a series of images and learns to incorporate the emotions into text generation. Such pre-training helps image-grounded emotional response generation (IgERG), where models generate responses given textual context and a sequence of speakers' images implying the emotions.

Particularly, to enhance the emotion perception, our model learns to model a series of emotions from a sequence of images via an image emotion sequentially labeling (IESL) and an image emotion classification (IEC) task. We obtain large-scale coarse data for those tasks via data augmentation, since those tasks have limited supervised data. To enhance the text generation, we apply BART's pre-training and masked language modeling (MLM) task to our pre-training. To enhance cross-modal interactions between emotion understanding and text generation, our model learns to incorporate emotions in the generated text by a controllable image-to-text generation (C-I2T) task. The text generation is controlled by the emotions detected from images. We construct a transformer-based framework carrying all pre-training tasks. Our contributions are as follows,

- We propose catering to object-independent emotional information in images during pre-training, which is a less explored topic.
- We propose several pre-training tasks in a unified framework that prompts image understanding, text generation, and cross-modal interactions.
- Our model obtains state-of-the-art performance on the IgERG task.
- We construct several multimodal pre-training datasets and will release them.

2   Related Work

Image grounding conversation has two categories. The first one, visual dialog, leads conversations to discuss the objects or events reflected in images [1,8]. Shuster et al. [46] assign specific styles to speakers in visual dialog. Agarwal et al. [1] study the effects of explicitly encoding historical contexts into visual dialog models. The second category captures speakers' emotions from images for an empathetic conversation. Poria et al. [39] and Hazarika et al. [14] discover speakers' emotion in conversations. Huber et al. [17] use a Seq2Seq [52] model to capture scene sentiment and use a facial encoding tool to extract the facial expressions from the images in conversation. Our task falls in this category. The above methods do not involve pre-training.

Emotional conversation systems can be categorized into two directions: expressing chatbot's emotions and catering to speakers' emotions. The first direction aims to enable chatbot to respond conditioned on a given emotion [7,50,68]. Colombo et al. [7] append the given emotion label on the input utterance. Song et al. [50] apply a lexicon-based attention to a Seq2Seq model and encourage Seq2Seq to implicitly express emotion. The second direction detects speakers' emotions and make empathetic responses according to speakers' emotion [12,38,42]. Skowron et al. [47] build a chatbot with the ability to detect user's emotion states. Rashkin et al. [42] propose a pipeline system that predicts emotion words and feed the explicit emotion words into a neural conversation model.

Lin et al. [28] propose an end-to-end neural conversation model. Lin et al. [29] and Zhong et al. [67] apply GPT [41] and BERT [9] to empathetic conversation. Another types of chatbots detect emotions from non-textual modalities [14], including tone or body languages. Some chatbots captures speakers' personalities in conversation [48,54,63].

Pre-trained language models significantly enhance natural language understanding (NLU) tasks [9,60], since language models contains commonsense knowledge [61] or language understanding abilities [33,37]. BERT [9] achieves state-of-the-art results on a wide range of NLU tasks. For natural language generation (NLG), GPT [41] trains to generate texts auto-regressively. BART [19] propose several text permutation techniques, such as text infilling and sentence shuffling. Multimodal (image-text) pre-training [23] mainly has four kinds of tasks as follows. 1. *object classification*: Su et al. [51] extract object representation with Faster-RCNN [43] model and apply object classification to pre-training. Li et al. [20] consider linguistic clues in object classification. 2. *object region modeling*: Li et al. [25] reconstruct the masked image regions by referring to the remaining part. Chen et al. [5] jointly train the masked region classification and the masked region modeling. 3. *image conditioned text generation*: Li et al. [26] pre-train with image captioning that typically describes the semantic information of image with a textual title. 4. *image-text matching*, encourages models to align texts and images in the semantic level [20]. Li et al. [22] learn to predict whether the given image-text pair are semantically aligned. Huang et al. [16] consider the alignment between multilingual texts and images. Those methods typically focus on pre-training to understand image objects, but our pre-training caters to object-independent information (e.g. emotions) in images.

3 Approaches

3.1 Overview

Our downstream task is the image-grounded emotional response generation (IgERG) that generates a response $\hat{S}_n$ given a series of context sentences $S = \{S_1, S_2, ..., S_{n-1}\}$ and the speakers' images corresponding to each sentence $I = \{I_1, I_2, ..., I_n\}$, where I_i, with speakers' facial expressions and gestures, reflects the emotional state of the speaker in the i-th turn. We construct a framework for the pre-training tasks and the fine-tuning task. As shown in Fig. 2, the framework mainly consists of four components:

- **Image encoder** E_{img} represents an image I_i with a vector $\boldsymbol{v}_i$.
- **Generator** is a transformer model [56] to generate text, where the pink and grey colored blocks in Fig. 2 indicate its encoder and decoder. Its input can be a sequence of sentences S, a sequence of images I, or the mixture of S and I; its output is the generated texts. If the input comes across an image I_i, the generator employs the image encoder to transfer the image into a vector $\boldsymbol{v}_i$ as its input. The generator's encoder output is defined as $\boldsymbol{e}_i$ at the i-th step.
- **Image emotion classifier** $D_{\text{img\_emo}}$ predicts emotion labels based on $\boldsymbol{e}_i$.

– **Text emotion evaluator** D_{txt_emo} evaluates texts generated by the generator.

Our tasks involve four types of datasets: 1.image-only datasets: each sample has one image or a sequence of images $I = \{I_1, I_2, ..., I_m\}$, $(m \geq 1)$; 2. text-only datasets: each sample is a raw sentence; 3. image-text datasets: each sample consist of several sentences $S = \{S_1, S_2, ..., S_n\}$ and each sentence has an aligned image $I = \{I_1, I_2, ..., I_n\}$; 4. textual emotion datasets: each sample has a sentence and its emotion label. We propose three kinds of pre-training tasks:

– **Pre-train for image emotion discovery** learns to detect image emotions on all datasets involving images, including image-only and image-text datasets.
– **Pre-train for text generation** learns to generate text on all datasets with texts, including text-only and image-text datasets.
– **Pre-train for cross-modal interaction** learns to generate text controlled by images on image-text datasets.

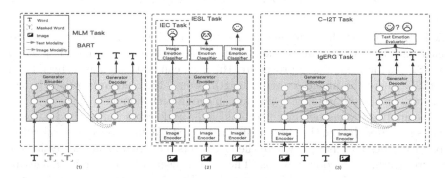

Fig. 2. Our model architecture with pre-training tasks (MLM, BART, IESL, IEC, and C-I2T) and fine-tuning tasks (IgERG). The generator, with an encoder (in pink) and a decoder (in grey), is shared among all subfigures. The image encoder and image emotion classifier are also shared in all time steps. Subfigure 1 shows the pre-training on text generation (MLM and BART) mentioned in Sect. 3.2; Subfigure 2 shows the pre-training of emotion discovery (IEC and IESL) mentioned in Sect. 3.2 and Subfigure 3 shows the pre-training of C-I2T and the fine-tuning of IgERG mentioned in Sect. 3.2 & 3.3. The orange and green arrows indicate information flows in image and text modality. (Color figure online)

3.2   Pre-training

Pre-training for Image Emotion Discovery. We propose two pre-training tasks: image emotion sequentially labeling (IESL) and image emotion classification (IEC) (Fig. 2.2), which enables our model to discover emotions from an image or a sequence of images. As a result, the image encoder and the generator's encoder represent each image I_i with a vector e_i that implies emotions.

IESL models a series of emotions reflected in a sequence of images $I = \{I_1, I_2, ..., I_m\}$, where the image sequence is consecutive screenshots from a video. Its output is a sequence of emotion labels. IESL involves the image encoder (E_{img}), generator's encoder (GE), and the image emotion classifier ($D_{\text{img\_emo}}$). The image encoder E_{img} first maps each image I_i into a vector $\boldsymbol{v}_i$. The generator's encoder GE takes the output vectors $\{\boldsymbol{v}_1, \boldsymbol{v}_2, ..., \boldsymbol{v}_m\}$ as the input and treats it as the sequence of "word embeddings" (As a transformer model, the generator's encoder inputs are word embeddings). Then, the generator's encoder obtains the output vectors $\{\boldsymbol{e}_1, \boldsymbol{e}_2, ..., \boldsymbol{e}_m\}$. To train the emotion labeling task, we feed each encoder output $\boldsymbol{e}_i$ to the image emotion classifier $D_{\text{img\_emo}}$ to predict the emotion at i-th step. Equation 1 and 2 show those operations, where E_{I_i} is the emotion label of I_i, K is emotion category number, $\mathbb{I}$ is an indicator function. We introduce the structure of the image encoder and emotion classifier in Sect. 3.4.

$$\boldsymbol{v}_i = E_{\text{img}}(I_i) \text{ for } i \in [1, m], \qquad \{\boldsymbol{e}_1, \boldsymbol{e}_2, ..., \boldsymbol{e}_m\} = GE(\{\boldsymbol{v}_1, \boldsymbol{v}_2, ..., \boldsymbol{v}_m\}), \quad (1)$$

$$L_{\text{IESL}} = \sum_{i=1}^{m} \sum_{k=1}^{K} \mathbb{I}(E_{I_i} = k) \log P(D_{\text{img\_emo}}(\boldsymbol{e}_i) = k). \quad (2)$$

IEC learns to detect the emotion reflected in a single image, where the input is an image and the output is an emotion label. This task shares the same components as IESL. As for the training procedure, this task can be treated as a special case of IESL, where the image sequence length is one. IEC's advantage is this task can work on the datasets with only one image in each sample.

We apply data augmentation to build large-scale coarse supervised datasets since supervised data in the above tasks are limited. We first collect several videos from TV series and movies, where the characters have rich facial expressions or gestures during the conversations. Then, we collect unlabeled raw images by taking screenshots from the TV series videos. Finally, we generate pseudo labels for the unlabeled images with Face++ API[2], which categorizes images into seven classes of emotion (e.g. fear). Therefore, each sample in those datasets has a sequence of images and labels, which is used for both IEC and IESL tasks.

In total, we obtain 5.9M training samples from six datasets for the above tasks. Except for one dataset that is an existing supervised image emotion dataset, we apply data augmentations on the other five since they do not have emotion labels. Among the five datasets, four come from the TV series and one is an image-text dataset [32].

Pre-training for Text Generation. To enhance text generation ability, our model applies the pre-training of BART [19] and pre-trains on the text data from the image-text datasets via masked language modeling (MLM) task (Fig. 2.1). BART pre-trains a model combining bidirectional and auto-regressive transformers. As rerunning BART's pre-training requires a great number of GPUs[3], we

---

[2] console.faceplusplus.com.cn.
[3] It roughly requires 64 GPUs for more than 2 weeks.

use the pre-trained BART released by FairSeq[4]. BART provides initial parameters for the generator. The motivation behind using BART is that BART has the same architecture as the generator and BART has achieved state-of-the-art performance on text generation (e.g. dialogue and summarization).

MLM learns to generate randomly masked tokens in a raw sentence [31]. MLM only employs our generator, where the encoder receives a masked sentence and the decoder reconstructs the original sentence. We train MLM on 6M sentences from the image-text dataset, and each sentence is a sample in the pre-training.

Pre-training for Cross-Modal Interaction. As shown in Fig. 2.3, we propose a pre-training task, controllable image-to-text generation (C-I2T), to enhance the interaction between images and texts. C-I2T learns to incorporate the emotions detected from the images into text generation and encourages emotions in texts and images to be consistent. The training samples consist of a series of images $I = \{I_a, I_{a+1}, ..., I_b\}$ and a series of sentences $S = \{S_a, S_{a+1}, ..., S_b\}$. The pre-training requires the model to generate a sentence $\hat{S}_b$ given the corresponding images $\{I_a, I_{a+1}, ..., I_b\}$ and the previous sentences $\{S_a, S_{a+1}, ..., S_{b-1}\}$.

We concatenate the sentences and images as a long sequence $\{I_a, S_a, I_{a+1}, S_{a+1}, ..., S_{b-1}, I_b\}$, where the sentence and image occurs alternately. Each utterance and its corresponding image gather together and the image is ahead of the utterance since we treat the image as the condition of its utterance. We use the image encoder to transfer each image I_i into a vector v_i, and generator's word embedding layer transfers a sentence with l words into a sequence of l word vectors $\{w_{i1}, w_{i2}, ..., w_{il}\}$. Then, we obtain a sequence of vectors $\{v_a, w_{a1}, w_{a2}, ..., v_i, w_{i1}, w_{i2}, ..., w_{(b-1)l}, v_b\}$ and feed the sequence into the generator to generate $\hat{S}_b$. Feeding a combined image-text sequence into the generator bridges the cross-modal attention to enhance image-text interaction, since every two inputs have a connection via self-attention in the generator (i.e. transformer) including the two inputs in different modalities. The text emotion evaluator $D_{\text{txt\_emo}}$ (See details in Sect. 3.4) predicts emotion of the generated text $\hat{S}_b$.

C-I2T's loss has two terms as Eq. 3: a negative likelihood loss $\mathcal{L}_{NLL}$ that encourages generating the ground truth and a cross-entropy loss $\mathcal{L}_{emo}$ that encourages the generated text $\hat{S}_b$ to have the same emotion as I_b, where E_{I_b} is I_b's emotion, $D_{\text{txt\_emo}}$ is the text emotion evaluator, and α balances the two losses.

$$\mathcal{L}_{\text{C-I2T}} = \mathcal{L}_{NLL} + \alpha\mathcal{L}_{emo}, \quad \mathcal{L}_{emo} = \sum_{k=1}^{K} \mathbb{I}(E_{I_b} = k) \log P(D_{\text{txt\_emo}}(\hat{S}_b) = k). \quad (3)$$

C-I2T uses an image-text dataset collected from the TV series. Each original sample is an image-text sequence consisting of a series of images $I = \{I_1, I_2, ..., I_n\}$ and sentences $S = \{S_1, S_2, ..., S_n\}$. The images I are from the screenshots of a video clip and the sentences S are the corresponding subtitles. All the sub-sequences in the image-text sequence act as training samples for

---

[4] dl.fbaipublicfiles.com/fairseq/models/.

C-I2T: each original sample (image-text sequence) is decomposed into multiple
C-I2T's samples: $\{I_a, I_{a+1}, ..., I_b\}$, $\{S_a, S_{a+1}, ..., S_b\}$ (for $\forall a, b \in (0, n], a+1 < b$).
As mentioned in Sect. 3.2, we generate pseudo emotion E_{I_i} for I_i with Face++.

Procedure of Pre-training. We conduct the above pre-training tasks in a unified framework. Our pre-training consists of two phases as mentioned in Fig. 2. In
the first phase, we load the pre-trained BART model to initialize the parameters
of our generator and obtain the text emotion evaluator $D_{\text{txt\_emo}}$ by fine-tuning
BERT on textual emotion datasets. In the second phase, based on the BART
and $D_{\text{txt\_emo}}$, we jointly pre-train the model on IEC, IESL, C-I2T, and MLM tasks.
To train the second phase, we mix up all the datasets and shuffle all batches of
samples (Samples in a batch come from one dataset). During the second phase
of pre-training, we pre-train IEC and IESL on the samples from datasets involving images; we pre-train MLM on samples involving texts and we pre-train C-I2T
on samples from image-text datasets. After all the pre-training, the pre-trained
model provides the initial parameters for the downstream IgERG task.

3.3   Fine-Tuning on the IgERG Task

IgERG is our downstream task. As shown in Fig. 2.3, most layers in our pre-
trained model provide the initial parameters for the downstream task, except
the classifier $D_{\text{img\_emo}}$ mentioned in Sect. 3.2 and the evaluator $D_{\text{txt\_emo}}$ mentioned in Sect. 3.2. Fine-tuning on the downstream task follows the pre-training
operations of C-I2T except for the use of L_{emo} loss. We concatenate the images
and context sentences into a sequence $\{I_1, S_1, I_2, S_2, ..., S_{n-1}, I_n\}$ as the input of
our model. The image encoder and generator's word embedding layer transfers
images and words into vectors. The generator's encoder receives those vectors
and the generator's decoder outputs the response $\hat{S}_n$.

3.4   Structure of Model Components

Our model (Fig. 2) consists of an image encoder, an image emotion classifier,
a text emotion evaluator, and a generator. The image encoder represents an
image I_i with a fixed dimensional vector v_i and feeds the vector to the generator.
The image encoder is ResNet-50 [15] without softmax layer. We replace the top
fully-connected layer with another fully-connected layer, in which the output
dimension equals the dimension of the generator's word embedding. Thus, the
image encoder's output $\{v_0, v_1, ..., v_m\}$ is fed into the generator by acting as its
input embeddings. The image encoder is trained by emotion discovery (IEC and
IESL) task (Sect. 3.2), and it is frozen in other pre-training tasks and fine-tuning.

   The image emotion classifier $D_{\text{img\_emo}}$ predicts the emotion of an image. Its
input is an image vector (i-th image's encoder output e_i). The image emotion
classifier consists of a fully-connected layer and a softmax layer. It transfers e_i
into a probability distribution. IEC and IESL train and use this classifier.

The text emotion evaluator D_{txt_emo} is a BERT-based classifier that predicts the emotion of the generated text in C-I2T task. It is pre-trained by BERT[5] and fine-tuned on text emotion datasets. Only C-I2T involves this D_{txt_emo}.

The generator is the core part that accomplishes the pre-training and fine-tuning tasks. It is trained in all pre-training tasks. The generator is the "transformer base" model in [56] with encoder (pink blocks in Fig 2) and a decoder (grey blocks in Fig 2). Both encoder and decoder have 6 stacked layers.

4 Experiments

4.1 Experimental Settings

Datasets and Hyper-parameters. Our pre-training tasks involve four types of datasets: 1. for image-only datasets, we use a supervised image emotion dataset, RAF-DB [24], with 15k samples. We collect 4.9M samples from four TV series (How I Met Your Mother, This Is Us, The Big Bang Theory, and Person of Interest). Each of the samples is a sequence of screenshots, and we assign emotional labels on those samples via Face++ API. 2. The text-only datasets are BART's datasets consisting of four datasets [13,35,55,69]. 3. For image-text datasets, we use OpenViD [32] dataset with 1M samples. It comes from the movies with subtitles, where the image comes from the sequence of screenshots and texts are the corresponding subtitles. Notice that, we split original OpenViD samples to obtain 21M samples for C-I2T as mentioned in Sect. 3.2. 4. We use three textual emotion datasets [4,27,39] with 0.92M samples in total, and they share the same emotion space with Face++ API. Our downstream task uses OpenViD [32] dataset with 0.9M/50k/50k samples for training/validation/testing. In all baselines and our model, following [32], we set the dimension of word embeddings and hidden layers to 512 and the dimension of image vector v to 1000. We set the dropout rate to 0.3 and the learning rate to 3e−5 during pre-training and fine-tuning. The factor α in Eq. 3 is 1. D_{txt_emo} is fine-tuned with a learning rate of 1e−5. We released our code[6].

Evaluation Metrics. We evaluate all methods with automatic and human evaluations. Our automatic metrics consists of: (1) Appropriateness, *Bleu-N* [21, 36] and *Nist-N* [10,40,53] measure N-gram match between outputs and ground truthes. *CiDEr* [57] is widely used in image captioning. (2) Informativeness, *Dist-N* [21,49] evaluates the response diversity via unique n-gram proportion in all responses. *Ent-N* [34,53] is the entropy on word count distribution. (3) Emotion Const. *E-Acc* and *E-F1* measure the accuracy and F1 score between the emotion of the last input image and the generated responses. Those metrics show the consistency of images and texts in terms of emotion.

---

[5] huggingface.co/bert-base-uncased.
[6] Our code is available at: github.com/stupidHIGH/MM-Pre-train.

We conduct human evaluations in three aspects: 1. overall quality *Qual* (fluency, relevance, and grammaticality) of the generated response, 2. informativeness and lexical diversity of results (*Info*). 3. *Emo* (in three factors: emotion consistency between input images and generated texts, emotion being well expressed in text, and whether the personas reflected in outputs and dialogue histories are consistent.) We hire five commercial annotators to annotate five copies of 300 randomly selected test samples with a 5-scale rating.

Baseline Methods. We verify our model by comparing following methods.

- No pre-training. `Trs` is the transformer model [56] without using images. `Trs+FV` and `Trs+CV` denote the transformer using images via Faster-RCNN [43] and ResNet [15], respectively. The above baselines come from Meng et al. [32] and do not involve pre-training. `Trs+CV` serves as the fine-tuning model of all the following methods for a fair comparison.
- Text-only pre-training. `BART` denotes the model pre-trained with BART [19] on text data and fine-tuned with `Trs+CV`'s model, where BART is widely-used pre-train model for text generation tasks (e.g. dialog and summarization).
- Image-Text pre-training. Oscar [24] is the multimodal pre-training model that pre-trains to align the object semantics.[7] `Oscar+BART`'s encoder comes from Oscar and its decoder comes from BART.

Table 1. Overall performance on automatic metrics and human evaluations (Results in rows 1 to 3 match the official github page (github.com/ShannonAI/OpenViDial), where the authors [32] publish the revised results of their paper. Kappa score [11] among annotators is 0.43 (moderate agreement among annotators)).

Pre-train type	Model	Appropriateness						Informativeness					Emotion const		
		Bleu3	Bleu4	Nist3	Nist4	CIDEr	Qual	Dist3	Dist4	Ent3	Ent4	Info	E-F1	E-Acc	Emo
No Pre-train	Trs	1.79	1.00	0.766	0.771	0.069	2.41	0.003	0.004	3.53	3.56	2.05	0.099	0.228	1.87
	Trs+FV	2.07	1.19	0.874	0.880	0.098	2.39	0.027	0.041	4.82	4.99	2.34	0.111	0.217	1.84
	Trs+CV	2.04	1.15	0.875	0.882	0.097	2.58	0.020	0.031	5.01	5.18	1.96	0.118	0.246	2.03
Text-only	BART	2.77	1.70	1.069	1.086	0.140	3.04	0.155	0.224	6.84	7.28	2.64	0.129	0.288	2.32
Image-Text	Oscar	2.18	1.21	0.865	0.870	0.098	2.88	0.021	0.034	5.78	6.07	2.35	0.121	0.248	2.21
	BART+Oscar	1.99	1.07	0.881	0.888	0.104	2.79	0.022	0.037	5.47	5.71	2.31	0.124	0.256	2.09
Image-Emotion-Text	Ours	**3.00**	**1.94**	**1.139**	**1.158**	**0.164**	**3.13**	**0.216**	**0.306**	**7.44**	**7.93**	**2.98**	**0.137**	**0.312**	**2.49**

4.2 Overall Results

Table 1 shows the performance of all methods. Among the baselines without pre-training (first 3 rows), `Trs+FV` and `Trs+CV` perform better than `Trs`, which verifies the image is helpful for our task. `Trs+CV`'s emotion consistency is a little higher than `Trs+FV`, so the following methods employ `Trs+CV` as the fine-tuning

---

[7] We choose Oscar as our baseline, since Cho et al. [6] and Li et al. [26] reported Oscar outperforms most existing multimodal pre-training models, including UNITER [5], XGPT [58], VL-BART [6] and VL-T5 [6] on VQA, NLVR, and image captioning.

model. Text pre-training `BART` obtains much higher performance than no pre-training models showing the power of pre-training. Text pre-training enhances *Dist* scores a lot. The reason is `BART` learned on a corpus with a large variety of sentences (160GB data) during its pre-training.

Image-text pre-training baselines (`Oscar` and `Oscar+BART`) outperform no pre-training models in the emotion consistency and get similar performances in other metrics. It shows image pre-training helps models to understand emotions. The performance of the two baselines is not as satisfactory as `BART`, showing the current image-text pre-training models do not fit for our task. As a model only pre-trained to align the object semantics between image and text, `Oscar` does not consider the object-independent information (e.g. emotion) during pre-training.

Considering object-independent information (emotion) in image-text pre-training, **Ours** surpasses all the baselines in all metrics. The improvements in emotion consistency show our pre-training enhances the cross-modal interaction that expresses image emotions in texts. The significant improvements on *Dist* (+39% in *Dist3* and +36% in *Dist4*) indicate suitable image-text pre-training can highly enlarge the informativeness of the generated text. The response generated by our model tends to have higher quality and appropriateness, as our model improves by 8% and 14% on *Bleu3* and *Bleu4*. Our promotion on emotion consistency verifies that considering the emotion reflected from the images during pre-training helps the model on understanding and expressing emotions.

4.3   Ablation Studies on Pre-training Tasks

Table 2. Ablation studies on the effectiveness of different pre-training tasks. "Ours − X" (row 2 to 5) indicates our full mode without pre-training of task X.

	Appropriateness					Informativeness				Emotion Const	
	Bleu3	Bleu4	Nist3	Nist4	CIDEr	Dist3	Dist4	Ent3	Ent4	E-F1	E-Acc
Ours	**3.00**	**1.94**	**1.139**	**1.158**	**0.164**	**0.216**	**0.306**	**7.44**	**7.93**	**0.137**	**0.312**
Ours − IESL	2.82	1.76	1.097	1.116	0.150	0.185	0.268	7.23	7.70	0.136	0.296
Ours − IEC	2.80	1.74	1.065	1.082	0.145	0.163	0.234	6.79	7.19	0.133	0.285
Ours − MLM	2.83	1.77	1.089	1.107	0.147	0.174	0.253	7.01	7.45	0.133	0.291
Ours − C-I2T	2.77	1.73	1.093	1.111	0.146	0.181	0.258	6.99	7.41	0.135	0.302

Table 2 shows the ablation studies on our proposed pre-training tasks. We construct a model variant by removing one specific pre-training task from our full model. As we propose four tasks, we obtain four variants and compare them to our full model **Ours**. For example, **Ours−IEC** denotes **Ours** without IEC.

Ours excels all model variants in row 2 to 5. It demonstrates all the proposed pre-training tasks are necessary and contribute to our full model. Among the rows 2 to 5, removing `IESL` from **Ours** (row 2) decrease the performance on all metrics. `IESL` is crucial since it models the change of emotions in a conversation session, which helps the downstream model capture the tendency of the speaker's emotion and generate suitable responses. Removing IEC from the full model causes the largest performance drop, which shows IEC is much important in pre-training. The reason is that it benefits from high-quality supervised training data with rich emotions (see dataset analysis in Sect. 4.5). Knowing that appropriateness measures the matching between ground truth and generated texts, it is reasonable to see that Ours − C-I2T performs the second-worst on appropriateness because C-I2T is the only task that pre-trains to make generated sentences and the ground truth similar.

4.4 Emotion Expression in Text Generation

To measure the ability to express emotions in the generated texts, we propose *%EW* and *%F1* for evaluation. *%EW* denotes the percentage of emotion words (occurs in an emotion word list[8] of each generated response. Considering the ground truth words, we measure the precision and recall of generated emotion words that match the ground truth as Eq. 4, where E denotes the emotion word list, $\hat{S}$ denotes a generated sentence, and S denotes a ground truth sentence. *%F1* is the harmonic average of the precision and recall.

$$\text{Precision} = \frac{|\hat{S} \cap E \cap S|}{|\hat{S} \cap E|}, \quad \text{Recall} = \frac{|\hat{S} \cap E \cap S|}{|E \cap S|} \tag{4}$$

The experimental results shown in Table 3 reflect the effectiveness of emotion expressions in two aspects: the quantity and quality of generated emotion words. As for the quantity, our model tends to generate far more emotion words in responses than the baselines (shown in the *%EW* of Table 3). This advantage mainly comes from the pre-training of *C-I2T*, since *C-I2T* learns to generate words to express emotions. As for the quality, the results on *%F1* verify the generated emotion words from our model are more likely to match the correct emotions (ground truth emotions). Our model works well owing to the pre-training of understanding emotions (IEC and IESL) and express emotions (C-I2T).

Table 3. The percentage of emotion words in generated responses from different methods, which shows the degree of emotion expressed by the model.

	Trs	Trs+CV	Trs+FV	BART	Oscar	Oscar+BART	Ours
%EW	1.5	3.7	2.7	4.3	4.3	3.5	**5.1**
%F1	3.1	15.3	10.8	60.2	12.1	12.0	**69.4**

---

[8] saifmohammad.com/WebPages/lexicons.html.

4.5   Studies on the Dataset Selection

Table 4. Comparisons among different types of image datasets on IEC. R is to conduct IEC on RAF-DB dataset with its original labels. R_FA, T_FA, and P_FA denote IEC on RAF-DB, TV, and Pose dataset with pseudo labels by Face++ API.

	Label	Emotion distribution	Image content	Appropriateness					Informativeness				Emotion const	
				Bleu3	Bleu4	Nist3	Nist4	CIDEr	Dist3	Dist4	Ent3	Ent4	E-F1	E-Acc
R	Real	Balanced	Facial Expressions	**2.20**	**1.27**	0.997	**1.009**	**0.107**	0.043	0.073	5.449	5.791	0.134	**0.325**
R_FA	Pseudo	Balanced	Facial Expressions	2.18	1.25	**1.005**	1.007	1.005	**0.052**	**0.087**	**5.707**	**6.105**	**0.136**	0.325
T_FA	Pseudo	Unbalanced	Face + Gestures	2.09	1.17	0.965	0.976	0.099	0.028	0.047	5.357	5.659	0.129	0.293
P_FA	Pseudo	Unbalanced	Few Facial Expressions	2.15	1.18	0.891	0.900	0.092	0.019	0.033	4.981	5.222	0.126	0.276

Most pre-training methods succeed owing to "big data" and our method also requires large-scale image datasets for IEC, IESL, and C-I2T. The qualities and characteristics of those datasets are crucial for our training. However, it's hard to select datasets from various image datasets. Here, we analyze the effectiveness of different types of image emotion datasets, verify the data augmentation, and give suggestions on dataset selection. In Table 4, we choose four datasets with different types and compare the performances of pre-training IEC on them.

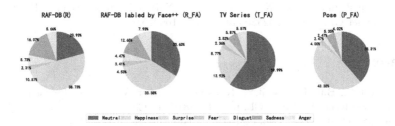

Fig. 3. Emotion distributions of four image datasets in Sect. 4.5.

We choose four datasets considering their image contents and emotion distributions: 1. RAF-DB (R) [24] is a supervised image emotion classification dataset, where most images describe facial expressions and the emotion distribution is well balanced. 2. We obtain a new dataset R_FA by relabeling the original RAF-DB via Face++. 3. TV dataset (T_FA) is a mixture of four TV series datasets (Sect. 4.1) labeled by Face++. The images contain speakers' facial expressions and gestures and its emotion distribution is unbalanced. 4. Pose dataset (P_FA) is a mixture of portrait photography datasets [30,64,66], where the face region covers a small part of each image. Its emotion distribution is unbalanced since the only two emotions cover 82% samples. We train all datasets with the same scale (15k samples) for a fair comparison. Figure 3 shows emotion distributions of the above datasets. The distributions of RAF-DB and RAF-DB labeled by

Face++ are similar. For emotion distribution of TV series, the "neutral" takes up more than half of the emotions. In Pose, "neutral" and "happiness" take up more than 82% of the emotions, indicating the distribution is quite unbalanced.

We conclude we should select datasets where 1. image contents are mainly about speakers' facial expressions, since T_FA surpasses P_FA; 2. the emotion distribution should be balanced, as R and R_FA excel T_FA and P_FA. Further, labels on the supervised data are useful according to the fact that R slightly outperforms R_FA in appropriateness. But the gap between R and R_FA is small, which shows the data augmentation by Face++ also constructs good datasets when the supervised label is not accessible.

5   Conclusion

In this paper, we propose a multimodal pre-training method that enhances the ability of text generation models to capture emotions from images and leverage the emotions for text generation. Our method pre-trains by discovering emotions in images, generating text, and incorporating the discovered emotions for text generation. We build a unified model carrying all pre-training tasks. Our pre-training method is applied to the IgERG task. The experimental results reveal our pre-training tasks enhance IgERG by a large margin.

Acknowledgement. Research on this paper was supported by Hong Kong Research Grants Council under grand No. 16204920 and National Natural Science Foundation of China under Grant No. 62025208 and No. 62106275.

References

1. Agarwal, S., Bui, T., Lee, J.Y., Konstas, I., Rieser, V.: History for visual dialog: Do we really need it? In: ACL, pp. 8182–8197 (2020)
2. Antol, S., et al.: VQA: visual question answering. In: ICCV (2015)
3. Chen, J., Zhang, L., Bai, C., Kpalma, K.: Review of recent deep learning based methods for image-text retrieval. In: MIPR, pp. 167–172. IEEE (2020)
4. Chen, S.Y., Hsu, C.C., Kuo, C.C., Ku, L.W., et al.: EmotionLines: an emotion corpus of multi-party conversations. In: ACL (2018)
5. Chen, Y.C., et al.: UNITER: universal image-text representation learning. In: ECCV (2020)
6. Cho, J., Lei, J., Tan, H., Bansal, M.: Unifying vision-and-language tasks via text generation. In: ICML 2021, pp. 1931–1942 (2021)
7. Colombo, P., Witon, W., Modi, A., Kennedy, J., Kapadia, M.: Affect-driven dialog generation. In: ACL, pp. 3734–3743 (2019)
8. Das, A., et al.: Visual dialog. In: CVPR, pp. 326–335 (2017)
9. Devlin, J., Chang, M.W., Lee, K., Toutanova, K.: BERT: pre-training of deep bidirectional transformers for language understanding. In: NAACL (2019)
10. Doddington, G.: Automatic evaluation of machine translation quality using n-gram co-occurrence statistics. In: HLT, pp. 138–145 (2002)
11. Fleiss, J.L.: Measuring nominal scale agreement among many raters. Psychol. Bull. **76**(5), 378 (1971)

12. Gao, J., et al.: Improving empathetic response generation by recognizing emotion cause in conversations. In: EMNLP (Finding), pp. 807–819 (2021)
13. Gokaslan, A., Cohen, V.: OpenWebText corpus. kylion007.github.io/OpenWebText tCorpus (2019)
14. Hazarika, D., Poria, S., Zadeh, A., Cambria, E., Morency, L.P., Zimmermann, R.: Conversational memory network for emotion recognition in dyadic dialogue videos. In: NAACL, vol. 2018, p. 2122. NIH Public Access (2018)
15. He, K., Zhang, X., Ren, S., Sun, J.: Deep residual learning for image recognition. In: CVPR, pp. 770–778 (2016)
16. Huang, H., et al.: M3P: learning universal representations via multitask multilingual multimodal pre-training. In: CVPR (2020)
17. Huber, B., McDuff, D., Brockett, C., Galley, M., Dolan, B.: Emotional dialogue generation using image-grounded language models. In: CHI, pp. 1–12 (2018)
18. Lee, D., Tian, Z., Xue, L., Zhang, N.L.: Enhancing content preservation in text style transfer using reverse attention and conditional layer normalization. In: ACL (2021)
19. Lewis, M., et al.: BART: denoising sequence-to-sequence pre-training for natural language generation, translation, and comprehension. In: ACL (2019)
20. Li, G., Duan, N., Fang, Y., Gong, M., Jiang, D.: Unicoder-VL: a universal encoder for vision and language by cross-modal pre-training. In: AAAI, vol. 34 (2020)
21. Li, J., Galley, M., Brockett, C., Gao, J., Dolan, B.: A diversity-promoting objective function for neural conversation models. In: NAACL, pp. 110–119 (2016)
22. Li, L., Chen, Y.C., Cheng, Y., Gan, Z., Yu, L., Liu, J.: HERO: hierarchical encoder for video+ language omni-representation pre-training. In: ACL (2020)
23. Li, L.H., Yatskar, M., Yin, D., Hsieh, C.J., Chang, K.W.: VisualBERT: a simple and performant baseline for vision and language. arXiv:1908.03557 (2019)
24. Li, S., Deng, W., Du, J.: Reliable crowdsourcing and deep locality-preserving learning for expression recognition in the wild. In: CVPR, pp. 2852–2861 (2017)
25. Li, W., et al.: UNIMO: towards unified-modal understanding and generation via cross-modal contrastive learning. arXiv preprint arXiv:2012.15409 (2020)
26. Li, X., et al.: Oscar: object-semantics aligned pre-training for vision-language tasks. In: Vedaldi, A., Bischof, H., Brox, T., Frahm, J.-M. (eds.) ECCV 2020. LNCS, vol. 12375, pp. 121–137. Springer, Cham (2020). https://doi.org/10.1007/978-3-030-58577-8_8
27. Li, Y., Su, H., Shen, X., Li, W., Cao, Z., Niu, S.: DailyDialog: a manually labelled multi-turn dialogue dataset. In: ACL (2017)
28. Lin, Z., Madotto, A., Shin, J., Xu, P., Fung, P.: MoEL: mixture of empathetic listeners. In: EMNLP, pp. 121–132 (2019)
29. Lin, Z., et al.: CAiRE: an end-to-end empathetic chatbot. In: AAAI, vol. 34, pp. 13622–13623 (2020)
30. Liu, Z., Luo, P., Wang, X., Tang, X.: Deep learning face attributes in the wild. In: ICCV (2015)
31. Logeswaran, L., Lee, H., Bengio, S.: Content preserving text generation with attribute controls. In: NeurIPS, pp. 5108–5118 (2018)
32. Meng, Y., et al.: OpenViDial: a large-scale, open-domain dialogue dataset with visual contexts. arXiv preprint arXiv:2012.15015 (2020)
33. Mikolov, T., Sutskever, I., Chen, K., Corrado, G.S., Dean, J.: Distributed representations of words and phrases and their compositionality. In: NeurIPS (2013)
34. Mou, L., Song, Y., Yan, R., Li, G., Zhang, L., Jin, Z.: Sequence to backward and forward sequences: a content-introducing approach to generative short-text conversation. In: COLING, pp. 3349–3358 (2016)

35. Nagel, S.: CC-news (2016). commoncrawl.org/2016/10/newsdataset-available
36. Papineni, K., Roukos, S., Ward, T., Zhu, W.: Bleu: a method for automatic evaluation of machine translation. In: ACL, pp. 311–318 (2002)
37. Pennington, J., Socher, R., Manning, C.: GloVe: global vectors for word representation. In: EMNLP, pp. 1532–1543 (2014)
38. Polzin, T.S., Waibel, A.: Emotion-sensitive human-computer interfaces. In: ITRW (2000)
39. Poria, S., Hazarika, D., Majumder, N., Naik, G., Cambria, E., Mihalcea, R.: MELD: a multimodal multi-party dataset for emotion recognition in conversations. In: ACL (2018)
40. Qin, L., et al.: Conversing by reading: contentful neural conversation with on-demand machine reading. In: ACL, pp. 5427–5436 (2019)
41. Radford, A., Narasimhan, K., Salimans, T., Sutskever, I.: Improving language understanding by generative pre-training (2018)
42. Rashkin, H., Smith, E.M., Li, M., Boureau, Y.L.: Towards empathetic open-domain conversation models: a new benchmark and dataset. In: ACL (2019)
43. Ren, S., He, K., Girshick, R., Sun, J.: Faster R-CNN: towards real-time object detection with region proposal networks. In: NeurIPS (2015)
44. Serban, I.V., et al.: A hierarchical latent variable encoder-decoder model for generating dialogues. In: AAAI, pp. 3295–3301 (2017)
45. Shang, L., Lu, Z., Li, H.: Neural responding machine for short-text conversation. In: ACL, pp. 1577–1586 (2015)
46. Shuster, K., Humeau, S., Bordes, A., Weston, J.: Image-chat: Engaging grounded conversations. In: ACL, pp. 2414–2429 (2020)
47. Skowron, M.: Affect listeners: acquisition of affective states by means of conversational systems. In: Esposito, A., Campbell, N., Vogel, C., Hussain, A., Nijholt, A. (eds.) Development of Multimodal Interfaces: Active Listening and Synchrony. LNCS, vol. 5967, pp. 169–181. Springer, Heidelberg (2010). https://doi.org/10.1007/978-3-642-12397-9_14
48. Song, Y., Liu, Z., Bi, W., Yan, R., Zhang, M.: Learning to customize model structures for few-shot dialogue generation tasks. In: ACL, pp. 5832–5841 (2020)
49. Song, Y., Tian, Z., Zhao, D., Zhang, M., Yan, R.: Diversifying neural conversation model with maximal marginal relevance. In: IJCNLP, pp. 169–174 (2017)
50. Song, Z., Zheng, X., Liu, L., Xu, M., Huang, X.J.: Generating responses with a specific emotion in dialog. In: ACL, pp. 3685–3695 (2019)
51. Su, W., et al.: VL-BERT: pre-training of generic visual-linguistic representations. In: ICLR (2019)
52. Sutskever, I., Vinyals, O., Le, Q.V.: Sequence to sequence learning with neural networks. In: NeurIPS, pp. 3104–3112 (2014)
53. Tian, Z., et al.: Response-anticipated memory for on-demand knowledge integration in response generation. In: Proceedings of the 58th Annual Meeting of the Association for Computational Linguistics, pp. 650–659 (2020)
54. Tian, Z., Bi, W., Zhang, Z., Lee, D., Song, Y., Zhang, N.L.: Learning from my friends: few-shot personalized conversation systems via social networks. In: AAAI, vol. 35, pp. 13907–13915 (2021)
55. Trinh, T.H., Le, Q.V.: A simple method for commonsense reasoning. arXiv preprint arXiv:1806.02847 (2018)
56. Vaswani, A., et al.: Attention is all you need. In: NeurIPS, pp. 5998–6008 (2017)
57. Vedantam, R., Lawrence Zitnick, C., Parikh, D.: CIDEr: consensus-based image description evaluation. In: CVPR, pp. 4566–4575 (2015)

58. Xia, Q., et al.: XGPT: cross-modal generative pre-training for image captioning. In: Wang, L., Feng, Y., Hong, Yu., He, R. (eds.) NLPCC 2021. LNCS (LNAI), vol. 13028, pp. 786–797. Springer, Cham (2021). https://doi.org/10.1007/978-3-030-88480-2_63

59. Yan, R., Song, Y., Wu, H.: Learning to respond with deep neural networks for retrieval-based human-computer conversation system. In: SIGIR, pp. 55–64 (2016)

60. Yang, Z., Dai, Z., Yang, Y., Carbonell, J., Salakhutdinov, R., Le, Q.V.: XLNet: generalized autoregressive pretraining for language understanding (2019)

61. Yu, C., Zhang, H., Song, Y., Ng, W.: CoCoLM: complex commonsense enhanced language model. arXiv preprint arXiv:2012.15643 (2020)

62. Zadeh, A., Zellers, R., Pincus, E., Morency, L.P.: Multimodal sentiment intensity analysis in videos: Facial gestures and verbal messages. IEEE IS **31**, 82–88 (2016)

63. Zhang, S., Dinan, E., Urbanek, J., Szlam, A., Kiela, D., Weston, J.: Personalizing dialogue agents: i have a dog, do you have pets too? In: ACL, pp. 2204–2213 (2018)

64. Zhang, Y., et al.: CelebA-spoof: large-scale face anti-spoofing dataset with rich annotations. In: Vedaldi, A., Bischof, H., Brox, T., Frahm, J.-M. (eds.) ECCV 2020. LNCS, vol. 12357, pp. 70–85. Springer, Cham (2020). https://doi.org/10.1007/978-3-030-58610-2_5

65. Zhang, Y., et al.: Multimodal style transfer via graph cuts. In: ICCV, pp. 5943–5951 (2019)

66. Zhang, Z., Luo, P., Loy, C.C., Tang, X.: Learning social relation traits from face images. In: ICCV (2015)

67. Zhong, P., Zhang, C., Wang, H., Liu, Y., Miao, C.: Towards persona-based empathetic conversational models. In: EMNLP, pp. 6556–6566 (2020)

68. Zhou, H., Huang, M., Zhang, T., Zhu, X., Liu, B.: Emotional chatting machine: emotional conversation generation with internal and external memory. In: AAAI, pp. 730–738 (2018)

69. Zhu, Y., et al.: Aligning books and movies: towards story-like visual explanations by watching movies and reading books. In: ICCV, pp. 19–27 (2015)

Information Networks Based Multi-semantic Data Embedding for Entity Resolution

Chenchen Sun[1,2,3]($\boxtimes$), Derong Shen[4], and Tiezheng Nie[4]

[1] School of Computer Science and Engineering, Tianjin University of Technology, Tianjin, China
suncc@email.tjut.edu.cn
[2] Engineering Research Center of Learning-Based Intelligent System (Ministry of Education), Tianjin University of Technology, Tianjin, China
[3] College of Intelligence and Computing, Tianjin University, Tianjin, China
[4] School of Computer Science and Engineering, Northeastern University, Shenyang, China
{shendr,nietiezheng}@mail.neu.edu.cn

Abstract. Entity resolution (ER) is an ongoing topic in data integration and data governance, which attracts considerable attention from multiple research fields. Recently, deep learning techniques have been substantially applied to entity resolution. We focus on entity resolution with graph based multi-semantic data embedding. In ER, data with attributes cannot be well represented by common word embeddings from natural language processing. In this work, data with attributes are modeled as a family of multitype bipartite information networks, each of which captures a specific type of semantics in data. Based on this, multi-semantic embeddings of data are collectively learned through the family of information networks. Particularly, a novel method is introduced to learn similarity based bipartite network embeddings. Generated tailored data embeddings are put into a flexible hierarchical ER framework, which outputs ER classification distributions. Our approach is comprehensively evaluated on a group of datasets, which presents its effectiveness.

Keywords: Entity resolution · Multi-semantic data embedding · Network embedding · Data with attributes · Data integration

1 Introduction

Data in organizations (such as enterprises, hospitals) are dispersed all over in various data formats. It is important to identify all records corresponding to the same real-world entity across data sources, which is a fundamental task called entity resolution (ER) in data integration and data governance [1, 2]. ER, also known as entity matching and record linkage, is a long-standing research topic in database, data mining and machine learning.

Recently deep learning (DL) has been extensively explored in ER research [3–6]. Although deep learning based ER (deep ER) research achieves remarkable progress, there still exists potential space for necessary improvements. A fundamental aspect

A. Bhattacharya et al. (Eds.): DASFAA 2022, LNCS 13247, pp. 20–35, 2022.
https://doi.org/10.1007/978-3-031-00129-1_2

of deep ER models is data embedding [3, 4]. The majority of existing works utilize universal word embeddings, such as word2vec [7, 8], GloVe [9] and FastText [10], all of which are pre-trained on large corpora of natural language processing (NLP). Despite their universality, ER tasks see deficiencies of pre-trained word embeddings. ER mainly processes data records from data with attributes (as illustrated in Fig. 1) [1], which are more rigorously organized than free texts in NLP. Data with attributes subsume structured data (like data in relational databases) and semi-structured data (like CSV data and JSON data). In the following, data refer to data with attributes if not particularly specified. Universal word embeddings fail to capture attribute-related semantics, which is detailed later. Also, ER datasets are often deeply domain-specific, like enterprise data and medical data. Such data commonly involve custom words out of universal vocabulary, which are missed by pre-trained word embeddings, called out-of-vocabulary (OOV) problem. Thus, tailored local embeddings of data with attributes are preferred in ER.

	Item	Model	Category
r_1	Lenovo ThinkPad X1 Carbon Gen 7	20QD001WUS	Laptops / Notebooks
r_2	Apple MacBook Air A1466,	MQD32LL/A, *Laptops / Notebooks*	

(a) Data source 1 (ds1)

	Title	Type	Brand
r_3	ThinkPad X1 Carbon (7th Gen), 20QD001WUS	Touchscreen Laptops	Lenovo
r_4	Latest ThinkPad X1 Carbon Gen 9, 20XW003GUS	Laptops	Lenovo

(b) Data source 2 (ds2)

Fig. 1. This is an example of data with attributes in our ER setting. In such data, records are annotated with attributes, such as ds1 and ds2. Each field is uniquely fixed by both a record and an attribute, such as $f_{11} = r_1[\text{Item}] = $ "Lenovo ThinkPad X1 Carbon Gen 7", $f_{12} = r_1[\text{Model}] = $ "20QD001WUS", and $f_{13} = r_1[\text{Category}] = $ "Laptops / Notebooks". There might be complex attribute associations (directed): from Item in ds1 to Title, Brand in ds2; from Title in ds2 to Item, Model in ds1. Also, there might be dirty data: in r_2, the field of Category should be "*Laptops/Notebooks*", which is however misplaced in the field of model.

Let us analysis data (as Fig. 1 shows) in details. These data are hybridly hierarchical. In the two-dimensional data hierarchy, there are four roles: attributes, records, fields and tokens. A schema consists of several non-ordered attributes, which shows its set nature. Each record is instantiated following the schema, where each field corresponds to a unique attribute. In such an organization, a field is horizontally related to other fields in the same record (row), and is vertically related to other fields in the same attribute (column). A field itself is a token sequence, a.k.a., a short text. Specifically, as shown in Fig. 2, there are three types of semantic relations: *attribute-field*, each field is annotated by a unique attribute; *record-field*, a record is composed of several fields; *field-token*, a field is composed of several tokens. Hence semantics in data is multitype, including schema semantics and hierarchical instance semantics (i.e., fields in a record and tokens in a field).

Traditional word or text embeddings [7–11], which originate from free texts in NLP, can neither well process the complex hierarchy nor fully capture multitype semantic relations in our data. In order to learn tailored local embeddings of data with attributes for ER, there are two major challenges. (1) The first challenge is how to build a unified data model to encode these multitype semantic relations. (2) The second challenge is how to learn effective data embeddings through the given unified data model for downstream ER tasks. On one hand, local data embeddings should fully capture both schema semantics and hierarchical instance semantics; on the other hand, local embeddings should be similarity driven, considering downstream ER tasks. EmbDI creates local embeddings for data integration tasks [5], and makes certain progress on our topic. Yet EmbDI is limited in following aspects. Its data model ignores the key role of fields, which reduces representation completeness of data with attributes. Its embedding method leverages vanilla random walks from DeepWalk [12], and does not fully exploit differences among objects of distinct types and heterogeneousness of semantic relations among such objects.

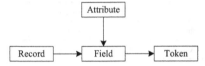

Fig. 2. Multitype semantic relations in data with attributes. Each arrowed line indicates a specific semantic relation type.

In this work, our goal is to generate tailored multi-semantic data embeddings for ER. First, we model data with attributes as a family of multitype bipartite information networks, including attribute-field network, record-field network and field-token network, each of which captures a specific type of semantic relations. Second, we learn multi-semantic local embeddings of data with attributes through these multitype information networks, which are tailored for ER. Since fields play key roles in both data organization and record comparisons, we choose to learn distributed representations of fields by collectively embedding the three bipartite information networks. In this way, field embeddings fully capture the hybrid hierarchy of schema semantics and instance semantics. Regarding similarity requirements of ER, we propose a similarity driven method for bipartite information network embedding, which maps vertices into a low dimensional space according to their effectively measured similarities. In essence, the embedding proximity distribution should be consistent with the measured similarity distribution. Third, we introduce a hierarchical representation-comparison-classification framework for ER. Schema mapping is not a priori in our ER setting, but can be inferred with field embeddings by the ER framework. Finally, we carry out comprehensive experiments over three types of datasets to evaluate the proposed approach. The evaluations present our improvements over previous works, and test different components of our approach.

Summary of contributions.

- We propose to represent data with attributes as a family of bipartite information networks, which fully preserve multitype semantic relations among records, attributes, fields and tokens in data.

- We propose to learn tailored multi-semantic distributed representations of data for ER tasks, by collectively embedding multitype information networks. Especially, we design a similarity based bipartite network embedding method.
- We propose a flexible representation-comparison-classification framework of deep ER, into which probabilistic schema mapping is integrated.
- We conduct extensive experimental evaluations on three types of seven datasets, which show effectiveness of our approach and effects of its components.

Organization of the rest paper. Section 2 formalizes the problem. Section 3 specifies multi-semantic data embedding through multitype information networks. Section 4 presents a representation-comparison-classification framework for ER. Section 5 conducts experimental evaluations over three types of datasets. Section 6 reviews related works. Section 7 concludes the whole work.

2 Problem Formalization

Entity resolution (ER) determines whether multiple records correspond to the same real-world entity. ER is actually a classification problem, which can be solved with deep neural networks [3, 4], a.k.a. deep ER. This work focuses on learning tailored local data embeddings for ER tasks, which is a fundamental problem in deep ER model building [4]. Basically, we model data as multitype information networks and generate multi-semantic distributed representations of data by networks embedding.

Multitype Information Networks for Data with Attributes. As Fig. 2 illustrates, there are three semantic relation types in data. To incorporate multitype semantic relations into a unified representation, we define a family of multitype information networks and Fig. 3 presents an example. We choose to model data as multitype bipartite information networks rather than a single heterogeneous information network, because different types of semantic relations are not comparable.

Definition 1. Attribute-Field Network. Attribute-field network is a weighted bipartite graph $G_{AF} = (A \cup F, E_{AF}, W_{AF})$, where A is a set of attributes, F is a set of fields, and E_{AF} is the set of edges between attributes and fields. The weight w_{ij} of the edge between attribute a_i and filed f_j is set to 1 uniformly.

Attribute-field network connects schema and instance, which is from abstract to instantiation. The attribute set size is usually small, and for instance, there are three attributes in ds1 or ds2 of Fig. 1. The number of fields corresponding to each attribute is up to record number, which is flexible. Thus the total number of fields can be very large.

Definition 2. Record-Field Network. Record-field network is a weighted bipartite graph $G_{RF} = (R \cup F, E_{RF}, W_{RF})$, where R is a set of records, F is a set of fields, and E_{RF} is the set of edges between records and fields. The weight w_{ij} of the edge between record r_i and filed f_j is set to 1 uniformly.

Record-field network is an affiliation from ensembles to components. A record has n fields, where n is the attribute set size.

Definition 3. Field-Token Network. Field-token network is a weighted bipartite graph $G_{FT} = (F \cup T, E_{FT}, W_{FT})$, where F is a set of fields, T is a set of tokens, and E_{FT} is the set of edges between fields and tokens. The weight w_{ij} of the edge between field f_i and token t_j is set to the number of times token t_j appears in field f_i, and is normalized into (0, 1] with the max number.

Field-token network captures token co-occurrences at the field level, which expresses fine-grained data semantics. Field semantics majorly stem from token level semantics, and so does for field similarity.

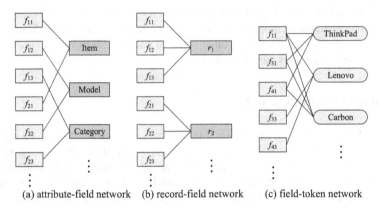

(a) attribute-field network   (b) record-field network   (c) field-token network

Fig. 3. A family of information networks constructed with data ds1 and ds2 in Fig. 1. All three information networks are partially illustrated, due to space limits.

Data with Attributes Embedding for Entity Resolution. With above three information networks, we formalize our problem. Our goal is to learn distributed representations of data with attributes, which is optimized for downstream ER tasks. Among four types of objects in the hybrid hierarchy, fields act a role of hubs, which connects attributes, records and tokens, as we see attribute-field network, record-field network and field-token network together. Fields, consisting of sequential tokens, are primary elements of records, and meanwhile, are semantically constrained by attributes. Since fields play the role of hubs across multitype information networks and are basic units for record comparisons, we choose to learn field embeddings for ER.

Definition 4. Data with Attributes Embedding (DAE) for Entity Resolution.
Given a collection of data with attributes from one or several data sources, the goal of DAE is to learn multi-semantic distributed representations of fields by embedding these multitype information networks built from the collection into a low dimensional vector space. Field embeddings should fully capture both schema semantics and hierarchical instance semantics. Also, field embeddings should be similarity orientated, where the proximity in the embedding space should be consistent with an effective similarity.

3 Multi-semantic Data Embedding Through Multitype Information Networks

Basically, we embed data with attributes through multitype information networks, and the output is tailored embeddings of fields, which cover all datasets to be resolved. An essential problem of ER is similarity computation, which calls for similarity based data embedding. Inspired by [13], we propose a common neighbor similarity based bipartite information network embedding method. Then we generate multi-semantic field embeddings by collectively leveraging the three information networks (of different types) constructed from data with attributes.

3.1 Similarity Based Bipartite Network Embedding

We embed bipartite networks with a novel common neighbor similarity.

Common Neighbor Similarity. For a bipartite network $G = (V_A \cup V_B, E_{AB}, W_{AB})$, V_A and V_B are two disjoint vertex sets of different types, and E_{AB} is the edge set between them. Generally, similarity between two vertices of the same type is indirectly indicated by their common neighbors of the other type, since such vertices are never linked directly. Given two vertices v_i and v_j from V_A, then their similarity can be measured as follows.

$$
sim_{cn}(v_i, v_j) = \frac{\sum_{v_k \in N(v_i) \cap N(v_j)} \frac{1}{d(v_k)}(w_{ik} + w_{jk})}{\sum_{v_m \in N(v_i)} \frac{1}{d(v_m)} w_{im} + \sum_{v_n \in N(v_j)} \frac{1}{d(v_n)} w_{jn}} \tag{1}
$$

$N(v_i)$ is the neighbor set of v_i; $d(v_i)$ is the degree of v_i. Our vertex similarity is a weighted variant of the dice similarity. We weight both edges and vertices in our similarity. For a vertex v_i and its neighbor v_k, their edge e_{ik} is naturally weighted as w_{ik} by network G. As a neighbor, vertex v_k is weighted by $1/d(v_k)$, which is inspired by classical IDF (Inverse Document Frequency). Thus, for vertex v_i, its neighbor v_k's importance is measured by both vertex v_k's weight and their edge weight, expressed as $(1/d(v_k))w_{ik}$. Finally, the denominator of the right part in formula 1 is v_i's weighted neighbors plus v_j's weighted neighbors, and the numerator is their weighted common neighbors. Note that each common neighbor is counted twice since the neighbor is linked to v_i and v_j separately.

Take the field-token network as an example. The more proportion of tokens two fields share, the more the two fields are similar; the more fields a token occurs in, the less the token contributes to field similarities.

Bipartite Network Embedding. Here we focus on embedding vertices of the same type in a bipartite network. Considering similarity computation desideration of ER, such as field similarities, the learned vertex feature representations are supposed to reflect distributions of a given vertex similarity $sim: V \times V \to \mathbb{R}$ over all vertices of the same type in a bipartite network.

We define neighborhoods in a bipartite network as follows: two vertices of one type belong to the same neighborhood(s) if they share at least one common neighbor vertex of the other type; otherwise, they are separated into different neighborhoods. Vertices sharing similar neighborhoods in networks should be mapped to be in close proximity in the embedding space.

Let $f: V \to \mathbb{R}^d$ be the mapping function from vertex v to its feature representation $f(v)$ (a d-dimensional vector), which we want to learn. To model proximity in the embedding space, we define the conditional probability of vertex v_j in set V_A given vertex v_i in set V_A, as shown in formula 2. This is actually a normalized proximity.

$$pxt(v_j|v_i) = \frac{\exp(f(v_j)^{\mathrm{T}} f(v_i))}{\sum_{v_k \in V_A} \exp(f(v_k)^{\mathrm{T}} f(v_i))} \tag{2}$$

Then given a bipartite network similarity $sim(\cdot, \cdot)$, we generate its similarity distribution. For instance, the conditional similarity of v_j given v_i is defined in formula 3.

$$p_{sim}(v_j|v_i) = \frac{sim(v_i, v_j)}{\sum_{v_k \in V_A} sim(v_i, v_k)} \tag{3}$$

We want the embedding proximity to be consistent with the given similarity. Thus, we define the objective function as Kullback-Leibler (KL) divergence between the embedding proximity distribution and the given similarity distribution, as formula 4 shows, and minimize it.

$$O_{emb} = \sum_{v_i \in V_A} KL(p_{sim}(\cdot|v_i) \| pxt(\cdot|v_i)) \tag{4}$$

Omitting some constants in formula 4, the objective function can be rewritten as formula 5, which is cross-entropy.

$$O_{emb} = -\sum_{v_i \in V_A} p_{sim}(\cdot|v_i) \log pxt(\cdot|v_i) \tag{5}$$

This objective can be optimized with asynchronous stochastic gradient descent (ASGD). But it is computationally expensive to calculate conditional probabilities $p_{sim}(\cdot|v_i)$ and $pxt(\cdot|v_i)$, which needs summation over the entire vertex set. To settle this issue, we adopt negative sampling (NEG) method [8], which, for each positive sample, selects K negative samples according to some noise distribution. A positive sample is defined as two vertices (of same type) sharing neighborhood(s), where they are neighbors to each other; a negative sample is defined as two vertices (of same type) sharing no neighborhood. Adoption of NEG makes our model scalable. Formally, we define negative sampling by the objective, as formula 6.

$$O_{NEG} = \sum_{\substack{v_i \in V_A \\ v_j \sim p_{sim}(\cdot|v_i)}} [\log \sigma(f(v_j)^{\mathrm{T}} f(v_i)) + \sum_{k \in [1,K]} \mathbb{E}_{v_k \sim P_n(v)} \log \sigma(-f(v_k)^{\mathrm{T}} f(v_i))] \tag{6}$$

$\sigma(x) = (1 + e^{-x})^{-1}$ is the sigmoid function. The first term models a positive sample, where v_j is sampled from neighborhoods of v_i, and their similarity by $sim(\cdot, \cdot)$ is positive; the second term models K negative samples randomly selected from the noise distribution $P_n(v)$, which is set by following [8].

Embeddings of attribute-field network, record-field network and field-token network can all be learned by the proposed model. In our model, $sim(\cdot, \cdot)$ is set to our proposed common neighbor similarity $sim_{cn}(\cdot, \cdot)$.

3.2 Multi-semantic Embedding for Data with Attributes

There are three information networks: attribute-field, record-field and field-token. As we see, fields occur in all multitype networks, and are also what we want to embed for downstream ER tasks. Each network indicates a unique affiliation relation, and has a particular semantic interpretation. Field-token networks contain token level semantics, which fundamentally contributes to field semantics, and plays a key role in field similarity computations. Record-field networks reflect record level semantics, where each small set of fields co-occur in the same record context. Attribute-field networks reflect schema semantics, where each (large) set of fields are constrained in the same attribute context.

Field embeddings should contain all multitype semantics. Therefore, field representations are collectively leaned through the three bipartite information networks. We define the collective objective function (formula 7) for multi-semantic field embedding, and minimize it.

$$O_{all} = \alpha O_{FT} + \beta O_{RF} + \gamma O_{AF} \tag{7}$$

$$O_{FT} = -\sum\nolimits_{v_i \in F \in G_{FT}} p_{sim}(\cdot|v_i) \log pxt(\cdot|v_i) \tag{8}$$

$$O_{RF} = -\sum\nolimits_{v_i \in F \in G_{RF}} p_{sim}(\cdot|v_i) \log pxt(\cdot|v_i) \tag{9}$$

$$O_{AF} = -\sum\nolimits_{v_i \in F \in G_{AF}} p_{sim}(\cdot|v_i) \log pxt(\cdot|v_i) \tag{10}$$

O_{FT} is the objective function for the field-token network embedding, O_{RF} is the objective function for the record-field network embedding, and O_{AF} is the objective function for the attribute-field network embedding. The hyperparameters α, β, γ ($\alpha + \beta + \gamma = 1$) are weights for different objectives, which control the contribution of each network embedding to overall field embedding.

We jointly train the model, where all the three types of networks are utilized. Since edges from different networks are not comparable, we choose to interleave updates of different network embeddings. Hence the model is updated network by network.

4   Flexible Entity Resolution with Multi-semantic Data Embeddings

There are two data sources S with schema $[a_1^s, \ldots, a_m^s]$ and T with schema $[a_1^t, \ldots, a_n^t]$. Entity resolution determines if two records r^s and r^t correspond to the same real-world entity, where $r^s = \{<a_1^s, f_1^s>, \ldots, <a_m^s, f_m^s>\}$ is from S and $r^t = \{<a_1^t, f_1^t>, \ldots, <a_n^t, f_n^t>\}$ is from T. Each field f_i is annotated by a unique attribute a_i, and is a token sequence.

With tailored multi-semantic field embeddings as the base, we propose a flexible representation-comparison-classification framework for ER. We integrate probabilistic schema mapping into ER. We adopt inter-attention to infer probabilistic attribute associations, and utilize intra-attention to arrange attribute weights.

Representation Layer. All dirty data to be resolved are put into DAE, and local field embeddings are generated. A field f_i^s from record r^s is represented as h_i^s. Then, r^s is represented as $H^S = [h_1^s, \ldots, h_m^s]$. [,] is a vector or matrix concatenation.

$$h_i^s = \text{DAE}(f_i^s) \tag{11}$$

Comparison Layer. This layer includes field alignment, comparison and weighting. Record comparisons are bidirectional, and we just specify $r^s \rightarrow r^t$ for simplicity. It aligns fields from r^s to r^t in probability, compares records in the field level, and arranges field weights. The output of this layer is a pair of directional record similarities.

Field Alignment. We build probabilistic schema mapping from r^s to r^t with inter-attention [14]. For each field representation h_i^s of record r^s, its soft-aligned representation is computed with all field representations of record r^t. Soft field alignment jointly analyzes two records, and results in pairwise field proximities from H^s to H^t, denoted as $\alpha^{s \rightarrow t}$. Field level inter-attention score from H^s to H^t is $(H^s)^T W^{s \rightarrow t} H^t$, where $W^{s \rightarrow t}$ is a trainable matrix. With softmax, attention scores are normalized into field alignment matrix $\alpha^{s \rightarrow t}$, where each entry $\alpha^{s \rightarrow t}(i, j)$ is proximity from h_i^s to h_j^t. $\widehat{H}^s$ is H^s's soft-aligned representation with H^t.

$$\alpha^{s \rightarrow t} = \text{softmax}((H^s)^T W^{s \rightarrow t} H^t) \tag{12}$$

$$\widehat{H}^s = H^t (\alpha^{s \rightarrow t})^T \tag{13}$$

Field Comparison. For each h_i^s and its soft-aligned representation $\widehat{h}_i^s$, we compute their element-wise absolute difference $|h_i^s - \widehat{h}_i^s|$ and Hadamard product $h_i^s \odot \widehat{h}_i^s$. The concatenation of the two interactions is put into a two-layer highway network, and a compact similarity representation $\widetilde{h}_i^s$ is generated. Formula 14, organized in the record way, presents initial field level similarity from r^s to r^t. So far, all fields play equally important roles in comparisons.

$$\widetilde{H}^s = \text{Highway}([|H^s - \widehat{H}^s|, H^s \odot \widehat{H}^s]) \tag{14}$$

Field Weighting. As is commonly known, different fields do not contribute equally to record similarities. We introduce an intra-attention mechanism [14] to capture field importances in similarity representations. $\widetilde{H}^s$'s intra-attention score is computed as product of $\widetilde{H}^s$ and a global context vector c^s, which is trainable. Attention scores are normalized with softmax into intra-attention β^s. Weighted similarity representation $s^{s \rightarrow t}$ from r^s to r^t is obtained by applying β^s on initial similarity $\widetilde{H}^s$.

$$\beta^s = \text{softmax}((\widetilde{H}^s)^T c^s) \tag{15}$$

$$s^{s \rightarrow t} = \widetilde{H}^s (\beta^s)^T \tag{16}$$

ER Classification Layer. We build a binary ER classifier with a highway network and softmax. The concatenation of similarities $s^{s \rightarrow t}$ and $s^{s \leftarrow t}$ is fed into a two-layer fully connected highway network, and the output is the aggregated similarity $s^{s\leftrightarrow t}$. $s^{s\leftrightarrow t}$ is put into a softmax classifier, and the final output is the ER distribution $P(y|r^s, r^t)$.

$$s^{s\leftrightarrow t} = \text{Highway}([s^{s\rightarrow t}, s^{s\leftarrow t}]) \tag{17}$$

$$P(y|r^s, r^t) = \text{softmax}(\boldsymbol{W}s^{s\leftrightarrow t} + b) \tag{18}$$

The ER model is trained by minimizing cross-entropy loss O_{ER}, where y_l is the label of ground truth and y_{pre} is the predicted label.

$$O_{ER} = crossEntropy(y_l, y_{pre}) \tag{19}$$

5 Experimental Evaluation

5.1 Experiments Setup

Datasets. As illustrated in Table 1, there are three groups of datasets for evaluations, including standard data and two types of hard data: dirty data and complex data. We implement an enhanced variant of UIS data generator [15] (eUIS for short) to help generate dirty data and complex data. We generate a standard person dataset Person-Person (PP), including two partitions with the same schema: name, telephone, address, city, state and zip code. In PP, there are duplicates between two partitions, but there is no duplicate inside each partition. Later, we construct a dirty version PP1 and a complex version PP2 with PP.

(1) Standard data. There are three standard datasets DBLP-Scholar (DS), DBLP-ACM (DA) and Fodors-Zagats (FZ) [4], which are well structured, are perfectly one-to-one aligned in schemas, and contain simple fields with few errors.

(2) Dirty data. Two dirty datasets PP1 and BR1 are derived from standard datasets PP and BeerAdvo-RateBeer (BR) [4] respectively. There are errors and value misplacements in dirty data. We generate a dirty dataset with a standard dataset in two steps: error injection and field misplacement. (a) With probability of 25%, a selected field in a record is injected into errors including edit errors (random character insertion, deletion, replacement and swap) and token errors (random token repeat, insertion, deletion, replacement and swap). (b) With probability of 40%, one field is randomly selected and moved into another attribute in the same record.

(3) Complex data. Two complex datasets PP2 and BR2 are derived from standard datasets PP and BR [4] respectively. There is at least one one-to-many attribute association between schemas of different data sources. We construct a complex dataset with a standard dataset in two steps: error injection and attribute merging. (a) Error injection here is similar to error injection of dirty data generation, except that probability is 20%. (b) Then, a subset of attributes is merged into a complex attribute given a schema. For PP2 from PP, name and address are merged into a complex attribute name-address in partition one; name and telephone are merged into a complex attribute name-telephone, and, address, city & zip code are merged

Table 1. Dataset descriptions.

Type	Dataset	Domain	Size	#Pos.	#Attr.
Standard	DBLP-Scholar (DS)	Citation	28,707	5,347	4-4
	DBLP-ACM (DA)	Citation	12,363	2,220	4-4
	Fodors-Zagats (FZ)	Restaurant	946	110	6-6
Dirty	Person-Person[1] (PP1)	Person	11,974	1,002	6-6
	BeerAdvo-RateBeer[1] (BR1)	Product	450	68	4-4
Complex	Person-Person[2] (PP2)	Person	11,974	1,002	5-3
	BeerAdvo-RateBeer[2] (BR2)	Product	450	68	3-2

into a second complex attribute address-city-zipcode in partition two. For BR2 from BR, Beer Name and Brew Factory Name are merged into a complex attribute BN-BFN in BeerAdvo; Beer Name, Style and ABV are merged into a complex attribute BN-style-ABV in RateBeer.

Metric. Our work focuses on resolution quality of ER. We choose the common metric F_1 measure for ER evaluation. $F_1 = 2PR/(P + R)$, P is precision and R is recall. P is the proportion of predicted matches that are truly matched, and R is the proportion of true matches that are correctly predicted.

Settings. Information networks based data embedding is implemented with C++, and the ER model is implemented with Python (PyTorch). All experiments are run on a server with 8 CPU cores (Intel(R) E5-2667, 3.2 GHz), 64G memory, and NVIDIA GeForce GTX 980 Ti.

Following previous works [13, 16–18], the network embedding dimensionality is set to $d = 128$. Each dataset is split into 3: 1: 1 for training, validation and testing of ER tasks. Numbers of epochs, mini-batch size and dropout rate are 15, 16 and 0.1. Adam is used as optimization algorithms.

5.2 Comparisons with Existing Works

We compare our approach DAE based ER (DAER) with existing graph based deep ER approaches EmbDI [5], GraphER [6] and two deep ER baselines DeepER [3], DeepMatcher [4] on three types of data.

Figure 4 illustrates overall performances of five ER approaches on three standard datasets: DS, DA and FZ. All approaches achieve relatively comparable (and good) performances on standard datasets, and F_1 gaps between different approaches are usually not big. Specifically, DAER outnumbers other approaches by 0.2% to 3% in ΔF_1 on DS and DA; all five approaches achieve the same F_1 on FZ. This is mainly due to that these standard datasets are easy to resolve.

Figure 5 illustrates overall performances of five ER approaches over two dirty datasets: BR1 and PP1. In general, DAER obviously outperforms the other four approaches on dirty data. On BR1, ΔF_1 between DAER and the others are at least 8.1%; on PP1, ΔF_1 between DAER and the others are at least 11.6%. There are many typos, token errors and even more value misplacements in dirty data, which make data hard to resolve. Our DAER's improvements majorly benefit from local field representations. Our tailored field representations capture multitype semantics, including token level (breaking attribute boundaries), record level & attribute level, and are learned based on similarities, both of which are essential for similarity computing in ER.

Figure 6 depicts overall performances of five ER approaches over two complex datasets: BR2 and PP2. Overall, DAER surpasses the other four approaches on complex data. On BR2, ΔF_1 between DAER and the others are at least 5.6%; on PP2, ΔF_1 between DAER and the others are at least 13.3%. There are complex attribute associations in schemas of complex data; also, complex data contains typos and token errors. Hence complex data are difficult to resolve. We think DAER's advantages over previous approaches come from following aspects: (1) tailored local field representations, which capture multitype semantics and are similarity driven, and (2) the proposed ER model, which integrates flexible schema mapping into ER.

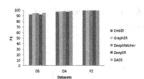

Fig. 4. General comparisons on standard data

Fig. 5. General comparisons on dirty data

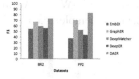

Fig. 6. General comparisons on complex data

5.3 Detailed Analysis

We evaluate key components of our proposed solution in detail.

Effect of Graph Embedding. Data embedding via multitype information networks is our major contribution. We compare different graph embedding methods for data embedding in ER. We use classical graph embedding methods PTE [18] and Node2Vec [16] for local field representations, and other parts stay the same, denoted as DAER-PTE and DAER-N2V respectively. In DAER-PTE, PTE is directly used for local field embedding. In DAER-N2V, Node2Vec replaces our bipartite network embedding method (for each information network embedding) in local field representations. On three standard datasets, they have comparable performances, as Fig. 7 shows. DAER overall outnumbers the other two approaches in F_1 on both dirty data and complex data. As Fig. 8 illustrates, ΔF_1 between DAER and the others are at least 7.7% on two dirty datasets. As Fig. 9 illustrates, ΔF_1 between DAER and the others are at least 9.1% on two complex datasets. The evaluation advantages show that our multitype information networks based data embedding is effective in ER. Our data embedding captures multitype semantics and considers object similarities, both of which are essential for similarity computations in ER.

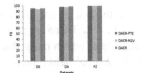
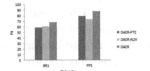
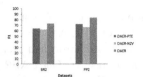

| **Fig. 7.** Graph embedding tests on standard data | **Fig. 8.** Graph embedding tests on dirty data | **Fig. 9.** Graph embedding tests on complex data |

Effect of ER Model. Probabilistic schema mapping (PSM) and field weighting (FW) are two key components of our ER model, and we test their effects on three types of data. DAER-[-FW] is DAER without FW, and DAER-[-PSM] is DAER with vanilla schema mapping instead of PSM. Figures 10, 11, 12 illustrate results on three standard datasets, on two dirty datasets and on two complex datasets respectively. On standard datasets, there are minor ΔF_1 between DAER and the other two approaches. On both dirty datasets and complex datasets, DAER obviously outperforms the others in F_1. Especially on complex datasets, loss of PSM reduces much more accuracies than loss of FW, due to existence of many complex attribute associations. The evaluation results testify that PSW and FW are effective components of our ER model, especially for dirty data and complex data, which commonly exist in the real world.

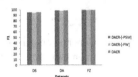
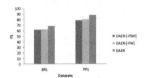
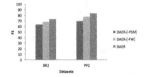

| **Fig. 10.** ER model tests on standard data | **Fig. 11.** ER model tests on dirty data | **Fig. 12.** ER model tests on complex data |

6 Related Work

Entity resolution attracts multiple research communities, such as database, data mining and machine learning [1, 19]. Currently, deep learning is strongly driving ER research. DeepER is a pioneer deep ER work [3], which builds an ER system with distributed representations of words and LSTM. Also, DeepER investigates DL based blocking for ER efficiency. DeepMatcher defines a design space of deep ER, including attribute embedding, attribute similarity representation and classifier [4]. DeepMatcher introduces four methods: heuristic-based, RNN-based, attention-based and hybrid. There are graph based deep ER works [5, 6]. GraphER is a token-centric approach, which utilizes GCN (graph convolutional network) to aggregate token-level comparisons [6]. EmbDI creates embeddings of relational data for data integration tasks, such as schema mapping and ER [5]. EmbDI constructs a graph with tokens, attributes & records, and run vanilla random walks over the graph to generate sentences to describe similarities across objects (like DeepWalk [12]). However, EmbDI disregards the key role of fields in graph construction, and does not fully utilize heterogeneousness of both objects (tokens, attributes & records) and their semantic relations in learning of embeddings.

Along with rapid DL developments, word embeddings have been widely used in NLP tasks. Trained over large NLP corpora, word embeddings map words into a compact vector space, which preserves syntactic and semantic word relationships. As a milestone, word2vec proposes two neural net language models skip-gram and CBOW [7, 8], which are able to learn high-quality word vectors with their simple but useful neural architectures. Word2vec produced profound influences on later word embeddings, and also triggered other embeddings, such as graph embeddings [20]. GloVe incorporates global information by matrix factorization and local information by context window into word representations [9], promoting its performance. Regarding unseen words, FastText extends the skip-gram model with character n-grams [10], where words are represented as sums of n-gram vectors.

Vertex embedding, as a core branch of graph embeddings, maps vertices into a low dimensional vector space by embedding graph structures [20]. Inspired by word2vec, DeepWalk captures the "context" of a vertex by running random walks, and utilizes skip-gram as the learning model [12], where generated walks act the role of sentences. Following DeepWalk, node2vec introduces biased random walks to diversify neighborhoods [16]. It guides random walks by configuring a mixture of BFS (breadth-first search) and DFS (depth-first search). LINE learns vertex embeddings by combining first-order and second-order proximities [17]. Incorporating both unlabeled and labeled information, PTE extends LINE for semi-supervised text data embedding [18]. As a versatile vertex similarity embedding framework, VERSE embeds graphs by reconstructing similarity distributions between vertices [13]. Our bipartite network embedding method is an improvement of VERSE that is adaptive to ER tasks. In heterogeneous information networks, metapath2vec defines meta-path based random walks and exploits a heterogeneous skip-gram model to learn vertex embeddings [21].

7   Conclusion

In this work, we study how to locally embed data with attributes for ER tasks. Data are modeled as a family of information networks, in which multitype semantic relations are preserved. Tailored multi-semantic distributed representations of fields are learned by collectively embedding these information networks. Particularly, a similarity driven method is proposed to embed each bipartite information network. With generated field embeddings, ER is carried out in a flexible representation-comparison-classification framework. Sufficient experimental evaluations over several datasets show that our approach is an effective solution. In future, an interesting potential research direction is how to apply our DAER approach to transfer learning of ER, which is meaningful for low-resource scenarios.

Acknowledgements. This work is supported by the National Natural Science Foundation of China (Grant Nos. 62002262, 62172082, 62072086, 62072084).

References

1. Köpcke, H., Thor, A., Rahm, E.: Evaluation of entity resolution approaches on real-world match problems. Proc. VLDB Endow. **3**(1–2), 484–493 (2010)

2. Hassanzadeh, O., Chiang, F., Lee, H.C., Miller, R.J.: Framework for evaluating clustering algorithms in duplicate detection. Proc. VLDB Endow. **2**(1), 1282–1293 (2009)
3. Ebraheem, M., Thirumuruganathan, S., Joty, S., Ouzzani, M., Tang, N.: Distributed representations of tuples for entity resolution. Proc. VLDB Endow. **11**(11), 1454–1467 (2018)
4. Mudgal, S., et al.: Deep learning for entity matching: a design space exploration. Proceedings of the 2018 International Conference on Management of Data, pp. 19–34 (2018)
5. Cappuzzo, R., Papotti, P., Thirumuruganathan, S.: Creating embeddings of heterogeneous relational datasets for data integration tasks. In: Proceedings of the 2020 ACM SIGMOD International Conference on Management of Data, pp. 1335–1349 (2020)
6. Li, B., Wang, W., Sun, Y., Zhang, L., Ali, M.A., Wang, Y.: GraphER: token-centric entity resolution with graph convolutional neural networks. In: Proceedings of the 29th International Joint Conference on Artificial Intelligence, pp. 8172–8179 (2020)
7. Mikolov, T., Chen, K., Corrado, G., Dean, J.: Efficient estimation of word representations in vector space. In: Proceedings of Workshop with the 2013 International Conference on Learning Representations (2013)
8. Mikolov, T., Sutskever, I., Chen, K., Corrado, G.S., Dean, J.: Distributed representations of words and phrases and their compositionality. In: Proceedings of the 2013 International Conference on Advances in Neural Information Processing Systems, pp. 3111–3119 (2013)
9. Pennington, J., Socher, R., Manning, C.D.: GloVe: global vectors for word representation. In: Proceedings of the 2014 Conference on Empirical Methods in Natural Language Processing (EMNLP), pp. 1532–1543 (2014)
10. Bojanowski, P., Grave, E., Joulin, A., Mikolov, T.: Enriching word vectors with subword information. Trans. Assoc. Comput. Linguist. **5**, 135–146 (2017)
11. Le, Q., Mikolov, T.: Distributed representations of sentences and documents. In: Proceedings of 2014 International Conference on Machine Learning, pp. 1188–1196 (2014)
12. Perozzi, B., Al-Rfou, R., Skiena, S.: DeepWalk: online learning of social representations. In: Proceedings of the 20th ACM SIGKDD International Conference on Knowledge Discovery and Data Mining, pp. 701–710 (2014)
13. Tsitsulin, A., Mottin, D., Karras, P., Müller, E.: VERSE: versatile graph embeddings from similarity measures. In: Proceedings of the 2018 World Wide Web Conference, pp. 539–548 (2018)
14. Hu, D.: An introductory survey on attention mechanisms in NLP problems. In: Proceedings of SAI Intelligent Systems Conference, pp. 432–448 (2019)
15. Hernández, M.A., Stolfo, S.J.: Real-world data is dirty: data cleansing and the merge/purge problem. Data Min. Knowl. Disc. **2**(1), 9–37 (1998)
16. Grover, A., Leskovec, J.: node2vec: scalable feature learning for networks. In: Proceedings of the 22nd ACM SIGKDD International Conference on Knowledge Discovery and Data Mining, pp. 855–864 (2016)
17. Tang, J., Qu, M., Wang, M., Zhang, M., Yan, J., Mei, Q.: LINE: large-scale information network embedding. In: Proceedings of the 24th International Conference on World Wide Web, pp. 1067–1077 (2015)
18. Tang, J., Qu, M., Mei, Q.: Pte: predictive text embedding through large-scale heterogeneous text networks. In: Proceedings of the 21th ACM SIGKDD International Conference on Knowledge Discovery and Data Mining, pp. 1165–1174 (2015)
19. Azzalini, F., Jin, S., Renzi, M., Tanca, L.: Blocking techniques for entity linkage: a semantics-based approach. Data Sci. Eng. **6**(1), 20–38 (2021)

20. Hamilton, W.L., Ying, R., Leskovec, J.: Representation learning on graphs: methods and applications. IEEE Data Eng. Bull. **40**(3), 52–74 (2017)
21. Dong, Y., Chawla, N.V., Swami, A.: metapath2vec: scalable representation learning for heterogeneous networks. In: Proceedings of the 23rd ACM SIGKDD International Conference on Knowledge Discovery and Data Mining, pp. 135–144 (2017)

Semantic-Based Data Augmentation for Math Word Problems

Ailisi Li[1], Yanghua Xiao[1,2(✉)], Jiaqing Liang[1], and Yunwen Chen[3]

[1] Shanghai Key Laboratory of Data Science, School of Computer Science,
Fudan University, Shanghai, China
{alsli19,shawyh}@fudan.edu.cn
[2] Fudan-Aishu Cognitive Intelligence Joint Research Center, Shanghai, China
[3] DataGrand Inc., Shanghai, China
chenyunwen@datagrand.com

Abstract. It's hard for neural MWP solvers to deal with tiny local variances. In MWP task, some local changes conserve the original semantic while the others may totally change the underlying logic. Currently, existing datasets for MWP task contain limited samples which are key for neural models to learn to disambiguate different kinds of local variances in questions and solve the questions correctly. In this paper, we propose a set of novel data augmentation approaches to supplement existing datasets with such data that are augmented with different kinds of local variances, and help to improve the generalization ability of current neural models. New samples are generated by knowledge guided entity replacement, and logic guided problem reorganization. The augmentation approaches are ensured to keep the consistency between the new data and their labels. Experimental results have shown the necessity and the effectiveness of our methods.

Keywords: Math word problem · Data augmentation

1 Introduction

Automatically solving Math Word Problem (MWP) has attracted more and more research attention in recent years. The MWP solvers are fed in with a natural language description of a mathematical question, and output a solution equation as the answer. In most cases, these questions are short narratives comprised of several known quantities and a query about an unknown quantity, whose value is the answer we desire. Table 1 shows a typical example of MWP, where x in the equation refers to the unknown quantity, and is calculated from the known quantities and specific constants such as π, 1, 2.

Numerous efforts have been devoted to solving this challenging task. Early studies relying on hand-crafted features [14,18,19] and predefined patterns [20] have limitations in generalization. Deep learning methods have become popular to solve the MWP task in recent years [22,24,28,32] due to their better capability

Table 1. An example of math word problem and new samples generated by semantic-based data augmentation approaches.

Original
Question 1: There are *390* kilograms of pears in the store, which is *40%* less than the weight of apples. x kilograms of apples are there in the store.
Equation 1: $x = 390 \div (1 - 40\%)$
Answer 1: 650
Knowledge guided entity replacement
Question 2: There are *390* kilograms of *bananas* in the *kitchen*, which is *40%* less than the weight of *watermelon*. x kilograms of *watermelon* are there in the *kitchen*.
Equation 2: $x = 390 \div (1 - 40\%)$
Answer 2: 650
Logic guided problem reorganization
Question 3: There are *390* kilograms of pears in the store, which is x less than the weight of apples. *650* kilograms of apples are there in the store.
Equation 3: $x = 1 - 390 \div 650$
Answer 3: 0.4

of generalization. [24] first modeled the MWP task as an equation generation task, and various works have followed this framework since then. Recent works in MWP mostly focus on designing complex generation models to capture more features from limited data. For example, [28] proposed a tree-structured decoder to imitate human behaviour, [32] utilized GCN, and [13] adopted the attention mechanism to better capture the relationship between quantities.

However, current MWP solvers still have weaknesses in terms of robustness and generalization. A superior MWP solver should understand a problem precisely in two ways. First, it is able to generate the same equation for a transformed question with only uninfluential entity replacement. For example, the equation generated for $Q1$ in Table 1 should not be changed when the *pears* in the question is replaced with *bananas*. Second, excellent models should be capable of generating a different equation when the logic of the question changes even if **the text of the question only changes slightly**. For example, in Table 1, the only difference between $Q1$ and $Q3$ is the position of token x. But their underlying logic is totally different and thus the corresponding equations are different. These two kinds of tiny local variances in questions lead to totally different results, one of which conserves the underlying logic, while the other one changes it completely. Humans are able to disambiguate these local variances easily while it's hard for most neural models to deal with discrete local variances. Previous works on MWP hardly consider this challenge.

The limitation of existing datasets for MWP is a main reason for the above-mentioned weaknesses. Since labeling MWP data is time-consuming, existing MWP datasets are all too small compared to datasets for other natural language processing tasks. Besides, few challenging samples with similar questions but different equations can be found in MWP datasets, which makes it hard for neural models to learn to deal with tiny local variances. The most popular and largest single-equation MWP dataset Math23k contains only 23,161 problems, which is rather limited compared to datasets in other field like sQuAD [17] with 150,000 questions. The weaknesses of existing datasets, especially the limited coverage of challenging samples, motivated us to augment the dataset with questions of minor variances leading to heterogeneous equations.

In this paper, we propose a set of semantic-based data augmentation approaches suitable for MWP task, namely knowledge guided entity replacement and logic guided problem reorganization. And two kinds of local variances are provided accordingly. Neural MWP solvers can benefit from our augmentation strategies in terms of generalization and the ability of dealing with tiny local variances. Unlike other popular augmentation approaches [25,27,30], which may cause inconsistency of the questions and equations in MWP task, our augmentation methods are carefully designed for MWP task to ensure consistency.

Knowledge Augmentation (Knowledge Guided Entity Replacement). As shown in Table 1, *pears* in the original question is replaced with another fruit *bananas*. And it is obvious that the replacement does not change the original logic, thus the new question conserves the original label. Our knowledge guided entity replacement method randomly replaces several entities in questions with other entities that belong to the same concept as the original ones. And we guide the replacement with knowledge base that contains much taxonomy knowledge.

Different from synonym replacement [4,25] which replaces random words with their synonyms, replacing entities could largely avoid semantic shifting since most entities in MWP are not crucial for the logic inference.

Logical Augmentation (Logic Guided Problem Reorganization). As the example shown in Table 1, quantities *390* and *40%* are known while quantity *650* is unknown in $Q1$. In the generated $Q3$, we let *40%* be the unknown one given *390* and *650*. The equation is changed accordingly to keep consistency. x in Eq. 3 is substituted for *650* and *40%* is replaced with x. Afterwards, we transform the new equation to its equivalent equation of the form $x = 1 - 390 \div 650$. To enrich the problem types of the training data, our logical augmentation iteratively set the known quantities in the original question and equation to the unknown. And the new equation is further transformed to its equivalent equation with mathematical properties for normalization.

It is worth mentioning that previous works only learn the equation of the unknown, while our augmentation method helps the neural models to make full use of the limited data by learning all the possible equations of the quantities in the question.

Our contributions are summarized as follows:

- To the best of our knowledge, this is the first systematical study of data augmentation for MWP task. And is easy to be applied to any neural models and extend to other math-related tasks.

- Our methods can generate coherent questions with consistent labels, which largely diversify both textual descriptions and equation templates. It also brings in massive challenging samples that existing datasets lack.
- Experimental results show the necessity and effectiveness of our methods. The performances of the additional evaluation also indicate that our methods largely enhances the generalization ability of neural models.

2   Related Work

2.1   Math Word Problem

Early works mostly utilized statistical methods or predefined rules. [18] utilized hand-crafted features to predict the lowest common ancestor operator for each quantity pair. [19] proposed a unit dependency graph based on [18]. [14] predefined a group of logic forms and converted the math question into them.

[24] first proposed to utilize a seq2seq model with recurrent neural network to generate equation template sequence. [21] proposed equation normalization to unify duplicated representations of equivalent expressions. [3] utilized a seq2seq model with the help of stack to align the semantic with the operator. [22] proposed a two-stage algorithm to predict a template tree. [13] applied group attention mechanism to extract more features.

[15,28,32] replaced the recurrent neural network based decoder with a tree-structured decoder and achieved satisfactory results. [31] leveraged the framework of knowledge distillation.

However, none of the works have tried to enhance the generalization ability of the neural models by data augmentation strategies.

2.2   Data Augmentation for Natural Language Processing (NLP)

We categorize data augmentation approaches for NLP into two types. The first type changes only the text while the second one changes both the text and labels.

Some research adds random noise to input data [29] or hidden states [12] making the models less sensitive to small perturbations. [25] systematically examined some basic augmentation methods including random synonyms replacement, word insertion, etc. [27] utilized tf-idf to help determine which words to replace. [23] adopted k-nearest neighbors to find synonyms in word embedding space. The Noise brought in by these methods could be tolerated in some tasks but not in MWP task due to the strict requirement of preciseness.

[9,26] fine tuned a pre-trained language model with text and label to generate new sentences given specific labels. [4] replaced words in the source and target sentences with rare words. Other approaches mainly applied generative models like VAE [7], seq2seq model [6], GPT [1], etc., to generate new text given a specific label. All of the methods above generating new text given a specific label are inappropriate to MWP task, because the equation is a sequence and is not enumerable.

Table 2. An example of the question generated with knowledge augmentation.

Q	*Ming Zhang ([PER]) bought n_0 apples ([FRU])*
Q_K	*Hong Li ([PER]) bought n_0 blueberries ([FRU])*

3 Methodology

3.1 Problem Statement

The MWP dataset contains training data D_{train} and testing data D_{test}, both of which are comprised of numerous questions Q and equations Eq. Neural models take Q as the input and generate the Eq sequence. In previous works, the neural model $\mathcal{M}$ is trained with the training data D_{train} and tested with the testing data D_{test}. Due to the limitations of D_{train} introduced in Sect. 1, we generate new samples D_{aug} from D_{train} by our augmentation approaches. And the neural model $\mathcal{M}$ is trained with both original and augmented data $D_{train} \cup D_{aug}$.

The question Q consists of a sequence of tokens $\mathcal{W} = \{w_i\}_{i=1}^{|\mathcal{W}|}$ including known quantities $\mathcal{N} = \{n_i\}_{i=1}^{|\mathcal{N}|}$ and entities $\mathcal{E} = \{e_i\}_{i=1}^{|\mathcal{E}|}$. Since quantities are fairly sparse, we replace quantities in Q with symbol n_i according to the occurrence order of the quantities during preprocessing phase. Quantities in Eq are replaced accordingly and we denote the answer of Eq as $\hat{n}$.

The target of MWP is to generate the Eq sequence which is composed of $\mathcal{N} \cup \mathcal{O} \cup \mathcal{C}$. Among them, $\mathcal{O}$ is the set of the operators (such as $\times$) and $\mathcal{C}$ is the constant set.

3.2 Knowledge Augmentation

The category information of entities is introduced to generate new questions. Given a sample (Q, Eq), we randomly choose θ entities mentioned in Q to be replaced with other entities belonging to the same concept as the original ones. If an entity e_i belonging to concept c is selected, the alternative entities are $e_j \in \mathcal{E}_c (i \neq j)$. Similar to [25], we set $\theta = \max(1, \alpha l)$, among which l refers to the length of Q and α is a hyper-parameter used to manage the replacement ratio. More entities are replaced for longer questions considering they tolerate noise better. We notice that an entity may appear more than once in a question, and all of them should be replaced if the entity is selected. For each question Q, the new questions generated by this means are denoted as Q_K. As an example in Table 2, entities *Ming Zhang* and *apples* are replaced with a random entity of *person* and *fruit* accordingly. Since entities are less informative for MWP inference, the generated questions conserve the logic of the original question. And the new equation matched with Q_K is still Eq.

Recognize Entities. There are two kinds of entities in MWP, namely **real-world entities** and **named entities**, and we take different strategies to recognize them. For detecting named entities like fake person names *Alice, Bob*, etc.

Table 3. An example of the question generated with logical augmentation.

Q	There are n_1 kilograms of pears in the store, which is n_2 less than the weight of apples. How many kilograms of apples are there in the store?
Q'	There are n_1 kilograms of pears in the store, which is n_2 less than the weight of apples. x kilograms of apples are there in the store
Q_L	There are n_1 kilograms of pears in the store, which is x less than the weight of apples. $\hat{n}$ kilograms of apples are there in the store

which are common in MWP, there are numerous models and tools work very well on this task. And we follow the method of [2] to recognize named entities. Another kind of entities is real-world entities like *apple, car*, etc. These entities can be easily recognized by referring to KGs. In this paper, we use WordNet [16] to guide our knowledge augmentation. Entities are linked to the WordNet syn-sets, and the direct hypernyms are viewed as their concepts. Besides, we restrict that only *physical entities* in questions could be replaced to avoid possible semantic shifting. *Abstract entities* like *time* should not be replaced in our case.

3.3 Logical Augmentation

Given a sample (Q, Eq), several (Q_L, Eq_L) pairs are generated by setting a known quantity in the original question to the unknown one in the new question. And the new equation is normalized with mathematical properties. Since the textual description of Q_L is similar to that of Q, neural models tend to generate similar equation sequences for them even though their ground truth solutions are totally different. With logical augmentation, neural models are forced to learn the different equations from the slight variations of input text, which to some extent enhances the inference ability of neural models.

Question Generation. In MWP, the letter x is usually assigned to represent the unknown. The value of it is the answer to the question Q (denoted as $\hat{n}$). The unknown quantities in questions are usually indicated by question words such as *how many*. Considering both question words and letter x refer to the same unknown, we replace the question words in Q with the letter x. And then the original question could be viewed as an assertive sentence about the logical relationship between the known and unknown quantities $\mathcal{N} \cup \hat{n}$. Previous training without augmentation only learns how to get $\hat{n}$ given $\mathcal{N}$, for Eq is in $x = f(\mathcal{N} \cup \mathcal{C})$ form, in which x refers to $\hat{n}$ as $\hat{n}$ is the unknown of Q. However, the logical relationship between other quantity pairs remains unlearned. To make full use of each data sample, we make the neural models additionally learn how to get $n_i \in \mathcal{N}$ given the other quantities $n_j \in \mathcal{N} \cup \hat{n}(i \neq j)$. As the letter x indicates the quantity we desire, quantities in Q are iteratively set to x. The generated questions are denoted as Q_L.

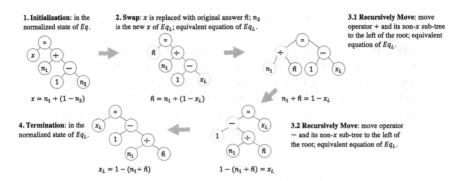

Fig. 1. The process of generating a new equation Eq_L. *As illustrated in Table 3, the new unknown quantity of Q_L and Eq_L is n_2 with n_1 and $\hat{n}$ known. The equation tree in the initialization step is built from Eq in Table 3. Equation trees from step 2 to step 4 are equivalent equations of Eq_L in different forms. And the tree in the termination step is the normalized form of Eq_L.*

Table 3 shows an example of questions generated with logical augmentation. Notably, when training with augmented data, original questions Q are replaced with Q' to ensure that all questions are expressed in an uniform form.

Equation Generation with Equation Tree Conversion. Since the known and unknown have been changed in the generated question Q_L, the corresponding equation Eq_L should be changed accordingly to keep consistency. To normalize the form of equations, the generated equations are transformed to their equivalent equations of a specific form based on mathematical properties with the help of equation trees. In the final normalized state, the term x is isolated on the left side of the equation.

Figure 1 shows an example of how a consistent and normalized equation Eq_L is generated from the original equation Eq.

Equation Tree. Equation tree is built from an equation whose root node is always the equal sign $=$. The left and right sub-trees of the root are expression trees of the left and right sides of an equation, thus the leaf nodes are operands and the inner nodes (except the root) are binary operators.

In the normalized state of an equation tree, the left sub-tree of the root only contains a leaf node x. The same equation tree in different states refers to equivalent equations of different forms. To normalize the equation, we transform the equation tree to its normalized state with two main actions based on the addition-subtraction property, the division property, and the multiplication property. 1) Move an operator node o (an inner node) and its non-x sub-tree (the sub-tree of o that does not contain x) to another side. 2) Switch o to its inverse operator, for example, switch $+$ to $-$ or $\times$ to $\div$.

The Process of Generating Eq_L. As shown in Fig. 1, there are four steps to generate Eq_L. 1) Initialization: a normalized equation tree is built from the original Eq. 2) Swap: x is substituted for $\hat{n}$, and the quantity n_i which is the new unknown is replaced with x_L (to distinguish with x). Since the new equation tree of Eq_L is not in normalized state, we take the available actions to transform the new equation tree to its normalized state. 3) Recursively Move: All of the operator nodes and their non-x sub-trees are recursively moved to the left of the root node in a top-down manner, leaving only x_L in the right. Different actions are chosen in different situations, all based on the natural mathematical properties. 4) Termination: Simply swapping the left and right sub-trees of the root node will make the equation tree normalized. And Eq_L could be restored from the equation tree.

4   Experiment

In this section, we conduct experiments to measure the scale of challenging samples in existing datasets. Besides, we evaluate our semantic-based augmentation strategies (denoted as s.based aug. for simplification) with three typical neural MWP solvers to show the improvements of the generalization ability brought in by s.based aug.. Extensive ablation studies are also conducted to verify the effectiveness of each augmentation strategy.

4.1   Dataset Analysis

Existing MWP datasets such as AI2 [5], SingleEQ [10], and AllArith [19] only have hundreds of samples. While relatively large-scale MWP datasets such as MAWPS [11] contains very few challenging samples as it is constructed with low lexical and template overlap. So we conduct our experiments on the largest and most popular MWP dataset Math23k [24], a Chinese MWP dataset with 23,161 pairs of (Q, Eq).

To make the neural models able to deal with discrete tiny local variances, it's necessary for the dataset to contain a great ratio of challenging samples that have similar questions but different equations. In this section, we will analyze Math23k from the amount and the quality of challenging samples in the training and the testing set.

Notably, the analysis in this section are all based on equation templates, which means only the structures of the equations are considered rather than the actual values. For example, equations $x = 3 + 2 + 1$ and $x = 5 + 4 + 2$ are viewed as the same in template.

The Amount of Challenging Samples in the Training Set. For a question $Q_i \in D_{train}$, if $\exists Q_j \in D_{train}(i \neq j)$ with different equation but similar text, Q_i and Q_j are considered as challenging samples. The similarity of two

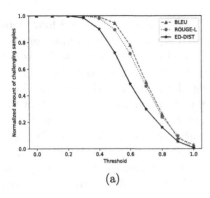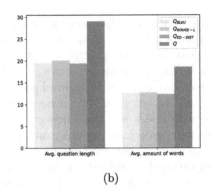

(a)   (b)

Fig. 2. (a) the amount of challenging samples filtered with different similarity score thresholds in the training set. (b) the quality of existing challenging samples.

questions are measured with three similarity scores, namely BLEU, ROUGE-L, and normalized reverse edit distance defined below (referred to as ED-DIST).

$$ED - DIST(Q_i, Q_j) = 1 - \frac{edit - dist(Q_i, Q_j)}{max(l_{Q_i}, l_{Q_j})}$$

l refers to the length of the question and $edit - dist(\cdot)$ is the Levenshtein distance of the given pair of questions. The more similar Q_i and Q_j are, the closer $ED-DIST(Q_i, Q_j)$ is to 1. Afterwards, we count the amount of challenging samples with the similarity thresholds set to different values as shown in Fig. 2(a). The amount of samples in Fig. 2(a) is normalized with respect to the size of the training set $|D_{train}|$. It's obvious that the amount of challenging samples in Math23k is rather limited as no more than 15% of the questions meet the condition when the threshold score is 0.9.

The Quality of the Challenging Samples. High-quality MWP questions are supposed to be longer sentences with rich background descriptions. Thus we analyze the average length and words diversity of challenging samples in D_{train}. The threshold of similarity scores is set to 0.9 to filter the challenging samples. In Fig. 2(b), Q refers to all questions in D_{train}, while Q_{metric} refers to the challenging samples filtered by certain metrics. e.g. Q_{BLEU} are challenging questions that have another similar question with BLEU score higher than 0.9. As shown in Fig. 2(b), the challenging samples filtered by all metrics are much shorter than the average question length, and so is the diversity of words. Table 4 shows some examples of existing challenging samples in D_{train}. The experimental results indicate that most of the challenging samples in D_{train} are too short and lack background descriptions.

Since both the amount and quality of existing challenging samples are not quite satisfactory, we propose knowledge and logical augmentation to generate questions with tiny local variances, which brings in massive high-quality challenging samples.

Table 4. Examples of challenging samples.

Challenging samples in D_{train}	Q_i	Q_j
	The divisor is 8 and the quotient is 2, how about the dividend?	The dividend is 24 and the divisor is 3, how about the quotient?
	3 times a number is 300, this number is equal to ?	A number is 7 times 21, this number is equal to ?
	Number A is equal to 150, and number B is 20% more than A. B = ?	Number A is 10.78, and number B is 3 more than B. B = ?
Challenging samples in D^*_{test}	In a parking lot, totally 48 cars and motorcycles are parked. Each car has 4 wheels, and each motorcycle has 3 wheels. If there are 20 motorcycles in the parking lot, how many wheels are there in total?	In a parking lot, totally 48 cars and motorcycles are parked. Each car has 4 wheels, and 172 wheels are there in total. If there are 20 motorcycles in the parking lot, how many wheels does a motorcycle have?

Testing Set Analysis. A good testing set requires low lexical and template overlap between the testing and training set [10]. Besides, as the ability to disambiguate tiny local variances in questions largely reflects the generalization ability of neural models, the testing set is supposed to have more challenging samples to better evaluate the ability of neural models. As shown in Table 5, we count the number of equation templates that only appeared in the D_{test} (New eq. template in Table 5) to measure the template overlap between the testing and training set. Meanwhile, the textual similarity between questions in the testing set and the training set is calculated with ED-DIST as introduced above. Specifically, for each Q_{te} in the testing set, we calculate a similarity score with $s = \max\{ED - DIST(Q_{te}, Q_{tr})\}, \forall Q_{tr} \in D_{train}$. The challenging samples in the testing set are counted as Sec. 4.1 with the threshold set to 0.9. Results in Table 5 are normalized with respect to the size of the testing set.

As shown in Table 5, the original testing set shares a high template (low new eq. templates) and lexical (high q. similarity) overlap with the training set, and contains limited challenging samples (low num. challenging samples). It indicates that the testing set is to some extent similar to the training set, making it hard to evaluate the actual inference ability of neural models perfectly. Not to mention the ability of neural models dealing with challenging samples, the limited scope of challenging samples indicates D_{test} is not able to evaluate neural models from this aspect.

Considering the weaknesses mentioned above, we manually labeled an additional testing set D^*_{test} with 380 samples in total which contains a great deal of **high-quality** challenging samples (Table 4 shows a pair of example challenging sample in D^*_{test}). Besides, the additional testing set D^*_{test} holds lower lexical

Table 5. Comparison of the two testing sets.

	New eq. template	Mean q. similarity	Num. challenging samples		
			BLEU	ROUGE-L	ED-DIST
D_{test}	0.0401	0.694	0.0211	0.0361	0.0201
D_{test}^*	0.155	0.593	0.574	0.282	0.0421

and template overlap with D_{train}, and also contains more challenging samples as shown in Table 5.

4.2   Experimental Setup

Dataset. We conduct our experiments on Math23k [24]. Besides training with the whole training set D_{train}, to evaluate the performance of our augmentation approaches on different data sizes, we randomly picked three training sets of sizes 500, 2k, and 20k (the whole training set) from the D_{train} as what have done in [25]. The neural models are evaluated on both original testing set D_{test} and the manually labeled D_{test}^*.

Neural Model $\mathcal{M}$. We evaluate s.based aug. strategies with three most typical MWP neural models.

- **Vanilla seq2seq** (marked as seq2seq). In this paper, we adopt a seq2seq model as [24] whose encoder is a 2-layer BiGRU and the decoder is a 2-layer LSTM. Besides, to help the model learn the slight variation of our augmented data, we utilize the attention mechanism before the feed-forward network.

$$\alpha_i = \frac{e^{V \tanh(W_1 h_{n_i} + W_2 \hat{h}_t + b)}}{\sum_{j=1}^{|\mathcal{N}|} e^{V \tanh(W_1 h_{n_j} + W_2 \hat{h}_t + b)}}$$

$$h_\mathcal{N} = \sum_{i=1}^{|\mathcal{N}|} \alpha_i h_{n_i}$$

$$P(y_t|y_0, ..., y_{t-1}) = f(\hat{h}_t \bigoplus h_\mathcal{N})$$

  V, W_1, W_2, b are all parameters, and h_{n_i} is the encoder hidden vector of the quantity n_i, while $\hat{h}_t$ refers to the decoder hidden vector of the time step t.
- **GTS** [28] is a tree-based neural model which has outperformed previous works significantly.
- **Graph2Tree** [32] is the state-of-the-art neural model for MWP.

Baseline Models for Comparison. We compare the performances of the three neural models trained with s.based aug. strategies with an extensive set of related work. **Math-EN** [21] proposed equation normalization based on a vanilla seq2seq model to effectively reduce the target space. **TRNN** [22] applied

Table 6. Answer accuracy (%) of neural models evaluated on the two testing sets. We evaluate s.based aug. on three most typical neural models. The improved performances w/ augmentation are shown in bold.

Model	D_{test}			D^*_{test}			
				All samples		Challenging samples	
	w/o	w/	w/	w/o	w/	w/o	w/
	aug.	s.based aug.	back trans.	aug.	s.based aug.	aug.	s.based aug.
Math-EN	66.7	–	–	–	–	–	–
TRNN	66.9	–	–	–	–	–	–
AST-Dec	69.0	–	–	–	–	–	–
GROUP-ATT	69.5	–	–	–	–	–	–
seq2seq	66.1	**71.2** ↑5.1	**66.4** ↑0.3	30.0	**53.2** ↑23.2	11.4	**27.3** ↑ 15.9
GTS	75.6	**76.1** ↑0.5	75.5 ↓0.1	33.2	**53.2** ↑20.0	11.4	**36.4** ↑ 25.0
Graph2Tree	77.4	77.0 ↓0.4	76.9 ↓0.5	30.3	**52.4** ↑22.1	11.4	**40.9** ↑ 29.5

a seq2seq model to predict a tree-structured template in a bottom-up manner. **GROUP-ATT** [13] proposed a group attention mechanism to extract intra-relation features. **AST-Dec** [15] proposed to generate abstract syntax tree of the equation in a top-down manner.

Baseline Augmentation Approach. We also compare the specifically designed s.based aug. with one of the most popular text augmentation strategy back translation [30], which is used to generate paraphrase of original sentence. The original sentence is first translated into a pivot language, and back to its original language. Other augmentation approaches such as synonymy replacement [25], generation-based methods [7], etc. are omitted here due to the intolerable noise for mathematical scenario.

Implementation Details. The baseline models are trained with D_{train} while the augmentation models are trained with $D_{train} \cup D_{aug}$. All models are evaluated on the same testing set D_{test}. To verify the ability of dealing with challenging samples, the three neural models trained with s.based aug. are additionally evaluated on D^*_{test}. Our evaluation metric is answer accuracy, which is calculated by comparing the answer of the predicted equation and the ground truth one rather than comparing the equation sequence. The parameters of the vanilla seq2seq model are set as [24]: The hidden units of both the encoder and decoder are 512. The word embedding dimension is set to 50 and the dropout for GRU and LSTM are set to 0.5. The number of epochs and mini-batch size are 80 and 32 respectively. We adopt an early stop policy after the accuracy of the validation set not increasing for 10 epochs. Adam optimizer [8] is used with learning rate set to 0.001, $\beta_1 = 0.9$ and $\beta_2 = 0.999$, and the learning rate is halved every 10 epochs. Settings of GTS and Graph2Tree are the same as that in [28] and [32].

4.3 Results

Comparison Results. Table 6 shows the answer accuracy of neural models trained with s.based aug. strategies compared with back translation and various baseline models. It's obvious that specifically designed s.based.aug. performs much better than back translation on MWP task. Since MWP has a strict requirement of precision, the noises brought in by back translation are sometimes unacceptable. We notice that many of the generated questions by the means of back translation are inconsistent with their equations. Trained with s.based aug., the accuracy of the seq2seq model increases by 5.1%, which is able to beat many other complex neural models with 71.2% answer accuracy. Performance gains of simple neural networks like seq2seq are much more than that of complex neural models like GTS and Graph2Tree. Our explanation is that complex models have better inference ability, making them able to learn more useful features from limited data, while simpler models have to learn these features from more diverse data. However, as analysed in Sect. 4.1, considering the original testing set D_{test} holds a high overlap with the training set, we reasonably suspect that the performance decrease is caused by the complex models likely to be overfitting and learns some dataset-specific features before augmentation. And the massive challenging samples brought in by our augmentation strategies may confuse the neural models, since none of them have specially designed to handle the tiny local variances. We'll further verify our hypothesis in Sect. 4.4.

As described in Sect. 4.1, the manually labeled testing set D_{test}^* holds lower overlap with the training set and contains 380 samples with a significant ratio of challenging samples. The three neural models trained w/ and w/o s.based aug. are tested on D_{test}^* to further evaluate the generalization ability of them. As shown in Table 6, we calculate the answer accuracy for all the questions and the challenging samples respectively. The challenging samples are filtered as 4.1 with the threshold set to 0.9. For the accuracy of the challenging samples, a challenging sample Q_i is viewed to be correctly answered only if all its similar questions Q_j has been correctly answered.

As shown in Table 6, s.based aug. strategies significantly boost the performances of all the three neural models with more than 20% increments. Besides, the ability of neural models dealing with discrete tiny local variances has also been largely improved. The results suggest that the s.based aug. strategies successfully benefit the generalization ability of the neural models. However, since existing MWP neural models hardly considered the discrete local variances which lead to respectively low accuracy, there is still a large space for future work to improve the ability of neural models dealing with such challenges.

Ablation Study. As illustrated in Table 7, both knowledge and logic guided augmentation methods contribute to the performance gains. And logical augmentation performs better than knowledge augmentation on all three datasets. This gap is more obvious on smaller datasets. We guess that on smaller datasets, the lack of problem types is more severe, thus enriching the equation templates

Table 7. The ablation study on datasets with different sizes. The best performance on each dataset is shown in bold.

Model	Training set size		
	500	2k	20k
seq2seq	9.52	31.6	66.1
seq2seq +knowledge	12.3 ↑2.78	33.5 ↑1.90	67.0 ↑0.90
seq2seq +logic	13.0 ↑3.48	34.1 ↑2.50	67.3 ↑1.20
seq2seq +s.based aug	**16.7** ↑7.18	**36.0** ↑4.40	**71.2** ↑5.10

Table 8. Examples of error cases that are predicted wrongly after augmentation.

$Q_{tr} \in D_{train}$	$Q_{te} \in D_{test}$	pred. Eq of Q_{te}	tgt. Eq of Q_{te}
A dictionary is priced at n_1 yuan, and after n_2 of the sale, the price is still n_3 higher than the purchase price. The purchase price of this dictionary is?	A dictionary is priced at n_1 yuan, and after n_2 of the sale, it will earn n_3. The purchase price of this dictionary is?	$x = n_1 \times (1 + n_3) \div n_2$	$x = n_1 \times n_2 \div (1 + n_3)$
A project can be completed in n_1 days if n_2 people come to do it. If n_3 people do it, how many days can it be done?	n_1 workers will complete a project within n_2 days. If it takes n_3 days to complete, how many workers are needed?	$x = 1 \div (n_1 \times n_2) \div n_3$	$x = (n_1 \times n_2) \div n_3$

is more needed. Besides, the results have shown that smaller datasets benefit more from the augmentation strategies than larger ones.

4.4   Case Study

In this section, we'll analyze the questions that are predicted correctly before the neural models trained with augmented data but are predicted wrongly afterwards. For each error case, we search for a most similar question in D_{train} to see whether these problems have occurred during training phase. And it turns out that most of the error cases have a nearly the same question in D_{train} as shown in Table 8. The $Q_{te} \in D_{test}$ are error predicted questions in the testing set. $Q_{tr} \in D_{train}$ are their similar questions found in the training set. In the first row of Table 8, the target Eq of the Q_{tr} is $x = n_1 \times n_2 \div (1 + n_3)$. According to the logical augmentation described in Sec. 3.3, one of the Eq_L could be $(\hat{n} \times (1 + n_3)) \div n_2$, which is equivalent to the mistakenly predicted equation in template. This result further support our hypothesis in Sect. 4.3 and explain the performance decreases of complex models on D_{test}.

5   Conclusion and Discussion

In this paper, we argue that discrete tiny local variances are a big challenge for neural models which previous works have ignored. And we propose a set of novel semantic-based data augmentation methods to supplement existing datasets with challenging samples. Both augmentation methods we proposed are able to generate coherent questions with consistent labels and largely diversify both textual descriptions and equation templates. Extensive experimental results have shown the necessity and effectiveness of the approaches we proposed. Besides, the idea we proposed could also be transferred to other math-related tasks like MWP generation.

Acknowledgment. This work was supported by National Key Research and Development Project (No. 2020AAA0109302), Shanghai Science and Technology Innovation Action Plan (No. 19511120400) and Shanghai Municipal Science and Technology Major Project (No. 2021SHZDZX0103).

References

1. Anaby-Tavor, A., et al.: Do not have enough data? Deep learning to the rescue! In: Proceedings of AAAI (2020)
2. Che, W., Feng, Y., Qin, L., Liu, T.: N-LTP: a open-source neural Chinese language technology platform with pretrained models. arXiv preprint arXiv:2009.11616 (2020)
3. Chiang, T.R., Chen, Y.N.: Semantically-aligned equation generation for solving and reasoning math word problems. arXiv preprint arXiv:1811.00720 (2018)
4. Fadaee, M., Bisazza, A., Monz, C.: Data augmentation for low-resource neural machine translation. arXiv preprint arXiv:1705.00440 (2017)
5. Hosseini, M.J., Hajishirzi, H., Etzioni, O., Kushman, N.: Learning to solve arithmetic word problems with verb categorization. In: Proceedings of EMNLP (2014)
6. Hou, Y., Liu, Y., Che, W., Liu, T.: Sequence-to-sequence data augmentation for dialogue language understanding. arXiv preprint arXiv:1807.01554 (2018)
7. Hu, Z., Yang, Z., Liang, X., Salakhutdinov, R., Xing, E.P.: Toward controlled generation of text. arXiv preprint arXiv:1703.00955 (2017)
8. Kingma, D.P., Ba, J.: Adam: a method for stochastic optimization. arXiv preprint arXiv:1412.6980 (2014)
9. Kobayashi, S.: Contextual augmentation: data augmentation by words with paradigmatic relations. arXiv preprint arXiv:1805.06201 (2018)
10. Koncel-Kedziorski, R., Hajishirzi, H., Sabharwal, A., Etzioni, O., Ang, S.D.: Parsing algebraic word problems into equations. Trans. Assoc. Comput. Linguist. **3**, 585–597 (2015)
11. Koncel-Kedziorski, R., Roy, S., Amini, A., Kushman, N., Hajishirzi, H.: MAWPS: a math word problem repository. In: Proceedings of ACL (2016)
12. Le, Q.V., Jaitly, N., Hinton, G.E.: A simple way to initialize recurrent networks of rectified linear units. arXiv preprint arXiv:1504.00941 (2015)
13. Li, J., Wang, L., Zhang, J., Wang, Y., Dai, B.T., Zhang, D.: Modeling intra-relation in math word problems with different functional multi-head attentions. In: Proceedings of ACL (2019)

14. Liang, C.C., Hsu, K.Y., Huang, C.T., Li, C.M., Miao, S.Y., Su, K.Y.: A tag-based statistical English math word problem solver with understanding, reasoning and explanation. In: Proceedings of IJCAI (2016)
15. Liu, Q., Guan, W., Li, S., Kawahara, D.: Tree-structured decoding for solving math word problems. In: Proceedings of EMNLP (2019)
16. Miller, G.A.: WordNet: a lexical database for English. Commun. ACM **38**, 39–41 (1995)
17. Rajpurkar, P., Zhang, J., Lopyrev, K., Liang, P.: SQuAD: 100,000+ questions for machine comprehension of text. arXiv preprint arXiv:1606.05250 (2016)
18. Roy, S., Roth, D.: Solving general arithmetic word problems. arXiv preprint arXiv:1608.01413 (2016)
19. Roy, S., Roth, D.: Unit dependency graph and its application to arithmetic word problem solving. arXiv preprint arXiv:1612.00969 (2016)
20. Shi, S., Wang, Y., Lin, C.Y., Liu, X., Rui, Y.: Automatically solving number word problems by semantic parsing and reasoning. In: Proceedings of EMNLP (2015)
21. Wang, L., Wang, Y., Cai, D., Zhang, D., Liu, X.: Translating a math word problem to an expression tree. arXiv preprint arXiv:1811.05632 (2018)
22. Wang, L., et al.: Template-based math word problem solvers with recursive neural networks. In: Proceedings of AAAI (2019)
23. Wang, W.Y., Yang, D.: That's so annoying!!!: a lexical and frame-semantic embedding based data augmentation approach to automatic categorization of annoying behaviors using #petpeeve tweets. In: Proceedings of EMNLP (2015)
24. Wang, Y., Liu, X., Shi, S.: Deep neural solver for math word problems. In: Proceedings of EMNLP (2017)
25. Wei, J., Zou, K.: EDA: easy data augmentation techniques for boosting performance on text classification tasks. arXiv preprint arXiv:1901.11196 (2019)
26. Wu, X., Lv, S., Zang, L., Han, J., Hu, S.: Conditional BERT contextual augmentation. In: Rodrigues, J.M.F., et al. (eds.) ICCS 2019. LNCS, vol. 11539, pp. 84–95. Springer, Cham (2019). https://doi.org/10.1007/978-3-030-22747-0_7
27. Xie, Q., Dai, Z., Hovy, E., Luong, M.T., Le, Q.V.: Unsupervised data augmentation for consistency training. arXiv preprint arXiv:1904.12848 (2019)
28. Xie, Z., Sun, S.: A goal-driven tree-structured neural model for math word problems. In: Proceedings of IJCAI (2019)
29. Xie, Z., et al.: Data noising as smoothing in neural network language models. arXiv preprint arXiv:1703.02573 (2017)
30. Yu, A.W., et al.: QANet: combining local convolution with global self-attention for reading comprehension. arXiv preprint arXiv:1804.09541 (2018)
31. Zhang, J., et al.: Teacher-student networks with multiple decoders for solving math word problem (2020)
32. Zhang, J., et al.: Graph-to-tree learning for solving math word problems. In: Proceedings of ACL (2020)

Empowering Transformer with Hybrid Matching Knowledge for Entity Matching

Wenzhou Dou[1], Derong Shen[1(✉)], Tiezheng Nie[1], Yue Kou[1], Chenchen Sun[2], Hang Cui[3], and Ge Yu[1]

[1] School of Computer Science and Engineering, Northeastern University, Shenyang, China
{shenderong,nietiezheng,kouyue,yuge}@cse.neu.edu.cn
[2] School of Computer Science and Engineering, Tianjin University of Technology, Tianjin, China
suncc@email.tjut.edu.cn
[3] University of Illinois at Urbana-Champaign, Champaign, USA
hangcui2@illinois.edu

Abstract. Transformers have achieved great success in many NLP tasks. The self-attention mechanism of Transformer learns powerful representation by conducting token-level pairwise interactions within the input sequence. In this paper, we propose a novel entity matching framework named GTA. GTA enhances Transformer for relational data representation by injecting additional hybrid matching knowledge. The hybrid matching knowledge is obtained via graph contrastive learning on a designed hybrid matching graph, in which the dual-level matching and multiple granularity interactions are modeled. In this way, GTA utilizes the prelearned knowledge of both hybrid matching and language modeling. This effectively empowers Transformer to understand the structural features of relational data when performing entity matching. Extensive experiments on open datasets show that GTA effectively enhances Transformer for relational data representation and outperforms state-of-the-art entity matching frameworks.

Keywords: Entity matching · Transformer · Pretrained language model · Hybrid matching graph · Graph contrastive learning

1 Introduction

Entity matching (EM), also known as entity resolution and record linkage, aims to identify records referring to the same real-world entity. Served as a long-standing critical problem [5] in data integration [9,24] and data cleaning [1], EM has been studied for many years [11] in plenty of fields such as e-commerce, medical treatment, etc. Recently, deep learning technologies achieved great success in database research, and have been a common way to solve EM tasks.

Figure 1 shows an example of EM tasks. Given a candidate record pair, the goal of EM is to determine whether they are referring to the same real-world

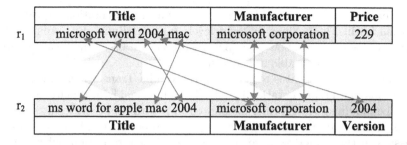

Fig. 1. An example of EM tasks. Records r_1 and r_2 can be seen as a matching pair according to dual levels—attribute level (*title, manufacturer*) and token level (*microsoft, word, 2004*, etc.).

entity. Due to the inherent hierarchical structure of the relational records, r_1 and r_2 can be compared at two levels—attribute level and token level. All the neural matching information between attribute values (e.g., values of *title* and *manufacturer*) and tokens (e.g., *microsoft, word, 2004*) jointly determines whether they match or not. Although the two values of attribute *title* are not exactly the same, the same attribute-aligned tokens (e.g., *word, 2004, mac*, etc.) and attribute-unaligned tokens (e.g., *microsoft* in attribute *title* and *manufacturer*), as well as the same values of aligned attributes (e.g., *manufacturer*), provide enough matching signals for the final result.

Depending on the level of comparisons, existing EM works can be divided into three categories: attribute-centric, token-centric, and hybrid-centric. Attribute-centric solutions [12,27] usually follow an alignment-comparison-summarization paradigm. They compare the aligned attributes and aggregate the similarity vectors to form the input to a binary classifier. However, these methods may fall flat when encountering situation like schema heterogeneity (e.g., attributes *price* and *version* in Fig. 1), which is a widespread scenario in real-world applications. Thus, recent works are mainly token-centric [22] or hybrid-centric [15,32,33], which additionally consider token-level matching information to provide EM signals.

Recently, Transformer-based works [2,23,34] have made great progress in EM tasks. They are token-centric solutions that cast EM as a sequence pair classification problem to leverage the pretrained language models (Pretrained LMs or PLMs) like BERT [8], RoBERTa [25] and DistilBERT [30], etc. By representing the candidate record pairs using BERT's [CLS] *RECORD1* [SEP] *RECORD2* [SEP] input schema, they conduct token-level comparisons between the two records via self-attention mechanism. However, vanilla Transformer is not quite suitable for EM tasks, for the reasons that (1) self-attention mechanism is originally designed for the semantic interaction of token level, and (2) the masked language model (MLM) training objective concerns on token-level prediction[1]. These two properties determine that Transformer is good at token-

---

[1] RoBERTa has proved that removing next sentence prediction (NSP) training objective can improve downstream task performance.

level interaction and less capable of relational data representation, especially for the scenarios in Fig. 1. Since different levels (i.e., attribute and token) contain various abstractions of knowledge [32], relational data's hierarchical feature is important and indispensable in EM tasks.

In this paper, our goal is to overcome the vanilla Transformer's insufficiency for relational data representation by injecting additional hybrid matching knowledge. The modified Transformer is adapted to regard input entries as relational records rather than natural language sentences to perform EM tasks. Thus, the key issues in this paper we need to solve are (1) how to obtain hybrid matching knowledge, i.e., Record-Token-Record (R-T-R) and Record-Attribute-Record (R-A-R), and embed the learned knowledge into representations of attributes and tokens, and (2) how to inject the above hybrid matching knowledge into Transformer for downstream EM tasks. Following this insight, we propose **GTA**, a novel EM framework that comprehensively integrates the hybrid matching knowledge of relational data and the tremendous language knowledge of pre-trained LM. GTA is a **G**raph-**T**ransformer-**A**ssembled architecture, which consists of two parts, namely Graph-based Hybrid Embedding (GHE for short) module and Adaptive Transformer-based Matching (ATM for short) module. The GHE module models both attribute-level and token-level matching into the graph topology, and then employs graph contrastive learning [26] to encode structural features into the embeddings of attributes and tokens. And then, the ATM module modifies Transformer's input format and embedding layer to absorb the prelearned knowledge in GHE module. In this way, the input embeddings of Transformer are structure-aware, which is to the benefit of the fine-tuning process of the pretrained LM for EM tasks.

The main contributions can be summarized as follows:

- We propose GTA, a novel EM framework that empowers Transformer with hybrid matching knowledge. GTA is a Graph-Transformer-Assembled architecture, which consists of GHE module for feature learning and ATM module for Transformer's fine-tuning for EM tasks.
- In GHE module, we model the hybrid matching (i.e., Record-Attribute-Record, Record-Token-Record) on the graph topology, ensuring the interactions between multiple granularities (i.e., Attribute-Attribute, Attribute-Token, Token-Token) simultaneously. We then design graph contrastive learning as a pretext task to conduct neural message passing to obtain structure-aware embeddings for attributes and tokens.
- In ATM module, we modify Transformer in two places—input format adaptation and prelearned knowledge injection, enabling Transformer to regard input entries as relational records rather than natural language sentences to perform EM tasks.
- We conduct extensive experiments on structured and dirty datasets and demonstrate that our proposed GTA framework effectively improves Transformer's representation for relational data, and outperforms state-of-the-art EM methods.

2   Related Work

Entity matching has attracted a lot of attention. Existing works can be classified into rule-based EM [7,13,31,37], crowdsourcing-based EM [16,18,36] and learning-based EM [3,12,15,20,23,27,33,34]. Learning-based EM has achieved great success recently. We then briefly introduce them according to the technologies (ML, DL, Transformer) they use.

Magellan [20] is a classical non-neural EM system based on machine learning (ML), which provides a variety of classifiers (decision tree, random forest, and SVM, etc.) to be trained on automatically generated features for EM. The tools it provides can significantly accelerate the entire EM pipeline.

With the rapid development of deep learning (DL) and its success in NLP, researchers introduce DL technologies into EM tasks to compare and aggregate information of relational data. DeepER [12] is one of the earliest methods to adopt word embeddings and LSTM neural networks to train an EM model. DeepMatcher [27] designs a space for DL-based EM, and proves that DL outperforms ML-based solutions on textual and dirty data. MCA [38] is an integrated multi-context attention framework for EM tasks that considers self-attention, pair-attention, and global-attention for three types of context. Both HierMatcher [15] and HAN [33] are end-to-end solutions for EM, which consider hybrid matching information (token and attribute levels). The difference is that HierMatcher performs token-level and attribute-level matching successively, and HAN solves the two in a two-tower mode.

Graph Neural Networks (GNNs) attract great attention due to their success in learning structural features in a lot of areas [14,17,29], thus recent works introduce GNNs to EM tasks. EMBDI [3] is a generic framework for obtaining local embeddings for data integration tasks, which leverages a compact tripartite graph to represent syntactic and semantic relationships between cell values. GraphER [22] encodes the semantic and structural features into an Entity Record Graph (ER-Graph) and trains an Entity Record GCN (ER-GCN) to obtain soft-structural embeddings for EM tasks. And GNEM [4] designs a record pair graph that allows each record pair to interact with relevant records and conducts the pairwise matching decision by borrowing valuable information from other pairs.

Recently, Transformer draws a great deal of concerns in both NLP [8] and CV [10] fields, due to the outstanding performance of self-attention mechanism in acquiring contextual information. Brunner et al. [2] proves the feasibility of adapting the pretrained LMs (e.g., BERT [8]) to EM tasks. DITTO [23] leverages pretrained LMs to solve EM tasks with three additional optimizations—domain knowledge injecting, text summarization, and data augmentation. BERT-ER [21] improves BERT-based EM model by delaying and enhancing BERT's interaction part, together with a blocking module to improve the EM efficiency. RPT [34] proposes a pretrained tuple-to-tuple model that supports several data preparation tasks like data cleaning, entity resolution, and information extraction, etc.

Our work is inspired by this paper [39] and Transformer-based related works [2,23,34]. The personalized dialogue generation model proposed in [39] enriches the dialogue context by additionally encoding the speakers' persona with dialogue histories, so as to enhance the Transformer's representation of dialogue context. So we carry on this idea to enhance Transformer for relational data representation and verify its feasibility in EM tasks. The difference is that the additional knowledge we inject into Transformer is obtained in a graph contrastive learning manner and meanwhile encoded into embeddings of attributes and tokens. The knowledge learning process requires no supervision and manual intervention. And then the modified Transformer is adapted to absorb these structure-aware embeddings to perform the downstream EM tasks.

3   Entity Matching via GTA Framework

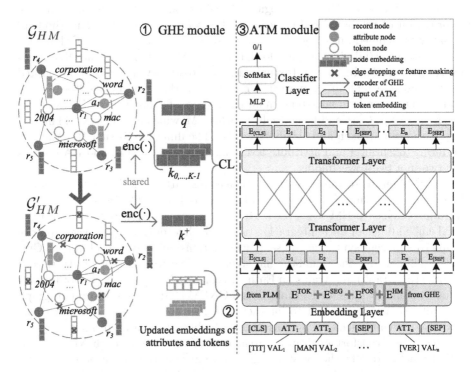

Fig. 2. The framework of GTA. This framework contains a Graph-based Hybrid Embedding (GHE) module for obtaining the hybrid matching knowledge, and an Adaptive Transformer-based Matching (ATM) module for the downstream EM tasks.

The framework of GTA is shown in Fig. 2. The entire workflow is ① GHE module first constructs the hybrid matching graph $\mathcal{G}_{HM}$ and its augmented view $\mathcal{G}'_{HM}$ to model the dual-level matching and multiple granularity interactions, and then performs contrastive learning (CL) on $\mathcal{G}_{HM}$ and $\mathcal{G}'_{HM}$ to obtain the updated embeddings. ② GTA extracts the updated embeddings of attributes and tokens and then maps them to the pretrained LM in ATM module. ③ ATM module modifies Transformer to absorb these structure-aware embeddings of attributes and tokens and finally fine-tunes the pretrained LM for EM tasks.

We first start with the preliminaries and then introduce the components of GTA framework, including a Graph-based Hybrid embedding (GHE) module and an Adaptive Transformer-based Matching (ATM) module.

3.1 Preliminaries

Let $r_1 = \{a_{11}, a_{12}, ..., a_{1m}\}$ and $r_2 = \{a_{21}, a_{22}, ..., a_{2n}\}$ be a candidate record pair from data sources S and S' separately. Each attribute value a_{1i} (or a_{2j}, $i \in [1, m]$ and $j \in [1, n]$) comprises a sequence of tokens like $\{t_1, t_2, ..., t_T\}$. Each token t_p ($p \in [1, T]$) in the sequence can be one of the string or numeric type.

Definition 1. *Entity matching. Given two relational data sources S and S', an entity matching framework takes as input a pair of records (e.g., r_1 from S and r_2 from S') and outputs the matching probability $P(y = 1|r_1, r_2)$.*

Definition 2. *Hybrid Matching Graph. A hybrid matching graph $\mathcal{G}_{HM} = \{\mathcal{V}, \mathcal{E}\}$, where $\mathcal{V}$ is the node set and $\mathcal{E}$ is the edge set. Each node $v \in \mathcal{V}$ can be one of record, attribute, or token. And each edge $e \in \mathcal{E}$ connects a record node with its contained attribute node or token node. $\mathcal{G}_{HM}$ is designed to model the dual-level matching and multiple granularity interactions of attributes and tokens. Each relevant record pairs are connected within two hops by their common attribute nodes or token nodes. And multiple granularity interactions of attribute-attribute, attribute-token, and token-token can also be conducted within two hops via the corresponding record nodes.*

3.2 Graph-Based Hybrid Embedding Module

The GHE module is designed to obtain hybrid matching knowledge in the manner of graph contrastive learning.

Hybrid Matching Graph Construction. As can be seen on the left side of Fig. 2, we first design a hybrid matching graph $\mathcal{G}_{HM}$ to model the relational data's dual-level matching process. In $\mathcal{G}_{HM}$, when two records have common attributes or tokens, they will be indirectly connected within two hops via the corresponding attribute nodes and token nodes, thus the neural matching information can be passed through both attribute level and token level. For instance, the dual-level matching information of r_1 and r_2 can be passed through both (1) attribute value a_1—*microsoft corporation* and (2) tokens—*word* and *mac*.

Beyond that, Fig. 3 shows more comprehensive analyses in attribute and token views. Since the record view in Fig. 3(a) has been analyzed before, we will not dwell on it. In Fig. 3(b) attribute view, the meta-path Attribute-Record-Attribute (e.g., a_1-r_1-a_2) implements the attribute-attribute interaction within a record (e.g., r_1). And the meta-path Attribute-Record-Token (e.g., a_1-r_1-*word*) implements the attribute-token interaction within a record (e.g., r_1). And in Fig. 3(c) token view, the meta-path Token-Record-Token (e.g., *word*-r_1-*mac*) implements the token-token interaction within a record (e.g., r_1). These

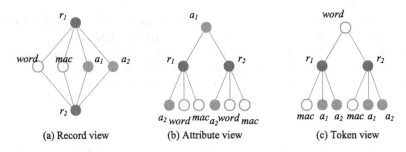

Fig. 3. The neural message passing process in different views—record view, attribute view, and token view.

meta-paths jointly ensure dual-level matching and multiple granularity interactions, enabling the attribute nodes and token nodes to perceive structural features for EM tasks.

Sampling and Training. We adopt contrastive learning as a pretext task to perform graph representation learning on $\mathcal{G}_{HM}$. First, we apply two ways of stochastic perturbation (edge dropping and node feature masking) on $\mathcal{G}_{HM}$, thus we get two views, one is an original view $\mathcal{G}_{HM}$, and the other is an augmented view $\mathcal{G}'_{HM}$. And then, we employ a 2-layer GCN with residual connections as an encoder to obtain node embeddings for both $\mathcal{G}_{HM}$ and $\mathcal{G}'_{HM}$:

$$\mathcal{F}(X^{(l)}) = \sigma(\widetilde{D}^{-\frac{1}{2}}\widetilde{A}\widetilde{D}^{-\frac{1}{2}}X^{(l)}W^{(l)}), \tag{1}$$

$$X^{(l+1)} = \mathcal{F}(X^{(l)}) + \sigma(X^{(l)}), \tag{2}$$

where $\widetilde{A} = A + I$ is the adjacency matrix with self-connections of the hybrid matching graph. $\widetilde{D}_{ii} = \sum_j \widetilde{A}_{ij}$ is the degree matrix. $X^{(l)} \in \mathbb{R}^{N \times D}$ is the node embedding matrix and $W^{(l)} \in \mathbb{R}^{D \times D}$ is a trainable weight matrix for the l-th layer. $\sigma(\cdot)$ is a nonlinear activation function, like ELU [6]. The GCN settings are designed based on the following considerations. Since all dual-level matching and multiple granularity interactions can be conducted within two hops on $\mathcal{G}_{HM}$ and $\mathcal{G}'_{HM}$, the 2-layer GCN is suitable for two-hop neural message passing. And the residual connections are designed to prevent the over-smoothing problem.

The sampling strategy for graph contrastive learning is as follows. For each record embedding $q \in X$ in the original view $\mathcal{G}_{HM}$ that acts as a query vector, we sample its corresponding augmented embedding $k_+ \in X'$ in $\mathcal{G}'_{HM}$ as a positive example. As for negative examples, we randomly select K record nodes which are far more than 2 hops (i.e., at least 4 hops) from the original record node, we denote each negative example as $k_i \in X$. The training objective is InfoNCE loss [28]:

$$\mathcal{L}_{HM} = -\log \frac{\exp(q^\top k_+/\tau)}{\sum_{i=0}^{K-1} \exp(q^\top k_i/\tau)}, \tag{3}$$

where τ is the temperature hyperparameter that acts as an adjusting factor to control the strength of penalties on hard negative examples [35].

By closing the distances of the original records with their augmented ones, and meanwhile pulling far from irrelevant negative records, the interactions within the candidate record pairs and their contained attributes and tokens are conducted in the graph contrastive learning process. And the updated attribute and token embeddings are endowed with structural features which help to improve the downstream Transformer-based EM performance.

3.3   Adaptive Transformer-Based Entity Matching

The above GHE module obtains hybrid matching information and multiple granularity interactions in a contrastive learning manner, we then extract and inject the knowledge into a Transformer-based pretrained language model to perform EM tasks. To absorb the prior knowledge, we modify the vanilla Transformer architecture in two places—input format adaptation and prelearned knowledge injection.

Input Format Adaptation. We first convert the raw record pair to a specific format that Transformer can absorb. By appending special tokens [CLS] at the beginning and [SEP] as a separator, we cast EM as a sentence pair classification task that can utilize pretrained knowledge in LMs (e.g., BERT, RoBERTa, and DistilBERT). Apart from that, we add a more fine-grained attribute-specific token before each attribute value to enable Transformer to conduct attribute-level interactions. For example, given a sequence as the value of attribute *title*, we add a special token [TIT] before it to get an attribute-aware entry. Take the record pair in Fig. 1 as an example, after the processing, the record pair can be converted into

[CLS] *RECORD1* [SEP] *RECORD2* [SEP],

where *RECORD1* refers to

[TIT] *microsoft word 2004 mac* [MAN] *microsoft corporation* [PRI] *229,*

and *RECORD2* refers to

[TIT] *ms word for apple mac 2004* [MAN] *microsoft corporation* [VER] *2004.*

So far, the record pairs as Transformer's inputs have been tokenized to accommodate structural perception. The reason for adding attribute-specific tokens (e.g., [TIT]) is that we can slightly enable self-attention mechanism to focus on the interactions of aligned attributes and meanwhile empower the attribute-specific tokens with the prior knowledge learned in the GHE module. The details will be illustrated in the next subsection.

Prelearned Knowledge Injection. After adapting the input format for Transformer, we then describe how to inject the prelearned knowledge into Transformer to perform EM tasks. Transformer-based LMs (e.g., BERT) design input

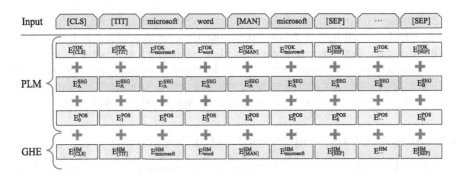

Fig. 4. ATM's input embeddings. The token, segment and position embeddings are inherent in pretrained language model (PLM), and the hybrid matching embeddings are extracted from the graph-based hybrid Embedding (GHE) module. Notice that, some tokens are omitted in this figure.

embedding E^{INPUT} as three components, they are token embedding E^{TOK}, segment embedding E^{SEG}, and position embedding E^{POS}. We retain the above three embeddings and additionally add the hybrid matching embeddings E^{HM} as the supplements to improve Transformer's EM performance.

In detail, we first extract all the updated attribute and token embeddings from GHE module. For different types of E^{HM}, we design different methods for knowledge injection. For example in Fig. 4, the token type embeddings (e.g., $E^{HM}_{microsoft}$ and E^{HM}_{word}) are directly mapped and injected without any additional processing. The attribute-specific embeddings (e.g., $E^{HM}_{[TIT]}$ and $E^{HM}_{[MAN]}$) are generated by averaging all corresponding attribute embeddings in GHE module. And the inherent [CLS] and [SEP] embeddings (i.e., $E^{HM}_{[CLS]}$ and $E^{HM}_{[SEP]}$) are copied from the PLM, which are equal to $E^{TOK}_{[CLS]}$ and $E^{TOK}_{[SEP]}$. Notice that, we do not add E^{HM} to the other three embeddings (i.e., E^{TOK}, E^{SEG} and E^{POS}) directly, but normalize and then scale E^{HM} using the scaling factor ψ before that. This helps to avoid the internal covariate shift problem and control the influence strength of GHE module by adjusting ψ, following the intuition that excessive knowledge injection may collapse the LM's inherence of language modeling.

Fine-Tuning Pretrained LM for EM Tasks. After hybrid matching knowledge injection, we fine-tune the pretrained LM on EM datasets to output the final decisions. The training objective is cross-entropy loss:

$$\mathcal{L}_{CLS} = -\frac{1}{|B|} \sum_{i=1}^{|B|} [y_i \log p_i + (1 - y_i) \log(1 - p_i)], \tag{4}$$

where $|B|$ is the size of training batch, $y_i \in \{0, 1\}$ is the label of i-th training pair, and p_i is the i-th output probability.

4   Experimental Evaluation

In this section, we evaluate our proposed GTA framework and compare it with existing works.

4.1   Experimental Settings

Datasets and Metric. We evaluate GTA framework on five open datasets[2] proposed in DeepMatcher. The datasets can be divided into two types (structured or dirty) and cover a variety of domains including software, beer, restaurant, and citation. The dataset sizes vary from 450 to 12, 363 to evaluate the scalability of GTA. Positive ratios (# Pos./Size) cover the range from 0.10 to 0.18 shows that EM is an unbalanced binary classification task. All EM datasets are split into 60%/20%/20% for training, validation, and test, which are the same as Deepmatcher [27]. The details of EM datasets are shown in Table 1. Following the previous works, we use F1 score as the metric to evaluate the EM performance.

Table 1. Details of EM datasets.

Dataset	Type	Domain	Size	# Pos.	# Att.
Amazon-Google (AG)	Structured	Software	11,460	1,167	3
BeerAdvo-RateBeer (BR)	Structured	Beer	450	68	4
Fodors-Zagats (FZ)	Structured	Restaurant	946	110	6
DBLP-ACM (DA_1)	Structured	Citation	12,363	2,220	4
DBLP-ACM (DA_2)	Dirty	Citation	12,363	2,220	4

Implementation Details and Training Settings. The GTA framework consists of two modules, GHE module and ATM module. For GHE module, we first separate each record into several attribute values. For each sequence of attribute value, we tokenize it with the pretrained tokenizer of the corresponding LM in ATM module. This helps a lot for the token mapping in knowledge injection process. For ease of knowledge injection, we adopt the same embedding dimension $D = 768$ for all the nodes in the hybrid matching graph $\mathcal{G}_{HM}$. We build $\mathcal{G}_{HM}$ and train the GHE module using Deep Graph Library[3]. The details of $\mathcal{G}_{HM}$ are shown in Table 2. The probabilities of stochastic perturbation (i.e., edge dropping and node feature masking) are both 20%, the number of negative samples is $K = 64$, and the temperature hyperparameter τ is 0.1 for the InfoNCE loss. For ATM module, we fine-tune the pretrained LMs (bert-base-uncased, roberta-base, distilbert-base-uncased) to perform EM using Hugging Face[4]. Experiments show that RoBERTa achieves the best EM performance, thus the experiment results we report are all using RoBERTa as the pretrained LM. The whole GTA framework is implemented using PyTorch as a backend.

---

[2] https://github.com/anhaidgroup/deepmatcher/blob/master/Datasets.md.
[3] https://www.dgl.ai/.
[4] https://huggingface.co/.

Table 2. Details of hybrid matching graph $\mathcal{G}_{HM}$.

Dataset	# Node	# Edge
Amazon-Google (AG)	17,549	50,999
BeerAdvo-RateBeer (BR)	25,969	100,038
Fodors-Zagats (FZ)	7,669	15,878
DBLP-ACM (DA_1)	19,069	98,427
DBLP-ACM (DA_2)	19,435	95,345

We train 500 epochs for GHE module and fine-tune 20 epochs for ATM module. The batch size $B = 64$ and learning rate $lr = 3e - 5$ are set for both GHE and ATM modules. Adam algorithm [19] with warming up and linear decay is used for optimization. The hybrid matching embeddings E^{HM} are scaled by multiplying a scaling factor $\psi = 0.2$ for Fodors-Zagats dataset and $\psi = 0.1$ for other datasets in knowledge injection. We conduct the experiments on a workstation with Intel Xeon W-2255 CPU @ 3.70 GHz and NVIDIA RTX A4000 with 16 GB memory.

The EM frameworks to be compared are as follows:

- **Magellan** [20]: A state-of-the-art ML-based (non-DL) system for EM tasks. It provides a variety of classifiers (decision tree, Naive Bayes, SVM, etc.) to be trained on automatically generated features.
- **RNN** [27]: A DL-based EM solution proposed in DeepMatcher that adopts Bi-GRU as an encoder to represent attribute values. Then it takes element-wise absolute difference as the comparison result to form the input of the classifier.
- **Attention** [27]: A DL-based EM solution proposed in DeepMatcher that adopts decomposable attention to implement attribute summarization and vector concatenation to perform attribute comparison.
- **Hybrid** [27]: A DL-based EM solution proposed in DeepMatcher that adopts Bi-GRU with decomposable attention to represent attribute values. Then it takes vector concatenation augmented with element-wise absolute difference as the input of the classifier.
- **MCA** [38]: A DL-based EM solution that designs multi-context attention network. MCA fully takes into account self-attention, pair-attention, and global-attention from three types of context.
- **HierMatcher** [15]: A DL-based EM solution that jointly considers hierarchical levels of matching granularity (token, attribute, and entity). It designs a cross-attribute token alignment module and attribute-aware attention mechanism that can solve EM in heterogeneous and dirty scenarios.
- **DITTO** [23]: A Transformer-based EM solution that leverages domain knowledge, text summarization, and data augmentation to improve pre-trained language models' ability for EM tasks.
- **BERT-ER** [21]: A Transformer-based solution that improves EM performance by delaying and enhancing BERT's interaction part, together with an adaptive blocking module to improve EM efficiency.

- **Baseline:** We directly fine-tune the pretrained LM as a baseline. This can be regarded the same as paper [2].

4.2   Main Results

Table 3 shows the results of GTA and existing works, including ML-based, DL-based and Transformer-based EM frameworks. We also set a Baseline which directly fine-tunes the pretrained LM to perform EM tasks without any optimization.

Table 3. EM performance of GTA and existing works. All the experimental results of the comparing methods are derived from the original papers, and the best results are bolded. We calculate $\Delta F1$ between our proposed GTA with Baseline which directly fine-tunes pretrained LM to perform EM tasks.

	AG	BR	FZ	DA$_1$	DA$_2$
Magellan	49.1	78.8	100	98.4	91.9
RNN	59.9	72.2	100	98.3	97.5
Attention	61.1	64.0	82.1	98.4	97.4
Hybrid	69.3	72.7	100	98.4	98.1
MCA	70.3	78.8	–	98.6	–
HierMatcher	74.9	–	–	98.8	98.1
DITTO	75.6	94.4	100	99.0	99.0
BERT-ER	75.3	87.5	–	98.7	–
Baseline	74.1	86.7	98.1	98.8	98.9
GTA	**76.2**	**96.3**	**100**	**99.1**	**99.0**
$\Delta F1$	+2.1	+9.6	+1.9	+0.3	+0.1

In detail, we can draw conclusions as below:

- GTA outperforms or reaches state-of-the-art results compared with existing EM frameworks. Compared to ML-based work (i.e., Magellan), GTA achieves 10.5 average F1 improvement, and compared to DL-based work (i.e., Hybrid), GTA achieves 6.4 average F1 improvement. And for HierMatcher, which is a hybrid-centric EM framework, GTA outperforms 1.3, 0.3, 0.9 F1 improvement on *Amazon-Google*, *DBLP-ACM$_1$* and *DBLP-ACM$_2$* dataset respectively. We can draw a conclusion that compared with these ML-based and DL-based works, GTA achieves improvement by additionally utilizing Transformer architecture and pretrained LM's prior knowledge, thus performs better in record sequence representation.
- GTA shows competitive performance compared to Transformer-based works (Baseline, DITTO and BERT-ER) and outperforms Baseline by an average of 2.8 F1 score. Due to the introduction of graph-based hybrid embedding (GHE) module, GTA encodes the dual-level matching information and multiple granularity interactions into a hybrid matching graph $\mathcal{G}_{HM}$. By conducting contrastive learning as a pretext task on $\mathcal{G}_{HM}$, structural features of

relational data can be obtained without any labeled data. Then the hybrid matching embeddings E^{HM} of attributes and tokens act as additional features to enhance Transformer's representation for relational data and improve the final EM performance.

- GTA shows robustness on dirty data. As can be seen in Table 2, GTA achieves state-of-the-art result on $DBLP\text{-}ACM_2$ dirty dataset. GTA's robust performance can be ascribed to two reasons: (1) both instance-level and hidden-level data augmentation on the hybrid matching graph $\mathcal{G}_{HM}$, this ensures that the prelearned hybrid matching embedding is generalized enough and not specialized to explicit attribute, token instances or feature dimensions, and (2) comparison and aggregation on both attribute-level and token-level, which ensures the attribute-unaligned tokens' interaction to perform comparison.

4.3 Detailed Analysis

Ablation Study. To evaluate the contribution of each component, we conduct an ablation study on GTA framework by ablating a specific component of GTA. GTA (-HM) refers to dropping hybrid matching knowledge injection. GTA (-AST) refers to dropping attribute-specific tokens. And GTA (-HM-AST) refers to dropping both of them. According to the results in Table 4, we can draw a conclusion that the additional attribute-specific tokens can accomplish finer separating to split various attributes, enabling Transformer to regard input entries as relational records. And the injected additional knowledge can significantly improve Transformer-based EM performance.

Table 4. Ablation study results compared with the full GTA framework.

	AG	BR	FZ	DA_1	DA_2
GTA	76.2	96.3	100	99.1	99.0
GTA (-HM)	74.3	87.3	98.1	98.9	98.9
	−1.9	−9.0	−1.9	−−0.2	−0.1
GTA (-AST)	75.6	95.4	99.8	98.9	99.0
	−0.6	−0.9	−0.2	−0.2	−0
GTA (-HM-AST)	74.1	86.7	98.1	98.8	98.9
	−2.1	−9.6	−1.9	−0.3	-0.1

Various LMs in GTA. We evaluate GTA using various BERT-like language models (i.e., BERT, RoBERTa, DistilBERT). As can be seen in Table 5, GTA (RoBERTa) shows the leading F1 score in our GTA framework. This can be attributed to that RoBERTa pretrains longer time on more data with bigger batch size than BERT and DistilBERT. RoBERTa also removes the next sentence prediction (NSP) objective of BERT, considering that the NSP objective may harm the downstream task performance. GTA (DistilBERT) also achieves good results, although the trainable parameter of DistilBERT (66M) is about half of RoBERTa (125M). For the challenging dataset *Amazon-Google*, GTA (RoBERTa) outperforms GTA (DistilBERT) by 4.8 F1 score, but for datasets like *Fodors-Zagats*, *DBLP-ACM$_1$* and *DBLP-ACM$_2$*, the margins of F1 score are 0.8, 0.4 and 0.6 separately.

Table 5. GTA's F1 score using various language models.

	AG	BR	FZ	DA$_1$	DA$_2$
GTA (BERT)	72.6	95.9	98.4	98.8	98.8
GTA (RoBERTa)	76.2	96.3	100	99.1	99.0
GTA (DistilBERT)	71.4	93.9	99.2	98.7	98.4

5   Conclusion and Outlook

In this paper, we propose a novel entity matching framework named GTA. GTA verifies the feasibility of knowledge injection for Transformer to perform EM tasks. By injecting additional hybrid matching knowledge, which is obtained via graph contrastive learning in a designed hybrid matching graph, GTA enhances Transformer for relational data representation. This enables Transformer to regard input entries as relational records to aggregate both attribute-level and token-level matching information. Compared with existing EM works, our proposed GTA framework effectively improves Transformer's representation for relational data and achieves state-of-the-art results on open datasets. Just like other related works, we hope to shed some light on this direction by conducting researches on AI4DB, and making contributions to both DB and AI communities.

Acknowledgements. This work is supported by the National Natural Science Foundation of China (62172082, 62072084, 62072086, U1811261).

References

1. Abedjan, Z., et al.: Detecting data errors: where are we and what needs to be done? Proc. VLDB Endow. **9**(12), 993–1004 (2016)
2. Brunner, U., Stockinger, K.: Entity matching with transformer architectures-a step forward in data integration. In: International Conference on Extending Database Technology, Copenhagen, 30 March–2 April 2020. OpenProceedings (2020)
3. Cappuzzo, R., Papotti, P., Thirumuruganathan, S.: Creating embeddings of heterogeneous relational datasets for data integration tasks. In: Proceedings of the 2020 ACM SIGMOD International Conference on Management of Data, pp. 1335–1349 (2020)
4. Chen, R., Shen, Y., Zhang, D.: GNEM: a generic one-to-set neural entity matching framework. In: Proceedings of the Web Conference 2021, pp. 1686–1694 (2021)

5. Christen, P.: Data matching: concepts and techniques for record linkage, entity resolution, and duplicate detection (2012)
6. Clevert, D.A., Unterthiner, T., Hochreiter, S.: Fast and accurate deep network learning by exponential linear units (ELUs). arXiv preprint arXiv:1511.07289 (2015)
7. Dalvi, N., Rastogi, V., Dasgupta, A., Das Sarma, A., Sarlós, T.: Optimal hashing schemes for entity matching. In: Proceedings of the 22nd International Conference on World Wide Web, pp. 295–306 (2013)
8. Devlin, J., Chang, M.W., Lee, K., Toutanova, K.: BERT: pre-training of deep bidirectional transformers for language understanding. In: Proceedings of the 2019 Conference of the North American Chapter of the Association for Computational Linguistics: Human Language Technologies, Volume 1 (Long and Short Papers), pp. 4171–4186 (2019)
9. Dong, X.L., Srivastava, D.: Big data integration. In: 2013 IEEE 29th international conference on data engineering (ICDE), pp. 1245–1248. IEEE (2013)
10. Dosovitskiy, A., et al.: An image is worth 16×16 words: transformers for image recognition at scale. In: International Conference on Learning Representations (2020)
11. Dunn, H.L.: Record linkage. Am. J. Public Health Natl. Health **36**(12), 1412–1416 (1946)
12. Ebraheem, M., Thirumuruganathan, S., Joty, S., Ouzzani, M., Tang, N.: Distributed representations of tuples for entity resolution. Proc. VLDB Endow. **11**(11), 1454–1467 (2018)
13. Elmagarmid, A., Ilyas, I.F., Ouzzani, M., Quiané-Ruiz, J.A., Tang, N., Yin, S.: NADEEF/ER: generic and interactive entity resolution. In: Proceedings of the 2014 ACM SIGMOD International Conference on Management of Data, pp. 1071–1074 (2014)
14. Fan, W., Ma, Y., Li, Q., He, Y., Zhao, E., Tang, J., Yin, D.: Graph neural networks for social recommendation. In: The World Wide Web Conference, pp. 417–426 (2019)
15. Fu, C., Han, X., He, J., 0001, L.S.: Hierarchical matching network for heterogeneous entity resolution. In: IJCAI, pp. 3665–3671 (2020)
16. Gokhale, C., et al.: Corleone: hands-off crowdsourcing for entity matching. In: Proceedings of the 2014 ACM SIGMOD International Conference on Management of Data, pp. 601–612 (2014)
17. Jin, W., et al.: Graph representation learning: foundations, methods, applications and systems. In: Proceedings of the 27th ACM SIGKDD Conference on Knowledge Discovery & Data Mining, pp. 4044–4045 (2021)
18. Marcus, A., Wu, E., Karger, D., Madden, S., Miller, R.: Human-powered sorts and joins. Proc. VLDB Endow. **5**(1) (2011)
19. Kingma, D.P., Ba, J.: Adam: a method for stochastic optimization. In: ICLR (Poster) (2015)
20. Konda, P., et al.: Magellan: toward building entity matching management systems. Proc. VLDB Endow. **9**(12), 1581–1584 (2016)
21. Li, B., Miao, Y., Wang, Y., Sun, Y., Wang, W.: Improving the efficiency and effectiveness for BERT-based entity resolution. In: Proceedings of the AAAI Conference on Artificial Intelligence, vol. 35, pp. 13226–13233 (2021)
22. Li, B., Wang, W., Sun, Y., Zhang, L., Ali, M.A., Wang, Y.: GraphER: token-centric entity resolution with graph convolutional neural networks. In: AAAI, pp. 8172–8179 (2020)

23. Li, Y., Li, J., Suhara, Y., Doan, A., Tan, W.C.: Deep entity matching with pre-trained language models. Proc. VLDB Endow. **14**(1), 50–60 (2020)
24. Li, Y., Li, J., Suhara, Y., Wang, J., Hirota, W., Tan, W.C.: Deep entity matching: challenges and opportunities. J. Data Inf. Qual. (JDIQ) **13**(1), 1–17 (2021)
25. Liu, Y., et al.: RoBERTa: a robustly optimized BERT pretraining approach. arXiv preprint arXiv:1907.11692 (2019)
26. Liu, Y., Pan, S., Jin, M., Zhou, C., Xia, F., Yu, P.S.: Graph self-supervised learning: a survey. arXiv preprint arXiv:2103.00111 (2021)
27. Mudgal, S., et al.: Deep learning for entity matching: a design space exploration. In: Proceedings of the 2018 International Conference on Management of Data, pp. 19–34 (2018)
28. Oord, A.V.D., Li, Y., Vinyals, O.: Representation learning with contrastive predictive coding. arXiv preprint arXiv:1807.03748 (2018)
29. Peng, Y., Choi, B., Xu, J.: Graph learning for combinatorial optimization: a survey of state-of-the-art. Data Sci. Eng. **6**(2), 119–141 (2021)
30. Sanh, V., Debut, L., Chaumond, J., Wolf, T.: DistilBERT, a distilled version of BERT: smaller, faster, cheaper and lighter. arXiv preprint arXiv:1910.01108 (2019)
31. Singh, R., et al.: Synthesizing entity matching rules by examples. Proc. VLDB Endow. **11**(2), 189–202 (2017)
32. Sun, C.C., Shen, D.R.: Mixed hierarchical networks for deep entity matching. J. Comput. Sci. Technol. **36**(4), 822–838 (2021)
33. Sun, C., Shen, D.: Entity resolution with hybrid attention-based networks. In: Jensen, C.S., et al. (eds.) DASFAA 2021. LNCS, vol. 12682, pp. 558–565. Springer, Cham (2021). https://doi.org/10.1007/978-3-030-73197-7_37
34. Tang, N., et al.: RPT: relational pre-trained transformer is almost all you need towards democratizing data preparation. Proc. VLDB Endow. **14**(8), 1254–1261 (2021)
35. Wang, F., Liu, H.: Understanding the behaviour of contrastive loss. In: Proceedings of the IEEE/CVF Conference on Computer Vision and Pattern Recognition, pp. 2495–2504 (2021)
36. Wang, J., Kraska, T., Franklin, M.J., Feng, J.: CrowdER: crowdsourcing entity resolution. Proc. VLDB Endow. **5**(11) (2012)
37. Wang, J., Li, G., Yu, J.X., Feng, J.: Entity matching: how similar is similar. Proc. VLDB Endow. **4**(10), 622–633 (2011)
38. Zhang, D., Nie, Y., Wu, S., Shen, Y., Tan, K.L.: Multi-context attention for entity matching. In: Proceedings of The Web Conference 2020, pp. 2634–2640 (2020)
39. Zheng, Y., Zhang, R., Huang, M., Mao, X.: A pre-training based personalized dialogue generation model with persona-sparse data. In: Proceedings of the AAAI Conference on Artificial Intelligence, vol. 34, pp. 9693–9700 (2020)

Tracking the Evolution: Discovering and Visualizing the Evolution of Literature

Siyuan Wu and Leong Hou U$^{(\boxtimes)}$

State Key Laboratory of Internet of Things for Smart City, Department of Computer and Information Science, University of Macau, Taipa, Macau SAR, China
yb87429@connect.um.edu.mo, ryanlhu@um.edu.mo

Abstract. A common task in research preparation is to survey related work in bibliographic databases. Scientists are finding the survey task notably difficult as the volume of the databases has been increased considerably over the past few decades. Making a good use of a survey paper of the research topic can vastly lower the difficulty but there may be no survey paper in some emerging research topics due to the rapid development. In this work, we propose a novel Literature Evolution Discovery (LED) process that aims to provide an explainable evolution structure of literature. The explainability is based on co-citation analysis and latent relationship extraction, which are done by Steiner tree algorithm and context-consistent factor graph model, respectively. The experiments show the superiority of our context-consistent factor graph model, compared with the state-of-the-art baselines. Our case studies and visualization results demonstrate the effectiveness and interpretability of our proposed algorithms in practice.

Keywords: Citation network · Factor graph · Steiner tree · Visualization

1 Introduction

Scientists are used to acquire new knowledge from literature in bibliographic databases. In the past, she might use a survey paper to understand the trend and challenges of a research topic. However, the technology life-cycle today in emerging areas is shorter than 1–2 decades ago so there may be no survey paper available in some fast changing areas. Another solution is to discover related work by navigating a large bibliographic database. The navigation may take a lot of time and probably miss some important references, especially for literature who cite hundreds of references. A similar finding is drawn in recent studies [4,19, 21,25]. Bibliographic search engines, e.g., Google Scholar, Microsoft Academic Graph, and CiteSeerX, etc., provide options to find related work of the literature. However, there is either no option to visualize the evolution structure of literature or is lack of explanation of the evolution structure. Master Reading Tree (MRT)

© The Author(s), under exclusive license to Springer Nature Switzerland AG 2022
A. Bhattacharya et al. (Eds.): DASFAA 2022, LNCS 13247, pp. 68–84, 2022.
https://doi.org/10.1007/978-3-031-00129-1_5

(https://www.aminer.cn/mrt) is a tool being developed recently in Aminer [23, 24] that helps scientists construct a functional citation flow map. However, the flow is *manually constructed* by experts so this feature is not available for every topic. In this work, we attempt to study an automatic discovery process for filling this important but missing feature in bibliographic databases. The literature discovery results can be visualized and interpreted which can help scientists acquire new knowledge more conveniently.

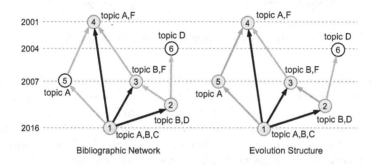

Fig. 1. A visualization example with evolution structure and its interpretability. (Color figure online)

Given a target literature p and a bibliographic database, we aim to extract the development history of p, so-called *evolution structure*, based on the literature relationship, e.g., topic similarity, co-authorship, co-citations, references, etc. The evolution structure should provide *explainable relationship* between literature in terms of both *topology* and *semantic context*. In Fig. 1, the left part is the original bibliographic network of target literature 1. Literature 2, 3, and 4 are highly cited and have overlapping topics of literature 1 (highlighted as blue nodes and black edges). The subgraph (of 4 nodes) can be extracted straightforwardly to understand the evolution of literature 1. While references 2 and 3 share a common topic B with literature 1, scientists may want to see more semantic explanation among these literature. Besides, the citation link (1, 4), which spans a long time, is less meaningful and there may be a stronger connection in between literature 1 and 4. The right part is an example of the evolution structure of literature 1 (i.e., blue nodes, black and green edges). It not only enriches the semantic explanation on topic B (for literature 1, 2, and 3) but also fills a missing reference (literature 5) based on topology structure.

Our Solution. In this work, we attempt to produce an interpretable evolution structure by two key processes, *latent relationship* estimation and *evolution structure* extraction, shown in the middle of Fig. 2. In particular, we estimate the latent relationship by a *factor graph* based optimization process, considering local and global *context consistency*. Our latent relationship estimation provides better context information since the estimation process is done by a recursively tracking (based on global factor graph configuration) instead of the local context similarity.

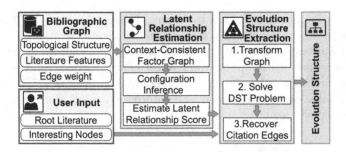

Fig. 2. Framework of literature evolution discovery.

With the *latent relationship* generated based on *context consistency*, our goal is to extract the evolution structure of a given literature such that the *total cost* is minimized. This problem is modeled as Literature Evolution Discovery (LED) in this work, which is a variant of the NP-hardness Directed Steiner tree problem (DST). Thereby, we approximately solve LED that finds DST on a transformed graph and our best algorithm performs in $O(\log^3 k)$ time, where k is the number of leaf nodes in the evolution structure.

Contributions. The main contribution of this work is threefold: (i) We propose to use a new concept, called *context consistency*, in a factor graph model. The consistent context configuration in the factor graph can be used as corresponding semantic interpretability; (ii) We define Literature Evolution Discovery problem (LED) for extracting evolution structure that preserves co-citation structures; (iii) Extensive experiments on real datasets and quantitative evaluations are provided, proving our method effective. Sufficient case studies demonstrate the interpretability of the discovered evolution structure.

2   Related Works

Latent Relationship Estimation. Relationship estimation between literature focuses on distance measurement. Conventional methods could be categorized into content-based (e.g., LDA [1]), structure-based [26], and hybrid methods ([5,13]). Recently, representation learning techniques [4] also received certain attention, e.g., Deepwalk [18], node2vec [8,14], LINE [22], GCN [15] and other variants [2,4]. These methods show strong capability in vector representation learning and the distance measurement. However, the distance calculated by the vectors is not interpretable as the training and the distance calculation are treated as black box processes.

Evolution Structure Extraction. Structure extraction in bibliographic databases has been studied based on various techniques, e.g., graphical model [28], information flow [6] and matrix decomposition [19], etc. Citation-LDA [28] aimed to capture the topic evolution without considering content/semantic information. Gordon et al. [6] generated a topic dependency graph but not the evolution structure of a specific literature. Vegas [19] focused on graph summarization by maximizing the overall influence flow and node summarization by matrix decomposition. Acemap [21] aimed to discover related paths and nodes in the database for visualization. RIDP [25] combined the content of papers and their citation papers, applying double-damping PageRank to imitate researcher reading behaviors to generate a study map, which shares the same intuition of ours. Although these work extract a substructure based on some objectives, the result is still lack of interpretability. As a comparison, the evolution structure in our work considers both context consistency and co-citation structure that are beneficial for interpretability.

3   Latent Relationship Learning

Inspired by the TPFG model [27], we model the latent relationship learning process as a factor graph configuration problem based on two aspects, (i) local context consistency and (ii) global context consistency.

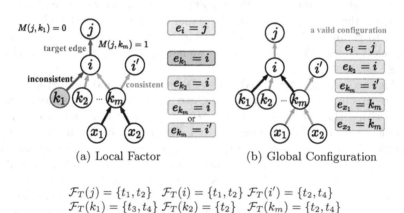

(a) Local Factor   (b) Global Configuration

$$\mathcal{F}_T(j) = \{t_1, t_2\} \quad \mathcal{F}_T(i) = \{t_1, t_2\} \quad \mathcal{F}_T(i') = \{t_2, t_4\}$$
$$\mathcal{F}_T(k_1) = \{t_3, t_4\} \quad \mathcal{F}_T(k_2) = \{t_2\} \quad \mathcal{F}_T(k_m) = \{t_2, t_4\}$$
(c) Literature Features

Fig. 3. Local factor and global configuration (with consistent context constraints) (Color figure online)

Notations. We consider a bibliographic graph $G = (\mathcal{P}, \mathcal{E}, \mathcal{W}, \mathcal{F}_\gamma)$, where $\mathcal{P}$ and $\mathcal{E}$ are literature node set and citation edge set, respectively. A citation edge $(i, j) \in \mathcal{E}$ indicates that literature j is cited by literature i. For the sake of presentation, we denote cited paper and citing paper nodes of p as $\mathcal{P}_\Uparrow(p)$ and $\mathcal{P}_\Downarrow(p)$ to distinguish hierarchical relationship. For each edge $(i, j) \in \mathcal{E}$, the initial weight $w_{i,j} \in \mathcal{W}$ indicates the proximity from i to j, that can be calculated by the degree and topic similarity [1], Personalized PageRank [12], or graph embedding [15,18]. In this work, we assume that the bibliographic network is a Directed Acyclic Graph (DAG). The features of a paper $p \in \mathcal{P}$ are denoted as $\mathcal{F}_\gamma(p)$. We consider four feature types $\mathcal{R} = \{T, A, \Uparrow, \Downarrow\}$ (topic, author, cited paper and citing paper) and $\gamma \in \mathcal{R}$. For example, literature i contains topic t_1, t_2 and literature i' contains topic t_2, t_4 (in Fig. 3(c)). These literature features constitute the semantic context in our context-consistent factor graph model (CCFG).

3.1 Context-Consistent Factor Graph

Local Context-Consistent Factor. Given any two literature i and j, we can find their local context-consistency $M(i, j)$ based their features $\mathcal{F}_\gamma(i)$ and $\mathcal{F}_\gamma(j)$,

$$M(i,j) = \begin{cases} 1 & \mathcal{F}_\gamma(i) \cap \mathcal{F}_\gamma(j) \neq \varnothing \\ 0 & otherwise \end{cases} \tag{1}$$

where $M(i, j) = 1$ if they have common features, otherwise is 0. As shown in Fig. 3(a), $M(k_1, j) = 0$ and $M(k_m, j) = 1$.

The next step is to find the *context consistency supports* from the neighbors. For simplicity, we denote an edge by a source-to-target variable e_i, where the source is i and the value of e_i (the target) can be any literature in $\mathcal{P}_\Uparrow(i)$. For instance, i is the source node of edge (i, j) and (i, j'). If $e_i = j$, the selected edge is (i, j). To estimate the latent relationship based on context consistency, we should select edges (set values) of a corresponding set of e_* who have consistent features.

In Fig. 3(a), given a target edge $e_i = j$, we aim to find context-consistent neighbors to support this relationship. **Case 1:** the k_2 and k_m share common features with j that $M(k_2, j) = 1$ and $M(k_m, j) = 1$ so that they are context-consistent, where the children edges $e_{k_2} = i$ and $e_{k_m} = i$ can be used to support the relationship $e_i = j$. **Case 2:** Since $M(k_1, j) = 0$, $e_{k_1} = i$ is an inconsistent case since the influence from j to i is not evidently transmits to k_1. **Case 3:** In particular, $e_{k_m} = i'$ is context-consistent with $e_i = j$, as the value of $M(k_m, j)$ is not related to $e_i = j$. It can be regard as a orthogonal influence, where k_m is influenced by another reference but not (i, j). By this analysis, we define the *local factor* f_i to calculate latent relationship score as follows[1],

$$f_i(e_i, \{e_k\}_{k \in \mathcal{P}_\Downarrow(i)}) = w_{i,e_i} \prod_k I(e_k, M(k, e_i)) \tag{2}$$

where the variables of f_i are the target edge(e_i) and a set of single edge from each literature citing i ($\{e_k\}$). It requires to select a specific edge from each literature k in $P_\Downarrow(i)$ to support the initial weight w_{i,e_i} for the target edge (i, e_i). Identity function I is to discriminate above consistent constraint, avoiding inconsistent context.

$$I(e_k, M(k, e_i)) = \begin{cases} 0 & e_k = i \wedge M(k, e_i) = 0 \\ 1 & otherwise \end{cases} \tag{3}$$

---

[1] Note that function $M()$ (cf. Eq. 1) considers only one type of features. For multi-feature cases, the local factor can be calculated by the average of $f_i(e_i,)$ with function $M()$ on each feature type γ.

where the only inconsistent context is as described in the Fig. 3(a) and **Case 2**, that selects the edge (k, i) to support target (i, j) and k has no common features with j.

Global Context-Consistent Function. The global context-consistent function F is then defined as the product of the local factors, where the dependency among e_* has been defined in $f()$ (cf. Eq. 2).

$$F(e_1, \ldots, e_n) = \frac{1}{Z} \prod_{i=1}^{|\mathcal{P}|} f_i(e_i, \{e_k\}_{\mathcal{P}_{\Downarrow}(i)}) \tag{4}$$

where Z is a normalizer and it is the value sum of $F()$ subject to $\{e_1, \ldots, e_n\}$. In particular, the configuration with any inconsistent context should result in $F = 0$. As shown in Fig. 3(b), a vaild configuration (green edges) is to find all values of e_* in the graph without inconsistent context. When a literature i has a null $\mathcal{P}_{\Uparrow}$, the $f_i(e_i = *)$ is initially set to 1, e.g., j and i' in Fig. 3(b). When a literature i cannot be used to support any relationships, $f_i(e_i = *)$ is set to 1 (not by Eq. 2) that indicates this edge has no effect on the consistent context of the target, e.g., k_1 in Fig. 3(a)).

Goal 1. Latent Relationship Estimation. With the global function F, the latent relationship score of an edge (i, j) is to fix $e_i = j$ and find the maximum global configuration (i.e., set values of $\{e_1, \ldots, e_n\}/e_i$). Therefore, the latent relationship score $s_{i,j}$ of edge (i, j) is the marginal probability of the maximum configuration of other literature as follows.

$$s_{i,j} = \max F(e_1, e_2, ..., e_n | e_i = j) \tag{5}$$

Goal 2. Interpretability. Another important goal is the interpretability for each latent relationship. Since we determine the latent relationship score with global context-consistent configuration, this is particularly helpful in the interpretation of the evolution structure. For instance, given an edge (i, j) in the evolution structure, the result is supported by the context-consistent edge set, e.g., $\{e_k\}_{k \in \mathcal{P}_{\Downarrow}}$. This can be viewed as a semantic explanation of the result.

Approximate Inference Process. The local context-consistent factor and global context-consistent function constitute a factor graph model [16]. It can be solved efficiently by the approximate inference process in [27].

4   Literature Evolution Structure Discovery

4.1   Literature Evolution Discovery Problem

Given a bibliographic network G and a root literature p, the evolution structure can be some reachable nodes from p in G (using a Markov process based on the estimated latent relationship). However, this naïve solution cannot generate satisfying result as it may only reserve references in one or two-hop neighbors or

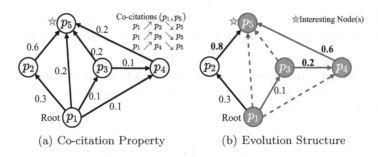

(a) Co-citation Property   (b) Evolution Structure

Fig. 4. Co-citation property and Evolution structure

the shortest paths between the references, which only collects paths to references together without structures.

To find a more meaningful evolution structure $ES(p)$, we make use of two concepts, *interesting nodes* and *co-citation*, that aims to provide better user experience in the visualization result.

Interesting Node Set. The interesting node set is a set of nodes $\mathcal{P}_I$ that are interested by the user and/or picked by a citation-based ranking approach. The goal of the discovery process is to find the evolution structure from the root node p to the nodes in $\mathcal{P}_I$. For instance, p_5 is an interesting node in Fig. 4.

Co-citation Property. Besides we use the co-citation concept in enriching the evolution structure. (1) If an edge (p_i, p_j) is added into the evolution structure $ES(p)$, then one of its co-citations is required to be added into $ES(p)$. (2) On the other hand, if an instance of a co-citation, denoted as $p_i \nearrow * \searrow p_j$, is picked into $ES(p)$, then the direct citation (p_i, p_j) will also be added into $ES(p)$. Shown in Fig. 4(a), the co-citation instance of (p_1, p_5) can be $p_1 \nearrow p_2 \searrow p_5$, $p_1 \nearrow p_3 \searrow p_5$ and $p_1 \nearrow p_4 \searrow p_5$.

Problem Definition. We are ready to present our Literature Evolution Discovery problem in Definition 1.

Definition 1 (Literature Evolution Discovery (LED)). *Given a bibliographic network G, a root literature $p \in \mathcal{P}$, interesting node set $\mathcal{P}_I \subseteq \mathcal{P}$, edge cost set C, LED returns an evolution structure $ES(p) \subseteq G$ such that the total edge cost is minimized, subject to (1) connectivity between p and interesting nodes and (2) co-citation property.*

The edge cost in G is derived from the latent relationship in Sect. 3. Mathematically, it is defined as $c(i, j) = 1 - s_{i,j}$ for an edge (i, j). Finding the evolution structure $ES(p)$ can be viewed as a directed Steiner tree problem [3,9] (see Definition 2). The DST problem is a well-known NP-hard problem, which can only be approximately solved by the algorithms based on shortest path [3,10] in quasi-polynomial time.

Algorithm 1. HideCite(Hcc)

Input: citation network G
Output: G', Co-Auxiliary $\mathcal{E}_{Aux}$
1: initialize $G' = G$, $\mathcal{E}_{Aux} = \varnothing$
2: **for** p_i = the next node in topological sorting **do**
3:     **for** $p_j \in \mathcal{P}_\Uparrow(p_i)$(in topological sorting) **do**
4:         find all $(p_i \nearrow * \searrow p_j)$
5:         **for** p_k in $\{*\}$ (in topological sorting) **do**
6:             mark (p_i, p_j) as hide; $\mathcal{E}_{Aux}(p_k, p_j)$.add$(p_i, p_k)$

Definition 2 (Directed Steiner Problem (DST)). *Given a directed graph* $G = (V, E)$, *a root node* $r \in V$, *a terminal (interesting) node set* $X \subseteq V$ *in size* $|X|$, *edge cost set* W, DST *is to find a tree* T^* *from* G *such that the total edge cost of the tree is minimized and root node have a path to each node in* X, *respectively.*

Graph Transformation. However, the evolution structure $ES(p)$ computed by DST does not reserve the co-citation property. For instance, DST may return an $ES(p_1)$ that contains (p_1, p_4) but not $p_1 \nearrow p_3 \searrow p_4$. To reserve the co-citation property in the evolution structure, we transform G into a new graph G' that *transforms* those directed citations if they are covered by at least one co-citations. The evolution structure extracted from G' (by DST) is able to reserve the co-citation property. We will show the algorithm details shortly in Sect. 4.2.

Example. As shown in in Fig. 4(b), three directed citations are removed (dashed lines) in the transformed graph G' as they are covered by at least one co-citation structure and the process is done in an inductive manner. We can then generate a Steiner tree ES' (blue nodes and blue solid lines) based on G'. Finally we recover the hidden directed citation if their co-citation is included in ES'. In Fig. 4(b), the solid lines of blue color are the result of ES' and the dashed lines of blue color are the recovered citations. All the blue color edges are the result of our evolution structure ES, that must obey the co-citation property.

Interpretability. LED generates an evolution structure that fulfills the co-citation property. In other words, a citation edge in the evolution structure, e.g., (p_1, p_5), can be explained by the co-citation structure. For instance, p_3 is a related work for citing p_5 in p_1. This interpretability also reveals the importance of edges semantically, i.e., some edges in the evolution structure are cornerstone citations and other edges help understand these core citation relationships.

4.2   Approximate Algorithm

In this section, we propose two methods to generate G' and the corresponding Steiner tree algorithms to extract ES'.

Hide Co-citation + Approx. SP (Hcc-ApSP). Hcc (Algorithm 1) generates G' by hiding direct citation edges based on the topological order of the citation

---

Algorithm 2. Co-citation Transformation (Tcc)

Input: citation network G, edge cost $\mathcal{C}$
Output: G', Co-Auxiliary $\mathcal{E}_{Aux}$, edge cost $\mathcal{C}'$
1: initialize $G' = G$, $\mathcal{C}' = \mathcal{C}$, $\mathcal{E}_{Aux} = \varnothing$
2: **for** $p_i =$ the next node in topological sorting **do**
3:     **for** $p_j \in \mathcal{P}_\Uparrow(p_i)$ (in topological sorting) **do**
4:         find all $(p_i \nearrow * \searrow p_j)$
5:         **for** p_k in $\{*\}$ (in topological sorting) **do**
6:             **if** $\mathcal{E}_{Aux}(p_k, p_j) \cap \mathcal{E}_{Aux}(p_i, p_j) = \varnothing$ **then**
7:                 $c'_{new}(k, j) = c'_{old}(k, j) + c(i, j)$
8:             **else**
9:                 $c'_{new}(k, j) = c'_{old}(k, j) + c'(i, j)$
10:            mark (p_i, p_j) as hide; $\mathcal{E}_{Aux}(p_k, p_j)$.add$(p_i, p_k)$

---

structure. To recover directed citation edge, we record each $p_i \nearrow * \searrow p_j$ into a set $\mathcal{E}_{Aux}$ that also helps to calculate the cost of an ES'. For instance, the ES in Fig. 4(a) can match the blue subgraph in Fig. 4(b).

Given G', the ES' can be generated by a trivial shortest path based algorithm that is used to solve DST [3,7]. Specifically, it searches the shortest paths from root p to each node in $\mathcal{P}_I$ and then combine them as a tree structure. The only adaption in ApSP is that the cost of a co-citation should also include the cost of the "embedded" directed citation. The cost recovery process can be done at the traversal process of Dijkstra algorithm. In Fig. 4(b), the shortest path from p_1 to p_5 is $p_1 \rightarrow p_3 \rightarrow p_4 \rightarrow p_5$. The shortest path cost is the sum cost of the path plus the costs of three hidden edges, (p_3, p_5), (p_1, p_4), and (p_1, p_5). The problem of this solution is the time complexity in recovering the edge costs, which is increasing with the length of shortest path.

Transform Co-citation + Approx. DST (Tcc-ApDST). The transformation algorithm Hcc does not pre-aggregate the cost so that the DST algorithm ApSP requires to recover the cost during the Dijkstra execution. In Tcc, our strategy is to construct a transformed graph G' that preserves corresponding citation edge costs. The pseudo code of the algorithm is shown in Algorithm 2. We first initiate edge cost $c' \in \mathcal{C}'$ in transformed G' with edge cost $c \in \mathcal{C}$ (line 1) and traverse all co-citation structure by a topological sorting (line 2–4). For a co-citation relation $p_i \nearrow p_k \searrow p_j$, we mark edge (p_i, p_j) as *hidden* in the graph G' (line 10) and transfer the cost $c(i, j)$ or $c'(i, j)$ to edge (p_k, p_j) as an accumulated cost, depending on whether (p_k, p_j) have a nested co-citation structure with (p_i, p_j) (line 5–9). These co-citation situations are shown in Fig. 5. The cases in Fig. 5(b)(c)(d) are co-citations about three edges in Fig. 5(a). We also record the auxiliary edge in $\mathcal{E}_{Aux}$, to make the judgment and recover the path (line 10).

After the graph transformation, all directed edges are *hidden*, but their costs are kept in the co-citation edges. In other words, the citation edges of a co-citation structure are likely embedded into the cost. We then apply the approximate algorithm in [3], denoted as ApDST, to generate the Steiner tree ES' from

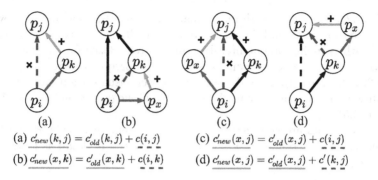

$$(a)\ c'_{new}(k, j) = c'_{old}(k, j) + c(i, j) \qquad (c)\ c'_{new}(x, j) = c'_{old}(x, j) + c(i, j)$$

$$(b)\ c'_{new}(x, k) = c'_{old}(x, k) + c(i, k) \qquad (d)\ c'_{new}(x, j) = c'_{old}(x, j) + c'(k, j)$$

Fig. 5. Co-citation transformation.

G'. With recovery of hidden edges by $\mathcal{E}_{Aux}$ in the generated ES', the result ES is not a tree structure but a subgraph confined by co-citation property. Note that this algorithm achieve $O(\log^3 |\mathcal{P}_I|)$ time complexity (same as ApDST), which is faster than ApSP.

5   Experiments

5.1   Experiment Setting

Datasets. Our bibliographic network dataset is collected from Aminer [20, 24]. The dataset mainly comes from 3 research areas, **AI, ICML, KDD**, described in Table 1. The bibliographic network G of each dataset is constructed based on two steps, (1) select a set of literature from source venues in 2010–2018 as seeds and (2) expand the relationship from the seed literature based on the citations. We keep the literature which is cited at least 10 times, and exclude the survey literature to avoid its disturbance.

Table 1. Datasets

Dataset	Source venue	# Nodes	# Edges
ICML	ICML	13,608	159,339
AI	NIPS, ICML, ICLR, CVPR, EMNLP, ACL, ECCV, IJCAI, AAAI	26,116	421,852
KDD	KDD	16,548	199,553

Ground Truth. In the discovery and visualization of the literature evolution, it is hard to find objective reference to judge the quality (especially for structure and interpretation). We extract 100 most popular Master Reading Trees (MRT) in Aminer [24] as ground truth for evaluation. Each MRT is a citation flow network with a root node, and the structure is manually generated by experts. The ground truth is mainly about literature in AI area. Based on citation network of **ICML, AI**, we generate comparison results for quantitative analysis (see Sect. 5.2, 5.3, and 5.4). In addition, we discover visualization cases for interpretability analysis about our model using the bibliographic network of **KDD**, (see Sect. 5.5).

All experiments are conducted on a Linux system with 3.60GHz CPU and 64G memory. The baselines and metric are introduced in detail as follows.

5.2   Latent Relationship Study

We conduct two groups of experiments for evaluating the effectiveness of CCFG in latent relationship mining. Considering the feature $\mathcal{F}_\gamma$ of literature nodes, we use topic, author, citation, reference and their combinations as feature γ respectively, to test the effect of feature selection on CCFG (denoted as CCFG-$\mathcal{F}_T$, CCFG-$\mathcal{F}_A$ etc. in Table 3). These features can be obtained in Aminer citation dataset [24]. To generate initial edge weight w of each edge, we utilized content-based and structure-based approaches. To evaluate the effectiveness of CCFG, we compare the basic weight of each baseline, with the latent relationship score further inferred by CCFG. They are denoted as *-Init and *-CCFG, respectively. The approaches (*) for generating the basic weight are listed as follows.

Content-Based Baseline: Topic similarity (t), which is contained in Aminer citation dataset [20].

Structure-Based Baseline: (1) Degree (d). (2) DeepWalk (dw) [18]. (3) GCN (gcn) [15]. Both of dw and gcn are implemented by Deep Graph Library (DGL) (https://www.dgl.ai/).

For fairness, we calculate the proximity between a root node and nodes of 1–2 hops by PageRank, and then select the top-k nodes to compare with nodes in MRT of the same root node.

In feature selection comparison, we adopt same weight generation method (i.e., deepwalk). In comparison between initial weight and CCFG, we adopt same feature, i.e., "Author"($\mathcal{F}_A$). Our comparisons and findings of this experiment are as follows.

Effectiveness of CCFG. For different input weights with/without CCFG (Table 2), our model CCFG improves the results by finding important relationships. (1) Each *-CCFG method has better results in Precision/Recall/F1 than its corresponding *-Init methods, both in top-20 of one hop (1hop@20) setting and top-50 of two hops (2hop@50) setting. *-CCFG methods could achieve best performance at most than baselines.

Effect of Feature Selection. The feature selection is not very sensitive in CCFG (Table 3). However, their targeting fields still show some differences. (1) Most of these features have a better performance compared with the basic weight by deepwalk (dw-Init) in both 1–2 hop. (2) "Author" feature is useful when characterizing close relationship, i.e., CCFG-$\mathcal{F}_a$ in 1hop@20 has better results than that of CCFG-$\mathcal{F}_t$. (3) The hybrid features, such as CCFG-$\mathcal{F}_{AT}$, CCFG-$\mathcal{F}_{AT⇑⇓}$, can also make better results than a single feature, where CCFG-$\mathcal{F}_{AT}$ is good at 2-hop and CCFG-$\mathcal{F}_{⇑⇓}$ and CCFG-$\mathcal{F}_{AT⇑⇓}$ is good at 1-hop.

Table 2. Comparisons of input weights. Feature is fixed as CCFG-$\mathcal{F}_A$

Methods	1hop@20			2hop@50		
	Pre	Rec	F1	Pre	Rec	F1
d-Init	.8416	.2431	.3722	.2714	.2780	.2680
d-CCFG-$\mathcal{F}_A$	**.8472**	**.2455**	**.3755**	**.2947**	**.3054**	**.2928**
t-Init	.8305	.2393	.3665	.2903	.2957	.2860
t-CCFG-$\mathcal{F}_A$	**.8500**	**.2464**	**.3768**	**.2947**	**.3054**	**.2928**
dw-Init	.8305	.2399	.3672	.2603	.2672	.2572
dw-CCFG-$\mathcal{F}_A$	**.8500**	**.2464**	**.3768**	**.2947**	**.3054**	**.2928**
gcn-Init	.8250	.2376	.3639	.2547	.2615	.2516
gcn-CCFG-$\mathcal{F}_A$	**.8500**	**.2464**	**.3768**	**.2947**	**.3054**	**.2928**

Table 3. Comparisons of feature selection. Weight is fixed as dw-CCFG.

Methods	1hop@20			2hop@50		
	Pre	Rec	F1	Pre	Rec	F1
dw-Init	.8305	.2399	.3672	.2603	.2672	.2572
dw-CCFG-$\mathcal{F}_A$	**.8500**	**.2464**	**.3768**	**.2947**	**.3054**	**.2928**
dw-CCFG-$\mathcal{F}_T$	.8361	.2419	.3701	.2592	.2708	.2583
dw-CCFG-$\mathcal{F}_{AT}$	.8472	.2455	.3755	.2914	.3032	.2901
dw-CCFG-$\mathcal{F}_{⇑⇓}$	**.8500**	**.2464**	**.3768**	.2836	.2929	.2812
dw-CCFG-$\mathcal{F}_{AT⇑⇓}$	**.8500**	**.2464**	**.3768**	.2847	.2937	.2821

5.3   Structure Study

We compare the proposed two algorithms for LED with a baseline solution, to prove the effectiveness and efficiency in structure extraction. (1) The baseline is mainly finding a Steiner tree with DST algorithm [3] in the original graph G (denoted as ApDST). And two algorithms we proposed are: (2) Hcc-ApSP, (3) Tcc-ApDST. Given a root in MRT, we generate the evolution structure with each method and compare returned nodes/edges with the nodes/edges in this MRT. To avoid the influence from the selection of interesting nodes $\mathcal{P}_I$, we select nodes in MRTs as interesting nodes.

Table 4. Comparisons of solutions for LED

Methods	Pre-p	Rec-p	F1-p	Pre-e	Rec-e	F1-e
ApDST	.4288	**.7993**	.5498	**.3867**	.2318	.2758
Hcc-ApSP	.4736	**.7993**	.5776	.3249	**.5264**	**.3898**
Tcc-ApDST	**.4949**	**.7993**	**.5943**	.3823	.4019	.3798

Table 5. The RunTime and structure costs of 10 cases

Structure cost									
Hcc-ApSP	13.1	40.6	24.0	19.3	40.0	32.1	33.4	20.2	30.3	21.5
Tcc-ApDST	11.8	27.9	18.0	16.6	30.4	19.4	22.1	15.7	23.2	16.9
Runtime (ms)									
Hcc-ApSP	187.7	438.5	3.0k	1.0k	82.0	4.0	11.5	707.3	3.0k	2.2k
Tcc-ApDST	0.4	0.5	0.5	0.5	0.3	0.1	0.2	0.4	0.4	0.5

Effectiveness of Co-citation. Both Hcc-ApSP and Tcc-ApDST consider co-citation structures when generating an ES, while ApDST ignores it. In Table 4, Hcc-ApSP and Tcc-ApDST have a higher Recall/F1 than that with ApDST-o. It demonstrates the co-citation transformation provides a balance between precision and recall. The co-citation structure is also beneficial for exploiting more latent nodes and matching more citations in MRTs (more details in Sect. 5.5).

Effectiveness of Hcc-ApSP **and** Tcc-ApDST. We compare them by Precision/Recall/F1 of nodes to evaluate the effectiveness of finding nodes in evolution structure, and by Precision/Recall/F1 of edges to evaluate the performance of finding structure. From Table 4, we could find Tcc-ApDST is superior than Hcc-ApSP, which demonstrates its performance on discovering meaningful literature. Considering Precision-e, Hcc-ApSP has a nearly good result with ApDST, and a larger improvement than Hcc-ApSP. It proves Tcc-ApDST could discover meaningful edges effectively. Besides, Tcc-ApDST is not good as Hcc-ApSP on Recall-e. Because Hcc-ApSP brings in much more but redundant edges than Tcc-ApDST. At last, on F1-e, by discovering much less but meaningful edges, Tcc-ApDST could achieve almost good performance as Hcc-ApSP, and are much better than ApDST.

Efficiency of Hcc-ApSP **and** Tcc-ApDST. As both of our methods are aimed at solving optimization problem LED, we randomly choose 10 cases to show the efficiency and optimality of them. In Table 5, we have findings: (1) Tcc-ApDST could save 99% of the running time of Hcc-ApSP. (2) The cost bound of Hcc-ApSP is loose. In the real testing, Tcc-ApDST could make a smaller cost result mostly, although an approximation of its cost with optimal value cannot be guaranteed theoretically.

5.4   Effectiveness of Our Solutions

To compare the effectiveness of our solutions for research structure discovery, we compare several baselines with traditional methods and state-of-art methods, on both **ICML** and **AI** datasets. We compare to the variants of structure expansion strategy in this work with different relationship mining approaches. Especially, RIDP [25] is a most related work with us, which aims to find a study map. We confine that the nodes are all in 2-hop for a root for equality. The metrics are precision/recall/F1 score for nodes and edges in MRTs, respectively.

The following are baselines and our methods. (1) Twohop (2H): The 2-hop subgraph from the root. (2) Hyrid (RIDP): The method in [25] which uses citation content-based LDA and damping PageRank, generating a study map by three steps: a) rank *topK* papers in damping PageRank as seed papers; b) construct subgraph from root with seed papers; c) remove isolated papers, called study map generation and denoted as SMG in our experiment. We set *topK* for each structure result same with its MRT in our experiment. (3) Random (Rand): Randomly choose *topK* nodes in 2-hops and construct the subgraph. (4) Citation Counts (CC-SMG): Choose *topK* nodes in 2-hop by citation counts and construct the subgraph by SMG. (5) Structure-based (DePR-SMG): Choose *topK* nodes by PageRank with degree as weights, and construct the subgraph by SMG. [11,17] (6) Content-based (Topic-SMG): Choose *topK* nodes by topic similarity with root and construct the subgraph by SMG [20,24]. (7) DeepWalk (DW-SMG):

Choose top-50 nodes by distance in deepwalk, and construct the subgraph by SMG [18]. (8) ApDST/Hcc-ApSP/Tcc-ApDST: The baseline without co-citation property described in Sect. 5.3, and our two algorithms for LED.

From Table 6, we can see that: (1) Our methods based on minimized structure cost have better results with **a large margin** than the variants of RIDP in most metrics, except DW-SMG is better than ApDST. Both Hcc-ApSP and Tcc-ApDST, which consider co-citation property, can generate structure more effectively. (2) The Tcc-ApDST is better than Hcc-ApSP in most of the metrics. Hcc-ApSP is better than Tcc-ApDST on Recall-e. The reason is that combining the shortest paths of root and each interesting node might bring more edges in the evolution structure. Though the Recall-p and Recall-e are very high in 2H, it returns all nodes as a baseline. Our methods based on minimized structure cost could still achieve close values with it.

Table 6. Comparisons of our solutions with baselines

Methods	Dataset: **ICML**						Dataset: **AI**					
	Pre-p	Rec-p	F1-p	Pre-e	Rec-e	F1-e	Pre-p	Rec-p	F1-p	Pre-e	Rec-e	F1-e
2H	.2811	**.7993**	.4031	.2004	**.6573**	.2954	.3374	**.6759**	.4219	.2310	**.5459**	.3008
Rand	.2777	.2777	.2777	.2046	.0878	.1171	.3365	.3244	.3300	.2446	.1681	.1857
CC-SMG	.3974	.3974	.3974	.2747	.2106	.2363	.4152	.4030	.4086	.2769	.2591	.2639
Topic-SMG	.2789	.2789	.2789	.2765	.0403	.0649	.3056	.2935	.2991	.1973	.1065	.1292
DePR-SMG	.2567	.2567	.2567	.1209	.0405	.0560	.3046	.2924	.2980	.3125	.0965	.1242
DW-SMG	.4642	.4642	.4642	.3972	.1255	.1841	.4367	.4079	.4199	.3517	.1441	.1896
RIDP	.3326	.4517	.3790	.2676	.1306	.1645	.3615	.4391	.3863	.2852	.1559	.1854
ApDST	.4288	**.7993**	.5498	**.3867**	.2318	.2758	.4298	.6012	.4838	.3561	.2055	.2444
CCFG-Hcc-ApSP	.4736	**.7993**	.5776	.3249	.5264	**.3898**	.5258	.6012	.5267	.3618	.3747	.3371
CCFG-Tcc-ApDST	**.4949**	**.7993**	**.5943**	.3823	.4019	.3798	**.5410**	.6012	**.5335**	**.4079**	.3343	**.3450**

5.5 Case Study

To show how the visualization of evolution structure gives advice for tracking the origin of a targeted literature in real scenario, we generate a case study from **KDD** dataset to discuss the effect of our solutions. The interesting nodes are top-20 nodes in PageRank with $s_{i,j}$ as input weight. Limited to space, we demonstrate a representative subgraph of their results in Fig. 6.

Given root node 1 and interesting nodes $\{4, 5, 6\}$, the evolution structure from LED is the connected subgraph with pink edges and nodes $\{1, 2, 3, 4, 5, 6\}$, where dashed edges are recovered edges (Fig. 6(a)). These direct citation links have candidate co-citations that should be hidden in Tcc stage. The result of ApDST is the connected tree with blue edges and nodes $\{1, 4, 5, 6\}$ (Fig. 6(b)). The grey edges are not contained in the result of ApDST but in that of Tcc-ApDST. An analysis of visualization stage is as follows.

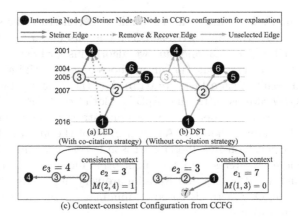

Fig. 6. A visualization example with interpretability. (a) Interesting nodes and Steiner nodes with pink edges are the result of Tcc-ApDST (dashed edges are recovered edges). (b) Interesting nodes and Steiner nodes with blue edges are the result of ApDST.

6   Conclusion

In this work, we propose a new framework to extract the literature evolution structure from a bibliographic databases. This task is challenging due to the diversification of the data and the collection of the latent information. To address these challenges, we adopt factor graph, co-citation based graph transformation, and Steiner tree. Our experimental results show the effectiveness of our proposed methods. An interesting future direction is to mine evolution history by different topics and develop an online visualization system.

Acknowledgements. This work was supported in part by the National Key Research and Development Plan of China (No. 2019YFB2102100), FDCT Macau (SKL-IOTSC-2021-2023, 0015/2019/AKP), and University of Macau (MYRG2019-00119-FST).

References

1. Blei, D.M., Ng, A.Y., Jordan, M.I.: Latent Dirichlet allocation. J. Mach. Learn. Res. **3**, 993–1022 (2003)
2. Cai, H., Zheng, V.W., Chang, K.C.: A comprehensive survey of graph embedding: problems, techniques, and applications. IEEE TKDE **30**(9), 1616–1637 (2018)
3. Charikar, M., et al.: Approximation algorithms for directed Steiner problems. In: Proceedings of the Ninth Annual ACM-SIAM Symposium on Discrete Algorithms, 25–27 January 1998, San Francisco, California, USA, pp. 192–200. ACM/SIAM (1998)
4. Dridi, A., Gaber, M.M., Azad, R.M.A., Bhogal, J.: Scholarly data mining: a systematic review of its applications. Wiley Interdiscip. Rev. Data Min. Knowl. Discov. **11**(2), e1395 (2021)

5. Effendy, S., Yap, R.H.C.: Analysing trends in computer science research: a preliminary study using the Microsoft academic graph. In: Proceedings of the 26th International Conference on World Wide Web Companion, Perth, Australia, 3–7 April 2017, pp. 1245–1250. ACM (2017)

6. Gordon, J., Zhu, L., Galstyan, A., Natarajan, P., Burns, G.: Modeling concept dependencies in a scientific corpus. In: Proceedings of the 54th Annual Meeting of the Association for Computational Linguistics, ACL 2016, 7–12 August 2016, Berlin, Germany, Volume 1: Long Papers (2016)

7. Grandoni, F., Laekhanukit, B., Li, S.: $O(\log^2 k/\log \log k)$-approximation algorithm for directed Steiner tree: a tight quasi-polynomial-time algorithm. In: Proceedings of the 51st Annual ACM SIGACT Symposium on Theory of Computing, STOC 2019, Phoenix, AZ, USA, 23–26 June 2019, pp. 253–264. ACM (2019)

8. Grover, A., Leskovec, J.: node2vec: scalable feature learning for networks. In: Proceedings of the 22nd ACM SIGKDD, San Francisco, CA, USA, 13–17 August 2016, pp. 855–864. ACM (2016)

9. Guan, S., Ma, H., Wu, Y.: Attribute-driven backbone discovery. In: Proceedings of the 25th ACM SIGKDD International Conference on Knowledge Discovery & Data Mining, Anchorage, AK, USA, 4–8 August 2019, pp. 187–195. ACM (2019)

10. Halperin, E., Krauthgamer, R.: Polylogarithmic inapproximability. In: Proceedings of the 35th Annual ACM Symposium on Theory of Computing, 9–11 June 2003, San Diego, CA, USA, pp. 585–594. ACM (2003)

11. Haveliwala, T.: Efficient computation of PageRank. Technical report, Stanford (1999)

12. Haveliwala, T., Kamvar, S., Jeh, G.: An analytical comparison of approaches to personalizing PageRank. Technical report, Stanford (2003)

13. Hoonlor, A., Szymanski, B.K., Zaki, M.J.: Trends in computer science research. Commun. ACM **56**(10), 74–83 (2013)

14. Kazemi, B., Abhari, A.: Content-based node2vec for representation of papers in the scientific literature. Data Knowl. Eng. **127**, 101794 (2020)

15. Kipf, T.N., Welling, M.: Semi-supervised classification with graph convolutional networks. In: 5th International Conference on Learning Representations, ICLR 2017, Conference Track Proceedings (2017)

16. Kschischang, F.R., Frey, B.J., Loeliger, H.: Factor graphs and the sum-product algorithm. IEEE Trans. Inf. Theory **47**(2), 498–519 (2001)

17. Page, L., Brin, S., Motwani, R., Winograd, T.: The PageRank citation ranking: bringing order to the web. Technical report, Stanford InfoLab (1999)

18. Perozzi, B., Al-Rfou, R., Skiena, S.: DeepWalk: online learning of social representations. In: The 20th ACM SIGKDD International Conference on Knowledge Discovery and Data Mining, KDD 2014, NY, USA, 24–27 August 2014 (2014)

19. Shi, L., Tong, H., Tang, J., Lin, C.: VEGAS: visual influence graph summarization on citation networks. IEEE Trans. Knowl. Data Eng. **27**(12), 3417–3431 (2015)

20. Sinha, A., et al.: An overview of Microsoft academic service (MAS) and applications. In: Proceedings of the 24th International Conference on World Wide Web, pp. 243–246. ACM (2015)

21. Tan, Z., Liu, C., Mao, Y., Guo, Y., Shen, J., Wang, X.: AceMap: a novel approach towards displaying relationship among academic literatures. In: Proceedings of the 25th International Conference on World Wide Web, WWW 2016, Montreal, Canada, 11–15 April 2016, Companion Volume, pp. 437–442. ACM (2016)

22. Tang, J., Qu, M., Wang, M., Zhang, M., Yan, J., Mei, Q.: LINE: large-scale information network embedding. In: Proceedings of the 24th International Conference on World Wide Web, Florence, Italy, 18–22 May 2015, pp. 1067–1077. ACM (2015)

23. Tang, J.: AMiner: toward understanding big scholar data. In: Proceedings of the Ninth ACM International Conference on Web Search and Data Mining, San Francisco, CA, USA, 22–25 February 2016, p. 467. ACM (2016)

24. Tang, J., Zhang, J., Yao, L., Li, J., Zhang, L., Su, Z.: ArnetMiner: extraction and mining of academic social networks. In: KDD 2008, pp. 990–998 (2008)

25. Tao, S., Wang, X., Huang, W., Chen, W., Wang, T., Lei, K.: From citation network to study map: a novel model to reorganize academic literatures. In: Proceedings of the 26th International Conference on World Wide Web Companion, Perth, Australia, 3–7 April 2017 (2017)

26. Valenzuela, M., Ha, V., Etzioni, O.: Identifying meaningful citations. In: Scholarly Big Data: AI Perspectives, Challenges, and Ideas, Papers from the 2015 AAAI Workshop, Austin, Texas, USA, January 2015

27. Wang, C., et al.: Mining advisor-advisee relationships from research publication networks. In: Proceedings of the 16th ACM SIGKDD International Conference on Knowledge Discovery and Data Mining, Washington, DC, USA, 25–28 July 2010 (2010)

28. Wang, X., Zhai, C., Roth, D.: Understanding evolution of research themes: a probabilistic generative model for citations. In: The 19th ACM SIGKDD International Conference on Knowledge Discovery and Data Mining, KDD 2013, Chicago, IL, USA, 11–14 August 2013 (2013)

Incorporating Commonsense Knowledge into Story Ending Generation via Heterogeneous Graph Networks

Jiaan Wang[1], Beiqi Zou[3], Zhixu Li[2], Jianfeng Qu[1(✉)], Pengpeng Zhao[1], An Liu[1], and Lei Zhao[1]

[1] School of Computer Science and Technology, Soochow University, Suzhou, China
jawang1@stu.suda.edu.cn, {jfqu,ppzhao,anliu,zhaol}@suda.edu.cn
[2] Shanghai Key Laboratory of Data Science, School of Computer Science,
Fudan University, Shanghai, China
zhixuli@fudan.edu.cn
[3] Department of Computer Science, Princeton University, Princeton, USA
bzou@cs.princeton.edu

Abstract. Story ending generation is an interesting and challenging task, which aims to generate a coherent and reasonable ending given a story context. The key challenges of the task lie in how to comprehend the story context sufficiently and handle the implicit knowledge behind story clues effectively, which are still under-explored by previous work. In this paper, we propose a Story Heterogeneous Graph Network (SHGN) to explicitly model both the information of story context at different granularity levels and the multi-grained interactive relations among them. In detail, we consider commonsense knowledge, words and sentences as three types of nodes. To aggregate non-local information, a global node is also introduced. Given this heterogeneous graph network, the node representations are updated through graph propagation, which adequately utilizes commonsense knowledge to facilitate story comprehension. Moreover, we design two auxiliary tasks to implicitly capture the sentiment trend and key events lie in the context. The auxiliary tasks are jointly optimized with the primary story ending generation task in a multi-task learning strategy. Extensive experiments on the ROCStories Corpus show that the developed model achieves new state-of-the-art performances. Human study further demonstrates that our model generates more reasonable story endings.

Keywords: Story ending generation · Heterogeneous graph network · Multi-task learning

1 Introduction

Story ending generation (SEG) is a natural language generation task, which aims at concluding a story ending given a context [33]. Generally, a story con-

---

J. Wang and B. Zou—Indicates equal contribution.

© The Author(s), under exclusive license to Springer Nature Switzerland AG 2022
A. Bhattacharya et al. (Eds.): DASFAA 2022, LNCS 13247, pp. 85–100, 2022.
https://doi.org/10.1007/978-3-031-00129-1_6

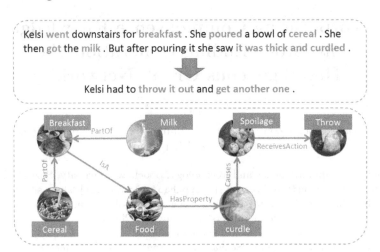

Fig. 1. The top graph shows an example of story ending generation. Orange words are entities and events. The bottom graph indicates the implicit knowledge behind the story. (Color figure online)

text contains a series of entities and events (known as story clues), each of which could have strong logical relationships with others, which leads to rich inter-active relations across the whole context. For humans, one may utilize his/her own commonsense knowledge to capture the story clues and conceive story end-ings. As shown in Fig. 1, the story clue of example story is: $went\_breakfast \Rightarrow poured\_cereal \Rightarrow got\_milk \Rightarrow it\_was\_thick\_and\_curdled$, which indicates the food may have gone bad. It is natural for humans to *throw the bad food and get another one* since we all know *spoiled food is harmful to health*. Therefore, in order to generate a coherent and reasonable ending, generative models should not only sufficiently comprehend the story context and further capture the story clues but also effectively handle the implicit knowledge behind them.

Most previous work views the story context as a linear sequence of words and ignores the rich relations among them. For example, Zhao et al. [33] and Li et al. [13] explore variant sequence-to-sequence (Seq2Seq) models encoding the story context in a left-to-right manner and decoding endings. They further utilize reinforcement learning to improve the rationality and/or diversity of generated endings. Gupta et al. [7] resort to an extra keywords extraction algorithm and model the keywords information in the Seq2Seq models. Luo et al. [15] make use of existing sentiment analyzers to consider the fine-grained sentiment in their Seq2Seq method. Guan et al. [6] design incremental encoding and multi-source attention mechanism to model the relation of adjacent sentences and incorpo-rate commonsense knowledge into contextual representation. These methods all adopt sequential modeling strategies for encoding the story context, hindering the exploration of inherently rich interactive relations across the whole context, which makes the story context and commonsense knowledge modeling inade-quate.

Recently, Huang et al. [10] suggest that the great importance of story clues hidden in the context and further propose multi-level graph convolutional networks over dependency parse (MGCN-DP) that models SEG task in a graph-to-text manner. The graph architecture better models the interactive relations across story context and result in more coherent endings compared to sequential methods. However, the relations of words from different context sentences cannot be explicitly captured by MGCN-DP which only contains word nodes from the same sentence in a graph. Besides, the MGCN-DP model does not consider commonsense knowledge behind the story, thus the generated endings could be suboptimal.

To remedy above issues, in this paper, we propose a **S**tory **H**eterogeneous **G**raph **N**etwork (SHGN) for SEG. The heterogeneous graph network shows its superiority in many tasks, such as recommender systems and summarization [2,29]. Specifically, three types of graph nodes are considered in our SHGN: *commonsense knowledge*, *words* and *sentences*. Besides, a *global* node is also introduced to the graph to aggregate the non-local information (see Fig. 3). To obtain contextualized representations for these nodes, large-scale pre-trained embedding models such as Sentence-BERT [22] and SimCSE [4] are used for contextual encoding. Then, a graph neural network is used to propagate message and update representations of nodes in the heterogeneous graph. The final node representations are passed through transformer decoders to generate story endings. In addition, to sufficiently comprehend the story context, we design two sub-tasks with special consideration for story ending generation: (1) the *sentiment prediction of story endings sub-task* uses representations of sentence nodes to predict the sentiment of corresponding story ending, which is constructed to push the model to capture the fine-grained sentiment trend; (2) the *clue words prediction sub-task* utilizes representations of word nodes to predict whether each word belongs to story clues, which is expected to force the model to identify the key events in the context. Such two sub-tasks can be coupled with the primary story ending generation task via multi-task learning strategy, resulting in our final model SHGN.

We conduct various experiments on the widely used ROCStories Corpus [19]. Experimental results show that our approach achieves state-of-the-art performances on SEG task. Human study indicates that our SHGN generates more coherent and reasonable story endings as compared to previous strong baselines.

Our main contributions in this paper are summarized as follows:

- We propose a Story Heterogeneous Graph Network for SEG, which explicitly models the information of story context at different granularity levels and the multi-grained interactive relations among them[1].
- We also design two auxiliary tasks (i.e., sentiment prediction of story endings and clue words prediction) to facilitate story comprehension. To the best of our knowledge, we are the first to apply multi-task learning strategy on SEG.

---

[1] We release our code and generated results at https://github.com/krystalan/AwesomeSEG.

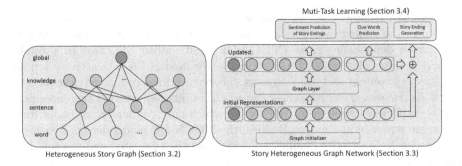

Fig. 2. Overview of our proposed model. For a story context, we first construct a heterogeneous story graph (Sect. 3.2). Then, we design our story heterogeneous graph network (SHGN) to initialize and update node representations in the graph (Sect. 3.3). Finally, auxiliary tasks are introduced to facilitate story comprehension, which are jointly optimized with the primary SEG task in the multi-task learning strategy (Sect. 3.4).

– Extensive experiments on widely used ROCStories Corpus show that our model achieves new state-of-the-art performances. Human study and case study further prove that our model could generate more coherent and reasonable story endings.

2   Related Work

Story Generation. Story Generation (SG), also known as storytelling, aims at generating a logical self-consistent story plot. Early SG work [5,23] mainly uses case-based or planning-based methods. Recently, researchers focus on generating stories with storyline or intermediate representations, such as skeletons [30], events [17], titles [12] and verbs [25]. In this way, they first generate intermediate representations, then rewrite and enrich them to obtain complete stories.

Story Ending Generation. Story Ending Generation (SEG) is a subtask of SG, which aims to understand the context and generate a coherent and reasonable story ending [33]. Li et al. [13] introduce Seq2Seq model with adversarial training to improve the rationality and diversity of the generated story endings. Similarly, Zhao et al. [33] employ Seq2Seq model based on reinforcement learning to generate more sensible endings. Gupta et al. [7] utilize an extra keywords extraction algorithm and model the keywords information in the proposed Seq2Seq model. Guan et al. [6] introduce a model which uses an incremental encoding scheme and commonsense knowledge to generate reasonable endings. Further, Huang et al. [10] propose a multi-level graph convolutional network to capture the dependency relations of input sentences. Although great progress has been made, the implicit knowledge and multi-grained interactive relations behind story context are still under-explored.

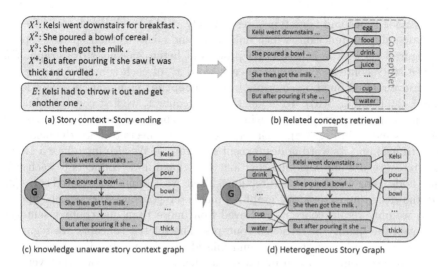

Fig. 3. Illustration of heterogeneous story graph construction process. Best viewed in color. Blue, orange, green and pink colors represent global, knowledge, sentence and word nodes, respectively.

3 Model

3.1 Overview

Story ending generation (SEG) task aims to generate a story ending conforming the corresponding context. Given a story context $X = \{X^1, X^2, ..., X^\mu\}$, where $X^k = x_1^k x_2^k ... x_l^k$ $(0 \leq k \leq \mu)$ represents the k-th sentence with l words. SEG aims at generating a story ending $E = y_1 y_2 ... y_m$ with m words.

Figure 2 shows the overview of our proposed model. To generate a coherent and reasonable story ending, we model the story context into a heterogeneous graph (Sect. 3.2). Based on the graph, we propose a Story Heterogeneous Graph Network (SHGN), which contains three components: (1) *graph initializer* is used to give each node an initial representation; (2) *graph layer* digests the structural information and gets updated node representations; (3) *transformer decoder* is used to generate the story endings according to final node representations (Sect. 3.3). Moreover, in order to sufficiently comprehend the story context, we design two auxiliary tasks, namely *sentiment prediction of story endings* and *clue words prediction*, which are expected to implicitly capture the sentiment trend as well as key events lie in context. These auxiliary tasks are jointly optimized with the primary story ending generation task in the multi-task learning strategy (Sect. 3.4).

3.2 Heterogeneous Story Graph Construction

We define the heterogeneous story graph as a directed graph $G = (\mathcal{V}, \mathcal{E}, \mathcal{A}, \mathcal{R})$, where $v \in \mathcal{V}$ represents each node and $e \in \mathcal{E}$ denotes each edge. $\mathcal{A}$ and $\mathcal{R}$ are the sets of node types and edge types, respectively. $\tau(v) : \mathcal{V} \to \mathcal{A}$ and $\phi(e) : \mathcal{E} \to \mathcal{R}$ are type mapping functions which link nodes and edges to their specific type.

Figure 3 shows the construction process of heterogeneous story graph. For a given story context X, we utilize each word x_t^k of each context sentence X^k to retrieve related one-hop concepts from ConceptNet commonsense knowledge graph[2] [24] (Cf. Fig. 3(b)). To model the story context, we first construct knowledge unaware story graph by viewing sentences and words as different types of nodes. As shown in Fig. 3(c), each sentence node connects to both the next sentence (if has) and its word nodes (except stopwords). We also introduce a global node to aggregate non-local information. The global node connects to each sentence node using edges in both directions. Then, we combine the related concepts and knowledge unaware story graph as our heterogeneous story graph. Specifically, we only retain concepts retrieved from more than one context sentence to control the quality of retrieved concepts. The related concepts are regarded as knowledge nodes. If there are multiple identical knowledge nodes or word nodes, we also combine them into a single one. Figure 3(d) shows the overview of our final heterogeneous story graph which contains four types of nodes (global, knowledge, sentence, word) and seven types of edges (global $\Rightarrow$ sentence, sentence $\Rightarrow$ global, sentence $\Rightarrow$ sentence, knowledge $\Rightarrow$ sentence, sentence $\Rightarrow$ knowledge, word $\Rightarrow$ sentence, sentence $\Rightarrow$ word) in total.

3.3 Story Heterogeneous Graph Network

The Story Heterogeneous Graph Network (SHGN) is used to initialize and update each node representation in the constructed graph (Sect. 3.2). Three components are introduced in SHGN: *graph initializer*, *graph layer* and *transformer decoder*.

Graph Initializer. The role of graph initializer is to give each node $v_i \in \mathcal{V}$ an initial representation $h_{v_i}^0$. For the global node, we randomly initialize its representation. For other nodes, the initial representations should contain the information of the corresponding documents, sentences, words or concepts. Owing to the emergence of large-scale pre-trained sentence embedding models [4,9,22], which show their superiority in many sentence-level NLP tasks, we decide to utilize SimCSE [4] (we also try different graph initializers in the experiments, and the SimCSE performs best. More details please refer to Sect. 4.4) to initialize representations. It is worth noting that the SimCSE model is also used to initialize knowledge and word nodes, instead of using geometric embedding methods (e.g., TransE) or word embedding methods (e.g., GloVe). In this way, there is no need to bridge the representation gap between different node types.

Graph Layer. Given a constructed graph $\mathcal{G}$ with node representations, we use Heterogeneous Graph Transformer (HGT) [8] as our graph layer, which models the heterogeneous graph by type-dependent parameters and can be easily applied to our heterogeneous story graph. Specifically, HGT includes:

---

[2] We only consider nouns, verbs, adjectives, and adverbs to retrieve related concepts from ConceptNet.

(a) *Heterogeneous Mutual Attention* is used to calculate the attention scores between source node s and target node t with the consideration of their edge $e = (s, t)$:

$$Attn^{(l)}(s, e, t) = \underset{\forall s \in N(t)}{softmax}(\alpha^{(l)}(s, e, t)) \qquad (1)$$

$$\alpha^{(l)}(s, e, t) = (k_s^{(l)} W_{att,\phi(e)}^{(l)} q_t^{(l)^{\top}}) \qquad (2)$$

$$k_s^{(l)} = W_{k,\tau(s)}^{(l)}(h_s^{(l-1)}) \qquad (3)$$

$$q_t^{(l)} = W_{q,\tau(t)}^{(l)}(h_t^{(l-1)}) \qquad (4)$$

where $N(t)$ denotes neighbors of target node t. $k_v^{(l)}/q_v^{(l)}$ represents the l-th layer key/query vector of node v. $h_v^{(l-1)}$ is the $(l-1)$-th layer representation of node v. $W_{att,\phi(e)}^{(l)}$, $W_{k,\tau(s)}^{(l)}$ and $W_{q,\tau(t)}^{(l)}$ are trainable parameters.

(b) *Heterogeneous Message Passing* is utilized to pass information from source nodes to target nodes:

$$Message(s, e, t) = \underset{i \in [1,h]}{\|} MSGHead^i(s, e, t) \qquad (5)$$

$$MSGHead^i(s, e, t) = W_{i,\tau(s)}^{(l)}(h_s^{(l-1)})W_{\phi(e)}^{MSG} \qquad (6)$$

where h is the number of heads in HGT, $W_{i,\tau(s)}^{(l)}$ and $W_{\phi(e)}^{MSG}$ are trainable parameters.

(c) *Target-Specific Aggregation* is used to aggregate information from the source nodes to the target node:

$$\tilde{h}_t^{(l)} = \underset{\forall s \in N(t)}{\oplus} (Attn^{(l)}(s, e, t) \cdot Message(s, e, t)) \qquad (7)$$

$$h_t^{(l)} = W_{\tau(t)}(\sigma(\tilde{h}_t^{(l)})) + h_t^{(l-1)} \qquad (8)$$

where $\oplus$ and σ represent addition operator and sigmoid function, respectively. $W_{\tau(t)}$ is trainable parameters.

After obtaining the output representation h_v^L for each node, we concatenate updated node representation h_v^L with corresponding initial representation h_v^0 and followed by a linear projection function to get the final node representation:

$$h_v^L = W_{final}[h_v^L, h_v^0] \qquad (9)$$

Transformer Decoder. We utilize the vanilla Transformer decoder [26] to decode the story endings. The inputs of multi-head attention in transformer decoder are node representations $H^L = [h_{v_0}^L; h_{v_1}^L; ...; h_{v_s}^L]$ (s denotes the total number of nodes in the graph) and decoder input D_{in}. This process is denoted as:

$$\tilde{D}_{in} = MultiHead(D_{in}, H^L, H^L) \qquad (10)$$

$$D_o = FFN(\tilde{D}_{in}) \qquad (11)$$

where FFN is two linear projections with a ReLU activation in the middle. D_o is the middle output of transformer decoder.

To generate next word, a linear projection and softmax function are used to predict the word probabilities:

$$\mathcal{P}(y_t|y < t, X) = softmax(W_o D_o) \qquad (12)$$

where W_o is trainable parameters, $\mathcal{P}(y_t)$ is the probability distribution over vocabulary.

Then, we calculate the negative data likelihood as loss function:

$$\mathcal{L}_{gen} = -\sum_t log\mathcal{P}(y_t = \tilde{y}_t|y < t, X) \qquad (13)$$

3.4   Auxiliary Tasks

To sufficiently comprehend the story context and generate more coherent story endings, we design two auxiliary tasks with the special consideration for SEG: *sentiment prediction of story endings* and *clue words prediction*. These two auxiliary tasks are jointly optimized with the SEG task in a multi-task learning strategy.

Sentiment Prediction of Story Endings. Generally, the sentimental trend of the story context plays a crucial role in the SEG [18]. In order to improve the sentimental consistency of the generated endings. We make use of representations of all sentence nodes to predict the sentiment of the corresponding ending:

$$Y_s = W_s(\sum_{v_s} h_{v_s}^L) \qquad (14)$$

where v_s denotes the sentence nodes and Y_s is the sentimental probability distribution.

We use the VADER toolkit [11] to construct sentiment labels of the gold story endings. The labels include "positive", "neutral" and "negative". During training, the cross entropy loss function can be defined as:

$$\mathcal{L}_{sen} = CE(Y_s, \tilde{Y}_s) \qquad (15)$$

Clue Words Prediction. The story clues that lie in context are also important for SEG. Huang et al. [10] find that the words of top-2-degree in the dependency tree of each sentence are similar to the story clues summarized by humans. In detail, the dependency tree model the word sequences in the directed graph architecture, whose edges represent the dependency relations between words, such as causal relation, modifier relation, etc. The words of top-2-degree have the

Table 1. Statistics of ROCStories Corpus. Average, minimum, maximum and 95th percentile of length of datasets in wordpieces.

ROCStories corpus	Samples	Story context				Story ending			
		avg.	min	max	95ptcl.	avg.	min	max	95ptcl.
Training	90,000	35.0	4	65	48	9.5	1	20	14
Validation	4,081	32.6	15	56	46	8.9	2	18	13
Testing	4,081	33.7	12	61	47	9.4	2	17	14

most dependency relations with others. Thus, we utilize Biaffine [1] to construct the dependency tree for each sentence in the context and further regard the words of top-2-degree as clue words. We use the representation of each word node to predict whether it is a clue word. The binary cross entropy loss function of *clue words prediction* is denoted by:

$$\mathcal{L}_{clu} = CE(Y_w, \tilde{Y}_w) \tag{16}$$

$$Y_w = W_w h_{v_w}^L \tag{17}$$

where v_w denotes the word nodes and Y_w is the clue words probability distribution.

Multi-Task Learning. In our SHGN model, all three tasks all jointly performed through multi-task learning. The final objective is defined as:

$$\mathcal{L} = \lambda_1 \mathcal{L}_{sen} + \lambda_2 \mathcal{L}_{clu} + (1 - \lambda_1 - \lambda_2)\mathcal{L}_{gen} \tag{18}$$

where λ_1 and λ_2 are hyper-parameters.

4 Experiments

4.1 Experiment Setup

Dataset. Following previous work [6,10], we evaluate SHGN on the ROCStories Corpus [19] which contains 98,162 five-sentence daily life stories collected from crowd-workers. Table 1 shows the detailed statistics of the Corpus. In SEG task, the first four-sentence of each story is regarded as story context while the last sentence is the ground truth ending.

Implementation Details. We implement our model based on `transformers` [28] and `PyTorch Geometric` [3] libraries. For graph initializer, we utilize the SimCSE [4] sentence embedding model released by original authors[3], which has a similar architecture with `Robera-base` (768 hidden size, 12 multi-head attention, 12 layers). Following Feng et al. [29], the number of graph layer is set to 1

---

[3] https://huggingface.co/princeton-nlp/sup-simcse-roberta-base.

and we set the hidden size to 768. The transformer decoder used in our experiments has 12 decoder layers, 12 multi-head attention and 768 hidden size. To construct labels for sentiment prediction of story endings, we use `VADER` toolkit[4]. The dependency parsing algorithm Biaffine [1] used in our experiments is implemented by `SuPar`[5] toolkit.

During training, we set the batch size to 64 and use linear warmup of 1,000 steps. We employ grid search of Learning Rate (LR) in [2e−5, 3e−5, 5e−5] and number epochs in [5, 10, 15]. The best configuration used LR = 5e−5, 15 epochs. During inference, we use beam search and the beam size is set to 5. The coefficient λ_1 and λ_2 used in multi-task learning strategy are both 0.1.

Automatic Evaluation. We make use of BLEU [20] and ROUGE [14] for our automatic evaluation metrics, and report BLEU-1,2,3,4 together with ROUGE-1,2,L scores. Following previous study [10], we utilize `nlg-eval`[6] and `pyrouge`[7] toolkits to calculate the scores. Note that the BLEU or ROUGE scores might vary with different toolkits.

Human Evaluation. Considering the limitation of automatic evaluation, it is necessary to conduct human evaluation. Specifically, three aspects are considered as the criteria: (1) Grammaticality evaluates correct, fluent and natural of the generated story endings; (2) Logicality is used to evaluate whether the generated endings are reasonable and coherent; (3) Relevance measures how relevant are the endings to the story context. For a model, We randomly choose 100 generated story endings and employ three NLP postgraduates to make the evaluation. The scoring adopts a 3-point scale, with 1 as the worst and 3 as the maximum.

4.2   Baseline Methods

We compare our model with several typical baselines and the state-of-the-art baselines:

- **Seq2Seq+Att** [16]: A LSTM-based Seq2Seq model with attention mechanism.
- **Transformer** [26]: Transformers is a parallel Seq2Seq model based on multi-head attention and feed forward networks.
- **HLSTM** [31]: A hierarchical LSTM utilizes word-level and sentence-level LSTM as its encoder, and uses vanilla LSTM as its decoder to generate text sequence.
- **IE+MSA** [6]: A SEG model which considers external commonsense knowledge and uses an incremental encoding scheme to generate endings.
- **T-CVAE** [27]: A conditional variational auto-encoder model based on transformers.

---

[4] https://github.com/cjhutto/vaderSentiment.
[5] https://github.com/yzhangcs/parser.
[6] https://github.com/Maluuba/nlg-eval.
[7] https://github.com/bheinzerling/pyrouge.

Table 2. Experiments on the ROCStories Corpus for the SEG task. The **bold** denote the best performance. For performances of baseline methods, [†] represents the reproducing results while [‡] denotes the results reported by Huang et al. [10].

Model	B1	B2	B3	B4	R1	R2	R-L
Seq2Seq+Att [16]	18.5[‡]	5.9[‡]	–	–	–	–	–
Transformer [26]	17.4[‡]	6.0[‡]	–	–	–	–	–
HLSTM [31]	22.1[‡]	7.1[‡]	–	–	–	–	–
IE+MSA [6]	24.3[†]	7.8[†]	3.9[†]	2.1[†]	17.5[†]	2.9[†]	20.8[†]
T-CVAE [27]	24.3[†]	7.7[†]	3.8[†]	2.0[†]	17.6[†]	3.0[†]	20.8[†]
Plan&Write [32]	24.4[†]	8.4[†]	4.1[†]	2.3[†]	18.1[†]	3.3[†]	21.4[†]
MGCN-DP [10]	24.5[†] (24.6[‡])	8.7[†] (8.6[‡])	4.3[†]	2.5[†]	18.4[†]	3.5[†]	21.9[†]
SHGN (Our)	**25.6**	**9.4**	**4.7**	**2.7**	**20.3**	**3.9**	**23.5**

- **Plan&Write** [32]: A story generation model, which first uses a given title (topic) to obtain several keywords, and then generates complete stories.
- **MGCN-DP** [10]: The state-of-the-art SEG model which utilizes the dependency parse tree to construct a graph for story context, and makes use of GCN to capture story clues. A transformer decoder is employed to generate final endings.

Table 3. ROCStories Corpus ablations. "w/o" means "without". glob.: global, know.: knowledge, init.: initializer, SESP.: sentiment prediction of story endings, CWP.: clue words prediction. The **bold** and underline denote the best and the second performances, respectively.

#	Model	B2	B4	R-L
1	SHGN	**9.4**	**2.7**	**23.5**
2	SHGN (w/o glob.)	9.2	**2.7**	22.7
3	SHGN (w/o know.)	9.0	2.6	22.4
4	SHGN (w/o word)	8.9	2.6	22.3
5	SHGN (init. Sentence-BERT)	9.3	**2.7**	23.2
6	SHGN (init. LSTM)	7.6	2.3	21.2
7	SHGN (w/o mutli-task)	9.1	2.5	22.7
8	SHGN (w/o SPSE.)	9.3	2.6	23.1
9	SHGN (w/o CWP.)	9.2	2.5	22.9

4.3 Main Results

Table 2 shows the results of automatic evaluation. The results show that our model significantly outperforms these baselines. Specifically, our model

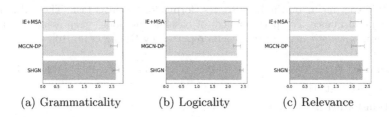

(a) Grammaticality   (b) Logicality   (c) Relevance

Fig. 4. Results on human evaluation, including means and variances. Our SHGN model outperforms IE+MSA and MGCN-DP on all three aspects.

achieves an improvement of 5.3%/5.3%/4.9%/4.5% over the IE+MSA/T-CVAE/Plan&Write/MGCN-DP in term of B1. As for B2, our model outperforms the IE+MSA/T-CVAE/Plan&Write/MGCN-DP by 20.5%/22.1% / 11.9%/8.0%, respectively. With respect to B4, our model implements an improvement of 28.6%/35.0%/17.4%/8.0%. And for R-L, our model achieves an improvement of 13.0%/13.0%/9.8%/7.3%. Other automatic metrics (i.e., B3, R1 and R2) also demonstrate the superiority of our SHGN. The results indicate that our model can comprehend the story context better based on the heterogeneous graph which model both the information of story context at different granularity (knowledge, sentence and word) levels and the multi-grained interactive relations among them.

4.4   Ablation Study

Effectiveness of Heterogeneous Graph. As described in Sect. 3.2, the constructed heterogeneous story graph contains four types of nodes: *global, knowledge, sentence* and *word*. We run 3 ablations, modifying various settings of our SHGN: (1) remove global node; (2) remove knowledge nodes; (3) remove word nodes. The effect of these ablations is shown in Table 3 (row 1 vs. row 2–4). In each case, the automatic evaluation scores are lower than our origin SHGN, which justifies the rationality of our model.

Effectiveness of Graph Initializer. We utilize SimCSE as our graph initalizer. We also run 2 ablations: (1) replace SimCSE [4] with Sentence-BERT [22]; (2) replace SimCSE with BiLSTM, where the forward and backward hidden states are concatenated as the initial node representations. GloVe.6B [21] word embedding (300 dimension) is used in the BiLSTM initializer. Table 3 (row 1 vs. row 5, 6) shows the effectiveness of SimCSE. Specifically, the results of BiLSTM initializer are dramatically dropped compared to pretrained sentence embedding models, which indicates the superiority of the pretrained models.

Effectiveness of Multi-Task Learning. In order to demonstrate the effectiveness of our auxiliary tasks, we remove each of them and all of them in ablation studies, respectively. The results are shown in Table 3 (row 1 vs. row 7–9), which indicates our designed auxiliary tasks can facilitate story context comprehension.

Table 4. Case Study on ROCStories Corpus. **Bold** words represent the key entities, events, or key words. *italic* words denote improper words.

	Case 1	Case 2
Context	Tim always had **stomach problems** He tried different things to **fix them** His doctors **couldn't** really **figure out** what was wrong It really **cut into** Tim's social **life**	Keith was **working at** a mechanic **shop** He had **given** a customer a **high quote** Keith **kept the difference** between the quote and the actual bill. Keith's boss **found out** what **he did**
IE+MSA	Tim was able to get a *new job*	Keith's boss was **mad** at him
MGCN-DP	Tim **felt better** after that	Keith *got* the **job**
SHGN	Tim eventually **got better**	Keith was **fired** from his **job**.
Gold	Eventually he learned to just **live with** the **discomfort**	Keith **lost** his **job**

4.5 Human Study and Case Study

We conduct human study on IE+MSA [6], MGCN-DP [10] and our SHGN. Figure 4 shows the results of human study. Our SHGN performs better than IE+MSA and MGCN-DP on all three aspects, which verifies that our SHGN performances better on generating reasonable and coherent story endings.

To provide a deeper understanding of generated endings, we show two examples from different models in Table 4. In Case 1, the IE+MSA model performs worst, since it generates a illogical story ending. In detail, the *"new job"* is irrelevant to the context. The MGCN-DP and SHGN output the endings with content about *"stomach problems"*, which more relevant to the context, but they miss the context clue *"couldn't figure out"*. This finding shows that: (1) The graph architecture could model the rich interactive relations across story and improve the relevance of generated endings; (2) SEG is still a challenging task. In Case 2, we know that there may be a bad ending for Keith since he did something illegal at work. However, MGCN-DP generates an unreasonable ending, which is contrary to the gold ending. IE+MSA and SHGN effectively capture the context clues and generate bad endings. Furthermore, the generated ending of SHGN is highly consistent with the gold ending, which shows the superiority of our model.

5   Conclusion

In this paper, we study SEG and propose Story Heterogeneous Graph Network (SHGN) which utilizes graph architecture and commonsense knowledge to comprehend the story context and handle the implicit knowledge behind the story. Besides, two auxiliary tasks are introduced to facilitate story comprehension. Extensive experiments on widely used ROCStories Corpus show that our SHGN achieves new state-of-the-art performances. Human study and case study further prove the effectiveness of our model.

Acknowledgement. This research is supported by the National Natural Science Foundation of China (Grant No. 62072323, 62102276), the Natural Science Foundation of Jiangsu Province (No. BK20191420, BK20210705, BK20211307), the Major Program of Natural Science Foundation of Educational Commission of Jiangsu Province, China (Grant No. 19KJA610002, 21KJD520005), the Priority Academic Program Development of Jiangsu Higher Education Institutions, and the Collaborative Innovation Center of Novel Software Technology and Industrialization. We would like to thank Duo Zheng and the members of the NLPTTG group, Netease Fuxi AI Lab for the helpful discussions and valuable feedback. We also thank anonymous reviewers for their suggestions and comments.

References

1. Dozat, T., Manning, C.D.: Deep biaffine attention for neural dependency parsing. In: 5th International Conference on Learning Representations, ICLR 2017, Toulon, France, 24–26 April 2017, Conference Track Proceedings. OpenReview.net (2017). https://openreview.net/forum?id=Hk95PK9le
2. Feng, X., Feng, X., Qin, B., Geng, X.: Dialogue discourse-aware graph model and data augmentation for meeting summarization. In: Proceedings of the Thirtieth International Joint Conference on Artificial Intelligence, IJCAI-21, pp. 3808–3814, August 2021. https://doi.org/10.24963/ijcai.2021/524. Main Track
3. Fey, M., Lenssen, J.E.: Fast graph representation learning with PyTorch geometric. ArXiv abs/1903.02428 (2019)
4. Gao, T., Yao, X., Chen, D.: SimCSE: simple contrastive learning of sentence embeddings. In: Proceedings of the 2021 Conference on Empirical Methods in Natural Language Processing, pp. 6894–6910. Association for Computational Linguistics, Punta Cana, November 2021. https://aclanthology.org/2021.emnlp-main.552
5. Gervás, P., Díaz-Agudo, B., Peinado, F., Hervás, R.: Story plot generation based on CBR. Know.-Based Syst. 18(4–5), 235–242 (2005). https://doi.org/10.1016/j.knosys.2004.10.011
6. Guan, J., Wang, Y., Huang, M.: Story ending generation with incremental encoding and commonsense knowledge. In: The Thirty-Third AAAI Conference on Artificial Intelligence, AAAI 2019, Honolulu, Hawaii, USA, 27 January–1 February 2019, pp. 6473–6480. AAAI Press (2019). https://doi.org/10.1609/aaai.v33i01.33016473
7. Gupta, P., Bannihatti Kumar, V., Bhutani, M., Black, A.W.: WriterForcing: generating more interesting story endings. In: Proceedings of the Second Workshop on Storytelling, pp. 117–126. Association for Computational Linguistics, Florence, August 2019. https://www.aclweb.org/anthology/W19-3413
8. Hu, Z., Dong, Y., Wang, K., Sun, Y.: Heterogeneous graph transformer. In: WWW 2020: The Web Conference 2020, Taipei, Taiwan, 20–24 April 2020, pp. 2704–2710. ACM/IW3C2 (2020). https://doi.org/10.1145/3366423.3380027
9. Huang, J., et al.: WhiteningBERT: an easy unsupervised sentence embedding approach. In: Findings of the Association for Computational Linguistics: EMNLP 2021, pp. 238–244. Association for Computational Linguistics, Punta Cana, November 2021. https://aclanthology.org/2021.findings-emnlp.23
10. Huang, Q., et al.: Story ending generation with multi-level graph convolutional networks over dependency trees. In: Proceedings of the AAAI Conference on Artificial Intelligence, vol. 35, no. 14, pp. 13073–13081 (2021). https://ojs.aaai.org/index.php/AAAI/article/view/17545
11. Hutto, C., Gilbert, E.: Vader: a parsimonious rule-based model for sentiment analysis of social media text. In: Proceedings of the International AAAI Conference on Web and Social Media, vol. 8, no. 1, pp. 216–225, May 2014. https://ojs.aaai.org/index.php/ICWSM/article/view/14550

12. Li, J., Bing, L., Qiu, L., Chen, D., Zhao, D., Yan, R.: Learning to write stories with thematic consistency and wording novelty. In: The Thirty-Third AAAI Conference on Artificial Intelligence, AAAI 2019, Honolulu, Hawaii, USA, 27 January–1 February 2019, pp. 1715–1722. AAAI Press (2019). https://doi.org/10.1609/aaai.v33i01.33011715

13. Li, Z., Ding, X., Liu, T.: Generating reasonable and diversified story ending using sequence to sequence model with adversarial training. In: Proceedings of the 27th International Conference on Computational Linguistics, pp. 1033–1043. Association for Computational Linguistics, Santa Fe, August 2018. https://www.aclweb.org/anthology/C18-1088

14. Lin, C.Y.: ROUGE: a package for automatic evaluation of summaries. In: Text Summarization Branches Out, pp. 74–81. Association for Computational Linguistics, Barcelona, July 2004. https://www.aclweb.org/anthology/W04-1013

15. Luo, F., et al.: Learning to control the fine-grained sentiment for story ending generation. In: Proceedings of the 57th Annual Meeting of the Association for Computational Linguistics, pp. 6020–6026. Association for Computational Linguistics, Florence, Jul 2019. https://www.aclweb.org/anthology/P19-1603

16. Luong, T., Pham, H., Manning, C.D.: Effective approaches to attention-based neural machine translation. In: Proceedings of the 2015 Conference on Empirical Methods in Natural Language Processing, pp. 1412–1421. Association for Computational Linguistics, Lisbon, September 2015. https://www.aclweb.org/anthology/D15-1166

17. Martin, L.J., et al.: Event representations for automated story generation with deep neural nets. In: Proceedings of the Thirty-Second AAAI Conference on Artificial Intelligence (AAAI-18), New Orleans, Louisiana, USA, 2–7 February 2018, pp. 868–875. AAAI Press (2018). https://www.aaai.org/ocs/index.php/AAAI/AAAI18/paper/view/17046

18. Mo, L., et al.: Incorporating sentimental trend into gated mechanism based transformer network for story ending generation. Neurocomputing **453**, 453–464 (2021)

19. Mostafazadeh, N., et al.: A corpus and cloze evaluation for deeper understanding of commonsense stories. In: Proceedings of the 2016 Conference of the North American Chapter of the Association for Computational Linguistics: Human Language Technologies, pp. 839–849. Association for Computational Linguistics, San Diego, June 2016. https://www.aclweb.org/anthology/N16-1098

20. Papineni, K., Roukos, S., Ward, T., Zhu, W.J.: Bleu: a method for automatic evaluation of machine translation. In: Proceedings of the 40th Annual Meeting of the Association for Computational Linguistics, pp. 311–318. Association for Computational Linguistics, Philadelphia, July 2002. https://www.aclweb.org/anthology/P02-1040

21. Pennington, J., Socher, R., Manning, C.: GloVe: global vectors for word representation. In: Proceedings of the 2014 Conference on Empirical Methods in Natural Language Processing (EMNLP), pp. 1532–1543. Association for Computational Linguistics, Doha, October 2014. https://www.aclweb.org/anthology/D14-1162

22. Reimers, N., Gurevych, I.: Sentence-BERT: sentence embeddings using Siamese BERT-networks. In: Proceedings of the 2019 Conference on Empirical Methods in Natural Language Processing and the 9th International Joint Conference on Natural Language Processing (EMNLP-IJCNLP), pp. 3982–3992. Association for Computational Linguistics, Hong Kong, November 2019. https://www.aclweb.org/anthology/D19-1410

23. Riedl, M.O., Young, R.M.: Narrative planning: balancing plot and character. J. Artif. Int. Res. **39**(1), 217–268 (2010)

24. Speer, R., Havasi, C.: Representing general relational knowledge in ConceptNet 5. In: Proceedings of the Eighth International Conference on Language Resources and Evaluation (LREC 2012), pp. 3679–3686. European Language Resources Association (ELRA), Istanbul, May 2012. http://www.lrec-conf.org/proceedings/lrec2012/pdf/1072_Paper.pdf
25. Tambwekar, P., Dhuliawala, M., Martin, L.J., Mehta, A., Harrison, B., Riedl, M.O.: Controllable neural story plot generation via reward shaping. In: Proceedings of the Twenty-Eighth International Joint Conference on Artificial Intelligence, IJCAI 2019, Macao, China, 10–16 August 2019, pp. 5982–5988. ijcai.org (2019). https://doi.org/10.24963/ijcai.2019/829
26. Vaswani, A., et al.: Attention is all you need. In: Advances in Neural Information Processing Systems, vol. 30. Curran Associates, Inc. (2017). https://proceedings.neurips.cc/paper/2017/file/3f5ee243547dee91fbd053c1c4a845aa-Paper.pdf
27. Wang, T., Wan, X.: T-CVAE: transformer-based conditioned variational autoencoder for story completion. In: Proceedings of the Twenty-Eighth International Joint Conference on Artificial Intelligence, IJCAI 2019, Macao, China, 10–16 August 2019, pp. 5233–5239. ijcai.org (2019). https://doi.org/10.24963/ijcai.2019/727
28. Wolf, T., et al.: Transformers: state-of-the-art natural language processing. In: Proceedings of the 2020 Conference on Empirical Methods in Natural Language Processing: System Demonstrations, pp. 38–45. Association for Computational Linguistics, October 2020. https://www.aclweb.org/anthology/2020.emnlp-demos.6
29. Xiachong, F., Xiaocheng, F., Bing, Q.: Incorporating commonsense knowledge into abstractive dialogue summarization via heterogeneous graph networks. In: Proceedings of the 20th Chinese National Conference on Computational Linguistics. pp. 964–975. Chinese Information Processing Society of China, Huhhot, August 2021. https://aclanthology.org/2021.ccl-1.86
30. Xu, J., Ren, X., Zhang, Y., Zeng, Q., Cai, X., Sun, X.: A skeleton-based model for promoting coherence among sentences in narrative story generation. In: Proceedings of the 2018 Conference on Empirical Methods in Natural Language Processing, pp. 4306–4315. Association for Computational Linguistics, Brussels, October-November 2018. https://www.aclweb.org/anthology/D18-1462
31. Yang, Z., Yang, D., Dyer, C., He, X., Smola, A., Hovy, E.: Hierarchical attention networks for document classification. In: Proceedings of the 2016 Conference of the North American Chapter of the Association for Computational Linguistics: Human Language Technologies, pp. 1480–1489. Association for Computational Linguistics, San Diego, June 2016. https://www.aclweb.org/anthology/N16-1174
32. Yao, L., Peng, N., Weischedel, R., Knight, K., Zhao, D., Yan, R.: Plan-and-write: towards better automatic storytelling. In: Proceedings of the AAAI Conference on Artificial Intelligence, vol. 33, no. 01, pp. 7378–7385, July 2019. https://ojs.aaai.org/index.php/AAAI/article/view/4726
33. Zhao, Y., Liu, L., Liu, C., Yang, R., Yu, D.: From plots to endings: a reinforced pointer generator for story ending generation. In: Zhang, M., Ng, V., Zhao, D., Li, S., Zan, H. (eds.) NLPCC 2018. LNCS (LNAI), vol. 11108, pp. 51–63. Springer, Cham (2018). https://doi.org/10.1007/978-3-319-99495-6_5

Open-Domain Dialogue Generation Grounded with Dynamic Multi-form Knowledge Fusion

Feifei Xu[1], Shanlin Zhou[1], Yunpu Ma[2(✉)], Xinpeng Wang[3], Wenkai Zhang[1], and Zhisong Li[4]

[1] School of Computer Science and Technology,
Shanghai University of Electric Power, Shanghai 201306, China
`xufeifei@shiep.edu.cn`, `zhoushanlin@mail.shiep.edu.cn`
[2] Chair of Database Systems and Data Mining, University of Munich,
80538 Munich, Germany
`cognitive.yunpu@gmail.com`
[3] School of Electronics and Information Engineering, Tongji University,
Shanghai 201804, China
[4] Data Intelligence Center, Alibaba Local Life Service Co., Ltd.,
Hangzhou 200062, China

Abstract. Open-domain multi-turn conversations normally face the challenges of how to enrich and expand the content of the conversation. Recently, many approaches based on external knowledge are proposed to generate rich semantic and information conversation. Two types of knowledge have been studied for knowledge-aware open-domain dialogue generation: structured triples from knowledge graphs and unstructured texts from documents. To take both advantages of abundant unstructured latent knowledge in the documents and the information expansion capabilities of the structured knowledge graph, this paper presents a new dialogue generation model, **D**ynamic **M**ulti-form **K**nowledge Fusion based Open-domain **C**hatting **M**achine (**DMKCM**). In particular, DMKCM applies an indexed text (a virtual Knowledge Base) to locate relevant documents as 1st hop and then expands the content of the dialogue and its 1st hop using a commonsense knowledge graph to get apposite triples as 2nd hop. To merge these two forms of knowledge into the dialogue effectively, we design a dynamic virtual knowledge selector and a controller that help to enrich and expand knowledge space. Moreover, DMKCM adopts a novel dynamic knowledge memory module that effectively uses historical reasoning knowledge to generate better responses. Experimental results indicate the effectiveness of our method in terms of dialogue coherence and informativeness.

Keywords: Conversation generation · Virtual knowledge base · Commonsense knowledge graph · Dynamic fusion

---

F. Xu and S. Zhou—Both—first authors with equal contributions.

1 Introduction

Open-domain conversation tries to meet human needs in terms of dialogue understanding and emotional resonance while keeping continuous. However, traditional merely data-driven multi-turn conversation models often generate simple and repetitive contents [15,19]. To address this issue, previous studies add additional persona information documents [16] or guide the conversation topic [22] to improve dialogue informativeness and diversity.

Notably, more recent studies investigate external knowledge as additional inputs of conversations [5,14,24,26,27], including knowledge graphs (denoted as KGs) [24,26,27], or unstructured texts [5,14]. Methods based on KGs show that KGs organize information around entities, making it easy to reason. Nevertheless, extracting relations for establishing the knowledge graph usually leads to the loss of information. More, it often generates less informative responses by simply applying and reformulating triples of KGs, e.g., KnowHRL [22] adds keywords from KGs using reasoning strategies to guide topics but the informativeness of the conversation has not increased significantly. Informative texts, e.g., comments about movies, can provide rich knowledge for the generation. However, their unstructured representation schemes require the language models to perform knowledge selection or attention from the knowledge texts, e.g., SKT [5] designs a complex screening process to use document knowledge. In general, these works are impossible to avoid the problem that the KGs are incomplete or the processing of documents is complicated. A very recent work, MKST [26] first attempts to apply different forms of knowledge in conversation. It extracts the entities mentioned in the sentences and links them to their corresponding entities in KGs as label knowledge. It designs a multi-knowledge-aware encoder to encode label, unstructured, and dialogue information together and get a generation by a knowledge-aware decoder. However, label knowledge is not achieved through reasoning which may not help the further expansion of dialogue. More, MKST just relies on dialogue data sets with background knowledge, e.g., Wizard-of-Wikipedia dataset.

To address above problems, we propose a new multi-turn dialogue generation model, **D**ynamic Multi-form **K**nowledge Fusion based Open-domain **C**hatting **M**achine (**DMKCM**). Its goal is to fuse abundant knowledge in an indexed corpus (a virtual Knowledge Base or a virtual KB) and information expansion capabilities of a commonsense knowledge graph (commonsense KG) simultaneously to enrich and expand informativeness in multi-turn conversation. The differences and functions of these two types of external knowledge can be summarized as follows:

- **Virtual KB:** This kind of knowledge base is usually an indexed corpus where each document link to its related documents with keywords. Each document in this base can express a complete meaning.
- **Commonsense KG:** This kind of knowledge graph includes the triples [*head_entity, relation, tail_entity*], whose entities are also called commonsense facts. These commonsense facts can enhance language representation in the commonsense aspect and even expand topics with reasoning by traversing entities and relations.

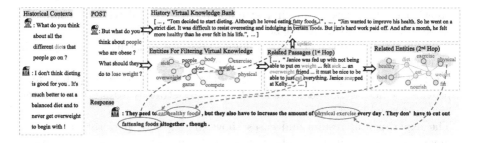

Fig. 1. An Example of Knowledge Fusion in a Conversation. Yellow indicate key words from 1st hop, and blue indicate entities of 2nd hop. Red indicates key words from history virtual knowledge. Different colored circles and dotted arrows point out the source of latent knowledge in the response. Black arrows indicate the flow of information. (Color figure online)

For DMKCM, we design two branches, including a dialogue branch (green blocks in left of Fig. 2) and a knowledge branch (orange blocks in left of Fig. 2). The dialogue branch generates responses by interchanging knowledge with the knowledge branch. On the knowledge branch, we separately take different reasoning strategies on virtual KB and commonsense KG to get passages (1st hop) and entities (2nd hop) that are related to the current dialog. 1st hop services for riching information of response. 2nd hop is to better capture concepts shifted in conversation structure which help generate more meaningful conversations, like "diet" and "weight" (concepts in "Historical context" and "POST" in Fig. 2) hop to related concepts, e.g., "healthy" and "exercise" elt., along the commonsense relations, in "Related Entities" of Fig. 1. This is a typical case in natural conversations. In addition, before 1st hop, we also expand concepts by commonsense KG to calculate the filtering scores, like "people" (from "POST" in Fig. 2) to "overweight" etl.). Using these filtering scores to select results inferred from Virtual KB helps to remove potential noise in reasoning results. When the topic is shifted, it is hard to find suitable knowledge from the current 1st hop to generate a response. Especially, we find history 1st hop (history virtual knowledge) can solve this issue, e.g., "fatty food" and "healthy" in history virtual knowledge bank related to response in Fig. 1. For this, history virtual knowledge is dynamically stored into the history virtual knowledge bank and provides knowledge support for the current turn. This helps the topic transition better in the current dialogue and also enriches the response to some extent. Our work improves the model explainability on knowledge extraction, helps to generate informative responses, and expands the topic of conversation to a certain extent. Explainability is important to dialogue in information-oriented scenarios, where a user needs to know how new knowledge in chatbot's responses is linked to the knowledge in their utterances, as Fig. 1 shows. Our experiments on two conversation datasets, including Persona-Chat [25] and DailyDialog [8], demonstrate the effectiveness of DMKCM.

In summary, the following contributions are made in this paper:

- This paper creatively proposes a novel dialogue generation model-DMKCM, to dynamically fuse multi-from knowledge into generation. To our best knowledge, this work is the first attempt to fuse virtual KB and common-sense KG into dialogue to get better responses.

- To adjust to the open domain dialogue task, we construct a new virtual knowledge base using the dataset of commonsense stories-ROCStory.
- We find that history virtual knowledge helps generate better responses and provides a new dynamically delayed updating strategy to store and filter history virtual knowledge.
- The experimental results and cases show the superior performance of our model. Various evaluating indicators indicate that DMKCM not only maximizes the advantages of achieved knowledge but also helps to generate more informative and coherent conversations.

2 Related Work

2.1 Dialogue Generation with External Knowledge

Many works have proved that external knowledge can facilitate dialogue generation. [27] presents a novel open-domain dialogue generation method to demonstrate how large-scale commonsense knowledge can facilitate language understanding and generation. [4] proposes a latent relation language model, a class of language models that parameterize the joint distribution over the words in a document and relevant entities via knowledge graph relations. For the use of the external documents, [10] incorporates external documents into the procedure of response generation in custom service dialogues. GLKS [14] adopts a global guide method to the local, and uses the dialogue contexts to filter out important n-gram information from the document to guide the generation process. However, the knowledge graphs lose facts, and external texts require complicated processing. These two forms of knowledge still have limitations in exerting external knowledge.

2.2 Virtual Knowledge Base

Virtual Knowledge Base (virtual KB) is an indexed corpus, which treats a corpus as a knowledge base containing entities and texts. It has been widely employed in open-domain Question Answer (QA) [2,3,11]. Virtual KB accomplishes the QA tasks by answering queries with spans from the corpus, ensuring that facts can be preserved in the relation extraction process. Whereas, to the best of our knowledge, virtual KB has not yet been mentioned in open-domain dialogue generation. DrKIT [2] is a state-of-the-art reasoning algorithm with QA on a virtual KB, which traverses textual data like a KB, softly following paths of relations between mentions of entities in the corpus. Inspired by this, we present a novel model, DMKCM, which includes a reasoning strategy based on DrKIT for getting more information related to our dialogue. To better fit our task, we convert a commonsense story corpus-the indexed ROCStories [12] as our virtual KB, instead of professional Wikipedia.

3   Model

3.1   Overview

The overview of DMKCM is shown in Fig. 2. DMKCM consists of two branches, dialogue branch and knowledge branch. **The dialogue branch** (green blocks in left of Fig. 2) aims to generate conversation based on an encoder-decoder model and interacts information with the knowledge branch to improve the informativeness expression of response. **The knowledge branch** (orange blocks in left of Fig. 2) is to reason, store, merge, and expand knowledge by Virtual Knowledge reasoning module (VK-reasoning), Dynamic Virtual Knowledge memory module (DVK-memory), Dynamic Virtual Knowledge selector module (DVK-selector), and Commonsense Knowledge expansion module (CK-expansion).

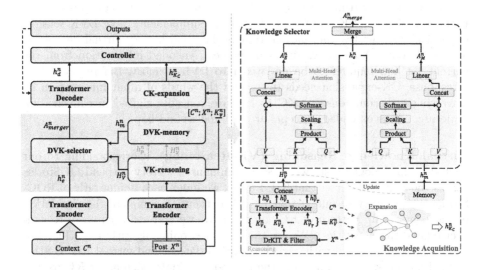

Fig. 2. The left is the overview of DMKCM, including the dialogue branch (green blocks) and knowledge branch (orange blocks). Right is details of the knowledge branch, which includes Knowledge Acquisition and Knowledge Selector. Knowledge Acquisition has three modules, including VK-reasoning, DVK-memory, and CK-expansion. Knowledge Selector represents the process of DVK-selector and the merge can be seen in Eq. 7. (Color figure online)

Before presenting our detail for the dialogue generation approach, we first introduce our notations and critical concepts. Formally, suppose we have a conversation dataset $D = \left\{ \left(C^i, X^i, R^i \right) \right\}_{i=1}^{N}$, where C^i represents conversational context before i-th turn. X^i is i-th user utterance. R^i is a response regarding to X^i. The final goal of this task is to estimate a generation probability distribution $P(R|[C,X],K)$ from D. Therefore, one can generate a response for $[C,X]$ following $P(R|[C,X],K)$, where $[C,X]$ means the concatenation for context C

and current user utterance X, K is the knowledge from knowledge branch, and R is the corresponding response. **Assume** that there are already $n - 1$ turns in a dialogue. We use Transformer encoder (T_enc) to encode X^n and C^n and get the last hidden state h_p^n and state h_e^n from X^n and $[C^n, X^n]$ separately. h_p^n and h_e^n represent encoded semantic information of X^n and $[C^n, X^n]$. Next, we elaborate on details for each of them.

3.2   Knowledge Branch

Firstly, VK-reasoning reasons and filters candidate documents related to the user utterance X^n as 1st hop from a virtual KB. We send X^n, candidate documents, and C^n to CK-expansion, aiming to expand commonsense concepts for better response. DVK-memory is a dynamic transfer module. It dynamically stores encoded vectors of 1st hop from previous $n - 1$ turns. We named these vectors as history virtual knowledge. Then, we filter and send related history virtual knowledge to DVK-selector as a knowledge supplement. Next, DVK-selector dynamically integrates information from 1st hop and history virtual knowledge for the decoder. Noticeably, after this process, encoded information from the current 1st hop requires to be updated into DVK-memory. Before generating a response, CK-expansion needs to expand the words of input information by traversing in a commonsense KG to capture concepts shifted in conversation structure. These extended concepts are denoted as 2nd hop.

Virtual Knowledge Reasoning (VK-Reasoning). Corresponding to our open-domain dialogue task, we select a commonsense story corpus-ROCStories [12] as the source of our virtual KB. Firstly, we index each unique title of ROC-Stories as an entity in our virtual KB. And then, to complete the simulation of the relationship pattern of KGs on the text, we traverse each story to connect related titles (entities). We use the latest reasoning algorithm DrKIT [2] to train a reasoning model with our conversation dataset and virtual KB. By this trained model, we reason and get the related candidate documents K_D^n to X^n. In particular, to obtain documents that are more relevant to X^n, we list related words of each word in X^n from a commonsense. The number of co-occurrence of each document and this list is regarded as this document's filtering score. We get top T candidate documents from K_D^n by this score. For convenience, these filtered candidate documents are named as K_V^n (1st hop).

As shown in right of Fig. 2, we encode K_V^n for DVK-selector. Concretely, these candidate documents are successively encoded by transformer encoder. Then, we get the last hidden state $h_{V_t}^n$ from the encoder layer, which represents encoding information of t-th candidate document and t is from 1 to T. T means the total number of candidate documents. Subsequently, we merge the last hidden states generated each time into a matrix called H_V^n. The process is as follows:

$$H_V^n = \left[h_{V_1}^n, h_{V_2}^n, \dots, h_{V_T}^n\right]^T,$$ (1)

$$h_{V_t}^n = T\_enc\left(K_{V_t}^n\right), (t = 1, \dots, T).$$ (2)

Dynamic Virtual Knowledge Memory (DVK-Memory). When the topic is shifted, e.g., "soda" in "A4" of Table 1, VK-reasoning may find documents about "soda" instead of **"weight"** and **"healthy"** which stand for the topic from context. This leads to little support for the current conversation because our model lacks practical knowledge to generate a response if only using the 1st hop knowledge.

Table 1. An Example of Conversation Topic Shift in DailyDialog. The topic words in the dialogue are emphasized in bold, and the words in red indicate the words that deviate from the current dialogue topic.

A1: What do you think about all the different **diets** that people go on ?
B1: ...a balanced diet and to never get **overweight**...
A2: But...What should they do to **lose weight** ?
B2: They need to eat **healthy foods**...cut out **fattening foods** altogether...
...
A4: How about drinking soda?
B4: ...**gain weight** by drinking far too much soda ... **no nutritional value**..

Thus, we design DVK-memory to dynamically store and filter history virtual knowledge representation, which provides a virtual knowledge supplement for the 1st hop and finally helps the topic transition has better smooth in the current dialogue. This process applies a dynamic delayed updating strategy (in Algorithm 1).

Algorithm 1: Algorithm of Dynamically Store and Filter History Virtual Knowledge

---

Input: The i-th (range from 1 to N) turn output of VK-reasoning: H_V^i
Output: Encoded vector of related history virtual knowledge: h_m^i

1 **for** *each* $i \in [1, N]$ **do**
2   **if** *the value of i is 1* **then**
3   Do DVK-selector;
4   Add H_V^i into Memory M^{i+1};
5   **else**
6   Extract historical knowledge representation of documents h_m^i related to h_p^i from M^i ;
7   Do DVK-selector;
8   Add H_V^i into Memory M^{i+1};

We denote $M = \{M^n\}_{n=2}^N$ as a set of history virtual knowledge representation, where M^n is historical knowledge representation for n-th turn of dialogue.

$$M^n = \left[H_V^1, \ldots, H_V^{n-1}\right]^T. \tag{3}$$

Then, We apply attention mechanism from [1] to calculate the extracting historical knowledge representation of documents h_m^n from M^n that is related to the representation of current user utterance h_p^n:

$$h_m^n = \sum_{i=2}^{n-1} \alpha_{w,k}^i M_k^i, \tag{4}$$

$$\alpha_{w,k}^i = \frac{exp\left(S_{w,k}^i\right)}{\sum_{i=1}^n exp\left(S_{w,k}^i\right)}, \tag{5}$$

$$S_{w,k}^i = V_a^T tanh\left(W_h\left[h_{p_w}^i; M_k^i\right]\right), \tag{6}$$

where M_k^i represents the k-th position hidden state of history virtual knowledge M^i and $h_{p_w}^i$ is the w-th token vector in h_p^i. V_a^T and W_h are trainable parameters. $S_{w,k}^i$ is the unnormalized attention weight by an attention neural network and $\alpha_{w,k}^i$ is the normalized attention weight from $S_{w,k}^i$.

Dynamic Virtual Knowledge Selector (DVK-Selector). We apply multi-head attention mechanism (MultiHead) [21] to extract features of current virtual knowledge H_V^n and historical knowledge h_m^n according to dialogue semantic information h_c^n. A gate is proposed for information fusion and its result is A_{merge}^n. Specifically,

$$A_{merge}^n = \begin{cases} \mu A_V^n + h_e^n, & n = 0 \\ \mu A_V^n + (1-\mu)A_M^n + h_e^n, & n > 0 \end{cases}, \tag{7}$$

$$A_M^n = MultiHead\left(h_e^n, h_m^n, h_m^n\right), \tag{8}$$

$$A_V^n = MultiHead\left(h_e^n, H_V^n, H_V^n\right), \tag{9}$$

$$\mu = sigmoid\left(W_g h_e^n\right). \tag{10}$$

Here, we use the sigmoid function to get a gating parameter μ for fusion, and the W_g are trainable parameters. A_V^n is the current virtual knowledge features related to h_e^n and A_M^n is the historical knowledge features related to h_e^n. Particularly, since DVK-memory takes a delayed updating strategy, it needs to remove the h_m^n when n is 0.

Commonsense Knowledge Expansion (CK-Expansion). To expand concepts and further enhance informativeness, we expand entities of K_V^n, C^n and X^n by searching neighbor nodes on a commonsense KG. We use $K_C^n = (k_h^n, k_r^n, k_t^n)$ to represent the knowledge triples, which connects the original entities and expanded entities. k_h^n is a set of words (entities) from K_V^n, C^n and X^n. k_t^n means expanded entities by the KG. k_r^n is the relation of k_h^n and k_t^n on the KG. Inspired by GCN that can encode graph structure well, we use Multi-layer

CompGCN (M_CompGCN) [20] to encode the knowledge triples by combining the node embedding and the relation embedding.

$$h_{K_h}^n, h_{K_r}^n, h_{K_t}^n = M\_CompGCN(K_C^n). \tag{11}$$

We use the dialogue context encoding h_e^n to compute the degree of attention β^i with the encoded head $h_{K_h}^n$ and the relation $h_{K_r}^n$, and then multiply with the encoded tail $h_{K_t}^n$. Finally, we get the representation of knowledge triples $h_{k_C}^n$.

$$h_{k_C}^n = \sum_{i=1}^{k} \beta^i h_{k_t}^i, \tag{12}$$

$$\beta^i = Softmax(h_e^i[h_{k_h}^i + h_{k_r}^i]), \tag{13}$$

where k is the number of the triples.

3.3 Generation

We use Transformer Decoder (T_dec) to generate words,

$$h_d^n = T\_dec(y_{t-1}^n, A_{merge}^n). \tag{14}$$

Then, a Controller is designed, in which the decoded hidden state h_d^n will be mapped into vocab size and outputs the probability of words P_v by Softmax function,

$$P_v = Softmax(W_v h_d^n). \tag{15}$$

In addition, we can also generate knowledgeable words by using knowledge expansion representation encoded in CK-expansion,

$$P_{K_C} = Softmax(\sum_{i=1}^{l} \gamma_i^n h_{k_C}^n), \tag{16}$$

$$\gamma_i^n = Softmax(h_d^n W_k h_{K_C}^n). \tag{17}$$

We get an attention weight γ_i^n by using the decoded hidden state h_d^n to focus on the $h_{k_C}^n$, which can make the model focus on the relative knowledge triples; then, we choose the knowledge entities according to entities probability P_{K_C} of relative weighted knowledge after Softmax function.

The final generated words will consider both the distribution of standard vocabulary and the distribution of knowledge entities. We use a soft gate probability g_t to choose the generated words from standard vocabulary or knowledge entities.

$$y_t = g_t \cdot P_v + (1 - g_t) \cdot P_{K_C} \tag{18}$$

$$g_t = \sigma(h_d^n) \tag{19}$$

3.4   Training

To train the proposed model, we minimize the negative log-likelihood

$$L_{NLL} = -\frac{1}{N}\sum_{i=1}^{N}\sum_{t=1}^{T} log P(y_t^{(n)}|y_{<t}^{(n)}, X^{(n)}, K^{(n)}), \qquad (20)$$

where N is the total number of the dataset, and T is the timestep of the n-th turn response sentence. $X^{(n)}$ represents the n-th turn user utterance in the dataset, and $K^{(n)}$ represents the n-th turn knowledge.

4   Experiments

4.1   Dataset

Conversation Corpus: We choose DailyDialog [8] and PersonaChat [25] as our datasets. In our work, four turns of dialogue are a unit of training sample and pre-processed statistics of the above datasets are shown in the Fig. 3.

Commonsense Knowledge Corpus: ConceptNet [17] is a semantic network designed to help computers understand the meanings of words that people use. Its English vocabulary contains approximately 1,500,000 nodes. **Source of Virtual KB:** The ROCStories [12] is a commonsense story corpus which contains 98,161 five-sentence stories.

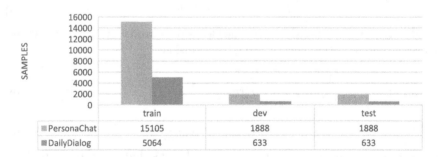

Fig. 3. Statistics of two datasets.

4.2   Comparison Method

We compare our model with representative baselines to investigate its effectiveness. The baselines are as follows: **(1) Attn-S2S** [18]: A classic method applies simple attention [1] to the input context based on the sequence-to-sequence model; **(2) Transformer** [21]: Transformer is a popular network architecture, based solely on attention mechanisms; **(3) Dir-VHRED** [23]: A recent work based on the latent variable hierarchical recurrent encoder-decoder model and characterizes the latent variables using Dirichlet distribution instead of traditional Gaussian distribution; **(4) GLKS** [14]: The newest dialogue generation model based on unstructured knowledge builds a global-to-local knowledge selection method to improve the quality of selected unstructured knowledge in background-based conversations; **(5) CCM** [27]: A commonsense knowledge-aware conversation model, which leverages commonsense knowledge from ConceptNet [17] through two graph attention mechanisms to facilitate informative response generation; **(6) MKST** [26]: The latest universal transformer-based architecture fuses label, background knowledge in open-domain conversation. Since MKST relies on datasets with background knowledge, we compare it with our model only on PersonaChat which has background information.

4.3   Implementation Details

We conduct the experiments with a Transformer structure (our baseline) with 8 heads, 6 layers, 512-dimensional hidden states, and a 2-layer GCN model. In the VK-reasoning, we set the number of reasoned candidate documents as 10 and filtered candidate documents as 5. During the CK-expansion, we search the neighbors of nodes and preserve the top 100 neighbor nodes. When processing datasets, the history context we choose is the previous turn of the current conversation. To train the model, we use the Adam optimizer [6] and use Adam-warmup to adjust the learning rate.

4.4   Evaluation Metrics

To analyze and evaluate our model more comprehensively, we use both automatic and human evaluations. **Automatic Evaluation**: Based on previous work, we apply several widely used automatic metrics. Specifically, we adopt PPL, BLEU-1,2,3,4) [13], and Distinct-1,2 (Dist-1,2) [7] to intuitively reveals quality, coherence and diversity of generated responses. PPL is the perplexity score that measures the quality of the language model. BLEU calculates word-based precision between a generated response and a gold response. Distinct evaluates the informativeness and diversity of the predicted responses. **Human Evaluation**: As known, automatic evaluation indicators have limitations in evaluating human conversations [9]. In our work, we randomly sample 200 test samples to conduct human evaluations. For response, we define three metrics: (1) Fluency (**Flu.**), i.e., degree of fluency and human readability; (2) Informativeness (**Inf.**), i.e., degree of knowledge for responses; (3) Appropriateness (**App.**), i.e., degree of relevance to the given context; Each response has 3 annotators to give a 3-graded whose rating range from 0 to 2. We take the average scores as the final results for each metric.

5   Results and Discussion

5.1   Performance Evaluation

Automatic Evaluation. Table 2 lists the automatic evaluation results for each model. Our model outperforms almost the baselines on two corpora. In the quality of the model, our PPL is the lowest, indicating that our generated responses Model is more grammatical. In the aspect of coherence, DMKCM has higher BLEU values, demonstrating our model tends to generate responses that are more similar to the gold responses than baselines in most cases. It can be inferred that our model can effectively obtain useful information from the historical context and historical knowledge in memory to help generate a response. On diversity, the Dist-1,2 metrics demonstrate that the models leveraging external knowledge achieve better performance than the knowledge-based model, e.g., GLKS, CCM, MKST, in generating meaningful and diverse responses. According to Table 2, in terms of indicators, DMKCM is better than MKST, which is the latest multi-knowledge based dialogue generation model. This signifies the effectiveness of our model on using structured knowledge or unstructured knowledge or the method of fusion.

Table 2. Automatic evaluation results of the proposed model and the baseline models. Numbers in bold indicate the best-performing model on the corresponding metrics.

Dataset	Model	PPL	BLEU-1	BLEU-2	BLEU-3	BLEU-4	Dist-1	Dist-2
PersonaChat	Attn-S2S	7.0079	0.4372	0.2525	0.1458	0.0842	0.0185	0.1208
	TRANSFORMER	6.5172	0.5023	0.2900	0.1675	0.0967	0.0425	0.2239
	Dir-VHRED	8.7117	0.4428	0.2557	0.1476	0.0852	0.0164	0.0795
	GLKS	7.5668	0.4684	0.2705	0.1562	0.0903	0.0420	0.1556
	CCM	7.3534	0.4703	0.2715	0.1568	0.0906	0.0596	0.2373
	MKST	7.1307	0.4460	0.2577	0.1491	0.0864	0.0808	0.2907
	DMKCM (our model)	**5.4549**	**0.5452**	**0.3149**	**0.1852**	**0.1087**	**0.0903**	**0.3328**
DailyDialog	Attn-S2S	7.7451	0.4166	0.2406	0.1390	0.0803	0.0281	0.2010
	TRANSFORMER	10.1355	0.4205	0.2428	0.1403	0.0811	0.0633	0.3096
	Dir-VHRED	9.3176	0.4276	0.2469	0.1426	0.0824	0.0399	0.2052
	GLKS	7.7635	0.4449	0.2570	0.1485	0.0859	0.0738	0.3496
	CCM	7.5592	0.4875	0.2773	0.1560	0.0860	0.0598	0.2350
	DMKCM (our model)	**5.7692**	**0.4928**	**0.2846**	**0.1645**	**0.0951**	**0.0801**	**0.3589**

Human Evaluation. Figure 4 clearly shows the human evaluation metrics results of DMKCM compared with the baselines through the radar chart. The three vertices of the radar chart respectively represent fluency, informativeness, and appropriateness. From the radar chart, DMKCM has the best performance on two datasets. Particularly, the informativeness has the most obvious advantage over other baselines, which indicates the effectiveness of our fusion of the multi-form knowledge and can generate coherent and informative responses.

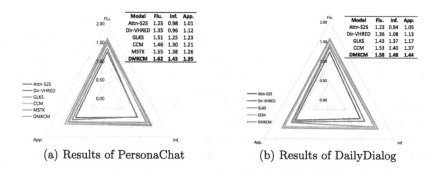

(a) Results of PersonaChat   (b) Results of DailyDialog

Fig. 4. Comparison of human Evaluation Results. The rating range from 0 to 2 and the bigger means the better.

5.2 Ablation Study

As shown in Table 3, we analyze the effectiveness of each module proposed in DMKCM through the following situations: **(1) 2Hop:** DMKCM drops CK-expansion; **(2) Mem and 2Hop:** DMKCM drops DVK-memory and CK-expansion; **(3) 1Hop and Mem:** DMKCM drops VK-reasoning and DVK-memory; **(4) 1Hop, Mem, and 2Hop:** This is baseline (Transformer). From the results, we can observe that the performance of situation (1) drops sharply from our model. This is within our expectations since CK-expansion helps to capture extra information from the post, which can improve the diversity of generated responses. This result can also show that fusing structured knowledge can effectively help dialogue generation. Metric results of the situation (1) are better than situation (2), which verifies that retaining history virtual knowledge using DVK-memory can effectively help dialogue generation. Situation (3) is related to unstructured knowledge and its poor results prove the effectiveness of VK-reasoning and DVK-memory, and the reasoned virtual knowledge affects generating responses. Situation (1) is better than situation (4) can also reflect all of the modules play important roles in our model. **In summary,** these modules of DMKCM designed for fusing structured and unstructured knowledge are helpful for response generating in terms of informativeness and coherence.

Table 3. Results of ablation study.

Dataset	Model	PPL	BLEU-1	BLEU-2	BLEU-3	BLEU-4	Dist-1	Dist-2	Flu./Inf./App.
PersonaChat	**DMKCM**	**5.4549**	**0.5452**	**0.3149**	**0.1852**	**0.1087**	**0.0903**	**0.3328**	**1.62/1.43/1.35**
	- 2Hop	6.5172	0.5023	0.2900	0.1675	0.0967	0.0838	0.3233	1.58/1.38/1.28
	- Mem and 2Hop	6.1340	0.5005	0.2890	0.1669	0.0964	0.0782	0.2712	1.53/1.31/1.18
	- 1Hop and Mem	6.7423	0.4843	0.2796	0.1615	0.0933	0.0862	0.3134	1.52/1.40/1.26
	- 1Hop, Mem and 2Hop	6.5172	0.5023	0.2900	0.1675	0.0967	0.0425	0.2239	1.46/1.22/1.08
DailyDialog	**DMKCM**	**5.7692**	**0.4928**	**0.2846**	**0.1645**	**0.0951**	**0.0801**	**0.3589**	**1.58/1.48/1.44**
	- 2Hop	7.6755	0.4643	0.2682	0.1550	0.0896	0.0744	0.3498	1.51/1.34/1.31
	- Mem and 2Hop	8.8904	0.4632	0.2677	0.1547	0.0894	0.0683	0.3414	1.48/1.32/1.29
	- 1Hop and Mem	8.0996	0.4572	0.2641	0.1526	0.0882	0.0792	0.3505	1.50/1.35/1.33
	- 1Hop, Mem and 2Hop	10.1355	0.4205	0.2428	0.1403	0.0811	0.0683	0.3414	1.44/1.21/1.25

5.3 Case Study

Sample conversations are shown in Table 4 which indicate that DMKCM can generate better responses than the baselines on two conversation datasets. Traditional end-to-end models, e.g., Attn-S2S and Dir-VHRED, tend to generate simple, safe, and incoherent responses without access to knowledge. Despite knowledge-based model, e.g., CCM, GLKS, and MKST, which generate informative responses, still produces irrelevant responses. Instead, DMKCM fuses knowledge from the knowledge graph and virtual knowledge as a whole and encodes more related information via DVK-selector and Controller modules, which supports generating more information and coherence response. In practice, the effect of these modules can be reflected, like Fig. 5. It thus generates more reasonable responses through better use of knowledge.

Table 4. Case study of generated responses.

-	PersonaChat	DailyDialog
-	**Historical Contexts:** A1: Hey, how's it going? I am eating pizza. B1: Not too bad. Pizza sounds great, I am on a diet. **Post:** A2: I'm a simple guy, I like pizza and trucks. But not diets	**Historical Contexts:** A1: Say, Jim, how about going for a few beers after dinner? B1: You know that is tempting but is really not good for our fitness **Post:** A2: What do you mean? It will help us to relax
Attn-S2S	I love the walking. I like spicy food	Do you really think it?
Transformer	That is cool. I like to eat healthy	I am sorry, but i am not sure that it will be a good idea
Dir-VHRED	I am always looking for ways to make money, I can make some money	But you know, it's not so much. Now that you have to do the drying up
GLKS	I love food too much. I am trying to open a restaurant next day	Do you really think so?
CCM	I am a good aggressive. I am a good teacher	I am like to get harvesting of the day
MKST	I like to work out	-
DMKCM	I am a waitress at a small restaurant. I am a vegan, but I like pizza	Do you think so? I do not really think we should get looking forward to it, it will just make us fat

   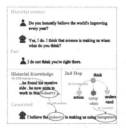

(a) A generated response example (knowledge from 1st hop)

(b) A generated response example (knowledge from history virtual knowledge and 2nd hop).

Fig. 5. Examples of DMKCM

6   Conclusion

To solve the challenge of lacking informative response in multi-turn dialogue generation, we propose a novel model, DMKCM. The existing methods of introducing knowledge into dialogue generation have some limits, so we combine virtual KB and commonsense KG to help generate better responses. In addition, we find that history virtual knowledge can improve responses and provide a new dynamically delayed updating strategy to store and filter history virtual knowledge. Experimental results on two datasets show that DMKCM can generate a more informative dialog with appropriate content ordering.

References

1. Bahdanau, D., Cho, K.H., Bengio, Y.: Neural machine translation by jointly learning to align and translate. In: 3rd International Conference on Learning Representations, ICLR 2015 (2015)
2. Dhingra, B., Zaheer, M., Balachandran, V., Neubig, G., Salakhutdinov, R., Cohen, W.W.: Differentiable reasoning over a virtual knowledge base. In: International Conference on Learning Representations (2020)
3. Godbole, A., et al.: Multi-step entity-centric information retrieval for multi-hop question answering. In: EMNLP 2019 MRQA Workshop, p. 113 (2019)
4. Hayashi, H., Hu, Z., Xiong, C., Neubig, G.: Latent relation language models. In: Proceedings of the AAAI Conference on Artificial Intelligence, pp. 7911–7918 (2020)
5. Kim, B., Ahn, J., Kim, G.: Sequential latent knowledge selection for knowledge-grounded dialogue. In: International Conference on Learning Representations (2020)
6. Kingma, D.P., Ba, J.: Adam: a method for stochastic optimization. arXiv preprint arXiv:1412.6980 (2014)
7. Li, J., Galley, M., Brockett, C., Gao, J., Dolan, B.: A diversity-promoting objective function for neural conversation models. In: HLT-NAACL (2016)
8. Li, Y., Su, H., Shen, X., Li, W., Cao, Z., Niu, S.: DailyDialog: a manually labelled multi-turn dialogue dataset. In: Proceedings of the Eighth International Joint Conference on Natural Language Processing (Volume 1: Long Papers), pp. 986–995 (2017)
9. Liang, W., Liang, K.H., Yu, Z.: Herald: an annotation efficient method to detect user disengagement in social conversations. In: ACL/IJCNLP (2021)
10. Long, Y., Wang, J., Xu, Z., Wang, Z., Wang, B., Wang, Z.: A knowledge enhanced generative conversational service agent. In: Proceedings of the 6th Dialog System Technology Challenges (DSTC6) Workshop (2017)
11. Moldovan, D., Pasca, M., Harabagiu, S., Surdeanu, M.: Performance issues and error analysis in an open-domain question answering system. In: Proceedings of the 40th Annual Meeting of the Association for Computational Linguistics, pp. 33–40 (2002)
12. Mostafazadeh, N., et al.: A corpus and cloze evaluation for deeper understanding of commonsense stories. In: Proceedings of the 2016 Conference of the North American Chapter of the Association for Computational Linguistics: Human Language Technologies, pp. 839–849 (2016)

13. Papineni, K., Roukos, S., Ward, T., Zhu, W.J.: Bleu: a method for automatic evaluation of machine translation. In: Proceedings of the 40th Annual Meeting of the Association for Computational Linguistics, pp. 311–318 (2002)

14. Ren, P., Chen, Z., Monz, C., Ma, J., de Rijke, M.: Thinking globally, acting locally: distantly supervised global-to-local knowledge selection for background based conversation. In: Proceedings of the AAAI Conference on Artificial Intelligence, pp. 8697–8704 (2020)

15. Shang, L., Lu, Z., Li, H.: Neural responding machine for short-text conversation. arXiv preprint arXiv:1503.02364 (2015)

16. Song, H., Zhang, W.N., Hu, J., Liu, T.: Generating persona consistent dialogues by exploiting natural language inference. In: Proceedings of the AAAI Conference on Artificial Intelligence, pp. 8878–8885 (2020)

17. Speer, R., Chin, J., Havasi, C.: ConceptNet 5.5: an open multilingual graph of general knowledge. In: Thirty-First AAAI Conference on Artificial Intelligence (2017)

18. Sutskever, I., Vinyals, O., Le, Q.V.: Sequence to sequence learning with neural networks. In: Advances in Neural Information Processing Systems, pp. 3104–3112 (2014)

19. Tang, J., Zhao, T., Xiong, C., Liang, X., Xing, E., Hu, Z.: Target-guided open-domain conversation. In: Proceedings of the 57th Annual Meeting of the Association for Computational Linguistics, pp. 5624–5634 (2019)

20. Vashishth, S., Sanyal, S., Nitin, V., Talukdar, P.: Composition-based multi-relational graph convolutional networks. In: International Conference on Learning Representations (2020)

21. Vaswani, A., et al.: Attention is all you need. In: Advances in Neural Information Processing Systems, pp. 5998–6008 (2017)

22. Xu, J., Wang, H., Niu, Z., Wu, H., Che, W.: Knowledge graph grounded goal planning for open-domain conversation generation. In: Proceedings of the AAAI Conference on Artificial Intelligence, pp. 9338–9345 (2020)

23. Zeng, M., Wang, Y., Luo, Y.: Dirichlet latent variable hierarchical recurrent encoder-decoder in dialogue generation. In: Proceedings of the 2019 Conference on Empirical Methods in Natural Language Processing and the 9th International Joint Conference on Natural Language Processing (EMNLP-IJCNLP), pp. 1267–1272 (2019)

24. Zhang, H., Liu, Z., Xiong, C., Liu, Z.: Grounded conversation generation as guided traverses in commonsense knowledge graphs. In: Proceedings of the 58th Annual Meeting of the Association for Computational Linguistics, pp. 2031–2043 (2020)

25. Zhang, S., Dinan, E., Urbanek, J., Szlam, A., Kiela, D., Weston, J.: Personalizing dialogue agents: I have a dog, do you have pets too? In: Proceedings of the 56th Annual Meeting of the Association for Computational Linguistics (Volume 1: Long Papers), pp. 2204–2213 (2018)

26. Zhao, X., Wang, L., He, R., Yang, T., Chang, J., Wang, R.: Multiple knowledge syncretic transformer for natural dialogue generation. In: Proceedings of The Web Conference 2020, pp. 752–762 (2020)

27. Zhou, H., Young, T., Huang, M., Zhao, H., Xu, J., Zhu, X.: Commonsense knowledge aware conversation generation with graph attention. In: IJCAI, pp. 4623–4629 (2018)

KdTNet: Medical Image Report Generation via Knowledge-Driven Transformer

Yiming Cao[1,2], Lizhen Cui[1,2(✉)], Fuqiang Yu[1,2], Lei Zhang[1,2], Zhen Li[3], Ning Liu[1], and Yonghui Xu[2(✉)]

[1] School of Software, Shandong University, Jinan, China
`clz@sdu.edu.cn`
[2] Joint SDU-NTU Centre for Artificial Intelligence Research (C-FAIR), Shandong University, Jinan, China
`xu.yonghui@hotmail.com`
[3] Department of Gastroenterology, Qilu Hospital of Shandong University, Jinan, China

Abstract. Writing medical image reports is an inefficient and time-consuming task for doctors. Automatically generating medical reports is an essential task of medical data mining, which can alleviate the workload of doctors and improve the standardization of reports. However, the existing methods mainly adopt the CNN-RNN structure to align image features with text features. This structure has difficulty dealing with the dependencies between distant text locations, leading to inconsistent context and semantics in the generated report. In this paper, we propose a knowledge-driven transformer (KdTNet) model for generating coherent medical reports. First, the visual grid and graph convolutional modules are devised to extract fine-grained visual features. Second, we adopt the transformer decoder to generate the hidden semantic states. Subsequently, a BERT-based auxiliary language module is employed to obtain the context language features of reports from the pre-defined medical term knowledge. We design a multimodal information fusion module to adaptively calculate the contribution of visual and linguistic features to the report for report generation. Extensive experiments on two real datasets explicate that our KdTNet model has achieved superior performance in captioning metrics and human evaluation compared with the state-of-the-art methods.

Keywords: Medical data mining · Medical report generation · Transformer model

1 Introduction

Medical images are widely-used in disease diagnosis and treatment [2]. The generation of clinical medical image reports mainly relies on doctors' medical knowledge and work experience. High-quality clinical medical image reports usually

A. Bhattacharya et al. (Eds.): DASFAA 2022, LNCS 13247, pp. 117–132, 2022.
https://doi.org/10.1007/978-3-031-00129-1_8

require detailed scrutiny and analysis by senior medical imaging experts. Due to differences in the experience of doctors, it is not easy to generate a clinical medical image report with medical images consistent with the text description. Given the precipitation of massive medical images, writing reports is a heavy burden for doctors. Besides, factors such as the doctor's work status can also cause problems such as time-consuming work and uneven report quality. In order to alleviate the heavy workload of doctors, automatic report generation [11,12,34] has become a critical task in clinical practice. It can significantly speed up the automation of workflows and enhance the quality and standardization of healthcare.

In recent years, deep learning have successfully generated accurate short text narratives for natural images in image captioning tasks [7,19,23], which makes medical image reports aroused widespread research interest. However, a significant challenge in medical report generation is that the report is a long narrative composed of multiple sentences. Therefore, applying traditional image captioning methods may not be sufficient to generate medical reports, as such methods usually generate brief descriptions for natural scenes. Most existing report generation methods [11,22,29] utilize the CNN-RNN based model with an attention mechanism to generate reports automatically. The image is encoded by a Convolutional Neural Network (CNN) to obtain visual features, which are then fed to a Recurrent Neural Network (RNN) to compile them into text. Some of the state-of-the-art works construct knowledge graphs to guide the image report generation, such as an abnormal graph [12], chest observation graph [34], and medical tag graph [13]. Although these works can generate textual narratives for medical images, most works typically suffer from textual incoherence and low linguistic diversity because that RNN-based models are limited by their inherent sequential nature that precludes parallelization and have difficulty handling dependencies between distant locations. Compared with RNN-based model, the transformer-based model presents more excellent performance in sequence modeling tasks. Chen et al. [1] proposed a memory-driven Transformer to record the key information of the generation process to generate a medical report. Liu et al. [17] proposed Posterior-and-Prior Knowledge Exploring-and-Distilling approach (PPKED) with three transformers to imitate the working mode of doctors. However, in these designs, Transformer-based models usually result in inconsistencies and redundancy in the generation of previously generated sentences. In addition, some essential medical terms and vocabulary in medical reports cannot be effectively generated.

To solve the above problems, we propose a knowledge-driven transformer (KdTNet) model to generate semantically coherent medical image reports. Specifically, we devise the visual grid module to extract the original grid features, We design a graph convolution module, including a pre-built medical disease graph and graph convolution network, which embeds grid features into prior medical knowledge to obtain node features. We employ a transformer-based encoder to embed node features to obtain fine-grained visual features. Subsequently, the transformer-based encoder transforms the visual features into hidden semantic features of the word sequence. In addition, we build an auxiliary language model composed of BERT and mask-multi-head-attention. The module is used to encode predefined medical descriptions to obtain the context language features, which can guide our KdTNet to focus on the context of medical reports information. Based on the above three features, we construct a multimodal information fusion module for the transformer architecture to perform sequence prediction, and calculate the contribution of visual and linguistic features to the generated sequence to generate a semantically coherent medical report.

The contributions are summarized as follows:

- We design the visualization grid module to extract the original grid features of medical images and devise the graph convolution module to combine grid features with medical disease knowledge to help transform visual features into linguistic features.
- We adopt the auxiliary language model composed of BERT and mask-multi-head-attention to encode predefined medical description knowledge into the context language features of medical reports.
- We build the multimodal information fusion module to weigh the contributions of visual and linguistic features to generate coherent medical reports.
- We evaluate our KdTNet on two real datasets. The results show our KdTNet obtains superior performance compared with the state-of-the-art methods.

2   Related Work

With the development of deep learning [7–9], text generation from images has been widely concerned. The current works mainly explore the image captioning and report generation for medical domain.

2.1   Image Captioning

The task of image captioning mainly applies the Encoder-Decoder models [23,25] to transform the image into a single descriptive sentence. The visual captioning models employ attention mechanisms [22,31] to improve performance. Besides, extra information is adopted to assist text generation for NLP [26] and image caption [11] tasks, such as the pre-trained embeddings [24,32,33], pre-built knowledge graphs [12], and pre-trained models [4]. However, these methods cannot be directly transferred to the task of generating medical reports. The medical report does not consist of only one sentence but a longer paragraph consisting of normal and abnormal descriptions. In the abnormal descriptions, each sentence focuses on a specific medical observation of a specific area in the image.

2.2   Report Generation

CNN-RNN Based Methods. Similar to image captioning, most existing works of report generation adopt the Encoder-Decoder frameworks to automatically generate a fluent report. Zhang et al. [35] employed a CNN model to obtain the image feature, and utilized LSTM to predict the diagnosis and generate the report of images. Some approaches [11,14] adopted extra information (such as context and topic representations) to assist report generation. Furthermore, generative models and reinforcement learning models were designed to analyze multiple features of image, text, and extra information. Other methods [6,13,29] appended auxiliary modules to CNN-RNN architecture, such as the Neural-Symbolic Learning (NSL) framework, recurrent generation model, and internal and external auxiliary signals. Judging from the results of the evaluation metrics, these auxiliary modules improved the performance to a certain extent.

Transformer Based Methods. The RNN-based model is limited by its sequence nature, and there is a dependency between remote locations. It is observed that the RNN model is easily misled by various abnormal findings in medical reports, making it unable to model abnormal descriptions effectively. Chen et al. [1] proposed a memory-driven Transformer to generate radiology reports, with a relational memory to record the previously generated information and a memory-driven conditional layer for the fusing the memory into Transformer. Liu et al. [17] designed a Posterior-and-Prior Knowledge Exploring-and-Distilling approach (PPKED) to imitate the radiologists' working patterns, and utilized PoKE, PrKE, and MKD modules to incorporate the prior medical knowledge into the report generation process.

Knowledge Based Methods. Many previous works [5,15,18,20,30] employed the graph neural network to aggregate neighbor information on graph architectures. Li et al. [12] proposed a KERP model to guide visual features into predefined abnormal graphs and the predefined medical templates for radiology reports generation. Zhang et al. [34] devised pre-constructed graph embedding modules on multiple disease findings to embed visual features into the node of the defined graph. However, these works focus on the generation of abnormal symptoms. Li et al. [13] utilized the medical tag and designed a graph encoder to fuse medical tag information and image features, and designed a GPT-based decoder to generate final reports.

Although these works have achieved good results on automatic evaluation metrics, they still utilize training models to forcibly align visual features of images with textual features of reports, which did not fully combine the image features and semantic features. Our work distinguishes itself from these works because we introduce fine-grained knowledge based on a transformer structure to guide the transformation from visual features to textual features. Different from these works, we utilize multimodal medical knowledge for information propagation and embedding. Medical disease graphs are designed to focus on different positions of images, and medical knowledge text fine-grained characterizes the contextual language features of the report.

3   Models

Medical image report generation is essentially an image-to-text generation task. In this paper, we follow the sequence-to-sequence paradigm. The input from the medical image is regarded as the source sequence $\mathbf{G} = \{\mathbf{g}_1, \mathbf{g}_2 \ldots, \mathbf{g}_N\}, \mathbf{g}_i \in \mathbb{R}^s$, where $\mathbf{g}_i$ is the grid feature of the image, and s donates the size of the feature vector. The corresponding report is the target sequence $T = \{t_1, t_2, ..., t_M\}, t_i \in \mathbb{V}$, where t_i is the generated tokens, and $\mathbb{V}$ is the corpus vocabulary composed of all candidate tokens.

The overview of our proposed KdTNet model is shown in Fig. 1. First, the visual grid module is applied to extract the original grid features. On this basis, we utilize our designed graph convolution module to combine the grid features with medical disease knowledge. Furthermore, we employ a transformer-based encoder and decoder to the visual feature of the image and hidden semantic features of the word sequence, respectively. Besides, the context language features are obtained by encoding predefined medical descriptions based on the auxiliary language model composed of BERT and mask multi-head attention. Based on the three obtained features, we propose a multi-modal information fusion module to perform sequence prediction and calculate the contributions of visual and linguistic features to obtain semantically coherent medical reports.

Our KdTNet can be divided into the following main parts: visual grid module, graph convolution module, auxiliary language module, and multi-module information fusion module. These several modules of the KdTNet and training objectives are illustrated in detail below.

3.1   Visual Grid Module

Given any medical image I, it is represented as a set of visual grid features $\mathbf{G} = \{\mathbf{g}_1, \mathbf{g}_2 \ldots, \mathbf{g}_N\}$ extracted from the pre-trained visual grid module, where N is the number of visual regions. Specifically, its grid features $\mathbf{G}$ are extracted by a pre-trained convolutional neural network (CNN), such as DenseNet and ResNet. The results of the encoding are adopted as the input sequences for other subsequent modules. The visual grid module can be expressed as:

$$\{\mathbf{g}_1, \mathbf{g}_2 \ldots, \mathbf{g}_N\} = F_g(I) \tag{1}$$

where $F_g(I)$ donates the visual grid module.

3.2   Graph Convolution Module

After obtaining the visual grid features of the medical image, we designed a graph convolution module to combine the image features with prior medical knowledge, which can guide the transformation of image features to medical text features. The graph convolution module mainly includes a predefined medical disease knowledge graph and graph convolution network.

Medical Disease Graph. The medical disease graph is devised to store common clinical diseases and their interactions. It is applied to train different diseases to learn to pay attention to different spatial positions of images. The nodes of

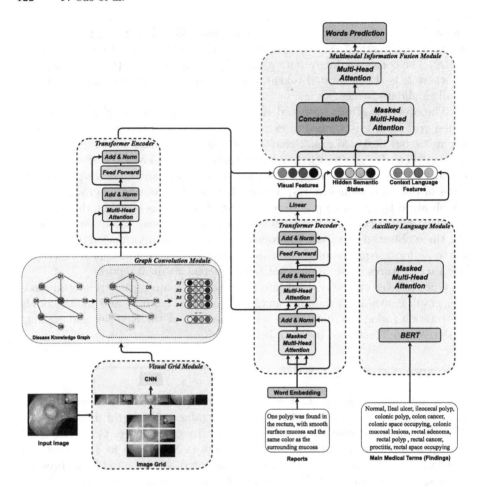

Fig. 1. An overview of our KdTNet architecture for medical report generation.

the medical disease graph represent major clinical diseases. The edges between nodes represent the co-occurrence relationship between diseases in the dataset or corpus.

In this paper, we construct two medical knowledge graphs for two classes of diseases, namely gastrointestinal and thoracic diseases. First, we assemble a medical disease graph for gastrointestinal diseases. Its nodes are typical gastrointestinal diseases, including 18 diseases such as ileocecal inflammation, colon polyps, and rectal cancer. The second is the medical disease graph for thoracic diseases. We reoptimized the disease graph according to previous research [12], covering 16 diseases, such as pleural effusion, pneumothorax, and pneumonia.

Graph Convolution Network. The graph convolution network (GCN) is employed to embed visual grid features into the medical disease graph and propagate and aggregate information in the graph. The initial features of the nodes in the medical disease graph come from the output **G** of the visual grid module. In particular, we adopt a Convolution layer with a 1*1 filter size, where the number of its channels is set to the number of nodes in the medical disease graph, inspired by the prior work [34]. In this way, the initial features of each node in the graph are obtained by calculating the attention weight of the corresponding channel.

After that, we utilize GCN to aggregate the information of disease nodes on the graph. In each convolutional layer, the information of neighbor nodes is aggregated by multiplying the current node features by the normalized adjacency matrix, and the aggregated features update the features of the nodes in the next layer. The GCN we adopted can be denoted as:

$$\mathbf{D}_{node}^{l+1} = F_{upda}(AF_{aggr}(\mathbf{D}_{node}^{l}, A)) \tag{2}$$

$$F_{aggr}(\mathbf{D}_{node}^{l}, A) = \sigma(A\mathbf{D}_{node}^{l}) \tag{3}$$

where $\mathbf{D}^l$ denotes the node features of the l-th layer, $\mathbf{D}^{l+1}$ is the node feature of the next layer, F_{upda} and F_{aggr} donate the activation function and aggregation function, respectively. A is the normalized Laplacian of the adjacency matrix.

The node features $\mathbf{f}_g$ output by GCN will be fed into the encoder of a transformer as the input to obtain the fine-grained visual feature.

3.3   Auxiliary Language Module

To make the generated report semantically coherent, we introduce the auxiliary language module to learn the contextual information between the text in medical reports. We first formed a corpus of main medical terms and descriptions for gastrointestinal diseases and thoracic diseases. These medical terms came from the "findings" part of the corresponding medical reports. For gastrointestinal diseases, we extract the medical findings or terms that often appear in reports, such as "smooth mucosa", "polypoid protrusion" and "tumor surface erosion". Besides, we also extracted common medical descriptions such as "no pneumothorax or plural effect", "biapical plural thickening" and "hyperexpanded lung" for thoracic diseases.

Furthermore, we employ the BERT-based model to extract linguistic features, appending a masked multi-head attention mechanism to prevent data leakage during the testing phase. The auxiliary language module generates the context language feature for a given corpus of medical terms. The process can be formalized as:

$$\mathbf{f}_B = BERT(Cor) \tag{4}$$

$$\mathbf{f}_m = MaskedAttention(FFN(\mathbf{f}_B) + pos) \tag{5}$$

$$\mathbf{f}_l = logSoftmax(FFN(\mathbf{f}_m)) \tag{6}$$

where $\mathbf{f}_B$ is the output of the BERT model, $\mathbf{f}_m$ represents the output of masked multi-head attention module, FFN is the fully connected feed-forward network, pos donates the position encoding of the input sequence, and $\mathbf{f}_l$ is the context language feature from the log softmax distribution of input sequence.

3.4   KdTNet Encoder

Our KdTNet model follows the classic transformer architecture, where the encoder takes node features extracted from images and medical disease knowledge as the input. Given the encoded features, the decoder generates the text sequence as output. The node features from the medical disease graph mentioned in Sect. 3.2 are first flattened, and embedded using a fully connected layer followed by a ReLU and a dropout layer. Subsequently, the embedded features are fed into the KdTNet encoder. The main component of the KdTNet encoder is Scaled Dot-Product Attention, which is expressed as:

$$Q = \mathbf{f}_e W_q, K = \mathbf{f}_e W_k, V = \mathbf{f}_e W_v \tag{7}$$

$$Z = softmax(\frac{QK^T}{\sqrt{d_k}})V \tag{8}$$

$$\mathbf{f}_v = \mathbf{f}_e + Z \tag{9}$$

where $\mathbf{f}_e$ is the embedding in the encoder, W_q, W_k, W_v are learnable weights for input metrics, d_k donates the scaling factor, and $\mathbf{f}_v$ represents the visual feature.

3.5   KdTNet Decoder

In the initial state, the text sequence features of the medical report are processed by word embedding, and then regarded as the input of the first decoder layer of the KdTNet decoder. The decoder can be formalized as:

$$\mathbf{f}_h = Decoder(\mathbf{f}_v, \mathbf{f}_w) \tag{10}$$

where $\mathbf{f}_h$ is the hidden semantic states from the decoder, $\mathbf{f}_v$ is the output of the encoder, and $\mathbf{f}_w$ donates the word sequence feature of the raw report.

3.6   Multimodal Information Fusion Module

We obtain three modal feature information through modules mentioned above: visual features embedded in medical disease knowledge, hidden semantic features, and context language features storing medical descriptions. To obtain a semantically coherent medical report, we designed a multi-modal information fusion module, attached to the upper layer of the KdTNet decoder. The module combines the feature information of the three modalities to calculate the contribution of visual features and linguistic features to each generated sequence, which is defined as follows:

$$q_{i,t} = \mathbf{f}_h W_i^Q, k_{i,t} = concat[\mathbf{f}_v; \mathbf{f}_l]W_i^Q, v_{i,t} = concat[\mathbf{f}_v; \mathbf{f}_l]W_i^V \quad (11)$$

$$head_{i,t} = softmax(q_{i,t}k_{i,t})v_{i,t} \quad (12)$$

$$head_i = concat[head_{i,1}, ..., head_{i,H}] \quad (13)$$

$$AR = concat[head_1, ..., head_N]W^A \quad (14)$$

where $q_{i,t}$ is the qurey vector of the t-th word in head i of multi-head attention, $k_{i,t}$ and $v_{i,t}$ represent the key and value matrix in head i of multi-head attention, $head_{i,t}$ donates the attention result for the t-th work in head i, $head_i$ is the attention result for the sequence in head i, and AR is the attention result of multi-head attention for the generated sequence.

3.7   Training

For each training sample, (I, r), where I is a group of images and r is the corresponding textual report composed of the ground truth sequence, the loss of report generation $\mathcal{L}$ is minimized by the cross-entropy loss:

$$\mathcal{L}(\theta) = -\sum_{i=1}^{N} \log(p_\theta(s_i|s_{1:i-1})) \quad (15)$$

where θ is the parameters of our KdTNet model, $s_{1:i-1}$ represents the ground truth sequence of the report r.

4   Experiment

In this section, we present the evaluation and analysis of the proposed methods.

4.1   Experimental Settings

Dataset. We conducted experiments on two medical image report datasets:

- Gastrointestinal Endoscope image dataset (**GE**). The GE is a private dataset collected from the Department of Gastroenterology in hospital, containing 15345 white light images of gastrointestinal endoscopy and 3069 Chinese reports. Each sample has multiple gastrointestinal endoscope images from different perspectives with their corresponding medical reports. In addition, 11 relative diseases are extracted from the "diagnosis" part of medical reports to construct the disease knowledge graph, and 89 abnormal findings are collected from the "finding" part of reports to compose the medical finding corpus.

– **IU-CX** [3]. The IU-CX is a widely-used public chest X-ray dataset, which
is used to verify the generality of our method. The original IU-CX consists
of 7,470 images associated with 3,955 radiology reports. We obtained 2896
radiology reports with frontal and lateral view chest X-ray images after data
clearing. Besides, we extracted 18 diseases for the thorax disease knowledge
graph and 70 abnormal findings for the medical finding corpus.

Parameter Settings. The method is implemented in Pytorch 1.7.1 based
on Python 3.8.5 and trained on a server with an Intel Core i9-10900K CPU,
and an Nvidia RTX 3090 GPU. We randomly split both datasets into 7:1:2
training:validation:testing data to train and evaluate our method. A pre-trained
DenseNet-121 is adopted to extract image features, where the grid size is set
to 7 * 7 The Chinese word segmentation module of Jieba [10] is employed for
processing the reports of GE. The number of heads in KdTNet is set to 8. The
dropout probability is 0.1. If not specifically specified, the hidden dimension of
KdTNet is 512. The ADAM optimizer with a batch size of 32 and a learning
rate of 1e−5 is employed to minimize the loss function.

Evaluation Metrics. We employed both automatic metrics and human evaluation to evaluate the performance for the task of medical report generation.

– Automatic Metrics, including CIDEr [27], ROUGE-L [16], BLEU [21].
– Human Evaluation: we randomly selected 50 samples from the test set and use
different baseline models to generate candidate reports to be evaluated. We
invite the hospital's gastroenterologists and graduate students to select the
report that best matches the ground-truth report from the candidate reports.

Baselines. We compared our proposed model to the following state-of-the-art
report generation and convention image captioning models: SaT [28], AAtt [19],
CoAtt [11], Transformer, M-dTransformer [1], RGKG [34] and PPKED [17]. For
IU-CX, the template retrieval methods such as HRGRA [14] and KER [12] are
also compared with our KdTNet for thoracic diseases. Note that due to the lack
of predefined templates on GE, these two methods are not used for comparison
on GE.

4.2   Results on Report Generation

Automatic Evaluation. To illustrate the effectiveness of our KdTNet, we
experiment with the baseline methods on both datasets for the report generation
task, with all performances reported in Table 1 on image captioning metrics. The
best and second best results are highlighted.

Referring to Table 1, Our KdTNet outperforms all baseline methods on the
BLEU-n and CIDEr scores of the GE dataset and the BLEU-n score of IU-CX,
which demonstrates the effectiveness and accuracy of our KdTNet in generating
medical reports. For BLEU-n, compared with the previous best method, KdTNet

Table 1. Comparison of report generation models on both image captioning metrics (CIDEr, ROUGE-L and BLEU-n) and human evaluation (HE) on GE and IU-CX dataset.

Dataset	Model	CIDEr	ROUGE-L	BLEU-1	BLEU-2	BLEU-3	BLEU-4	HE
GE	SaT	0.557	0.613	0.643	0.552	0.506	0.414	0.028
	AAtt	0.579	0.617	0.649	0.549	0.491	0.419	0.078
	CoAtt	0.674	**0.748**	0.774	0.654	0.618	0.575	0.124
	Transformer	0.604	0.691	0.689	0.572	0.584	0.521	0.116
	M-dTransformer	0.679	0.736	0.779	0.677	0.619	0.574	0.128
	PPKED	0.683	0.742	<u>0.789</u>	<u>0.682</u>	<u>0.621</u>	<u>0.580</u>	<u>0.192</u>
	RGKG	<u>0.684</u>	0.726	0.752	0.676	0.609	0.554	0.118
	KdTNet(ours)	**0.692**	<u>0.748</u>	**0.792**	**0.686**	**0.624**	**0.583**	**0.216**
IU-CX	SaT	0.288	0.307	0.204	0.125	0.088	0.064	–
	AAtt	0.29	0.311	0.211	0.129	0.091	0.069	–
	CoAtt	0.274	0.372	0.447	0.286	0.207	0.157	–
	HRGRA	**0.341**	0.320	0.436	0.295	0.209	0.152	–
	Transformer	0.283	0.363	0.397	0.251	0.176	0.133	–
	M-dTransformer	0.328	0.371	0.461	0.302	0.213	0.164	–
	PPKED	<u>0.332</u>	**0.376**	<u>0.472</u>	<u>0.311</u>	<u>0.216</u>	<u>0.167</u>	–
	KER	0.314	0.342	0.451	0.298	0.209	0.159	–
	RGKG	0.316	0.370	0.437	0.285	0.203	0.149	–
	KdTNet(ours)	**0.341**	<u>0.375</u>	**0.474**	**0.316**	**0.225**	**0.169**	–

has achieved 1.13%-3.31% improvement on the GE data set and 0.43%-4.17% on IU-CX, which shows that the report generated by our KdTNet is smoother and more realistic as the higher the BLEU-n scores, the more pronounced the overlap between the generated n-grams and the ground-truth reports. In terms of CIDEr, the performance of KdTNet on the IU-CX dataset is not as good as HRGRA, because HRGRA directly employed CIDEr as the reward of its reinforcement learning algorithm in the training process. As for ROUGR-L, KdTNet scores lower than CoAtt on the IU-CX data set and PPKED on GE. CoAtt and PPKED achieve the best performance on the IU-CX and GE, respectively. A reasonable explanation is that they both use additional semantic information as to their inputs, making the longest subsequence in their generated reports most consistent with the ground truth reports.

In addition, there are several observations by comparing the results of different methods. Firstly, the methods adopted prior knowledge (such as PPKED, KER, RGKG, and our KdTNet) outperform most report generation, models. The observation explicates that it is necessary to incorporate prior knowledge information into the generative model for guiding the transformation of visual to linguistic features. Compared with the CNN-RNN based models (such as SaT and AAtt), the Transformer-based model is more excellent on most metrics, indicating that the text generated by the Transformer structure is more similar to the ground-truth report than the vanilla CNN-RNN structure. In particular, PPKED ranks second in performance on BLEU-n, which points that this model can generate more accurate n-grams than other models. Besides, models that use additional information also affect the performance of specific metrics, such as CoAtt and PPKED models that use additional semantic information, and KER and HRGRA models that use predefined report templates.

Human Evaluation. We randomly selected 50 samples from GE and invited three gastroenterologists and seven graduate students to evaluate the reports generated by KdTNet and baseline methods. Each method generates a corresponding report for each sample, and we ask each expert to select the one that best matches the ground truth report from the generated reports. The human evaluation metric is the proportion of each method selected by experts in the total number. The result is shown in the last column of Table 1. The reports generated by our KdTNet account for 21.6% of the reports selected by experts. The results demonstrate that our KdTNet is capable of generating semantically consistent and reliable reports.

Qualitative Analysis. To further investigate the effectiveness of our KdTNet, we conduct qualitative analysis on both datasets with their ground-truth and generated reports. The visualization results of KdTNet on both datasets are presented in Fig. 2. The first two rows depict two samples from the GE dataset (note that gastroenterologists translate the reports of GE from Chinese to English), and the last two rows are from the IU dataset. It can be observed that KdTNet is capable of generating reports consistent with the description written by doctors. For example, the generated report accurately reported the location (such as descending colon, sigmoid colon, and transverse colon) and type (congestion and edema, tumor, and polyp) of the lesion in the GE sample. Similarly, in the IU-CX sample, KdTNet also accurately describes most types of lesions. It is worth noting that KdTNet not only generates abnormal descriptions, but also the descriptions of normal regions. For example, "Smooth mucosa" and "Normal peristalsis" in the GE, and "No focal consolidation" and "The lungs are clear" in the IU-CX. For the necessary medical terms in the ground-truth reports, KdTNet covers almost all of these terms in its generated reports. In addition, the reports generated by KdTNet are also consistent with the position indicated by the attention maps. The result explicates that KdTNet can generate accurate reports and interpretable attention areas.

4.3   Ablation Studies

Effect of Components. We conducted ablation studies on GE and IU-CX to investigate the effectiveness of each module in KdTNet. Specifically, \GCM excludes the graph convolution module from KdTNet, \ALM, and \MIF drops the auxiliary language module and multimodal information fusion module. As shown in Fig. 3, KdTNet\GCM has the worst performance, revealing that the introduction of related diseases and the combination of prior knowledge and image features can effectively improve the accuracy of report generation. On the

Image	Attention Maps	Ground Truth Report	Generated Report

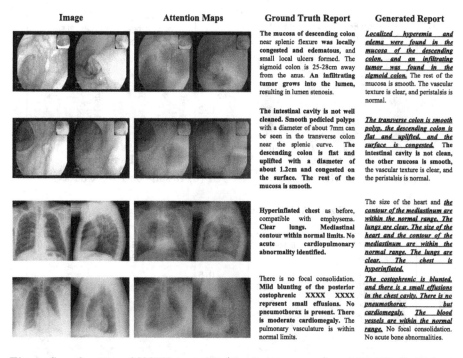

The mucosa of descending colon near splenic flexure was locally congested and edematous, and small local ulcers formed. The sigmoid colon is 25-28cm away from the anus. An **infiltrating tumor grows into the lumen,** resulting in lumen stenosis.

Localized hyperemia and edema were found in the mucosa of the descending colon, and an infiltrating tumor was found in the sigmoid colon. The rest of the mucosa is smooth. The vascular texture is clear, and peristalsis is normal.

The intestinal cavity is not well cleaned. Smooth pedicled polyps with a diameter of about 7mm can be seen in the transverse colon near the splenic curve. The **descending colon is flat and uplifted** with a diameter of about 1.2cm and **congested on the surface.** The rest of the mucosa is smooth.

The transverse colon is smooth polyp, the descending colon is flat and uplifted, and the surface is congested. The intestinal cavity is not clean, the other mucosa is smooth, the vascular texture is clear, and the peristalsis is normal.

Hyperinflated chest as before, compatible with emphysema. **Clear lungs. Mediastinal contour within normal limits.** No acute cardiopulmonary abnormality identified.

The size of the heart and *the contour of the mediastinum are within the normal range. The lungs are clear. The size of the heart and the contour of the mediastinum are within the normal range. The lungs are clear. The chest is hyperinflated.*

There is no focal consolidation. Mild **blunting of the posterior costophrenic** XXXX XXXX represent small **effusions.** No pneumothorax is present. There is moderate **cardiomegaly.** The pulmonary vasculature is within normal limits.

The costophrenic is blunted, and there is a small effusions in the chest cavity. There is no pneumothorax but cardiomegaly. The blood vessels are within the normal range. No focal consolidation. No acute bone abnormalities.

Fig. 2. Sample cases of KdTNet on GE (the first two rows) and IU-CX. **Bold text** indicates consistency between the generated reports and ground truth. Underlined text indicates the correspondence between the generated reports and the attention maps.

other hand, the performance of KdTNet\MIF is poor, which determines that the multimodal information fusion module helps reduce the distance between visual features and language features. The performance of KdTNet\ALM is similar to that of KdTNet, indicating that the auxiliary language module has played a specific role in performance improvement. However, the auxiliary language module enhances the interaction between the image and the generated text.

Effect of Embedding Size. We analyze the impact of embedding size in KdT-Net on model performance in the section. Figure 4 displays the performance comparison of the embedding size of our proposed model on the GE and IU-CX. In general, as the embedding size increases, the performance first increases and then decreases. Increasing the embedding size from 64 to 128 can significantly improve performance. However, when the embedded size is 1024, the performance of KdTNet decreases. It demonstrates that the use of a large number of embedding sizes has a powerful representation ability. However, if the embedding size is too large, the complexity of our model will increase significantly. Therefore, we need to find the appropriate embedding size to strike a balance between performance and complexity.

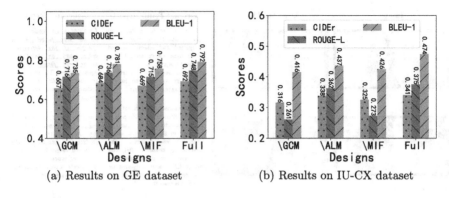

Fig. 3. Ablation study for different designs

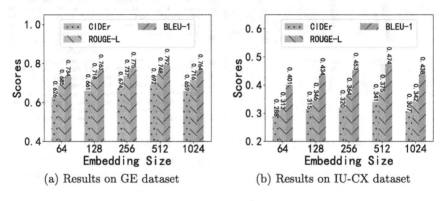

Fig. 4. Ablation study for different embedding sizes

5   Conclusion

In this paper, we propose a knowledge-driven transformer (KdTNet) model for semantically consistent and coherent medical reports generation. The visual grid module is devised to extract fine-grained visual features. We utilize a graph convolutional module to incorporate visual features into predefined medical disease graphs and adopt transformer-based encoders for embedding node features to fine-grained visual features. The transformer decoders the visual features to generate the hidden semantic states, and employ a bert-based auxiliary language model to the context language features of reports. Finally, we designed the multimodal information fusion module (MIF) to adaptively calculate the contribution of visual and language features to the report for a complete medical report generation. Extensive experiments on two real datasets demonstrate our KdTNet model achieves superior performance compared with mainstream approaches. For future work, it is essential to explore the relevant relationship between image region and main medical terms. It is interesting to adopt other GNN architectures to combine the transformer-based models for feature learning.

Acknowledgements. The research work was supported by the National Key R&D Program of China (No. 2021YFF0900800), the major Science and Technology Innovation of Shandong Province grant (No. 2021CXGC010108), the Shandong Provincial Key Research and Development Program (Major Scientific and Technological Innovation Project) (No. 2019JZZY011007), and partially supported by the NSFC (No. 91846205).

References

1. Chen, Z., Song, Y., Chang, T., Wan, X.: Generating radiology reports via memory-driven transformer. In: Proceedings of the 2020 Conference on Empirical Methods in Natural Language Processing, EMNLP 2020, pp. 1439–1449 (2020)
2. Delrue, L., et al.: Difficulties in the interpretation of chest radiography. In: Comparative Interpretation of CT and Standard Radiography of the Chest, pp. 27–49 (2011)
3. Demner-Fushman, et al.: Preparing a collection of radiology examinations for distribution and retrieval. Med. Inform. Assoc. **23**(2), 304–310 (2016)
4. Devlin, J., et al.: BERT: pre-training of deep bidirectional transformers for language understanding. In: NAACL-HLT 2019, pp. 4171–4186 (2019)
5. Du, G., Zhou, L., Yang, Y., Lü, K., Wang, L.: Deep multiple auto-encoder-based multi-view clustering. Data Sci. Eng. **6**(3), 323–338 (2021)
6. Han, Z., et al.: Unifying neural learning and symbolic reasoning for spinal medical report generation. Med. Image Anal. **67**, 101872 (2021)
7. He, K., Zhang, X., Ren, S., Sun, J.: Deep residual learning for image recognition. In: IEEE CVPR, pp. 770–778 (2016)
8. Hochreiter, S., Schmidhuber, J.: Long short-term memory. Neural Comput. **9**(8), 1735–1780 (1997)
9. Huang, G., Liu, Z., van der Maaten, L., Weinberger, K.Q.: Densely connected convolutional networks. In: IEEE CVPR, pp. 2261–2269 (2017)
10. Jieba: "Jieba" Chinese text segmentation: built to be the best python Chinese word segmentation module (2018). https://github.com/fxsjy/jieba
11. Jing, B., Xie, P., Xing, E.P.: On the automatic generation of medical imaging reports. In: ACL, pp. 2577–2586 (2018)
12. Li, C.Y., Liang, X., Hu, Z., Xing, E.P.: Knowledge-driven encode, retrieve, paraphrase for medical image report generation. In: AAAI, pp. 6666–6673 (2019)
13. Li, M., Wang, F., Chang, X., Liang, X.: Auxiliary signal-guided knowledge encoder-decoder for medical report generation (2020)
14. Li, Y., Liang, X., Hu, Z., Xing, E.P.: Hybrid retrieval-generation reinforced agent for medical image report generation. In: NeurIPS, pp. 1537–1547 (2018)
15. Liang, X., Hu, Z., Zhang, H., Lin, L., Xing, E.P.: Symbolic graph reasoning meets convolutions. In: NeurIPS, pp. 1858–1868 (2018)
16. Lin, C.Y.: Rouge: a package for automatic evaluation of summaries. In: Text Summarization Branches Out, pp. 74–81 (2004)
17. Liu, F., Wu, X., Ge, S., Fan, W., Zou, Y.: Exploring and distilling posterior and prior knowledge for radiology report generation. In: IEEE Conference on Computer Vision and Pattern Recognition, CVPR 2021, pp. 13753–13762. IEEE (2021)
18. Liu, Y., et al.: Pre-training graph transformer with multimodal side information for recommendation. In: MM 2021, pp. 2853–2861. ACM (2021)
19. Lu, J., Xiong, C., Parikh, D., Socher, R.: Knowing when to look: adaptive attention via a visual sentinel for image captioning. In: IEEE CVPR, pp. 3242–3250 (2017)

20. Norcliffe-Brown, W., Vafeias, S., Parisot, S.: Learning conditioned graph structures for interpretable visual question answering. In: NeurIPS 2018, pp. 8344–8353 (2018)
21. Papineni, K., Roukos, S., Ward, T., Zhu, W.: Bleu: a method for automatic evaluation of machine translation. In: ACL, pp. 311–318 (2002)
22. Rennie, S.J., Marcheret, E., Mroueh, Y., Ross, J., Goel, V.: Self-critical sequence training for image captioning. In: IEEE CVPR, pp. 1179–1195 (2017)
23. Shin, H., et al.: Learning to read chest X-rays: recurrent neural cascade model for automated image annotation. In: IEEE CVPR, pp. 2497–2506 (2016)
24. Song, Y., Shi, S.: Complementary learning of word embeddings. In: IJCAI 2018, pp. 4368–4374 (2018)
25. Sutskever, I., Vinyals, O., Le, Q.V.: Sequence to sequence learning with neural networks. In: NIPS, pp. 3104–3112 (2014)
26. Tian, Y., et al.: Joint Chinese word segmentation and part-of-speech tagging via two-way attentions of auto-analyzed knowledge. In: ACL 2020 (2020)
27. Vedantam, R., Zitnick, C.L., Parikh, D.: Cider: consensus-based image description evaluation. In: IEEE CVPR, pp. 4566–4575 (2015)
28. Vinyals, O., Toshev, A., Bengio, S., Erhan, D.: Show and tell: a neural image caption generator. In: IEEE CVPR, pp. 3156–3164 (2015)
29. Xue, Y., et al.: Multimodal recurrent model with attention for automated radiology report generation. In: Frangi, A.F., Schnabel, J.A., Davatzikos, C., Alberola-López, C., Fichtinger, G. (eds.) MICCAI 2018. LNCS, vol. 11070, pp. 457–466. Springer, Cham (2018). https://doi.org/10.1007/978-3-030-00928-1_52
30. Yao, T., Pan, Y., Li, Y., Mei, T.: Exploring visual relationship for image captioning. In: Ferrari, V., Hebert, M., Sminchisescu, C., Weiss, Y. (eds.) Computer Vision – ECCV 2018. LNCS, vol. 11218, pp. 711–727. Springer, Cham (2018). https://doi.org/10.1007/978-3-030-01264-9_42
31. You, Q., Jin, H., Wang, Z., Fang, C., Luo, J.: Image captioning with semantic attention. In: IEEE CVPR, pp. 4651–4659 (2016)
32. Zhang, H., et al.: Multiplex word embeddings for selectional preference acquisition. In: Proceedings of the 9th International Joint Conference on Natural Language Processing, EMNLP-IJCNLP 2019, pp. 5246–5255 (2019)
33. Zhang, W., Ying, Y., Lu, P., Zha, H.: Learning long- and short-term user literal-preference with multimodal hierarchical transformer network for personalized image caption. In: AAAI 2020, pp. 9571–9578. AAAI Press (2020)
34. Zhang, Y., Wang, X., Xu, Z., Yu, Q., Yuille, A.L., Xu, D.: When radiology report generation meets knowledge graph. In: AAAI, pp. 12910–12917 (2020)
35. Zhang, Z., Xie, Y., Xing, F., McGough, M., Yang, L.: MDNet: a semantically and visually interpretable medical image diagnosis network. In: IEEE CVPR 2017, pp. 3549–3557 (2017)

Fake Restaurant Review Detection Using Deep Neural Networks with Hybrid Feature Fusion Method

Yifei Jian, Xingshu Chen, and Haizhou Wang[✉]

School of Cyber Science and Engineering, Sichuan University, Chengdu, China
whzh.nc@scu.edu.cn

Abstract. Online review websites all over the world have developed by leaps and bounds in the past decade, followed by more and more fake reviews that are very difficult to identify. This not only affects the decision-making of consumers, but also seriously undermines the fairness of the consumption market. Most existing research ignored the abundant metadata in reviews and its effective combination with the review text. In this paper, we firstly collect reviews from one of the most popular Chinese review platforms Dianping, and release two fake review datasets we construct, which makes up for the relative lack of datasets in this field. Afterward, we present some novel and effective features for fake review detection by analyzing the characteristics of the collected reviews. In addition, we propose a multimodal fake review detection model BAM (BERT + Attention + MLP), which takes the review text and the extracted features as input, and uses neural networks as well as multimodal fusion technology to realize the recognition of fake reviews. The experimental results show that BAM has low dependence on the size of the dataset. Compared with baseline models, it can be found that BAM has a better detection performance.

Keywords: Fake reviews · Abundant metadata · Labeled dataset · Extracted features · Multimodal model

1 Introduction

With the rapid development of e-commerce platforms and online review websites, more and more consumers tend to refer to the reviews on review websites as a basis for their purchase decisions. In addition, with the expansion of this trend, more and more consumers are also willing to post online reviews to share their real consumption experiences with others. Online reviews have played an important role in guiding the consumption decisions of consumers [4]. However, the authenticity of online reviews has become a worrying problem, and the accurate identification of fake reviews has become one of the most urgent network security issues that need to be resolved [9].

© The Author(s), under exclusive license to Springer Nature Switzerland AG 2022
A. Bhattacharya et al. (Eds.): DASFAA 2022, LNCS 13247, pp. 133–148, 2022.
https://doi.org/10.1007/978-3-031-00129-1_9

Fake review, also known as deceptive review, is regarded as a kind of opinion spam [12], which refers to reviews that do not have any useful reference value for the decision-making of consumers, or even play a role in misleading guidance. Their authors, the publishers of fake reviews, are also known as spammers. Positive fake reviews are helpful for merchants to earn a good reputation to promote their products, while negative fake reviews will degrade their image among consumers [34]. Therefore, driven by huge economic interests, more and more merchants choose to post fake online reviews to promote their products or slander other merchants in a competitive relationship, so as to make a profit [26,36]. At the same time, ordinary consumers may also post fake reviews because of receiving rewards from merchants or having conflicts of interest with merchants.

Nowadays, many online review websites as well as e-commerce platforms provide sound review functions, the most famous ones are Amazon [12], Yelp [24], and Dianping [19]. Among them, Dianping[1] is the earliest independent third-party online review website in the world, and is also one of the most popular review platforms in China. Compared with other platforms, the review data provided by Dianping is very abundant.

In order to identify fake reviews, various detection methods have been proposed in recent years. However, most of these methods focus on the English review platforms, rarely on the Chinese review platforms [33]. Moreover, most of them are combined with common feature extraction methods and traditional machine learning classifiers, and the classification effect is not very ideal [7].

The focus of this paper is on the detection of Chinese fake reviews in restaurants. Firstly, since the collection and labeling of fake reviews is very difficult, recent researches [28,33] show that there are currently only a few labeled datasets of fake reviews, which largely hinders the development of this field. To our best knowledge, there is currently no publicly labeled dataset of Chinese fake reviews with rich metadata. Secondly, most of the existing research focuses on review text and user behaviors, while ignoring the feature extraction of the abundant metadata in reviews [23], which is partly due to the lack of rich metadata in the existing datasets. Thirdly, the current deep learning detection methods for fake review detection are still relatively lacking [1], and there are only a few fake review detection models that can effectively combine multimodal input. In view of the above challenges, this paper mainly proposes three contributions.

- We collect reviews from the most popular Chinese review platform (i.e. Dianping) by our developed crawler. According to our strict annotation rules, two labeled datasets of fake reviews are constructed and published. Compared with other datasets, the sources of reviews in ours are more extensive and they contain richer metadata.
- A total of 30 different emotion indicator features and statistic features for reviews are extracted to help fake review detection. Among them, 10 features are newly defined, and 9 features are redefined by us.
- We construct a multimodal fake review detection model based on deep learning, which can take the review text, emotion indicator features, and statistic

---

[1] https://www.dianping.com.

features as input. Experimental results prove that our model, namely BAM (BERT+Attetion+MLP), can effectively identify fake reviews and significantly outperform baselines.

2   Related Work

2.1   Traditional Machine Learning Approaches

Fake review detection was first officially proposed by Jindal et al. [12]. They considered that fake review detection is a two-class classification problem, and used Logistic Regression, Support Vector Machine (SVM), Decision Tree, and Naive Bayes to classify reviews. Subsequently, Ott et al. [22] used Logistic Regression and SVM to conduct supervised learning detection on their dataset, which contains 400 truthful and 400 gold-standard deceptive reviews. In the same year, Li et al. [13] followed the Naive Bayes method, and added the experiments of semi-supervised methods. Soon after, Li et al. [16] analyzed the performance of linear SVM on the extended dataset based on [22]. Afterward, Barbado et al. [3] defined a feature framework and used Logistic Regression, Decision Tree, Random Forest, and Gaussian Naive Bayes to achieve fake review detection. Recently, Gutierrez-Espinoza et al. [10] used Decision Tree, Random Forest, SVM, Extreme Gradient-Boosting Tree (XGBT) to detect fake reviews, and also added the ensemble learning method to explore its improvement in detection effect.

2.2   Deep Learning Approaches

Since the traditional machine learning methods cannot learn the inner meaning of the review text, Wang et al. [32] and Li et al. [17] proposed approaches based on Convolutional Neural Networks (CNN) to learn the representation of reviews. Since feature extraction is very important in the direction of fake review detection [34], they also successfully improved the detection performance by combining behavioral information and selected features. Moreover, Ren et al. [25] designed a model, which uses CNN to learn the representation of sentences in reviews, and then uses gated Recurrent Neural Networks (RNN) to combine the sentence representations. Based on these studies, Hajek et al. [11] proposed two advanced models based on Deep Feed-Forward Neural Network (DFFNN) and CNN respectively, which integrate the word context as well as consumer emotions. In recent years, with the rapid development of Natural Language Processing (NLP) technology, the drawbacks of traditional text embedding methods have been exposed. Kennedy et al. [29] proposed to use Bidirectional Encoder Representations from Transformers (BERT) [6] for contextualized fake review detection. They showed that the contextualized vector representations extracted by BERT can better learn the meaning of the text. As a branch of machine learning algorithms, deep learning has strong self-learning ability, which makes it able to learn diverse hierarchical representations of data [35]. However, in the existing detection methods based on deep learning, there is a lack of methods that can effectively combine the extracted features and contextualized vector representations. Hence, motivated by the research status, we design a novel model BAM (BERT+Attention+MLP), which takes the review text, emotion indicators, and statistic features as input, and uses neural networks as well as multimodal fusion technology to realize the detection of fake reviews.

3   Methodology

3.1   Review Collection and Labeling

Compared with datasets in other fields (such as fake news detection), the number of fake review datasets is rare [28]. In addition, the majority of existing studies on fake reviews are aimed at English text [33]. Because of linguistic reasons, most of them cannot be directly applied to the detection of Chinese fake reviews. To our best knowledge, there is currently almost no publicly labeled Chinese fake review dataset with rich metadata available. Therefore, we have to collect and label the data by ourselves. Compared with other review platforms such as Amazon and Yelp, the review data provided by Dianping is the most comprehensive and abundant. Besides the review text, it also provides a lot of metadata (e.g. five different types of ratings and interactive information, etc.), which is helpful to explore the characteristics of fake reviews. Based on the above reasons, we select Dianping as the review collection platform and the object of our research.

According to the world city ranking[2], we randomly select 10 developed cities in China as the scope of the review collection. Furthermore, we select the restaurants at the top of the recommendation list of each city as the sources of review crawling, and the time frame for the post of all reviews is from January 2019 to December 2020.

In order to label the reviews more accurately and effectively, we analyze the reviews in detail and list 11 characteristics that fake reviews always contain. Then, according to these characteristics, the collected reviews are labeled.

(1) The review text looks like an advertisement, overly narrating objective facts.
(2) Too many photos, likes, interactions, or recommended foods are included.
(3) The text description is exaggerated, and the rating is too high or too low.
(4) To praise or disparage a restaurant, the review text compares it with other restaurants in competitive relationships.
(5) The review text contains too many consumption details.
(6) The review text describes a certain consumption experience from the perspective of others, making it more authentic.
(7) There are sentences in the review text that encourage and induce other consumers to come and consume.
(8) Reviewers use default usernames, or even are anonymous.

---

[2] https://www.lboro.ac.uk/gawc/world2020t.html.

(9) The review only shows dissatisfaction with certain aspects of the consumption experience, but all the ratings given are very low.

(10) The review contains a lot of courtesy text.

(11) The sentence structure of the review is uniform, and a fixed template is obviously applied.

When a review sample meets more than two characteristics above, it will be regarded as a fake review. Each review is labeled by two researchers, and the conflict is resolved by the third one [13]. In the end, we get the 10-city-review dataset (CR-Dataset) from 10 different developed cities in China. Besides, we also use the same approach to obtain the Chengdu-review dataset (CR-Dataset-CD), to verify the generalization ability of detection models. The introduction of the two datasets[3] is shown in Table 1.

Table 1. Dataset introduction.

Dataset name	Number of reviews in the dataset	Source
CR-Dataset	2,220 fake reviews and 10,286 truthful reviews (12,506 reviews in total)	Ten developed cities in China (exclude Chengdu)
CR-Dataset-CD	833 fake reviews and 4,767 truthful reviews (5,600 reviews in total)	Chengdu

3.2 Review Feature Extraction

In our work for fake review detection, a total of 30 features, including 10 new features and 9 redefined features (* indicates a feature that already has been defined in other references, but is redefined by us), are identified. The feature set is classified into two categories, namely emotion indicator features and statistic features. Furthermore, the statistic features are divided into four subcategories, namely text-based, information-based, reviewer-based, and interaction-based features.

Emotion Indicator Features. This type of feature is mainly obtained by analyzing the review text through NLP technology. Since many fake reviews contain strong emotional tendencies, emotional features can help distinguish fake reviews from truthful reviews [11].

1. *Maximum, minimum, average, sum, and variance of emotional words score (MAE, MIE, AE, SE, and VE)**. The text of a review often consists of several sentences, and each sentence contains several words. Using word segmentation technology and HowNet [13,36], the weight of each sentiment word can be calculated. Furthermore, the maximum, minimum, average, summary, and variance of the score weights of emotional words in the review text (five features in total) can be calculated.

---

[3] https://github.com/FakeReview/Dataset.

2. *Comprehensive emotional score (CE)*. We use SnowNLP[4], a popular natural language processing library for Chinese, to score text sentiment [5]. It implements a Bayesian classifier for the sentiment of Chinese consumer reviews. The value range of $CE(r)$ of review r is $\{CE(r)|0 \leq CE(r) \leq 1\}$.
3. *Ratio of emotional words (RE)*. By analyzing the difference between fake reviews and truthful reviews in our dataset, 32 emotional words with distinguishing significance are selected, and their proportion in each text is calculated [12,13,36].
4. *Temptation of reviews (TR) [New]*. We judge whether the review text entices other users to come and consume, and take it as one of the emotional indicator features to identify fake reviews. If a review r has words of temptation, then $TR(r) = 1$, otherwise 0.

Statistic Features. Each review is composed of text and metadata, which contain abundant statistic features that can be used to help identify fake reviews. In our work, statistic features are divided into text-based features, information-based features, reviewer-based features, and interaction-based features.

Text-Based Features. Review text has always been the research focus of fake review detection. Existing work has proved that the text-based features can help distinguish between fake and truthful reviews.

1. *Ratio of restaurant names (RRN)*. In order to achieve the effect of advertising or to compare different restaurants, some fake reviews will mention the name of the restaurant many times. This makes the name of the restaurant appear more often in fake reviews than in truthful reviews [12,13].
2. *Ratio of numerals (RN)*. The frequent appearance of numbers can be considered as a manifestation of more details in the review text. And this feature is believed to help distinguish fake reviews from all the reviews [12,29].
3. *Ratio of first-person pronouns (PP1)*. The use of the first person represents the expression of subjective consciousness, which is considered as an important feature of deceptive reviews and is widely used in fake review detection [17,19,21,24,27].
4. *Ratio of exclamation, question, and tilde mark (EQT)**. Many researchers have believed that the number of question marks ("?") and exclamation marks ("!") in reviews is an important feature for detecting fake reviews [13,19,24, 27]. And through the analysis of our dataset, we find that tilde marks ("~") can also contribute to distinguishing fake reviews.
5. *Length of review text (LR)*. Many research results have shown that the average length of fake reviews is greater than that of truthful reviews, so this feature is extracted in our work [12,13,19,24].
6. *Review content template (RCT) [New]*. Reviews that apply a fixed template have extremely high similarities and are difficult to provide effective advice to other customers, so they are classified as fake reviews [15]. If a template is applied to the content of a review r, then $RCT(r) = 1$, otherwise 0.

---

[4] https://github.com/isnowfy/snownlp.

Information-Base Features. All reviews in our dataset contain many fixed additional information, which will not change once they are posted. This kind of information contains a wealth of features that can help recognize fake reviews.

1. *Overall rating (OR).* The rating of reviews is considered to be an important factor affecting consumers' decision-making [20]. And the overall rating of fake reviews shows a phenomenon of polarization, either too high or too low [12]. Therefore, this feature is extracted to identify fake reviews.

2. *Extremity of rating (ER)*.* Positive fake reviews tend to have extremely high ratings, while negative fake reviews tend to have extremely low ratings. Many researchers have considered this feature when studying fake reviews [12,19, 21,24]. There are five different kinds of ratings, namely overall rating, taste rating, environment rating, service rating, and ingredient rating included in our dataset. Hence, the ER feature of review r is defined as

$$ER(r) = \begin{cases} 0 & \exists Rating(i) \in (2.5, 4) \\ 1 & otherwise \end{cases} \tag{1}$$

   where the value range of $Rating(i)$ of is $\{0, 0.5, 1.0, 1.5, 2.0, 2.5, 3.0, 3.5, 4.0, 4.5, 5.0\}$, and $i \in \{overall, taste, environment, service, ingredient\}$.

3. *Number of photos (NP).* The number of photos contained in fake reviews is often different from truthful reviews [30], so this feature is considered to identify fake reviews in our work.

4. *Number of recommended foods (NF) [New].* Dianping allows users to list recommended foods under the reviews. Similar to the feature of NP, this feature can also reflect the difference between fake reviews and truthful reviews.

5. *Certification tag (CT) [New].* Dianping provides certification tags for a small number of reviews, which means the users indeed posting reviews after consumption. It is difficult for ordinary fake reviews to get such certification. If a review r has a certification tag, then $CT(r) = 1$, otherwise 0.

6. *Variance of ratings (VR)*.* The five kinds of ratings given by many fake reviews show consistency, that is, all of them are very high or very low. Taking this characteristic into account, we calculate the variance of the five ratings of reviews and used it to distinguish fake reviews. In our work, the VR feature of review r is defined as

$$VR(r) = \frac{1}{5} \sum_{i} (Rating(i) - avgRating)^2 \tag{2}$$

   where $avgRating$ refers to the average of the five kinds of ratings, and $i \in \{overall, taste, environment, service, ingredient\}$.

7. *Deviation of ratings (DR)*.* For truthful reviews, the overall rating is generally determined according to the average value of four sub ratings, so the deviation between them is very small, while fake reviews have no obvious rules. Therefore, we consider the DR feature of review r to detect fake reviews, which is defined as

$$DR(r) = Rating(overall) - \frac{1}{4} \sum_{j} Rating(j) \tag{3}$$

   where $j \in \{taste, environment, service, ingredient\}$.

8. *Consumption amount (CA)* [*New*]. Dianping supports users to give the specific consumption amount in reviews. As an option, most truthful reviews do not include it through our observations. While fake reviews are likely to use it to increase or decrease the average consumption amount, thereby affecting customers' decisions. If a review r contains the consumption amount, then $CA(r) = 1$, otherwise 0.

9. *Review time (RT)* [*New*]. The restaurants in the dataset we constructed will not be open late at night, so the post time of truthful reviews should be between 7 am and 24 pm. Very few people will choose to post reviews in the early morning or late at night, while the post time of fake reviews does not necessarily follow this rule. If a review r is posted within this time interval, then $RT(r) = 1$, otherwise 0.

Reviewer-Based Features. Fake reviews are posted by spammers, so the characteristics of review publishers are an important direction to consider when detecting fake reviews.

1. *Registration on the main site (RMS).* Whether a user is registered on the main site of Dianping is believed to be able to effectively identify spammers [14]. Similarly, we consider using this feature to identify fake reviews.

2. *Anonymous review (AR).* Spammers tend to post reviews without using a username, which can help them hide themselves, especially for malicious spammers to avoid disputes with merchants [21]. If a user u is anonymous, then $AR(u) = 1$, otherwise 0.

3. *Whether the user is a VIP (VIP)* [*New*]. When a user meets certain conditions, he/she can apply to become a VIP member of Dianping. For most users, reviews posted by VIP members are more likely to be accepted, that is, they are more likely to be considered truthful. If a user u is a VIP, then $VIP(u) = 1$, otherwise 0.

4. *Default username (DU)* [*New*]. Conceiving a plausible username for each user requires some time cost for spammers. Typical spammers tend to use the default username given by the platform. If a user u uses the default username, then $DU(u) = 1$, otherwise 0.

Interaction-Based Features. In addition to the characteristics of the reviews themselves, fake reviews can also be identified from the interactive information.

1. *Number of likes (NL).* In order to better disguise themselves, fake reviews often collect a large number of likes to enhance their credibility. Therefore, some seemingly ordinary reviews sometimes have an unexpected number of likes, which can effectively help distinguish fake reviews [12,21].

2. *Number of interactions (NI)* [*New*]. Users can interact with a review by posting opinions, that is, commenting on other reviews. In [30], the authors have used the relationship between them to build heterogeneous information network to detect spammers. In our work, the number of interactions between a review and other users is counted as a feature for detecting fake reviews.

3. *Reply of restaurant merchant (RRM)* [*New*]. Dianping supports the person in charge of a restaurant (merchant) to reply to reviews, and this reply is different from the interaction of ordinary users. Some fake reviews tend to use the merchant's reply to enhance their authenticity. If the restaurant merchant replies to a review r, then $RRM(r) = 1$, otherwise 0.

3.3   Fake Review Detection

The BAM model proposed in this paper for fake review detection is shown in Fig. 1. Each of its components and their functions and principles are as follows.

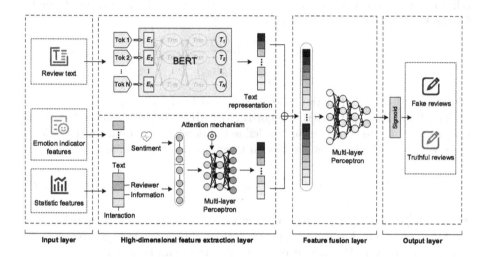

Fig. 1. BAM fake review detection model.

Input Layer. In the BAM model, any review sample (including review text and metadata) will be processed to extract the emotion indicator features E and statistic features S. After that, the two types of extracted features together with the review text T will be used as the input of the model.

High-Dimensional Feature Extraction Layer. The layer is mainly divided into two parts. In the first part, we use a state-of-the-art technology called BERT [6], which is a deep bidirectional, unsupervised language representation that only uses plain text corpus for pre-training. BERT is very useful to extract high-quality and high-dimensional language features from the text, and has been proven to help fake review detection [29].

Each word in the review text is replaced with a unique token, then they will be fed into a bidirectional transformer structure. Finally, the text T of each review is processed into a high-dimensional vector, that is, the BERT text representation B, which can be represented as

$$B = BERT(T) = \{b_1, b_2, ..., b_n\} \tag{4}$$

where n represents the dimension of the text representation vector.

In the second part, the emotion indicator features E and the statistic features S will undergo early fusion [2]. After that, the fused features $C = \{c_1, c_2, ..., c_l\}$ will be sent to an MLP with an Attention mechanism [31], where l is the length of the features.

The Attention mechanism can help increase the weight of important features and reduce the weight of unimportant features, thereby helping the model to achieve better detection performance [9,17]. We define the calculation formula for the weight α_i of each feature c_i as

$$\alpha_i = \frac{\exp(g(c_i))}{\sum\limits_{i=1}^{l} \exp(g(c_i))} \tag{5}$$

where g is a learnable function of the MLP, and $i \in \{1, 2, ..., l\}$.

The Back Propagation (BP) algorithm of MLP [10] is based on the chain rule. In the forward propagation, we define the number of neurons in the hidden layer as p, then the input of the hidden layer can be expressed as

$$h_j = \sum\limits_{i=1}^{l} w_{ij}\alpha_i + b_{ij} \tag{6}$$

where $j \in \{1, 2, ..., p\}$, w_{ij} represents the weight matrix from the input layer to the hidden layer, and b_{ij} represents the optional bias term from the input layer to the hidden layer. The output of the hidden layer can be given by

$$h'_j = f(h_j) \tag{7}$$

where f is a nonlinear activation function.

Through the MLP with an Attention mechanism, the emotion indicator features and statistic features are finally transformed into high-dimensional features A with the same dimensions as the BERT text representation B.

Feature Fusion Layer. This layer is mainly responsible for fusing the BERT text representation B with the features A extracted by the MLP with an Attention mechanism, so as to realize the expansion of the feature space in the fake review detection. This can describe the distribution of data in the feature space to a greater extent, and further achieve the purpose of improving the classification performance. The fused feature F, which can also be called multimodal representation [19], can be expressed as

$$F = B \oplus A \tag{8}$$

After that, the feature F is sent to another MLP, and the output vector β of the whole neural network is obtained.

Output Layer. In the output layer, a Fully Connected (FC) layer with a Sigmoid function is used to perform two-class classifications and output the final classification results. The representation vector β of each review can be used to calculate the probability $\hat{y}$ that the review is fake, which is given by

$$\hat{y}(\beta) = \frac{1}{1 + e^{-\beta}} \tag{9}$$

4   Experiments and Results

In this section, we evaluate the performance of the proposed BAM model for fake review detection. The experimental datasets are CR-Dataset and CR-Dataset-CD constructed by us. Based on the number of fake reviews, we randomly sampled the truthful reviews in the datasets in a stratified manner, so that the number of truthful reviews from each city is balanced with the number of fake reviews. In the analysis of the numerical results, we use the stratified 10-fold cross-validation method during the experiment and show the average value of evaluation metrics.

4.1   Different Dataset Sizes

In order to explore the impact of different dataset sizes on BAM, and to verify the low dependence of BAM on the dataset size, we carry out a comparison experiment of model detection performance for different dataset sizes. Specifically, CR-Dataset is randomly divided in this experiment, resulting in eight different dataset sizes. The number of fake reviews and truthful reviews contained in them are 800:800, 1,000:1,000, 1,200:1,200, 1,400:1,400, 1,600:1,600, 1,800:1,800, 2,000:2,000, and 2,200:2,200 respectively. The performance evaluation metrics of BAM under different dataset sizes are shown Table 2.

Table 2. The performance of BAM under different dataset sizes.

Metric	Size							
	800 × 2	1,000 × 2	1,200 × 2	1,400 × 2	1,600 × 2	1,800 × 2	2,000 × 2	2,200 × 2
Accuracy	0.9393	0.9460	0.9512	0.9575	0.9586	0.9597	0.9600	0.9602
Precision	0.9463	0.9596	0.9550	0.9660	0.9631	0.9614	0.9618	0.9611
Recall	0.9325	0.9440	0.9487	0.9488	0.9547	0.9585	0.9586	0.9596
F1-score	0.9388	0.9461	0.9514	0.9573	0.9588	0.9596	0.9600	0.9603
ΔF1-score	—	0.0073	0.0053	0.0059	0.0015	0.0008	0.0004	0.0003

It can be seen that as the number of review samples in the dataset increases, the overall evaluation metrics of BAM show an upward trend. Here we focus on the F1-score, which is the comprehensive evaluation metric. When the number of review samples is less than 1,400:1,400, the F1-score change significantly with the dataset size. The larger the dataset size, the greater the F1-score. However, when the number of review samples is higher than 1,400:1,400, the F1-score no longer increase significantly with the increase of the dataset size. This shows that when the dataset meets a certain size, the detection performance of BAM will no longer be significantly affected by it, which verifies the low dependence of the BAM model on dataset size.

4.2   Ablation Study

The BAM detection model adopts multiple inputs from the review data. In order to prove that the input of different aspects is effective, and explore which aspect of input features has a greater impact on the detection performance, the feature ablation experiment is carried out for BAM on CR-Dataset. The subsets of the feature set can be represented by the set-difference function given as

$$F \backslash F_1 = \{x | x \in F \wedge x \notin F_1\} \tag{10}$$

where F is the set with all features, F_1 is the subset of F with a particular category of features. Firstly, the performance of BAM with F, $F \backslash BERT$, $F \backslash Emotion$, and $F \backslash Statistic$ are compared. Secondly, to explore the impact of statistic features on detection performance, the performance of BAM with F, $F \backslash Text$, $F \backslash Information$, $F \backslash Reviewer$, and $F \backslash Interaction$ are compared. The evaluation metrics of BAM with different feature sets are show in Fig. 2.

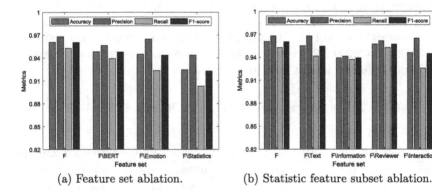

(a) Feature set ablation.   (b) Statistic feature subset ablation.

Fig. 2. The performance of BAM in feature ablation test.

It can be seen from the comparative experimental results that the BERT text representation features, emotion indicator features, and statistic features all have a great impact on the detection performance. Among them, the contribution of statistic features is obviously the largest, the contribution of emotion indicator features is the second, and the contribution of BERT text representation features is relatively the least. For the four types of statistic features, the information-based features have the largest contribution, followed by the interaction-based features. The text-based feature, and the reviewer-based features, rank third and fourth respectively.

4.3   Model Performance Comparison

In order to better highlight the performance of BAM in our work, we select seven state-of-the-art detection methods that have achieved good results in the field of fake review detection as the comparison baselines. These methods include Logistic Regression (LR) [3,12], Support Vector Machine (SVM) [4,14,16], Naive Bayes (NB) [13,18,22], Decision Tree (DT) [3,10], Random Forests (RF) [3,8], Convolutional Neural Network (CNN) [11], and Deep Feed-Forward Neural Network (DFFNN) [11]. Afterward, we carry out the experiment of the performance comparison on CR-Dataset and CR-Dataset-CD. Since the review source of CR-Dataset-CD is completely different from that of CR-Dataset, the review samples in them are subject to different feature distributions. In other words, the samples in CR-Dataset-CD are completely unknown to the classifier trained on CR-Dataset. Therefore, in addition to the general model performance comparison on CR-Dataset, in order to further verify the generalization ability of models in the real environment, we also test on CR-Dataset-CD. The final performance of BAM and seven different baseline models is shown in Fig. 3.

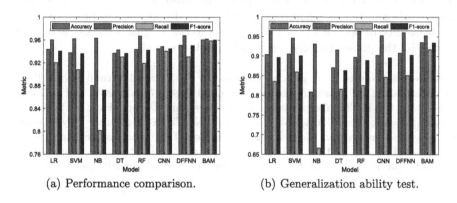

(a) Performance comparison.   (b) Generalization ability test.

Fig. 3. The performance of different models in comparison.

Through the comparative experiment results, it can be found that the performance (accuracy and F1-score) of BAM is the best in either scenario. For all models, few truthful reviews are misjudged as fake reviews, but many fake reviews are misjudged as truthful reviews. This is reflected in the fact that the precision of all models is higher than the recall, which proves that the confusion of fake reviews makes it difficult for detection models to effectively distinguish them from truthful reviews. Although there is a gap in the accuracy and recall of all models, the difference between the two is the smallest in the result of BAM, which verifies the stability of BAM. Moreover, it is also worth noting that the precision of most baseline models is not much different from that of BAM, and some models even surpass BAM in precision. However, their recall is far lower than that of BAM, which makes their F1-score inferior to that of BAM.

5   Conclusion

This paper collects a large number of reviews from the Dianping platform, and formulates a set of strict rules to label all of them. Afterward, two labeled Chinese fake review datasets are constructed, which are CR-Dataset and CR-Dataset-CD respectively. Referring to the previous work, a total of 30 emotion indicator features and statistic features are extracted. Among them, 10 features are newly defined, and 9 features are redefined by us in the field of fake review detection. Moreover, we propose a multimodal model BAM (BERT + Attention + MLP) for fake review detection, which can take both the review text and the extracted features as input. The BAM model has low dependence on the size of the training dataset, and significantly outperforms the state-of-the-art baselines in detecting fake reviews under different scenarios.

Nevertheless, our current work is only to undertake research on Chinese restaurant reviews. If enough corresponding metadata is obtained, it can also be extended to other fields. For example, in movie reviews, the ratio of movie titles in the text can be used to replace the ratio of restaurant names. This will be taken into account in future work. In addition, we will further explore the feasibility of applying our detection method in English reviews, as it could emphasize the presence of the features in English as well.

Acknowledgements. This work is supported by the National Natural Science Foundation of China (NSFC) under grant nos. 61802270, 61802271, 81602935, and 81773548. In addition, this work is also partially supported by Joint Research Fund of China Ministry of Education and China Mobile Company (No. CM20200409), the Key Research and Development Program of Science and Technology Department of Sichuan Province (No. 2020YFS0575), and the Sichuan University and Yibin Municipal People's Government University and City Strategic Cooperation Special Fund Project (Grant No. 2020CDYB-29).

References

1. Archchitha, K., Charles, E.: Opinion spam detection in online reviews using neural networks. In: Proceedings of the 19th International Conference on Advances in ICT for Emerging Regions, Colombo, Sri Lanka, pp. 1–6 (2019)
2. Baltrusaitis, T., Ahuja, C., Morency, L.P.: Multimodal machine learning: a survey and taxonomy. IEEE Trans. Pattern Anal. Mach. Intell. **41**(2), 423–443 (2019)
3. Barbado, R., Araque, O., Iglesias, C.A.: A framework for fake review detection in online consumer electronics retailers. Inf. Process. Manag. **56**(4), 1234–1244 (2019)
4. Byun, H., Jeong, S., Kwon Kim, C.: SC-Com: spotting collusive community in opinion spam detection. Inf. Process. Manag. **58**(4), 102593 (2021)

5. Chang, W., Xu, Z., Zhou, S., Cao, W.: Research on detection methods based on Doc2vec abnormal comments. Futur. Gener. Comput. Syst. **86**, 656–662 (2018)

6. Devlin, J., Chang, M.W., Lee, K., Toutanova, K.: BERT: pre-training of deep bidirectional transformers for language understanding. In: Proceedings of the 17th Annual Conference of the North American Chapter of the Association for Computational Linguistics: Human Language Technologies, Minneapolis, Minnesota, USA, pp. 4171–4186 (2019)

7. Fahfouh, A., Riffi, J., Mahraz, M.A., Yahyaouy, A., Tairi, H.: PV-DAE: a hybrid model for deceptive opinion spam based on neural network architectures. Expert Syst. Appl. **157**, 113517 (2020)

8. Fontanarava, J., Pasi, G., Viviani, M.: Feature analysis for fake review detection through supervised classification. In: Proceedings of the 4th International Conference on Data Science and Advanced Analytics, Tokyo, Japan, pp. 658–666 (2017)

9. Gao, Y., Gong, M., Xie, Y., Qin, A.K.: An attention-based unsupervised adversarial model for movie review spam detection. IEEE Trans. Multimed. **23**, 784–796 (2021)

10. Gutierrez-Espinoza, L., Abri, F., Namin, A.S., Jones, K.S., Sears, D.R.W.: Ensemble learning for detecting fake reviews. In: Proceedings of the IEEE 44th Annual Computers, Software, and Applications Conference, Madrid, Spain, pp. 1320–1325 (2020)

11. Hajek, P., Barushka, A., Munk, M.: Fake consumer review detection using deep neural networks integrating word embeddings and emotion mining. Neural Comput. Appl. **32**(23), 17259–17274 (2020). https://doi.org/10.1007/s00521-020-04757-2

12. Jindal, N., Liu, B.: Analyzing and detecting review spam. In: Proceedings of the 7th IEEE International Conference on Data Mining, Omaha, Nebraska, USA, pp. 547–552 (2007)

13. Li, F., Huang, M., Yang, Y., Zhu, X.: Learning to identify review spam. In: Proceedings of the 22nd International Joint Conference on Artificial Intelligence, Barcelona, Catalonia, Spain, pp. 2488–2493 (2011)

14. Li, H., Chen, Z., Mukherjee, A., Liu, B., Shao, J.: Analyzing and detecting opinion spam on a large-scale dataset via temporal and spatial patterns. In: Proceedings of the 9th AAAI International Conference on Web and Social Media, Oxford, England, pp. 634–637 (2015)

15. Li, J., Zhang, P., Yang, L.: An unsupervised approach to detect review spam using duplicates of images, videos and Chinese texts. Comput. Speech Lang. **68**, 101186 (2021)

16. Li, J., Ott, M., Cardie, C., Hovy, E.: Towards a general rule for identifying deceptive opinion spam. In: Proceedings of the 52nd Annual Meeting of the Association for Computational Linguistics, Baltimore, Maryland, USA, pp. 1566–1576 (2014)

17. Li, L., Qin, B., Ren, W., Liu, T.: Document representation and feature combination for deceptive spam review detection. Neurocomputing **254**, 33–41 (2017)

18. Ligthart, A., Catal, C., Tekinerdogan, B.: Analyzing the effectiveness of semi-supervised learning approaches for opinion spam classification. Appl. Soft Comput. **101**, 107023 (2021)

19. Liu, Y., Pang, B., Wang, X.: Opinion spam detection by incorporating multimodal embedded representation into a probabilistic review graph. Neurocomputing **366**, 276–283 (2019)

20. Liu, Y., Zhou, W., Chen, H.: Efficiently promoting product online outcome: an iterative rating attack utilizing product and market property. IEEE Trans. Inf. Forensics Secur. **12**(6), 1444–1457 (2017)

21. Noekhah, S., Binti Salim, N., Zakaria, N.H.: Opinion spam detection: using multi-iterative graph-based model. Inf. Process. Manag. **57**(1), 102140 (2020)

22. Ott, M., Choi, Y., Cardie, C., Hancock, J.T.: Finding deceptive opinion spam by any stretch of the imagination. In: Proceedings of the 49th Annual Meeting of the Association for Computational Linguistics: Human Language Technologies, Portland, Oregon, USA, pp. 309–319 (2011)

23. Radovanovic, D., Krstajic, B.: Review spam detection using machine learning. In: Proceedings of the 23rd International Scientific-Professional Conference on Information Technology, Piscataway, New Jersey, USA, pp. 1–4 (2018)

24. Rayana, S., Akoglu, L.: Collective opinion spam detection: bridging review networks and metadata. In: Proceedings of the 21st ACM SIGKDD International Conference on Knowledge Discovery and Data Mining, Sydney, New South Wales, Australia, pp. 985–994 (2015)

25. Ren, Y., Ji, D.: Neural networks for deceptive opinion spam detection: an empirical study. Inf. Sci. **385–386**, 213–224 (2017)

26. Ruan, N., Deng, R., Su, C.: GADM: manual fake review detection for O2O commercial platforms. Comput. Secur. **88**, 101657 (2020)

27. Shehnepoor, S., Salehi, M., Farahbakhsh, R., Crespi, N.: NetSpam: a network-based spam detection framework for reviews in online social media. IEEE Trans. Inf. Forensics Secur. **12**(7), 1585–1595 (2017)

28. Stanton, G., Irissappane, A.A.: GANs for semi-supervised opinion spam detection. In: Proceedings of the 28th International Joint Conference on Artificial Intelligence, Macao, China, pp. 5204–5210 (2019)

29. Sun, Y., Loparo, K.: Contextualized opinion spam detection. In: Proceedings of the 57th Annual Meeting of the Association for Computational Linguistics: Student Research Workshop, Florence, Italy, pp. 344–350 (2019)

30. Sun, Y., Loparo, K.: Opinion spam detection based on heterogeneous information network. In: Proceedings of the IEEE 31st International Conference on Tools with Artificial Intelligence, Portland, Oregon, USA, pp. 1156–1163 (2019)

31. Tan, J.H., Chan, C.S., Chuah, J.H.: Comic: toward a compact image captioning model with attention. IEEE Trans. Multimed. **21**(10), 2686–2696 (2019)

32. Wang, X., Liu, K., Zhao, J.: Handling cold-start problem in review spam detection by jointly embedding texts and behaviors. In: Proceedings of the 55th Annual Meeting of the Association for Computational Linguistics, Vancouver, Canada, pp. 366–376 (2017)

33. Wu, Y., Ngai, E.W., Wu, P., Wu, C.: Fake online reviews: literature review, synthesis, and directions for future research. Decis. Support Syst. **132**, 113280 (2020)

34. Xie, S., Wang, G., Lin, S., Yu, P.S.: Review spam detection via temporal pattern discovery. In: Proceedings of the 18th ACM SIGKDD International Conference on Knowledge Discovery and Data Mining, Beijing, China, pp. 823–831 (2012)

35. Yang, W., Zhang, X., Tian, Y., Wang, W., Xue, J.H., Liao, Q.: Deep learning for single image super-resolution: a brief review. IEEE Trans. Multimed. **21**(12), 3106–3121 (2019)

36. You, L., Peng, Q., Xiong, Z., He, D., Qiu, M., Zhang, X.: Integrating aspect analysis and local outlier factor for intelligent review spam detection. Futur. Gener. Comput. Syst. **102**, 163–172 (2020)

Aligning Internal Regularity and External Influence of Multi-granularity for Temporal Knowledge Graph Embedding

Tingyi Zhang[1], Zhixu Li[2], Jiaan Wang[1], Jianfeng Qu[1(✉)], Lin Yuan[1], An Liu[1], Lei Zhao[1], and Zhigang Chen[3]

[1] School of Computer Science and Technology, Soochow University, Suzhou, China
{tyzhang1,jawang1,lyuan}@stu.suda.edu.cn, {jfqu,anliu,zhaol}@suda.edu.cn
[2] Shanghai Key Laboratory of Data Science, School of Computer Science,
Fudan University, Shanghai, China
zhixuli@fudan.edu.cn
[3] iFLYTEK Research, Suzhou, China
zgchen@iflytek.com

Abstract. Representation learning for the Temporal Knowledge Graphs (TKGs) is an emerging topic in the knowledge reasoning community. Existing methods consider the internal and external influence at either element level or fact level. However, the multi-granularity information is essential for TKG modeling and the connection in between is also under-explored. In this paper, we propose the method that **A**ligning-internal **R**egularity and external **I**nfluence of **M**ulti-granularity for **T**emporal knowledge graph **E**mbedding (ARIM-TE). In particular, to prepare considerate source information for alignment, ARIM-TE first models element-level information via the addition between internal regularity and the external influence. Based on the element-level information, the merge gate is introduced to model the fact-level information by combining their internal regularity including the local and global influence with external random perturbation. Finally, according to the above obtained multi-granular information of rich features, ARIM-TE conducts alignment for them in both structure and semantics. Experimental results show that ARIM-TE outperforms current state-of-the-art KGE models on several TKG link prediction benchmarks.

Keywords: Temporal knowledge graph · Representation learning · Multi-granularity information alignment

1 Introduction

Knowledge Graph (KG) has proved its powerful strength in various downstream tasks, such as recommender systems [18,44], question answering [6,42] and relation extraction [29,37]. Despite its great benefits, most of the KGs suffer from

---

T. Zhang and Z. Li—Equal contribution.

© The Author(s), under exclusive license to Springer Nature Switzerland AG 2022
A. Bhattacharya et al. (Eds.): DASFAA 2022, LNCS 13247, pp. 149–164, 2022.
https://doi.org/10.1007/978-3-031-00129-1_10

incompleteness [36] due to the emergence of new facts. To alleviate this problem, Knowledge Graph Embedding (KGE) is often regarded as an effective approach for Knowledge Graph Completion (KGC) which focuses on deriving new facts based on existing ones. Specifically, KGE methods embed KG elements (i.e., entities and relations) into a low-dimensional vector space while preserving the original semantic and structural information [13]. Recently, Temporal Knowledge Graphs (TKGs) have gained increasing attention. In order to accurately present the temporal knowledge, TKGs include the valid time of each fact, e.g., the fact that Pierre Curie won the Nobel Prize is presented as (*Pierre Curie, wonPrize, Nobel Prize*, 1903).

Generally, Temporal Knowledge Graph Embedding (TKGE) methods can consider the important information from internal and external perspectives. The internal information models the regularity of evolution in TKG while the external information takes the randomness into consideration. Existing methods try to model the information from either element-level or fact-level: (i) Element-level modeling methods [7,8,10,41] focus on learning the evolving characteristics of the entities and relations in TKG. The external influence is usually modeled as element-level uncertainty. (ii) Fact-level modeling methods [4,20,32,45] aim at incorporating the regularity information or randomness of facts in structure and semantics along the time. However, the information from multi-granularity is integral for TKGE. Specifically, on the one hand, only modeling the important evolution characteristics of the elements, the vital influence from the neighboring facts would be neglected, e.g., the Tokyo 2020 Olympic Games postponed to 2021 due to the occurrence of COVID-19 pandemic. On the other hand, only considering the fact-level modeling would weaken the element-level evolution characteristics in representation, e.g., the Olympics are held every four years rather than a coincidence. Therefore, the alignment should be adopted to explore the semantic and structural characteristics of multi-granular information. Based on the well-modeled features in multi-granularity, alignment could effectively establish an informative and interactive mechanism in between to promote the modeling of multi-granular information and enhance the representation of the TKG. Although some recent methods [15,21] consider the element-level evolution characteristics and the fact-level concurrent or temporally adjacent facts. These methods aim at acquiring better representations in a single granularity. Besides, the external influence in multi-granularity is also ignored. To address the above issues, we propose the method that **A**ligning internal **R**egularity and external **I**nfluence of **M**ulti-granularity for **T**emporal knowledge graph **E**mbedding (ARIM-TE). ARIM-TE leverages the multi-granular information and explores consistence in between through semantic and structural alignment. In order to achieve fully alignment with abundant features, ARIM-TE models the internal regularity with external influence at both element-level and fact-level. Specifically, ARIM-TE first models element-level information as the addition between the internal regularity and external perturbation. Secondly, based on the information flowing from the element level, fact-level information is modeled as the fusion of the internal and external information by introducing the merge gate.

On the one hand, supported by the internal regularity at element level, the modeling of fact-level internal regularity includes the local property of temporal adjacent facts and the global structure revealed by the evolving local property in sequence. On the other hand, the external influence at fact level is modeled as a modification of its respective element-level external influence through introducing Gaussian distributed random perturbation so as to tackle with occasional facts. Finally, ARIM-TE conducts the information alignment in structure and semantics between the multi-granular information.

In summary, our main contributions are as follows:

- We introduce a new TKGE model, namely ARIM-TE, which is the first to simultaneously consider element-level and fact-level information with their respective internal regularity and external influence.
- We propose the merge gate to learn the fact-level information by fusing the internal information with external gaussian distributed random perturbations based on corresponding element-level information.
- We craft the alignment between multi-granular information in structure and semantics to achieve an informative TKGE.
- Experimental results on link prediction task show that our ARIM-TE outperforms the state-of-the-art KGE models on ICEWS14 and GDELT.

2   Related Work

2.1   Static Knowledge Graph Embedding

Traditional KGE methods can be generally divided into two categories: translational distance methods and semantic matching methods [34].

Translational distance methods assume the relation as the translation from the head entity to the counterpart of the tail entity in embedding space. TransE [2] is a typical translation-based model, which cannot precisely deal with complex relations (e.g., 1-to-N, N-to-1 and N-to-N relations) due to its strong assumption. TransH [35] utilizes the hyperplanes to divide the facts based on different relations. TransR [22], TransD [11] and TranSparse [12] construct different projection functions to learn a better representation of KG elements.

Semantic matching methods such as tensor factorization approaches [17,27, 28,33,43] model the KG as a three-way tensor and learn the representation of each element by tensor-decomposition. Other approaches [1,5,9,23,26,31] conduct semantic matching through various kinds of neural networks.

Different from the above methods which ignore the temporal information of facts, our paper mainly focuses on TKGE.

2.2   Temporal Knowledge Graph Embedding

TKGE methods model the information from internal and external perspectives. Some methods only pay attention to characteristics of elements. TA-DistMult [7] focuses on modeling time-aware representations of relations with a recurrent neural network. By Graph Convolution Neural Network, TeMP [38] models the structural information of entities within graph and integrates the information across time. TeRo [40] embeds the evolution of entities as rotations. ChronoR [30] models time and relation as the rotation transformation from head entity to tail entity. DE-SimplE [8] obtains the diachronic embeddings for relations and entities. DyERNIE [10] and HERCULES [25] pay attention to hyperbolic embeddings for TKG. DyERNIE leverages velocity vectors to learn dynamic entity representations on Riemannian manifolds. HERCULES utilizes temporal relations as the curvature of Riemannian manifolds. Rather than methods mentioned above, which only consider the internal information, ATiSE [41] reckons the information from external through including the randomness in additive time series decomposition.

Other methods pay attention to fact-level information. HyTE [4] projects each triple into its corresponding time hyperplane. ConT [24] learns a new core tensor for each timestamp. TTransE [20] models the temporal information as a translational vector in score function. TNTComplEx [19] and TeLM [39] utilize discrete timestamps and conduct 4th-order tensor factorization to obtain embeddings. CygNet [45] models the whole fact with copy mode and generation mode. The fact-level external influence is usually modeled with probabilistic, e.g. Know-Evolve [32] leverages temporal point process to model the occurrence of each fact.

Recent methods consider the importance of information in multi-granularity. Re-Net [15] models the joint probability for each fact based on the evolution characteristics of each element and the information on neighborhood aggregator. RE-GCN [21] embeds elements with evolved representations by considering the concurrent or temporally adjacent facts and the static property of entities in their name strings. However, they ignore the external influence in multi-granularity and the lack of alignment in structure and semantics would lead inconsistency in final representation. Different from the existing methods, ARIM-TE conducts semantic and structural alignment in multi-granularity with information which is riched of both internal and external features.

3   Problem Formulation

The notations in this paper are as follows: the lower-case letters denote the scalars, the boldface lower-case letters denote the vectors, and the boldface upper-case letters denote the matrices. Additionally, $\sigma(\cdot)$ is the sigmoid function, tanh is an activation function, $\circ$ denotes the Hadamard product. For vectors $v_1 \in \mathbb{R}^{d_1}$ and $v_2 \in \mathbb{R}^{d_2}$, $[v_1; v_2] \in \mathbb{R}^{d_1+d_2}$ is the concatenation operation.

The TKG $\mathcal{G}$ consists of temporal facts in the form of (h, r, t, τ), where $\tau \in \mathcal{T}$ represents the valid time of the fact and $\mathcal{T}$ is the set of timestamps. $h, t \in \mathcal{E}, r \in \mathcal{R}$, where $\mathcal{E}, \mathcal{R}$ denote the set of entities and relations, respectively. The TKG can be divided into several sub-KGs: $\mathcal{G} = \mathcal{G}_1 \cup \mathcal{G}_2 \cup \cdots \cup \mathcal{G}_K$ based on the valid time of the fact, K denotes the number of time steps in TKG. The sub-KG $\mathcal{G}_k$ consists of facts that are valid at time step k.

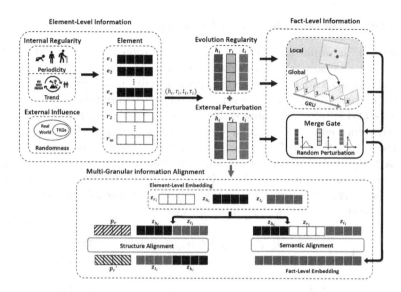

Fig. 1. The overview of our ARIM-TE, consists of element-level information modeling (Sect. 4.1), fact-level information modeling (Sect. 4.2) and multi-granular information alignment (Sect. 4.3).

Temporal Knowledge Graph Embedding aims at learning the low-dimensional representation of head entity $h \in \mathbb{R}^{d \times 1}$, tail entity $t \in \mathbb{R}^{d \times 1}$ and relation $r \in \mathbb{R}^{d \times 1}$. The task of Temporal Knowledge Graph Completion is to derive new facts based on existing ones. In this paper, we focus on interpolation problem [16] which only considers the facts that are valid on the timestamp $\tau \in \mathcal{T}$ during link prediction. We adopt entity prediction task for evaluation.

4 Methodology

We propose a three-step method, which **A**ligning internal **R**egularity and external **I**nfluence of **M**ulti-granularity for **T**emporal knowledge graph **E**mbedding (ARIM-TE). Figure 1 illustrates the overview of our ARIM-TE. As the source of alignment, the multi-granular information is fully modeled with features of both internal regularity and external influence. In the first step, ARIM-TE obtains the periodicity and trend as internal evolution regularity, together with the external perturbation as external influence for each element (Sect. 4.1). In the second step, ARIM-TE utilizes merge gate to model the fact-level information as the fusion of internal regularity and external influence (Sect. 4.2). In the last step, ARIM-TE conducts the information alignment in structure and semantics with informative features in multi-granularity. Considering the form of the fact, ARIM-TE aligned element-level information based on their structure (i.e., (head, relation, tail)). Additionally, the information in multi-granularity is aligned semantically so as to enhance the representation of the TKG (Sect. 4.3).

4.1   Element-Level Information Modeling

The semantic information of each entity and relation in TKG is varied along the time. Some characteristics evolve regularly. For example, the growth of a person shows a trend and the scenic spots usually have peak season and off season. Compared to entities, the semantic information of relations evolves at a lower rate and is relatively more stable during a short period of time [8]. Consequently, representing the regular semantic evolution of each relation with a time-agnostic vector is sufficient. In addition to the internal evolution characteristics, ARIM-TE also measures the external influence of each element to alleviate the incompleteness problem of the TKG. The information out of TKG is modeled as the external perturbation. In this paper, we utilize addition operation to combine the characteristics in learning the element-level information.

$$\mu = \begin{cases} \alpha_e sin(\rho_{s_e}\varphi_\tau + \nu_{s_e}) + \rho_{u_e}\varphi_\tau + \nu_{u_e} & \text{if entity,} \\ \alpha_r & \text{if relation.} \end{cases} \tag{1}$$

$$\delta = \begin{cases} \beta_e \zeta_{\tau_e} & \text{if entity,} \\ \beta_r \zeta_{\tau_r} & \text{if relation.} \end{cases} \tag{2}$$

$$z = \mu + \delta, \tag{3}$$

where φ_τ represents the time embeddings, μ denotes the evolution regularity of entity or relation. For entity e, α_e denotes the periodicity feature, $sin(\cdot)$ models the periodic activation function which is parameterized by ρ_{s_e} and ν_{s_e}. ρ_{u_e} and ν_{u_e} fit the semantic evolution trend for entities. Because the semantic information of each relation is relatively stable, μ simplifies into α_r when representing the relation r. Considering different semantic information stability of relation and entity, the external perturbation δ for entity and relation includes the separate time variable ζ_{τ_e}, ζ_{τ_r} and element-related characteristics β_e, β_r. Finally, the element-level embedding z is obtained by the combination of evolution regularity μ and external perturbation δ. All of the vectors mentioned above have d dimensions.

After training with the score function $score_e$, the fact $f = (h, r, t, \tau)$ in TKG is initialized with element-level information.

$$score_e = \|z_h + z_r - z_t\|. \tag{4}$$

$$z_f = \mu_f + \delta_f = [\mu_h; \mu_r; \mu_t] + [\delta_h; \delta_r; \delta_t] = [z_h; z_r; z_t]. \tag{5}$$

where μ_f and δ_f denotes the fact that initialized by evolution regularity and external perturbation, respectively.

4.2   Fact-Level Information Modeling

In TKG, the knowledge is presented in the form of fact. Though the important evolution characteristics of the elements is obtained after element-level information modeling, the truth that some facts may have correlations or causation relationships in between is still under-explored. The external influence out of the TKG should also be concerned at fact level. Therefore, the merge gate is proposed to combine the external influence with internal local property under the guidance of the global structure. For internal regularity modeling at fact level, ARIM-TE considers the local property of neighboring facts at corresponding time step based on the element-level evolution regularity. The evolving property along the time facilitates the modeling of global structure in TKG which supervise the local property at each time step. The external influence at fact-level is modeled as Gaussian distributed random perturbation based on the corresponding element-level information.

Local Property Encoding. Indeed, the occurrence of each fact would somehow influence the upcoming events. The ignored relevance between facts would lead the imprecise representation modeling. Though some events would have a long-lasting influence on the others, such as the advent of electricity which changed the way people live. The facts that happened within a short time period are more likely to have strong dependencies. Consequently, ARIM-TE adopts local property sequence to aggregate the information of neighboring facts at different time steps. According to the valid time τ, the TKG is divided into sequence $G_k(k \in \{1, 2, 3, ..., K\})$ of length K without overlaps. The specific semantic information of neighboring facts at time step K is modeled on a local hyperplane o_k with its corresponding property projection vector ω_k. For a fact (h, r, t, τ) initialized with element-level regularity characteristics μ at time step k, the local property is encoded as:

$$
\begin{aligned}
o_k(h) &= \mu_h - (\omega_k^T \mu_h \omega_k), \\
o_k(r) &= \mu_r - (\omega_k^T \mu_r \omega_k), \\
o_k(t) &= \mu_t - (\omega_k^T \mu_t \omega_k),
\end{aligned}
\tag{6}
$$

Global Structural Modeling. The modeling of local influence only focuses on the information of neighboring facts at corresponding time step, separately. However, some facts would have long-lasting effects. Therefore, ARIM-TE adopts Gated Recurrent Unit (GRU) [3] to obtain the global structure information of the TKG across the time. During global structure learning, ARIM-TE models the local property at each time step with new-coming facts as well as the vital information in history.

$$
h_k = \text{GRU}(o_k, h_{k-1}).
\tag{7}
$$

With the property projection ω_k at time step k, the hidden representation is updated to h_k which contains structural information up to time step k. In order to keep a consecutive sequence, the hidden representation h_k should be close to the property projection vector ω_{k+1} on the hyperplane o_{k+1} of the time step $k+1$. ARIM-TE introduces auxiliary loss to guide the modeling of local property in sequence with TKG's global structure.

$$loss_{aux} = \frac{1}{T-1} \sum_{k=1}^{T-1} \|\boldsymbol{h}_k - \boldsymbol{\omega}_{k+1}\|_2^2. \tag{8}$$

Internal and External Influence Merging. Because of the limited coverage of the TKG, the external information should also be considered in order to supplement the information internal. The external influence $\boldsymbol{n}_f$ for fact is modeled as Gaussian distributed random perturbation ϵ based on the element-level external influence $\boldsymbol{\delta}_f$.

$$\boldsymbol{n}_f = \boldsymbol{\delta}_f \epsilon = [\boldsymbol{\delta}_h \epsilon_h; \boldsymbol{\delta}_r \epsilon_r; \boldsymbol{\delta}_t \epsilon_t] \tag{9}$$

Through the merge gate, the internal regularity in local and global would further merged with external influence through gate mechanism, e.g. a fact(h, r, t, τ):

$$\begin{aligned}
\boldsymbol{a}_f &= [\boldsymbol{o}_k(h); \boldsymbol{o}_k(r); \boldsymbol{o}_k(t)] \\
\boldsymbol{m}_n &= \sigma(\boldsymbol{U}_n \boldsymbol{a}_f + \boldsymbol{W}_n \boldsymbol{n}_f), \\
\boldsymbol{m}_a &= \sigma(\boldsymbol{U}_a \boldsymbol{a}_f + \boldsymbol{W}_a \boldsymbol{n}_f), \\
\tilde{\boldsymbol{m}}_h &= \tanh(\boldsymbol{U}_h \boldsymbol{a}_f + \boldsymbol{W}_h(\boldsymbol{m}_a \circ \boldsymbol{n}_f)), \\
\boldsymbol{m}_f &= (1 - \boldsymbol{m}_n) \circ \tilde{\boldsymbol{m}}_h + \boldsymbol{m}_n \circ \boldsymbol{n}_f.
\end{aligned} \tag{10}$$

$\boldsymbol{a}_f$ represents internal regularity of the fact which encoded with corresponding local property. Specifically, the local property in sequence is originally fused with global structural of TKG through training with auxiliary loss. $\boldsymbol{m}_n$ focuses on keeping the necessary external influence of the fact. $\boldsymbol{m}_a$ decides the requirement of internal information. Similar to GRU, we introduce a hidden state $\tilde{\boldsymbol{m}}_h$ in merge gate. Finally, the fact-level representation $\boldsymbol{m}_f$ is calculated by mixing internal regularity and external influence of TKG. $\boldsymbol{U}_n, \boldsymbol{U}_a, \boldsymbol{U}_h, \boldsymbol{W}_n, \boldsymbol{W}_a, \boldsymbol{W}_h$ are the weight matrices of the merge gate.

4.3   Multi-granular Information Alignment

After element-level information learning, ARIM-TE learns the element-level internal evolution regularity and measures the external influence as external perturbation. Further, fact-level information is obtained through the merge gate which fuses the internal regularity with the external random perturbation based on the element-level information within facts. In order to explore the consistence of the information in multi-granularity, ARIM-TE conducts interactive information alignment in structure and semantics.

Considering the facts are denoted with relations and entities in order (i.e., the head entity, the relation and the tail entity), the structure of the facts with different types of relations are ignored through the simple combination of element-level information. Consequently, ARIM-TE aligns each type of relation in the form of fact with its related entity pairs in order (h, t) and in reverse (t, h), respectively:

$$q = p_r[z_h; z_t], \qquad (11)$$

$$q' = p_r'[z_t; z_h], \qquad (12)$$

$$s_f = [p_r; p_r'; q; q - q']. \qquad (13)$$

The entity pairs are presented with element-level information z. q, q' represents the corresponding relation-specific structural information with entity pairs in order and in reverse. The information in multi-granularity should be semantic consistence since the fact is the combination of its elements. To align the semantic information of the fact at element-level z_f and fact-level m_f, ARIM-TE conducts the introspective alignment instead of simple concatenation.

$$c_f = [z_f; z_f \circ m_f; z_f - m_f; m_f]. \qquad (14)$$

The operations $z_f - m_f$ and $z_f \circ m_f$ are element-wise subtraction and multiplication which are targeted at capturing contradiction and amplifying signals, respectively. With the concatenation of structure alignment s_f and semantic alignment c_f information, the plausibility of each potential fact is measured through a multi-layer perceptron (MLP) with the learnable parameters θ.

$$score_f = MLP([c_f; s_f]; \theta). \qquad (15)$$

4.4   Model Learning

We adopt negative sampling strategy which randomly replace the head entity or tail entity for each positive fact $f = (h, r, t, \tau) \in \mathcal{G}$. In model learning, we build the query of head entity $(?, r, t, \tau)$ and tail entity $(h, r, ?, \tau)$, then construct the candidate set $S_{f,h}$ and $S_{f,t}$, respectively. Each candidate set consists of the target key and a number of entities selected by negative sampling. Finally, the cross-entropy loss for fact $f \in \mathcal{G}$ with head query and tail query is formulated as:

$$\begin{aligned} loss_{ce} = -(\sum_{f \in \mathcal{G}} log \frac{exp(score(f))}{\sum_{h' \in S_{f,h}} exp(score(h', r, t, \tau))} \\ + log \frac{exp(score(f))}{\sum_{t' \in S_{f,t}} exp(score(h, r, t', \tau))}). \end{aligned} \qquad (16)$$

where $score(\cdot)$ could be changed in different learning steps. ARIM-TE adopts $score_e$ in element-level information learning. With $score_f$ the total loss in latter learning process is formulated as:

$$loss = loss_{ce}(score_f) + \gamma loss_{aux}. \qquad (17)$$

Specifically, $loss_{aux}$ is utilized to assist the modeling of local property with global structure, γ denotes the trade-off hyper-parameter for the auxiliary loss.

5   Experiments

5.1   Datasets

We evaluate our model on two public datasets, i.e., ICEWS14 [7] and GDELT [32]. ICEWS14 is a common benchmark in TKG evaluation which is selected from the public dataset Integrated Crisis Early Warning System (ICEWS). GDELT is the subset of the Global Database of Events, Language, and Tone which was extracted by Trivedi. ICEWS14 includes the facts that happened in 2014 and GDELT retains the facts that occurred from April 1, 2015 to March 31, 2016. Each fact in datasets is annotated with its corresponding valid timestamp. e.g., ($Barack\,Obama,\ make\,a\,visit,\ France,$ 2014.02.12). The statistics of each dataset is shown in Table 1.

Table 1. Statistics of experimental datasets.

	$\mathcal{E}$	$\mathcal{R}$	$\mathcal{T}$	Train	Valid	Test
ICEWS14	7,128	230	365	72,826	8,941	8,963
GDELT	500	20	366	2,735,685	341,961	341,961

5.2   Evaluation Metrics and Baselines

We evaluate our ARIM-TE on link prediction task. The link prediction task refers to answer two kinds of queries (i.e., $(?,\ r,\ t,\ \tau)$ and $(h,\ r,\ ?,\ \tau)$) generated from each fact $(h,\ r,\ t,\ \tau)$ in the test set. Take the head entity prediction $(?,\ r,\ t,\ \tau)$ as an example, we score and rank all potential entities in the filtered setting [2] which filters the entities according to the facts in TKG, since other entities except for the target head entity h may also link the query as a valid fact in TKG. We follow a similar approach for the tail entity prediction $(h,\ r,\ ?,\ \tau)$ to get the rank of the target entity. We report the Hit@n to show the proportion that the target entity in test set is included in the top-n of the filtered candidate list. Usually, the n is set to 1, 3 and 10. We also provide the Mean Reciprocal Rank (MRR) calculated by averaging the reciprocated rank of the target entity for each query. We compare our ARIM-TE with previous state-of-the-art methods including three static KGE models which ignore the temporal information: TransE [2], DistMult [43], SimplE [17]. We also select several competitive temporal KGE methods: ConT [24], TTransE [14], HyTE [4], TA-DistMult [7], DE-SimplE [8], ATiSE [41], TNTComplEx [19], TeRo [40] and TeMP [38].

5.3   Implementation Details

Our ARIM-TE is implemented using PyTorch. The time granularity in the experiment is set to month so as to alleviate the unbalance issue on each time step. Following the experimental set-up in DE-SimplE, the dimension of embeddings

Table 2. Comparison of different models on ICEWS14 and GDELT. The best results among all models are in bold.

Model	ICEWS14				GDELT			
	MRR	Hit@1	Hit@3	Hit@10	MRR	Hit@1	Hit@3	Hit@10
TransE	0.280	9.4	–	63.7	0.113	0.0	15.8	31.2
DistMult	0.439	32.3	–	67.2	0.196	11.7	20.8	34.8
SimplE	0.458	34.1	51.6	68.7	0.206	12.4	22.0	36.6
ConT	0.185	11.7	20.5	31.5	0.144	8.0	15.6	26.5
TTransE	0.255	7.4	–	60.1	0.115	0.0	16.0	31.8
HyTE	0.297	10.8	41.6	65.5	0.118	0.0	16.5	32.6
TA-DistMult	0.477	36.3	–	68.6	0.206	12.4	21.9	36.5
DE-SimplE	0.526	41.8	59.2	72.5	0.230	14.1	24.8	40.3
ATiSE	0.545	42.3	63.2	75.7	–	–	–	–
TNTComplEx	0.607	51.9	65.9	77.2	–	–	–	–
TeRo	0.562	46.8	62.1	73.2	–	–	–	–
TeMP-GRU	0.601	47.8	68.1	82.8	0.275	19.1	29.7	43.7
TeMP-SA	0.607	48.4	**68.4**	**84.0**	0.232	15.2	24.5	37.7
ARIM-TE	**0.624**	**56.3**	65.1	74.1	**0.503**	**42.9**	**53.1**	**64.6**

d is 100. We choose the Adam optimizer in the training process. In GDELT, the negative ratio is 5 along the whole process. The batch size is 4096 at element level, then drops to 1024 with the purpose of improving the generalization performance. The learning rate is 1e-3 at element level, then drops to 3e-4. Considering the datasets with different sizes, in ICEWS14 the negative ratio is 500 at element level, then increase to 1000. The batch size is 512 at element level. Same as in GDELT, we adopt a smaller batch size of 128 in latter learning. The learning rate for ICEWS14 is 3e-4 in the whole training process. The trade-off hyper-parameter γ for global structural loss is set to 10 for the two datasets. The two-layer MLP with 1024 and 512 hidden sizes is chosen in information alignment. The dropout rate is tuned from {0.0, 0.2, 0.4}.

5.4   Results

In this section, we report the performance of our ARIM-TE and compare it with previous state-of-the-art models on two TKG datasets: ICEWS14 and GDELT. The best link prediction evaluation results of each baseline model and our ARIM-TE are shown in Table 2. Table 2 shows that ARIM-TE outperforms the baseline methods on two datasets and achieves SOTA performance on GDELT. On GDELT, ARIM-TE has great improvements in link prediction of 22.8% in MRR, 23.8% in Hit@1, 23.4% in Hit@3 and 20.9% in Hit@10 over the best baseline method. On ICEWS14, ARIM-TE gets the improvement of 4.4% in Hit@1 and

1.7% in MRR compared to the best baseline method, which confirms the importance of multi-granular information in getting accurate representation. Compared to ICEWS14, GDELT has denser training data on each snapshot. Our ARIM-TE fully learns the local property and the global structure on GDELT with its denser training data. Additionally, with more informative interactions between entities in GDELT, the element-level information is explored adequately. The alignment of expressive multi-granularity information significantly enhances the representation of the TKG and greatly improves the link prediction performance on GDELT compared to all the baselines. Though the data in ICEWS14 is relatively sparse, ARIM-TE still gets more accurate link prediction performance which implies that our ARIM-TE could effectively model the information internal and external with alignment of multi-granular information.

5.5   Ablation Study

To better understand our ARIM-TE, we run experiments on GDELT with several variants. The power of multi-granular information is measured through the variant ARIM-TE-E which only considers the element-level information. With the purpose of validating the effectiveness of two different components in element-level evolution regularity modeling, we construct the variants which only models trend ARIM-TE-ET or periodic characteristics ARIM-TE-ES. The performance of global structure and external influence is measured in the variant ARIM-TE-FG and ARIM-TE-FO, respectively. We also test the effectiveness of merge gate in variant ARIM-TE-FM which simply models the combination of the time property internal and external with sum. Finally, we measure the power of alignment in variant ARIM-TE-A which simply combine the multi-granularity information through concatenation. The variant ARIM-TE-AS only considers the alignment in semantics to test effectiveness of the structure alignment.

Table 3. Results for different variants of our ARIM-TE on GDELT. The best results among all models are in bold.

Variant	MRR	Hit@1	Hit@3	Hit@10
ARIM-TE	**0.503**	**42.9**	**53.1**	**64.6**
ARIM-TE-E	0.166	0.0	26.8	42.0
ARIM-TE-ET	0.484	40.8	51.1	63.0
ARIM-TE-ES	0.483	40.8	51.0	62.9
ARIM-TE-FG	0.482	40.6	50.9	62.8
ARIM-TE-FO	0.461	38.3	48.8	61.2
ARIM-TE-FM	0.448	36.9	47.4	60.1
ARIM-TE-A	0.431	35.1	45.6	58.4
ARIM-TE-AS	0.472	39.6	49.7	61.9

The results in Table 3 show that the multi-granular information greatly improves the performance of our ARIM-TE. ARIM-TE gains 42.9% Hit@1 and 33.7% MRR improvements over the variant ARIM-TE-E. The alignment in semantics and structure would significantly improve the performance with 7.8% in Hit@1. The external information at fact level would improve the performance with 4.2% in MRR and 4.6% in Hit@1 than ARIM-TE-FO. The well-combination of the fact-level information internal and external would boost its performance. The merge gate effectively fuses the information with the improvements of 6.0% in Hit@1 and 5.5% in MRR than simply add them together in ARIM-TE-FM. The structure alignment is also important and improves the performance of TKG with 3.3% in Hit@1. The lack of either trend or seasonal characteristics in evolution regularity modeling would weaken the performance of TKGE. Besides, ARIM-TE could better learn the evolving structure of TKG across time with more accuracy through considering the global influence in TKG.

6   Conclusion

We propose a novel model ARIM-TE based on multi-granular information with alignment in structure and semantics to enhance the performance of representation for TKGs. To prepare for an effective alignment, we consider element-level and fact-level information from both perspectives of internal regularity and external influence. Moreover, fact-level information is modeled by the message flow from its respective elements and is further fused by an elaborated merge gate. Experimental results indicate the effectiveness and superior performance of our ARIM-TE on several TKG benchmarks.

Acknowledgement. This research is supported by the National Key R&D Program of China (No. 2018AAA-0101900), the National Natural Science Foundation of China (Grant No. 62072323, 62102276), the Natural Science Foundation of Jiangsu Province (No. BK20191420, BK20210705, BK20211307), the Major Program of Natural Science Foundation of Educational Commission of Jiangsu Province, China (Grant No. 19KJA610002, 21KJD520-005), the Priority Academic Program Development of Jiangsu Higher Education Institutions, and the Collaborative Innovation Center of Novel Software Technology and Industrialization.

References

1. Bordes, A., Glorot, X., Weston, J., Bengio, Y.: A semantic matching energy function for learning with multi-relational data. Mach. Learn. **94**(2), 233–259 (2013). https://doi.org/10.1007/s10994-013-5363-6
2. Bordes, A., Usunier, N., Garcia-Duran, A., Weston, J., Yakhnenko, O.: Translating embeddings for modeling multi-relational data. In: Neural Information Processing Systems (NIPS), pp. 1–9 (2013)
3. Chung, J., Gulcehre, C., Cho, K., Bengio, Y.: Empirical evaluation of gated recurrent neural networks on sequence modeling. In: NIPS 2014 Workshop on Deep Learning (2014)

4. Dasgupta, S.S., Ray, S.N., Talukdar, P.: HyTE: hyperplane-based temporally aware knowledge graph embedding. In: Proceedings of the 2018 Conference on Empirical Methods in Natural Language Processing, pp. 2001–2011 (2018)
5. Dettmers, T., Minervini, P., Stenetorp, P., Riedel, S.: Convolutional 2D knowledge graph embeddings. In: Thirty-Second AAAI Conference on Artificial Intelligence, pp. 1811–1818 (2018)
6. Dong, L., Wei, F., Zhou, M., Xu, K.: Question answering over Freebase with multi-column convolutional neural networks. In: Proceedings of the 53rd Annual Meeting of the Association for Computational Linguistics and the 7th International Joint Conference on Natural Language Processing (Volume 1: Long Papers), pp. 260–269 (2015)
7. Garcia-Duran, A., Dumančić, S., Niepert, M.: Learning sequence encoders for temporal knowledge graph completion. In: Proceedings of the 2018 Conference on Empirical Methods in Natural Language Processing, pp. 4816–4821 (2018)
8. Goel, R., Kazemi, S.M., Brubaker, M., Poupart, P.: Diachronic embedding for temporal knowledge graph completion. In: Proceedings of the AAAI Conference on Artificial Intelligence, pp. 3988–3995 (2020)
9. Guo, L., Sun, Z., Hu, W.: Learning to exploit long-term relational dependencies in knowledge graphs. In: International Conference on Machine Learning, pp. 2505–2514 (2019)
10. Han, Z., Chen, P., Ma, Y., Tresp, V.: DyERNIE: dynamic evolution of Riemannian manifold embeddings for temporal knowledge graph completion. In: Proceedings of the 2020 Conference on Empirical Methods in Natural Language Processing (EMNLP), pp. 7301–7316 (2020)
11. Ji, G., He, S., Xu, L., Liu, K., Zhao, J.: Knowledge graph embedding via dynamic mapping matrix. In: Proceedings of the 53rd Annual Meeting of the Association for Computational Linguistics and the 7th International Joint Conference on Natural Language Processing (Volume 1: Long Papers), pp. 687–696 (2015)
12. Ji, G., Liu, K., He, S., Zhao, J.: Knowledge graph completion with adaptive sparse transfer matrix. In: Thirtieth AAAI Conference on Artificial Intelligence, pp. 985–991 (2016)
13. Ji, S., Pan, S., Cambria, E., Marttinen, P., Yu, P.S.: A survey on knowledge graphs: representation, acquisition, and applications. IEEE Trans. Neural Netw. Learn. Syst. 1–21 (2021). https://doi.org/10.1109/TNNLS.2021.3070843
14. Jiang, T., et al.: Towards time-aware knowledge graph completion. In: Proceedings of COLING 2016, the 26th International Conference on Computational Linguistics: Technical Papers, pp. 1715–1724 (2016)
15. Jin, W., Qu, M., Jin, X., Ren, X.: Recurrent event network: autoregressive structure inference over temporal knowledge graphs. In: Proceedings of the 2020 Conference on Empirical Methods in Natural Language Processing, pp. 6669–6683 (2020)
16. Kazemi, S.M., et al.: Representation learning for dynamic graphs: a survey. J. Mach. Learn. Res. 21, 1–73 (2020)
17. Kazemi, S.M., Poole, D.: Simple embedding for link prediction in knowledge graphs. In: Advances in Neural Information Processing Systems 31: Annual Conference on Neural Information Processing Systems 2018, pp. 4289–4300 (2018)
18. Kumar, S., Zhang, X., Leskovec, J.: Predicting dynamic embedding trajectory in temporal interaction networks. In: Proceedings of the 25th ACM SIGKDD International Conference on Knowledge Discovery & Data Mining, pp. 1269–1278 (2019)
19. Lacroix, T., Obozinski, G., Usunier, N.: Tensor decompositions for temporal knowledge base completion. In: 8th International Conference on Learning Representations (2020)

20. Leblay, J., Chekol, M.W.: Deriving validity time in knowledge graph. In: Companion Proceedings of the the Web Conference 2018, pp. 1771–1776 (2018)
21. Li, Z., et al.: Temporal knowledge graph reasoning based on evolutional representation learning. In: The 44th International ACM SIGIR Conference on Research and Development in Information Retrieval, SIGIR 2021, pp. 408–417 (2021)
22. Lin, Y., Liu, Z., Sun, M., Liu, Y., Zhu, X.: Learning entity and relation embeddings for knowledge graph completion. In: Proceedings of the AAAI Conference on Artificial Intelligence, pp. 2181–2187 (2015)
23. Liu, Q., et al.: Probabilistic reasoning via deep learning: neural association models. arXiv preprint arXiv:1603.07704 (2016)
24. Ma, Y., Tresp, V., Daxberger, E.A.: Embedding models for episodic knowledge graphs. J. Web Semant. 100490 (2019)
25. Montella, S., Rojas-Barahona, L.M., Heinecke, J.: Hyperbolic temporal knowledge graph embeddings with relational and time curvatures. In: Findings of the Association for Computational Linguistics: ACL/IJCNLP 2021, pp. 3296–3308 (2021)
26. Nathani, D., Chauhan, J., Sharma, C., Kaul, M.: Learning attention-based embeddings for relation prediction in knowledge graphs. In: Proceedings of the 57th Annual Meeting of the Association for Computational Linguistics, pp. 4710–4723 (2019)
27. Nickel, M., Rosasco, L., Poggio, T.: Holographic embeddings of knowledge graphs. In: Proceedings of the AAAI Conference on Artificial Intelligence, pp. 1955–1961 (2016)
28. Nickel, M., Tresp, V., Kriegel, H.P.: A three-way model for collective learning on multi-relational data. In: Proceedings of the 28th International Conference on International Conference on Machine Learning, pp. 809–816 (2011)
29. Riedel, S., Yao, L., McCallum, A., Marlin, B.M.: Relation extraction with matrix factorization and universal schemas. In: Proceedings of the 2013 Conference of the North American Chapter of the Association for Computational Linguistics: Human Language Technologies, pp. 74–84 (2013)
30. Sadeghian, A., Armandpour, M., Colas, A., Wang, D.Z.: ChronoR: rotation based temporal knowledge graph embedding. In: Proceedings of the AAAI Conference on Artificial Intelligence, pp. 6471–6479 (2021)
31. Schlichtkrull, M., Kipf, T.N., Bloem, P., van den Berg, R., Titov, I., Welling, M.: Modeling relational data with graph convolutional networks. In: Gangemi, A., et al. (eds.) ESWC 2018. LNCS, vol. 10843, pp. 593–607. Springer, Cham (2018). https://doi.org/10.1007/978-3-319-93417-4_38
32. Trivedi, R., Dai, H., Wang, Y., Song, L.: Know-evolve: deep temporal reasoning for dynamic knowledge graphs. In: Proceedings of the 34th International Conference on Machine Learning, pp. 3462–3471 (2017)
33. Trouillon, T., Welbl, J., Riedel, S., Gaussier, E., Bouchard, G.: Complex embeddings for simple link prediction. In: Proceedings of The 33rd International Conference on Machine Learning, pp. 2071–2080 (2016)
34. Wang, Q., Mao, Z., Wang, B., Guo, L.: Knowledge graph embedding: a survey of approaches and applications. IEEE Trans. Knowl. Data Eng. **29**(12), 2724–2743 (2017)
35. Wang, Z., Zhang, J., Feng, J., Chen, Z.: Knowledge graph embedding by translating on hyperplanes. In: Proceedings of the AAAI Conference on Artificial Intelligence, pp. 1112–1119 (2014)
36. West, R., Gabrilovich, E., Murphy, K., Sun, S., Gupta, R., Lin, D.: Knowledge base completion via search-based question answering. In: Proceedings of the 23rd International Conference on World Wide Web, pp. 515–526 (2014)

37. Weston, J., Bordes, A., Yakhnenko, O., Usunier, N.: Connecting language and knowledge bases with embedding models for relation extraction. In: Proceedings of the 2013 Conference on Empirical Methods in Natural Language Processing, pp. 1366–1371 (2013)
38. Wu, J., Cao, M., Cheung, J.C.K., Hamilton, W.L.: TeMP: temporal message passing for temporal knowledge graph completion. In: Proceedings of the 2020 Conference on Empirical Methods in Natural Language Processing (EMNLP), pp. 5730–5746 (2020)
39. Xu, C., Chen, Y.Y., Nayyeri, M., Lehmann, J.: Temporal knowledge graph completion using a linear temporal regularizer and multivector embeddings. In: Proceedings of the 2021 Conference of the North American Chapter of the Association for Computational Linguistics: Human Language Technologies, pp. 2569–2578 (2021)
40. Xu, C., Nayyeri, M., Alkhoury, F., Yazdi, H.S., Lehmann, J.: TeRo: a time-aware knowledge graph embedding via temporal rotation. In: Proceedings of the 28th International Conference on Computational Linguistics, pp. 1583–1593 (2020)
41. Xu, C., Nayyeri, M., Alkhoury, F., Yazdi, H., Lehmann, J.: Temporal knowledge graph completion based on time series gaussian embedding. In: International Semantic Web Conference, pp. 654–671 (2020)
42. Xu, K., Reddy, S., Feng, Y., Huang, S., Zhao, D.: Question answering on Freebase via relation extraction and textual evidence. In: Proceedings of the 54th Annual Meeting of the Association for Computational Linguistics (Volume 1: Long Papers), pp. 2326–2336 (2016)
43. Yang, B., Yih, S.W.T., He, X., Gao, J., Deng, L.: Embedding entities and relations for learning and inference in knowledge bases. In: Proceedings of the International Conference on Learning Representations (ICLR) 2015 (2015)
44. You, J., Wang, Y., Pal, A., Eksombatchai, P., Rosenberg, C., Leskovec, J.: Hierarchical temporal convolutional networks for dynamic recommender systems. In: The World Wide Web Conference, pp. 2236–2246 (2019)
45. Zhu, C., Chen, M., Fan, C., Cheng, G., Zhang, Y.: Learning from history: modeling temporal knowledge graphs with sequential copy-generation networks. In: Proceedings of the AAAI Conference on Artificial Intelligence, pp. 4732–4740 (2021)

AdCSE: An Adversarial Method for Contrastive Learning of Sentence Embeddings

Renhao Li, Lei Duan$^{(\boxtimes)}$, Guicai Xie, Shan Xiao, and Weipeng Jiang

School of Computer Science, Sichuan University, Chengdu, China
{lirenhao,guicaixie,shanxiao,weipengjiang}@stu.scu.edu.cn,
leiduan@scu.edu.cn

Abstract. Due to the impressive results on semantic textual similarity (STS) tasks, unsupervised sentence embedding methods based on contrastive learning have attracted much attention from researchers. Most of these approaches focus on constructing high-quality positives, while only using other in-batch sentences for negatives which are insufficient for training accurate discriminative boundaries. In this paper, we demonstrate that high-quality negative representations introduced by adversarial training help to learn powerful sentence embeddings. We design a novel method named AdCSE for unsupervised sentence embedding. It consists of an untied dual-encoder backbone network for embedding positive sentence pairs and a group of negative adversaries for training hard negatives. These two parts of AdCSE compete against each other mutually in an adversarial way for contrastive learning, obtaining the most expressive sentence representations while achieving an equilibrium. Experiments on 7 STS tasks show the effectiveness of AdCSE. The superiority of AdCSE in constructing high-quality sentence embeddings is also validated by ablation studies and quality analysis of representations.

Keywords: Sentence embedding · Contrastive learning · Adversarial training

1 Introduction

Sentence embeddings are used successfully for a variety of NLP applications such as semantic similarity comparison [10], sentence clustering [26], and information retrieval [23]. As a result, plenty of methods have been proposed and obtained high-quality sentence representations with additional supervision [8,16,25]. However, it is costly with human annotation and unavailable in real-world applications.

This work was supported in part by the National Key Research and Development Program of China (2018YFB0704301-1), the National Natural Science Foundation of China (61972268), the Sichuan Science and Technology Program (2020YFG0034).

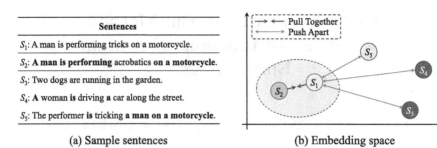

Sentences
S_1: A man is performing tricks on a motorcycle.
S_2: **A man is performing** acrobatics **on a motorcycle.**
S_3: Two dogs are running in the garden.
S_4: A woman **is** driving **a** car along the street.
S_5: The performer **is** tricking **a man on a motorcycle.**

(a) Sample sentences   (b) Embedding space

Fig. 1. A toy example of positives and negatives in contrastive learning where we bolded the same words as S_1 in the other sentences. S_1 is the input sentence; S_2 is a positive sentence of S_1 obtained by replacing synonyms; S_3, S_4 and S_5 are negatives of S_1, respectively. Among them, S_3 and S_4 are randomly sampled sentences while S_5 has a higher word coverage with S_1 compared to them. Thus, we take S_5 as a high-quality negative of S_1 for example.

Existing unsupervised sentence embedding methods with a contrastive objective have drawn much attention from researchers due to their impressive results on the semantic textual similarity (STS) tasks [6,12,17,29]. The main idea of contrastive learning is to pull semantically close neighbors (or 'positives') together and push apart non-neighbors (or 'negatives') [13].

How to construct positives and negatives for the given sentences is the key point of using contrastive learning in an unsupervised manner. Following this idea, recently developed methods, including ConSERT [29] and SimCSE [12], focus on constructing high-quality positives for the input sentences. ConSERT explores four different data augmentation strategies to generate positive views. SimCSE with an unsupervised manner applies the standard dropout twice as minimal data augmentation to compose positive pairs. For negatives, they only use all other sentences from the same batch where sentences are randomly sampled. This ignores that the quality of negatives also plays an important role in contrastive learning. Taking sentences in Fig. 1 as an example.

Example 1. It is easy for sentence embedding models to distinguish S_3 and S_4 as the negatives of S_1. However, S_5 is difficult to distinguish from S_1 due to its high word coverage with S_1. It is referred as a **hard negative** of S_1.

In general, hard negatives are more related to the input sentence in semantics compared to randomly sampled negatives. To build expressive sentence embeddings, we do not consider generating or sampling negative sentences from the input sentences. Instead, we directly obtain negative representations in embedding space by adversarial training.

In this paper, we design AdCSE: <u>Ad</u>versarial Method for <u>C</u>ontrastive Learning of <u>S</u>entence <u>E</u>mbeddings, which consists of an untied dual-encoder backbone network and a group of negative adversaries:

- *Backbone network*: most of the existing methods utilize two encoders with shared parameters as their backbone for embedding. Instead, we adopt an untied dual-encoder as the backbone network to embed the input sentences and their corresponding positives.

- *Negative adversaries*: for unsupervised training hard negatives, negative adversaries are utilized to challenge the discriminative ability of backbone network by adversarial training.

With a contrastive learning objective, these two parts of AdCSE alternately update their parameters through adversarial training. When they reach equilibrium, the most expressive sentence embeddings will be obtained.

Our main contributions can be summarized as follows.

- We design a novel unsupervised method, named AdCSE, to build high-quality sentence embeddings with a contrastive learning objective.
- We improve the quality of negatives by introducing adversaries to an untied dual-encoder in contrastive learning framework. Expressive sentence embeddings are obtained by adversarial training between hard negatives and positives.
- We evaluate AdCSE on 7 STS tasks. Empirical results demonstrate the effectiveness of AdCSE over many competitive baselines. Additionally, fine-grained analysis such as embedding quality analysis and case study further validates its superiority in constructing powerful sentence embeddings.

The rest of this paper is organized as follows. Sect. 2 presents a comprehensive review of the related work. In Sect. 3, we discuss the critical techniques of the proposed model AdCSE. By comparing with many competitive baseline methods on STS tasks, the superior performance of AdCSE is demonstrated in Sect. 4. In Sect. 5, we get deep insight into AdCSE with further analysis. Finally, we conclude our work in Sect. 6.

2   Related Work

Our work is related to the existing research on sentence embedding and contrastive learning. We introduce the related work briefly below.

2.1   Sentence Embedding

Previous methods for sentence embedding include two main categories: (1) supervised learning with labeled sentences, and (2) unsupervised sentence embedding with unlabeled sentences, while a few of them adapt for both of the settings.

Supervised Approaches. To preserve the original information from sentences as much as possible, most of the early works focus on the fusion of multi-grained sentence features by CNNs or RNNs [8,16]. Since BERT [11] showed advanced performance on a variety of NLP downstream tasks, some attempts of generating sentence embedding using pre-trained language models have been applied to sentence-pair regression tasks. However, the native derived sentence representations from BERT are proved to be collapsed. To make full use of pre-trained language model in sentence-level tasks, Reimers *et al.* [25] first designed a sentence-BERT to derive semantically meaningful sentence embeddings. With siamese and triplet network structures, sentence-BERT is able to tackle semantic similarity search using cosine similarity.

Unsupervised Approaches. To further adapt sentence embeddings to downstream tasks like STS, a series of works are proposed for the anisotropy problem brought by BERT-based sentence representations. Li et al. [19] proposed a flow-based model by mapping embeddings to a standard Gaussian latent space. While BERT-whitening introduced by Su et al. [27] is another effective way to enhance the isotropy of sentence representations, which applies whitening operation to BERT and achieves competitive results. Another line of works are based on the distributional hypothesis (Mikolov et al. [22]), where context information of the sentences is considered adequately. For instance, Skip-thought (Kiros et al. [18]) utilizes an encoder-decoder framework to sequentially predict the words of adjacent sentences. Instead of training a model to reconstruct the surface form of the input sentence or its neighbors, Logeswaran et al. [20] designed quick thoughts (QT) to predict the adjacent sentences by the current sentence.

2.2   Contrastive Learning

Contrastive learning is a kind of self-supervised technique to learn powerful representation by distinguishing samples generated by the same object against the different. Based on this intuition, approaches with contrastive learning are enabled to achieve impressive results in unsupervised visual representation learning [9,14,15].

Recently, contrastive learning has been widely applied in NLP tasks for its strong ability to train the model in an unsupervised manner. Zhang et al. [30] proposed a CNN-based model IS-BERT, which constructs positive sample pairs by maximizing the mutual information between the global sentence embedding and its corresponding local contexts embeddings. Yan et al. [29] explored four kinds of data augmentation methods for sentence-level contrastive learning in both unsupervised and supervised settings. Instead of using a siamese network with shared parameters, Carlsson et al. [6] employed an untied dual-encoder framework to counter the task bias on final layers of models imposed by pre-training objectives. To make full use of embeddings of different layers in BERT, Kim et al. [17] designed a self-guided contrastive approach which fine-tunes the BERT by making the [CLS] representation of the last layer close to its hidden states. SimCSE proposed by Gao et al. [12] applies dropout to contrastive learning of sentence embeddings which acts as minimal data augmentation and performs effectively.

3   The Design of AdCSE

In this section, we present the details of AdCSE. We first introduce the problem formulation of unsupervised sentence embedding based on contrastive learning in Sect. 3.1. Then the backbone network for embedding positive sentence pairs

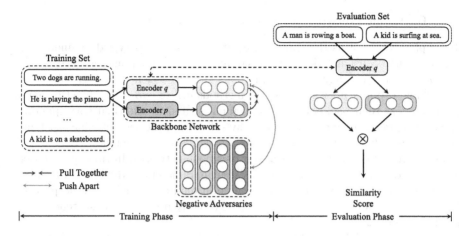

Fig. 2. Illustration of the proposed model AdCSE (Best viewed in color).

is presented in Sect. 3.2. In Sect. 3.3, we describe the negative adversaries for training hard negatives in detail. At last, the learning objective and algorithm of AdCSE are elaborated in Sect. 3.4. Figure 2 shows the architecture of AdCSE, which includes a dual-encoder backbone network along with the negative adversaries. These two parts of AdCSE interact with each other adversarially with a contrastive learning objective.

3.1 Problem Formulation

Given a set of input sentences $\mathcal{X}$, for each sentence $x_i \in \mathcal{X}$, the goal of unsupervised sentence embedding is to learn a representation $\mathbf{h}_i \in \mathbb{R}^d$ in embedding space $\mathcal{H}$. This process is abbreviated as $\mathcal{F} : \mathcal{X} \to \mathcal{H}$.

The goal of contrastive learning is to learn expressive representations by pulling semantically similar neighbors together and pushing non-neighbors apart. Specifically, for each sentence $x_i \in \mathcal{X}$, a function $\varphi(\cdot)$ is designed to map x_i to a semantically similar sentence $x_i^+ = \varphi(x_i)$ in order to compose a positive sentence pair (x_i, x_i^+). We adopt the normalized temperature-scaled cross-entropy loss (NT-Xent) as the contrastive objective. Denoting $\mathbf{h}_i \in \mathbb{R}^d$ and $\mathbf{h}_i^+ \in \mathbb{R}^d$ as the embeddings of x_i and x_i^+ mapped by the encoder in AdCSE, the training objective for (x_i, x_i^+) with a mini-batch of N sentence pairs is:

$$\mathcal{L}_i = -\log \frac{e^{\text{sim}(\mathbf{h}_i, \mathbf{h}_i^+)/\tau}}{\sum_{j=1}^{N} e^{\text{sim}(\mathbf{h}_i, \mathbf{h}_j^+)/\tau}} \tag{1}$$

where τ represents the temperature hyperparameter to control the scale of samples and $\text{sim}(\cdot)$ calculates the cosine similarity of two embeddings.

3.2   Backbone Network for Contrastive Learning

To employ NT-Xent as the training objective in an unsupervised manner, some existing works construct the positive sample x_i^+ from the input sentence x_i with sentence-level data augmentation methods such as token shuffling, token cutoff [29] and dropout masks [12]. These methods utilize two encoders with shared parameters as their backbone to embed input sentences and their corresponding enhancements respectively, therefore may suffer from the semantic inconsistency issue between the input sentence and its positive. Instead, we follow He *et al.* [14] to utilize an untied dual-encoder as our backbone networks. In this case, we take advantage of two encoders with inconsistent parameters to distinguish between x_i and x_i^+ by themselves rather than an additional data augmentation process. In other words, for a given sentence x_i, we directly use the same sentence as its positive sample x_i^+.

In AdCSE, BERT is used for the two untied encoders: encoder q and encoder p. θ_q and θ_p are denoted as their corresponding learnable parameters. Then for a positive sentence pair (x_i, x_i^+), we apply two independent BERT encoders followed by pooling layers to map the sentences to representations $(\mathbf{h}_i, \mathbf{h}_i^+)$:

$$\mathbf{h}_i = g_{\theta_q}(f_{\theta_q}(x_i)) \tag{2}$$

$$\mathbf{h}_i^+ = g_{\theta_p}(f_{\theta_p}(x_i^+)) \tag{3}$$

where f_{θ_q} and f_{θ_p} are the [CLS] representations in the last layer of the two untied BERT. g_{θ_q} and g_{θ_p} stand for two independent pooling layers which consist of linear projection and the activation function $\tanh(\cdot)$. Note that θ_q and θ_p are updated in different ways in AdCSE for the dual-encoder to learn $\mathbf{h}_i$ and $\mathbf{h}_i^+$ respectively from the inconsistent parameters. To this end, gradient descent is adopted for optimizing θ_q while a momentum update is used to smooth the evolving process of θ_p which is proved to be effective by He *et al.* [14] in the field of computer vision. With $k \in \{1, 2, ..., K\}$ where K is the total steps of training, encoder parameters after the k-th step of training are denoted as $\theta_q^{(k)}$ and $\theta_p^{(k)}$. Given a momentum m, θ_p is updated as follows:

$$\theta_p^{(k)} = m\theta_p^{(k-1)} + (1-m)\theta_q^{(k-1)} \tag{4}$$

3.3   Adversaries for Hard Negatives Training

Inspired by Hu *et al.* [15], we adopt adversarial training to unsupervised construct hard negatives for contrastive sentence embedding. To challenge the ability of distinguishing between positives and negatives in the backbone network, negative adversaries $\mathcal{N} = \{\mathbf{n}_j | \mathbf{n}_j \in \mathbb{R}^d, 1 \leq j \leq M\}$ with M negatives are randomly initialized at first. Then they keep up with $\mathbf{h}_i$ in every sample batch by iteratively updating the learnable parameters of itself θ_n through adversarial training.

To be more specific, the backbone network tends to minimize the contrastive loss by making $\mathbf{h}_i$ close to $\mathbf{h}_i^+$ while pulling apart $\mathbf{h}_i$ and $\mathbf{n}_j$. In the meanwhile,

Algorithm 1. Pseudocode of AdCSE
___
Input: $\mathcal{X}$: training set; K: the total number of training steps; α_q: learning rate of encoder q; α_n: learning rate of adversaries; m: momentum for updating encoder p; τ: temprature

Output: θ_q: learning parameters of encoder q; θ_p: learning parameters of encoder p; θ_n: learning parameters of negative adversaries

1: initialize θ_q, θ_p and θ_n;
2: shuffle samples in $\mathcal{X}$;
3: **for** $k = 1 \rightarrow K$ **do**
4:     sample a batch $\mathcal{X}_{batch}^{(k)}$ from $\mathcal{X}$ without repetition;
5:     momentum update $\theta_p^{(k)}$ using Equation 4;
6:     **for** each sample x_i in batch $\mathcal{X}_{batch}^{(k)}$ **do**
7:         obtain query embedding $\mathbf{h}_i$ by encoding x_i with encoder q;
8:         obtain positive embedding $\mathbf{h}_i^+$ by encoding x_i with encoder p;
9:     **end for**
10:    obtain embeddings of hard negatives in $\mathcal{N}$;
11:    compute the contrastive loss $\mathcal{L}$ using Equation 5;
12:    update $\theta_q^{(k)}$ using Equation 6;
13:    update $\theta_n^{(k)}$ using Equation 7;
14: **end for**
15: **return** θ_q, θ_p and θ_n
___

the negative adversaries tend to confuse the discriminator with the updated hard negatives by maximizing the contrastive loss. We believe the joint training of the backbone network and negative adversaries benefits the performance of AdCSE on evaluation.

3.4 Learning Objective and Algorithm

Based on the contrastive learning strategy, we present the following loss function of AdCSE which is derived from Equation 1:

$$\mathcal{L} = -\frac{1}{N} \sum_{i=1}^{N} \log \frac{e^{\text{sim}(\mathbf{h}_i, \mathbf{h}_i^+)/\tau}}{e^{\text{sim}(\mathbf{h}_i, \mathbf{h}_i^+)/\tau} + \sum_{j=1}^{M} e^{\text{sim}(\mathbf{h}_i, \mathbf{n}_j)/\tau}} \tag{5}$$

where N is the number of positive sample pairs in a mini-batch while M is the number of adversaries for training hard negatives. Intuitively, the above objective tends to push the negative sample embedding $\mathbf{n}_j$ closer towards the input sentence embedding $\mathbf{h}_i$ from the current minibatch. Therefore, harder negatives will be trained by the adversaries with the parameters updated. In this way, gradient decent and ascent are respectively applied to update parameters θ_q and θ_n for adversarial training:

$$\theta_q^{(k)} = \theta_q^{(k-1)} - \alpha_q \frac{\partial \mathcal{L}(\theta_q^{(k-1)}, \theta_n^{(k-1)})}{\partial \theta_q^{(k-1)}} \qquad (6)$$

$$\theta_n^{(k)} = \theta_n^{(k-1)} + \alpha_n \frac{\partial \mathcal{L}(\theta_q^{(k-1)}, \theta_n^{(k-1)})}{\partial \theta_n^{(k-1)}} \qquad (7)$$

where α_q and α_n are learning rates of encoder q and the negative adversaries respectively. With the contrastive loss of the model $\mathcal{L}$ in Eq. 5, the adversarial target mentioned in Sect. 3.3 is presented as:

$$\theta_q^\star, \theta_n^\star = \arg \min_{\theta_q} \max_{\theta_n} \mathcal{L}(\theta_q, \theta_n) \qquad (8)$$

where $\theta_q^\star$ and $\theta_n^\star$ are the parameters to equilibrate the two parts of AdCSE. We train this model in an adversarial way in the hope that the best performance of the model could be reached with the saddle point $(\theta_q^\star, \theta_n^\star)$ for this minimax problem. Based on the procedures above, we present the pseudo-code of AdCSE in Algorithm 1.

4   Experiments

We trained AdCSE on unlabeled Wikipedia corpus and evaluated its performance on 7 semantic textual similarity (STS) tasks. All experiments were conducted on a server with an RTX3090 and 24 GB memory. The model AdCSE was implemented by Python 3.6.2 with Pytorch 1.7.1 based on CUDA 11.0.

4.1   Experimental Setup

Datasets. Following Gao *et al.* [12], we used a million sentences randomly sampled from Wikipedia for our self-supervised training[1]. For evaluation, 7 STS tasks were utilized to conduct our experiments, including STS tasks 2012–2016 (STS12–STS16) (Agirre *et al.* [1–5], STS Benchmark (STS-B) (Cer *et al.* [7]) and SICK- Relatedness (SICK-R) (Marelli *et al.* [21]). Each sample in these datasets contains a pair of sentences together with a gold score between 0 and 5, indicating their ground-truth semantic similarities. We obtained all these datasets through the SentEval toolkit[2]. Please note that we only used development sets and test sets of STS tasks for evaluation so that all of the STS experiments were fully unsupervised.

Evaluation Metrics. We followed the evaluation metrics of SimCSE [12] to measure the semantic similarity of sentences. For sentence pairs in the evaluation set, we obtained their embeddings through $\mathcal{F}$ and calculated the set of

---

[1] https://huggingface.co/datasets/princeton-nlp/datasets-for-simcse/resolve/main/wiki1mforsimcse.txt.

[2] https://github.com/facebookresearch/SentEval.

predicted similarities $\hat{\mathcal{Y}}$ by cosine($\cdot$) function. Denoting the given set of ground-truth semantic scores as $\mathcal{Y}$, the ranks of $\hat{\mathcal{Y}}$ and $\mathcal{Y}$ are respectively acquired with the ranking function r($\cdot$). To assess how well the relationship between these two variables are described using a monotonic function, the spearman correlation was applied to evaluate the correlation between them, which is defined as follows:

$$\rho = \frac{\text{cov}(\text{r}(\hat{\mathcal{Y}}), \text{r}(\mathcal{Y}))}{\sigma_{\text{r}(\hat{\mathcal{Y}})}\sigma_{\text{r}(\mathcal{Y})}} \tag{9}$$

where cov($\cdot$) calculates the covariance of two variables while σ represents the standard deviations of them. The closer spearman correlation is to 1, the more similar the predicted ranked similarities from AdCSE and the ranked ground-truth are. To facilitate comparison with other baselines, we report $\rho \times 100$ as the spearman correlation in the rest of this paper.

Baselines. In our experiments, several state-of-the-art unsupervised sentence embedding methods were selected as baselines.

- **GloVe embeddings** [24] is an unsupervised learning algorithm to obtain vector representations for words. By performing aggregated global word-word co-occurrence statistics on a corpus, the method is able to generate sentence embeddings using the averaging word vectors.
- **BERT** [11] is a pre-trained language model using self-attention mechanism. Benifiting from both mask language model and next sentence prediction tasks, the model applies high-quality embeddings for various NLP tasks in a self-supervised manner.
- **BERT-flow** [19] maps embeddings to a standard Gaussian latent space to solve the anisotropy problem for sentence representations.
- **BERT-whitening** [27] enhances the isotropy of sentence representations by applying whitening operation to BERT.
- **IS-BERT** [30] is a CNN-based model which maximizes the mutual information to optimize sentence embeddings.
- **CT-BERT** [6] utilizes a dual-encoder framework together with contrastive loss to counter the task bias on final layers of models imposed by pre-training objectives.
- **ConSERT** [29] explores four kinds of data augmentation methods for sentence-level contrastive learning.
- **SG-BERT** [17] is a self-guided contrastive approach which fine-tunes the BERT by making the [CLS] output of the last layer close to its hidden states.
- **SimCSE** [12] applies dropout inside BERT as the minimal sentence-level data augmentation method in contrastive learning and acquires state-of-the-art performance on STS tasks.

Table 1. Evaluation results on the test set of STS tasks. We report the spearman correlation $\rho \times 100$ and bolded the best results. ♣: results from Reimers *et al.* [25]; ◇: results from Gao *et al.* [12]; ♡: results from Zhang *et al.* [30]; ♠: results from Yan *et al.* [29]; ★: results from Kim *et al.* [17]; baseline results without labels were implemented by ourselves.

Model	STS12	STS13	STS14	STS15	STS16	STS-B	SICK-R	Avg.
GloVe embeddings (avg.)♣	55.14	70.66	59.73	68.25	63.66	58.02	53.76	61.32
BERT$_{base}$ (cls before pooler)	21.53	32.11	21.28	37.89	44.24	20.29	42.42	31.39
BERT$_{base}$ (first-last avg.)	39.69	59.37	49.67	66.03	66.19	53.88	62.06	56.70
BERT$_{base}$-flow◇	58.40	67.10	60.85	75.16	71.22	68.66	64.47	66.55
BERT$_{base}$-whitening◇	57.83	66.90	60.90	75.08	71.31	68.24	63.73	66.28
IS-BERT$_{base}$♡	56.77	69.24	61.21	75.23	70.16	69.21	64.25	66.58
CT-BERT$_{base}$◇	61.63	76.80	68.47	77.50	76.48	74.31	69.19	72.05
ConSERT$_{base}$♠	64.64	78.49	69.07	79.72	75.95	73.97	67.31	72.74
SG-BERT$_{base}$★	68.49	80.00	71.34	81.71	77.43	77.99	68.75	75.10
SimCSE-BERT$_{base}$ (unsup.)◇	68.40	82.41	**74.38**	80.91	**78.56**	76.85	72.23	76.25
AdCSE-BERT$_{base}$ (Ours)	**70.52**	**84.10**	74.18	**82.15**	78.42	**78.32**	**73.16**	**77.26**

Implementation Details. For pure BERT, we adopted model weights released by Huggingface's Transformers[3] for evaluation. [CLS] output from BERT (cls before pooler) and the average embedding of the first and last layers (first-last avg.) are reported in this paper. For GloVe embeddings, we used averaging word vectors as sentence embeddings and report the result from Reimers *et al.* [25]. For BERT-flow, BERT-whitening and CT-BERT, we report the results reproduced by Gao *et al.* [12] which share the same evaluation setting with us. For IS-BERT and ConSERT, we report the results under unsupervised settings from their original paper (Zhang *et al.* [30], Yan *et al.* [29]). In addition, results evaluated by model named Contrastive (BT + SG-OPT) in Kim *et al.* [17] are reported as our baseline for SG-BERT.

Our implementation is based on SimCSE (Gao *et al.* [12]) and AdCo (Hu *et al.* [15])[4]. For AdCSE reported here, the max sequence length is set to 32 and dropout rate of both encoder q and encoder p are set to 0.1 just like the BERT defaults. We set learning rates for encoder q and negative adversaries to 3e–5 and 3e–3 respectively. Momentums of encoder p and negative adversaries are set to 0.995 and 0.9 respectively. The temperature τ of NT-Xent loss is set to 0.05. Besides, both batch size and the number of negatives are set to 64. We removed the projection layer in the evaluation phase to make the model more generalizing. Following SimCSE (Gao *et al.* [12]), we evaluated on the development set of STS-B every 125 steps during training and save the best model checkpoint for testing.

---

[3] https://github.com/huggingface/transformers.

[4] Our code is publicly available at https://github.com/lirenhao1997/AdCSE.

4.2   Main Results

Evaluation results on 7 STS tasks of AdCSE and other baselines are presented in Table 1, where the best results are in bold. We have the following observations:

– AdCSE yielded the best performance on most of the STS tasks. Specifically, it outperformed the previous state-of-the-art models on STS12, STS13, STS15, STSB, and SICK-R tasks, while having a small gap to them on STS14 and STS16. Taking STS12 as an example, AdCSE improved over the strongest baseline by 3.1%. Overall, AdCSE improved the previous best-averaged spearman correlation from 76.25% to 77.26%. This verifies the significance of untied dual-encoder networks and negative adversaries.
– In addition, compared with pure BERT, methods introducing contrastive learning performed better on 7 STS tasks. We attribute the collapse of pure BERT to their limitations on sentence embedding. The introduction of comparative learning can alleviate this collapse of pure BERT.

4.3   Ablation Study

To get deep insight into AdCSE and verify the validity of its two components separately, we conducted an ablation study of AdCSE.

– w/o negative adversaries: We removed the negative adversaries in AdCSE and only kept the dual-encoder backbone network for embedding. In this case, only different in-batch samples were used as negatives for contrastive learning.
– w/o untied dual-encoder: the untied dual-encoder backbone network was replaced by two encoders with shared parameters while negative adversaries were kept in this model.
– w/o both: a model without neither negative adversaries nor the untied dual-encoder (instead, using two encoders with shared parameters as the backbone) were evaluated as the baseline.

Table 2. Ablation study on AdCSE.

Model	STS12	STS13	STS14	STS15	STS16	STS-B	SICK-R	Avg.
AdCSE	**70.52**	**84.10**	**74.18**	82.15	**78.42**	78.32	**73.16**	**77.26**
- w/o negative adversaries	69.51	82.13	73.34	**82.49**	78.32	**78.60**	72.00	76.63
- w/o untied dual-encoder	67.75	77.22	70.01	80.46	77.59	76.48	68.86	74.05
- w/o both	66.02	79.83	69.90	76.42	75.54	74.33	70.07	73.16

With other settings held constantly, evaluation results of the ablated models on test set of STS tasks are shown in Table 2. According to the results, we observe that the best results of most evaluation tasks were obtained by complete AdCSE.

Removal of the untied dual-encoder and the negative adversaries separately led to a decrease in model performance, which indicates the contribution of both parts of AdCSE. It is worth noting that, the negative adversaries had about the same improvement in model performance with shared-parameter encoders (73.16 → 74.05) and the untied dural-encoder (76.63 → 77.26) as backbone, respectively, which verified the stability of adversarial training in AdCSE.

5   Further Analysis

We further validate AdCSE by analyzing sentence representations as well as the real cases in semantic similarity comparison. Impacts of both batch size and temperature are also investigated in this section.

5.1   Analysis of Embedding Space

To evaluate the quality of the embedding space of the model, we employed two metrics proposed by Wang *et el.* [28] for contrastive learning which called *alignment* and *uniformity*. For the embedding process $\mathcal{F}$, alignment ℓ_{align} is defined with the expected distance between positive pairs, while uniformity ℓ_{uniform} is the logarithm of the average pairwise Gaussian potential:

$$\ell_{\text{align}}(\mathcal{F}) \triangleq \mathbb{E}_{(x,y) \sim p_{\text{pos}}} \left[\|\mathcal{F}(x) - \mathcal{F}(y)\|_2^2 \right] \tag{10}$$

$$\ell_{\text{uniform}}(\mathcal{F}) \triangleq \log \mathbb{E}_{x,y \overset{i.i.d}{\sim} p_{\text{data}}} \left[e^{-2\|\mathcal{F}(x) - \mathcal{F}(y)\|_2^2} \right] \tag{11}$$

where p_{pos} denotes the distribution of positive pairs and p_{data} denotes the data distribution. In Fig. 3, we showed alignment and uniformity of the sentence representations from some sentence embedding methods, where the averaged evaluation results on STS tasks were also reported along with the scatter points.

It can be seen that, embeddings from BERT encoders were better in alignment compared to the contrastive-based methods SimCSE and AdCSE, while their uniformity was worse which is the main reason for their poor evaluation results on STS tasks. The uniformity of AdCSE was slightly worse than that of unsupervised SimCSE, while it had a better alignment. In general, the contrastive learning effectively improves uniformity of pre-trained embeddings whereas keeping a good alignment. Moreover, the addition of negative adversaries further improves alignment, resulting in further model performance improvements.

5.2   Case Study on Semantic Similarity

Besides the performance evaluated by spearman correlation, the discrimination ability of models could be presented through a case study on semantic similarity calculation in an intuition way. Table 3 shows real cases from the development set of STS-B task where BERT, SimCSE and AdCSE were evaluated by measuring

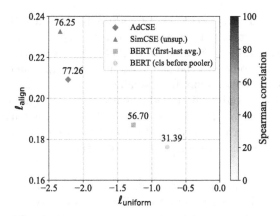

Fig. 3. Alignment and uniformity of some sentence embedding methods along with their averaged evaluation results on STS tasks. For both ℓ_{align} and ℓ_{uniform}, the lower the better.

Table 3. Case study on semantic similarity measuring, where GT is the ground truth similarity relatedness of the sentence pair, scored in [0, 5]. The predicted scores of models were obtained by mapping cosine similarities to the range of ground truth score. For each sentence pair, we bolded the predicted score closest to its ground truth.

	Sentence pair	GT	BERT (First-last avg.)	SimCSE (unsup.)	AdCSE
#1	A person drops a camera down an escelator. A man tosses a bag down an escalator.	2.75	3.75	3.30	**3.21**
#2	A woman is cutting some herbs. A woman is chopping cilantro.	2.80	4.22	3.49	**3.26**
#3	Five kittens are eating out of five dishes. Kittens are eating food on trays.	2.75	4.33	3.83	**3.10**

the cosine similarity of their output embeddings. Accroding to the results, BERT was far from the ground truth while SimCSE and AdCSE were able to measure the relatedness of sentence pairs more precisely. When it came to the hard case, SimCSE failed to handle the sentence pairs with high word coverage, such as (*Five kittens are eating out of five dishes, Kittens are eating food on trays*). In contrast, AdCSE could deal with this situation better thanks to the adversarial training with respect to hard negatives.

5.3   Influence of Batch Size and Temperature

We investigated the impact of batch size for training and temperature of $\mathcal{L}$ on model performance. Where we reported the averaged spearman correlation of the test set of all STS tasks as the model performance.

Batch Size. In some previous works of contrastive learning [9], a larger batch size may result in better performance. Thus, we experimented with different batch sizes of AdCSE on STS tasks. Note that we adjusted the number of negative adversaries accroding to the batch size in these experiments. As shown in Fig. 4(a), AdCSE benefited more from smaller batch sizes (32, 64, 96) compared to SimCSE, and achieved its best performance when the batch size was set to 64. A possible reason for this phenomenon is that a larger batch size together with more adversaries are in need of adjusting the corresponding learning rates α_q and α_n, which are hard to control for adversarial training.

Temperature. The hyperparameter temperature τ in Eq. 5 is used to control the smoothness of the distribution normalized by softmax operation. The distribution is smoothed by a large temperature while sharpened by a small one. Thus, an appropriate temperature can help the model learn from hard negatives by influencing its gradients during backpropagation. In our experiments, we explored the influence of temperature to AdCSE. As Fig. 4(b) shows, the best performance of AdCSE was reached with $\tau = 0.05$. Either too small or too large temperature affected the model's ability to learn from negative samples.

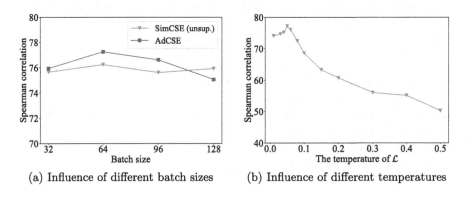

(a) Influence of different batch sizes   (b) Influence of different temperatures

Fig. 4. Performance analysis of batch size and temperature on STS tasks.

6   Conclusion

In this paper, we design a novel unsupervised sentence embedding method, named AdCSE, which consists of an untied dual-encoder backbone network and a group of negative adversaries. Employing contrastive learning as an objective, AdCSE is able to learn expressive sentence representations by adversarial training. Evaluation results on 7 STS tasks indicate that AdCSE is competitive compared with state-of-the-art methods. With ablation empirical evidence and in-depth analysis, we show the importance of each part of AdCSE and validate its effectiveness from different perspectives.

In the future, we plan to take advantage of information from different layers in BERT to improve the performance of AdCSE.

References

1. Agirre, E., et al.: SemEval-2015 Task 2: semantic textual similarity, English, Spanish and pilot on interpretability. In: SemEval@NAACL-HLT, pp. 252–263 (2015)
2. Agirre, E., et al.: SemEval-2014 Task 10: multilingual semantic textual similarity. In: SemEval@COLING, pp. 81–91 (2014)
3. Agirre, E., et al.: SemEval-2016 Task 1: semantic textual similarity, monolingual and cross-lingual evaluation. In: SemEval@NAACL-HLT, pp. 497–511 (2016)
4. Agirre, E., Cer, D.M., Diab, M.T., Gonzalez-Agirre, A.: SemEval-2012 Task 6: a pilot on semantic textual similarity. In: SemEval@NAACL-HLT, pp. 385–393 (2012)
5. Agirre, E., Cer, D.M., Diab, M.T., Gonzalez-Agirre, A., Guo, W.: *SEM 2013 shared task: semantic textual similarity. In: *SEM, pp. 32–43 (2013)
6. Carlsson, F., Gyllensten, A.C., Gogoulou, E., Hellqvist, E.Y., Sahlgren, M.: Semantic re-tuning with contrastive tension. In: ICLR (2021)
7. Cer, D.M., Diab, M.T., Agirre, E., Lopez-Gazpio, I., Specia, L.: SemEval-2017 Task 1: semantic textual similarity multilingual and crosslingual focused evaluation. In: SemEval@ACL, pp. 1–14 (2017)
8. Chen, Q., Zhu, X., Ling, Z., Wei, S., Jiang, H., Inkpen, D.: Enhanced LSTM for natural language inference. In: ACL, pp. 1657–1668 (2017)
9. Chen, T., Kornblith, S., Norouzi, M., Hinton, G.E.: A simple framework for contrastive learning of visual representations. In: ICML, pp. 1597–1607 (2020)
10. Conneau, A., Kiela, D., Schwenk, H., Barrault, L., Bordes, A.: Supervised learning of universal sentence representations from natural language inference data. In: EMNLP, pp. 670–680 (2017)
11. Devlin, J., Chang, M., Lee, K., Toutanova, K.: BERT: pre-training of deep bidirectional transformers for language understanding. In: NAACL-HLT, pp. 4171–4186 (2019)
12. Gao, T., Yao, X., Chen, D.: SimCSE: simple contrastive learning of sentence embeddings. In: EMNLP (2021)
13. Hadsell, R., Chopra, S., LeCun, Y.: Dimensionality reduction by learning an invariant mapping. In: CVPR, pp. 1735–1742. IEEE Computer Society (2006)
14. He, K., Fan, H., Wu, Y., Xie, S., Girshick, R.B.: Momentum contrast for unsupervised visual representation learning. In: CVPR, pp. 9726–9735 (2020)
15. Hu, Q., Wang, X., Hu, W., Qi, G.: AdCo: adversarial contrast for efficient learning of unsupervised representations from self-trained negative adversaries. In: CVPR, pp. 1074–1083 (2021)
16. Kalchbrenner, N., Grefenstette, E., Blunsom, P.: A convolutional neural network for modelling sentences. In: ACL, pp. 655–665 (2014)
17. Kim, T., Yoo, K.M., Lee, S.: Self-guided contrastive learning for BERT sentence representations. In: ACL/IJCNLP, pp. 2528–2540 (2021)
18. Kiros, R., et al.: Skip-thought vectors. In: NeurIPS, pp. 3294–3302 (2015)
19. Li, B., Zhou, H., He, J., Wang, M., Yang, Y., Li, L.: On the sentence embeddings from pre-trained language models. In: EMNLP, pp. 9119–9130 (2020)
20. Logeswaran, L., Lee, H.: An efficient framework for learning sentence representations. In: ICLR (2018)

21. Marelli, M., Menini, S., Baroni, M., Bentivogli, L., Bernardi, R., Zamparelli, R.: A SICK cure for the evaluation of compositional distributional semantic models. In: LREC, pp. 216–223 (2014)
22. Mikolov, T., Sutskever, I., Chen, K., Corrado, G.S., Dean, J.: Distributed representations of words and phrases and their compositionality. In: NeurIPS, pp. 3111–3119 (2013)
23. Palangi, H., et al.: Deep sentence embedding using long short-term memory networks: analysis and application to information retrieval. IEEE ACM Trans. Audio Speech Lang. Process. **24** (2016)
24. Pennington, J., Socher, R., Manning, C.D.: Glove: global vectors for word representation. In: EMNLP, pp. 1532–1543 (2014)
25. Reimers, N., Gurevych, I.: Sentence-BERT: sentence embeddings using Siamese BERT-networks. In: EMNLP-IJCNLP. pp. 3980–3990 (2019)
26. Reimers, N., Schiller, B., Beck, T., Daxenberger, J., Stab, C., Gurevych, I.: Classification and clustering of arguments with contextualized word embeddings. In: ACL, pp. 567–578 (2019)
27. Su, J., Cao, J., Liu, W., Ou, Y.: Whitening sentence representations for better semantics and faster retrieval. CoRR abs/2103.15316 (2021)
28. Wang, T., Isola, P.: Understanding contrastive representation learning through alignment and uniformity on the hypersphere. In: ICML, pp. 9929–9939 (2020)
29. Yan, Y., Li, R., Wang, S., Zhang, F., Wu, W., Xu, W.: ConSERT: a contrastive framework for self-supervised sentence representation transfer. In: ACL/IJCNLP, pp. 5065–5075 (2021)
30. Zhang, Y., He, R., Liu, Z., Lim, K.H., Bing, L.: An unsupervised sentence embedding method by mutual information maximization. In: EMNLP, pp. 1601–1610 (2020)

HRG: A Hybrid Retrieval and Generation Model in Multi-turn Dialogue

Deji Zhao[1], Xinyi Liu[1], Bo Ning[1(✉)], and Chengfei Liu[2]

[1] School of Information Science and Technology, Dalian Maritime University,
Dalian, China
{dejizhao,lxy869281725,ningbo}@dlmu.edu.cn
[2] Swinburne University of Technology, Melbourne, Australia
cliu@swin.edu.au

Abstract. In multi-turn dialogue generation, the generated response should consider the content before the current turn of dialogue. Due to multiple turns, it is difficult to maintain the context consistency by using only a few previous turns of the dialogue indiscriminately. Except for the context information, we can retrieve additional candidates from historical contexts, according to semantic similarity. Therefore, in this paper, we integrate the historical information into the generative model called HRG. The HRG model can generate a response by using both context information and retrieved historical candidates, which contain richer information such as theme and latent information. We encode contexts, current turn and historical information separately to find the most important turns and give the current turn a higher level of attention. Then we propose a hierarchical fusion encoder to integrate the retrieval information through a KL divergence gate dynamically. Finally, we conduct experiments on the Ubuntu large-scale English multi-turn dialogue community dataset and Daily dialogue dataset. The results show that our hybrid model performs well on both automatic evaluation and human evaluation compared with the existing baseline models.

Keywords: Hybrid retrieval method · Multi-turn dialogue · Dialogue system · Text generation

1 Introduction

Dialogue systems intend to use big data to provide users with fast and concise answers. They can be divided into single-turn dialogue systems and multi-turn dialogue systems. Because multi-turn dialogue is more popular and widely used in actual situations, it has been widely used in real-world applications, including customer service systems, personal assistants, and chatbots.

In order to make a dialogue system behave more like humans, two methods are mainly used: one is based on retrieval and the other is based on generation. The retrieval method is widely used in single-turn dialogues to quickly respond

A. Bhattacharya et al. (Eds.): DASFAA 2022, LNCS 13247, pp. 181–196, 2022.
https://doi.org/10.1007/978-3-031-00129-1_12

to user requests and get answers. As a multi-turn dialogue is much more complicated than a single-turn dialogue, it needs to handle more contextual information and extract effective information from context for generating a better response for the multi-turn dialogue. So how to effectively use complex context to generate a more appropriate response is an important problem in the multi-turn dialogue generation task.

In multi-turn conversation model, it is very difficult to distinguish which turn is important in the context dialogue. Hierarchical Recurrent Encoder-Decoder (HRED) [14,17] is widely used in multi-turn dialogue generation. This method captures context information through a hierarchical encoder and decoder structure. HRED adds an additional encoder to the traditional encoder-decoder model. Compared with ordinary recurrent neural network (RNN) language model, it can capture the context, reduce the calculation steps between adjacent sentences, and realize multiple turns of dialogue generation. However, the performance of the model will be damaged if the context is handled indiscriminately, and the model often fails to grasp the specific sentence information. Then in order to solve this problem, the recent relevant context with self-attention (ReCoSa) [25] model uses long-distance self-attention mechanism to model the context and response separately, so as to find out which word is important. It uses the self-attention to encode and decode the information, the parameters of both encoder and decoder are learned by maximizing the averaged likelihood of the training data. However, due to the introduction of long attention to calculate the weight of the context only, the model tends to lose the valid information of the current turn and often replies to repeated information.

Consider more specific information is proved very effective on multi-turn conversation model in previous research. Short-text Topic-level Attention Relevance with Biterm Topic Model (STAR-BTM) [24] integrates the topic information in the dialogue generation. However, the topic information is an implicit representation in a conversation, few valid information can be learned. Conditional Historical Generation (CHG) [26] focuses more on the integration of historical information. But CHG can't learn additional information and repeated useless historical dialogue information of a session can't improve the performance of the model.

Table 1. An example in the Daily dialogue dataset.

Context	Examples
Utterance 1	What do you think of our price ?
Utterance 2	Your price has gone up sharply
Utterance 3	Prices of the raw materials have been raised, we have to adjust the price.
Current turn	I agree with you there, but your price is unreasonable.
Retrieval 1	The cost of production has been skyrocketing in recent years.
Retrieval 2	Our price depends on the quantity of the order.
Response	Moderate prices will bring about large sales and more profit.

On the other hand, it is also a big challenge to use valid information in context to retrieve relevant historical information and integrate it with the model. There are many attempts in the integration of both the retrieval methods and generative methods in single-turn dialogue generation. Zhu et al. [28] use adversarial training methods to integrate retrieval information in single-turn dialogue. They propose a Retrieval-Enhanced Adversarial Training approach to make better use of N-best response candidates. However, using adversarial training in text generation tasks will greatly increase the difficulty of the model training. So far, no one has done the combination of the retrieval methods and generative methods in multi-turn dialogues, possibly the complexity caused by multiple turns.

Table 1 is an example in the Daily dialogue dataset. People say more than one sentence in a turn. This makes the multi-turn dialogue models more difficult to capture the most effective information. Utterance 1, utterance 2 and utterance 3 are former turns, the current turn is the asking question, retrieval 1 and retrieval 2 are the answers obtained by the retrieval method according to the contextual information. The words in red color represent the response by integrating important words of different turns and historical information in word level. We argue that the current turn is the most important and applying an attention mechanism on the current turn can get different weights of previous turns in sentence level. From Table 1, we can see the response is related to the current turn information, the context information and historical information can be used as a supplement. Utterance 1 and utterance 2 are talking about the price of the product. Utterance 3 is talking about the reasons for the rise in commodity prices. In current turn the customer wants a better price. Retrieval 1 and retrieval 2 serve as supplements to context, finding more background knowledge and hoping to reach a deal with customers in terms of quantity. The response can give the right answer according to context and historical topic, like 'scale' and 'profit' information. Due to the employment of retrieving historical information, the response is more diverse. The historical information can give response a correct direction. Without introducing historical information, it is difficult to guarantee the diversity of the generated dialogue and the consistency of the context.

In this paper, we propose the hybrid retrieval and generation Model (HRG) model, which is a multi-turn dialogue generation model that combines generated information and retrieved historical information. The model can obtain the information of the same scene according to the semantic similarity. We separate the contexts into two parts, one is current turn information and the other is previous turn information. We argue the current turn information is the most important context. The motivation of this paper is that the combination of generative and retrieval methods will make the response generated by the model more in line with the actual context, and the model can capture more information and find the latent features to maintain context semantic consistency. Compared with the existing multi-turn dialogue methods, we use the retrieved historical information and propose a novel fusion method to integrate historical information hierarchically. Compared with the existed single-turn dialogue method, our model can fit multi-turn tasks. In our proposed HRG model, we use KL divergence to measure the difference between the retrieval information and the context information and give different weights to different contexts, and finally use hierarchical fusion encoder to dynamically integrate the retrieved information. KL divergence is widely used to generate image and increase generalization ability in computer vision.

In the retrieval stage, we first use the semantic similarity method to encode different sentences and then find the most relevant answer based on the given sentence. At the same time, due to the very large amount of data, our retrieval method uses the distilled robustly optimized BERT pretraining approach (RoBERTa) [10]. In the generative stage, we use the Transformer encoder block to capture context information, current turn information, and retrieved historical information separately and then use the hierarchical fusion encoder to integrate the retrieval information, finally, we send the fusion vectors into the decoder to get the response. In our experiment, we use two public datasets, the Daily dialogue dataset, and the English Ubuntu community dataset to evaluate our model. The results show that our model can produce more flexible answers and more appropriate responses than existing baseline models. The contextual consistency of the dialogue is maintained after incorporating retrieval information, which shows that our method is effective and reasonable.

The contributions of this paper are summarized as follows:

- We propose the HRG model, which integrates the retrieved historical information on the basis of the generative model. Due to the employment of historical information, our model can find latent information and maintain the consistency of the dialogue context.
- We propose a novel hierarchical fusion encoder to integrate the retrieved historical information through a KL divergence gate dynamically. Using hierarchical fusion encoder can effectively utilize historical information.
- We conduct experiments on the Ubuntu large-scale English multi-turn dialogue community dataset and Daily dialogue dataset. The experimental results show that our hybrid model performs well on both automatic evaluation and human evaluation compared with the existing baseline models.

2 Related Work

Most of the existing dialogue systems are based on the retrieval method. They use this method to find rich information and respond smoothly. They choose the information and discourse in the previous turn as input and choose the context-sensitive natural response. However, in the generative stage, the answers generated by the generative-based method are more flexible and can cope with complex contexts.

2.1 Single-Turn Response Matching

Retrieval-based methods choose a response from candidate responses. Retrieval-based methods focus more on message-response matching. Matching algorithms have to overcome semantic gaps between messages and responses [3]. Early studies of retrieval-based chatbots focus on response selection [20,21], where only the single-turn message is used to select a proper response. They calculate the similarity between the context and the answer vector which is encoded by long short-term memory (LSTM). Recent semantic method calculate the similarity of sentences in pre-train model, such as [2,10]. Azzalini et al. [1] try to use entity linkage to improve semantic quliaty.

2.2 Multi-turn Response Matching

In multi-turn retrieval matching, the existing work is to splice the utterances in context and match the final response. In multi-turn response selection, current message and previous utterances are taken as input. The model selects a response from the repositories which is the most relevant to the whole context. Identifying important information in the previous contexts is very important. Lowe et al. [11] encoded the context and candidate response into a context vector and a response vector through RNN and then computed the similarity between the two vectors. Selecting the previous utterances in different strategies and combining them with current messages is proved effective [23]. The next improvement is to establish a strategy to select context information. Yan et al. [23] select the previous utterances in different strategies and combined them with current message to get the answer.

2.3 Single-Turn Response Generation

Early end-to-end open domain dialogue generation work was inspired by neural machine translation [13,16]. A good generator often can cope with complex contexts and produce a fluent, grammatical answer. The widely used method is sequence to sequence (Seq2Seq) model [18], it uses RNN or LSTM as encoder and decoder. Li et al. [6] try to use meta-learning to train a better encoder. Given a context sequence, a recurrent neural network-based encoder is first utilized to encode each message, and then in the decoding stage, another RNN decoder is used to generate the response. The parameters of both encoder and decoder are learned by maximizing the averaged likelihood of the training data. Due to the introduction of the attention mechanism, the performance of seq2seq model is better, and the decoder can use attention mechanism to utilize sentence information differently, which improves the accuracy of generation.

2.4 Multi-turn Response Generation

Despite many existing research works on single-turn dialogue generation, multi-turn dialogue generation has gained increasing attention [9,12,27]. One reason is that it is more accordant with the real application scenario, such as customer services chatbot. Because more information is considered, customers usually change topics in multi-turn dialogue, which has brought huge challenges to researchers in this field. In the multi-turn dialogue generation task, hierarchical recurrent encoder-decoder architectures (HRED) [14] are proposed to capture context information. Later, Serben propose HRED model with hidden variables, called VHRED [15]. This method introduces hidden variables into the intermediate state in the previous HRED to improve the diversity of the generated dialogue. ReCoSa [25] model can find the most important information of word-level in multi-turn dialogue. ReCoSa use long self-attention mechanism [19] to model multi-turn dialogues. Hierarchical self-attention network (HSAN) [5] can find the most important words and utterances in the context simultaneously.

However, although these methods fully model the context information, they have not considered the diversity of the generated sentences and the consistency of the context information. Now, the data-driven single-turn dialogue system can be roughly divided into two categories, one is generative tasks, and the other is retrieval tasks. In generative tasks, the same as the multi-turn dialogue generation method, it has improved on the basic Seq2Seq model. Recently, using the decoder in transformer block [19] is widely used because of its speed advantage, but this method often generates repetitive and meaningless responses.

There are some works that integrate retrieval and generative methods. Zhu et al. [28] used adversarial training methods to combine generative sentences with sentences obtained from retrieval to get good results, but this method is based on a single-turn of dialogue. And using generative adversarial networks (GAN) is hardly training. Similar to the fusion retrieval information, STAR-BTM [24] integrates the topic information in the dialogue generation. CHG [26] proposes a merchant history conversation selection module which can copy words directly from the relevant history conversations. However, it is difficult to retrieve valid information and fuse them into the context, no one has merged the retrieved information in multi-turn of dialogue according to our current research.

3   HRG Model

In this section, we illustrate our model in detail, whose architecture is depicted in Fig. 1. The red bar represents the distribution of retrieved historical information. The knowledge base contains all candidate context information.

3.1   Current Turn and Context Attention

To find the most important turns in contexts, we get the weights of the contexts through the attention mechanism. Given contexts:

$$Contexts = \{utterance_1, utterance_2, ..., utterance_n\},$$

each $utterance_i$ in contexts can be represented as $utterance_i = \{w_1, w_2, ..., w_n\}$, where w_i represents a word in an utterance. All sentences in one session are encoded by Bi-GRU to get the representation of the sentence separately, then get the word level importance coefficient and obtain every sentence representation through current turn and context attention. In addition, we integrate the

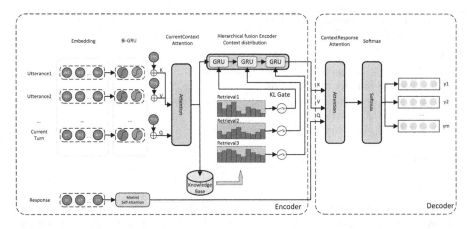

Fig. 1. The architecture of Hybrid Generation and Retrieval model. Each turn passes the Bi-GRU to get the sentence representation and add the turn level position encoding.

positional embedding to the contexts so as to find the importance of different contexts at the sentence level. The final contexts representation is shown as follow:

$$Contexts = \left\{ \left[\overrightarrow{h}_1, \overleftarrow{h}_1 \right] + TOS_1, ..., \left[\overrightarrow{h}_T, \overleftarrow{h}_T \right] + TOS_t \right\},$$

where TOS_i is the turn of sentence embedding, which represents positional embedding to indicate the order of turns, $\overrightarrow{h}$ and $\overleftarrow{h}$ express Bi-directional sentence vector respectively. The current turn representation is:

$$CurrentTurn = \left\{ \left[\overrightarrow{h}, \overleftarrow{h} \right] \right\},$$

The context attention layer and attention formula are calculated as follows,

$$Attention = softmax \left(\frac{QK^T}{\sqrt{D_K}} \right) V \tag{1}$$

$$h^{CurCon} = Attention \left(CurrentTurn, Contexts \right), \tag{2}$$

where h^{CurCon} is the current turn and contexts attention representation.

3.2 Retrieval

In order to obtain the supplement information of related contexts, we use distilled RoBERTa [10] language model. First, we set up message-response pairs, and use RoBERTa to encode each message into a 768-dimension vector. Then we use each message as a query to calculate the semantic similarity in the vector library, to search the most matching message in the range k. Finally, during the training process, according to the matching score, the semantic algorithm

Algorithm 1. Retrieval algorithm

Input: h^{Con}, h^{Cur}
Output: h^{CurRet}

1: $h^{CurCon} = Attention\,(CurrentTurn, Contexts)$
2: **for** epoch in number of epochs **do**
3:     $h^{Ret} = SemanticSimilarity(h^{ConCur}, Knowledgebase)$
4:     $h^{Ret} = SelfAttention(h^{Ret})$
5:     $h^{CurRet} = Attention(h^{Cur}, h^{Ret})$
6: **end for**
7: **return** h^{CurRet}

returns the most matching message called candidate context. In addition, our retrieval model is also trained with the model, which can not only improve the quality of model generation but also improve the retrieval performance of similar sentences. So our model can enrich the context and response fields.

Since the training set used in the retrieval is the same as the training set used in the model training, in order to avoid the retrieval method from seeing the ground truth, we exclude the retrieved sentences from the ranking. The retrieval function between retrieval and attention vectors similarity is calculated by cosine similarity.

Algorithm 1 presents how to find semantic retrieval vectors, h^{Con}, h^{Cur} represent context vector and current turn vector. And input the retrieval knowledge, through our algorithm, we can finally get the most relevant retrieval vectors. The similarity formula mentioned in the algorithm is pre-encoded as shown in the above formula.

3.3   Hierarchical Fusion Encoder

In order to solve the semantic consistency of context and response, it is very important to integrate the information obtained from semantic retrieval into context. Based on this principle, we propose a hierarchical divergence fusion encoder and use KL divergence to measure the word distribution between context and retrieval context. KL divergence can measure the difference between two distributions. Then we use a hierarchical encoder to fuse retrieval and context information differences.

Through the retrieval part, we get the context candidate and response candidate. Taking context level fusion as an example, in order to detect the difference between the retrieved candidate context information and context information, we let the candidate context pass KL gate, and the formula is as follows:

$$D_{KL}\left(h^{Con}|h^{Ret}\right) = \sum_i h_i^{Con} \log \frac{h_i^{Con}}{h_i^{Ret}} \qquad (3)$$

$$Gate = D_{KL}\left(h^{Con}|h^{Ret}\right), \qquad (4)$$

where i represents the ith utterance in the context. From formula 2 we get the context's attention weights, we use it to control the amount of information in each context. Our hierarchical fusion encoder can dynamically fuse context in each time step in order to get the final contexts vector. The representation of fusion $context_i$ is calculated as follows:

$$h^{Fus} = h^{Con} * W_{Con} + (1 - Gate) * h^{Ret}, \tag{5}$$

where W_{Con} represent the attention weights in context attention, h^{Fus} represents the final fusion retrieval vector.

Each fusion context vector should pass the GRU to get the final contexts representation. Figure 2 shows fusion information is auto-regressive and gets the final time step representation. Then we concatenate contexts representation and current turn representation. The response representation is similar to context, each word can be seen as the context in contexts, and get the fusion representation as to the decoder input.

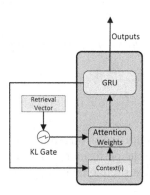

Fig. 2. Context hierarchical fusion encoder.

Table 2. Human evaluation results of mean score, proportions of three levels (+2, +1, and 0 represent excellent, good and average respectively).

	Appropriateness				Informativeness				Grammaticality			
	Means	+2	+1	0	Means	+2	+1	0	Means	+2	+1	0
Seq2Seq	0.58	24.7	8.6	66.7	0.42	8.5	24.8	66.7	1.49	73.1	3.1	23.8
ReCoSa	0.71	30.5	10.2	59.3	0.50	10.8	28.5	60.7	1.71	82.6	5.5	11.9
STAR-BTM	0.72	28.2	15.1	56.7	0.56	12.1	31.7	56.2	1.74	84.9	4.2	10.9
CHG	0.80	32.1	15.6	52.3	0.61	14.1	32.8	53.1	1.77	85.3	6.7	8.0
HSAN	0.85	33.5	17.8	48.7	0.64	15.9	32.2	51.9	1.74	86.8	2.0	13.0
HRG	**0.88**	**34.8**	**18.1**	47.1	**0.77**	**21.8**	**33.4**	44.8	**1.88**	**90.1**	**7.8**	2.1

3.4   Decoder

Finally, the values in the encoder and the decoder are jointly trained. At this time, we combine the fused contexts vectors and current turn vectors together and sent them into the decoder. The context response attention can be expressed by the following equation:

$$h^{CR} = Attention\left(h_{Con}^{Fus}, h_{Res}\right), \tag{6}$$

where h_{Con}^{Fus} represents the hierarchical fusion of contexts and current turn, h_{Res} represents response vector. Given an input response: $Response = \{y_1, y_2, \ldots, y_m\}$, the likelihood of the corresponding response sequence is:

$$P\left(Y|C;\theta\right) = \prod_{t=1}^{T'} P\left(y_t|C, y_1, \ldots, y_{t-1}; \theta\right) \tag{7}$$

After passing the context and response attention, we can generate words of response through softmax:

$$P\left(y_t|C, y_1, \ldots, y_{t-1}; \theta\right) = P\left(y_t|h^{CR}; \theta\right) \tag{8}$$

4   Experiment

4.1   Datasets

We use the Ubuntu community multi-turn dialogue dataset [11] and Daily dialogue dataset [8] to evaluate the performance of our proposed model. We use the official script to mark positive sample training.

4.2   Baselines

Seq2Seq: Sequence to sequence model with attention mechanism [18].

HRED: Hierarchical Recurrent Encoder-Decoder [14]. Using this method, multiple dialogue turns are modeled separately.

VHRED: VHRED is a variant of HRED. In order to increase robustness, implicit variable information is added [15].

ReCoSa: Relevant context with self-attention [25]. Use long distance attention method can capture important word information.

STAR-BTM: Multi-turn dialogue generation integrate the topic information [24].

CHG: Utilizing historical dialogue representation learning and historical dialogue selection mechanism [26].

HSAN: A hierarchicalself-attention network, which attends to find the important words and utterances in context simultaneously [5].

4.3 Experiment Settings

In order to make a fair comparison between all baseline methods, the hidden layer size is set to 512, the batch size is set to 32 and8 heads attention is used. We use Pytorch to run all models on three Tesla T4 GPUs.

4.4 Human Evaluation

We randomly sampled 200 messages from the Ubuntu test set to conduct the human evaluation as it is extremely time-consuming. We recruit 5 evaluators to judge the response from three aspects [4].

- Appropriateness: a response is logical and appropriate to its message.
- Informativeness: a response has meaningful information relevant to its message.
- Grammaticality: a response is fluent and grammatical.

4.5 Automatic Evaluation

We use perplexity [14], BLEU [22] and Dist-1, Dist-2 [7] to evaluate the diversity of our responses, where Dist-k is the number of different k-grams after normalization of the total number of words in the response.

   We have done a lot of experiments on both datasets to verify the effectiveness of our retrieval model. In order to ensure the fairness and consistency of the baseline model, we conducted several groups of experiments, which are: comparison between our complete model (with retrieval information) and other baseline models without retrieval information (as shown in Table 3), in order to prove the importance of introducing semantic retrieval information; Comparison between our complete model (with retrieval information) and other baseline models with simple fusion retrieval information (as shown in Table 4), in order to ensure the fairness between our model and baseline model and prove the effectiveness of using hierarchical fusion encoder; A simple way to introduce retrieval information is to use the same semantic retrieval mechanism to find the most relevant retrieval sentence of the current turn and context attention features and directly spliced together. What's more, we also conduct an ablation experiment on the Daily dialogue dataset (as shown in Table 5) to prove our hierarchical encoder is efficient. The KL gate is replaced by a simple attention.

Table 3. Performance of different models on Ubuntu dataset and daily dialogue dataset.

Dataset	Ubuntu				Daily dialogue			
Model	PPL	BLEU	Dist-1	Dist-2	PPL	BLEU	Dist-1	Dist-2
Seq2Seq	110.384	0.5026	0.903	1.091	170.228	0.7923	0.814	0.886
HRED	180.528	0.6045	1.021	2.112	160.540	0.7206	0.972	1.132
VHRED	151.610	0.5991	1.125	2.132	154.681	0.8315	0.815	1.117
ReCoSa	96.057	1.2485	1.718	3.768	97.174	1.3111	1.443	3.156
STAR-BTM	104.893	1.3303	1.601	4.525	99.221	1.2875	1.330	3.326
CHG	98.532	1.2814	1.625	4.438	104.258	1.2832	1.448	3.143
HSAN	95.225	1.3087	1.670	4.547	89.144	1.3015	1.291	3.112
HRG	**89.735**	**1.3401**	**1.781**	**4.558**	**77.243**	**1.4353**	**1.512**	**3.361**

Table 4. Comparison of our complete model and simple fusion retrieval information model on Ubuntu and daily dialogue dataset.

Dataset	Ubuntu				Daily dialogue			
Model	PPL	BLEU	Dist-1	Dist-2	PPL	BLEU	Dist-1	Dist-2
Seq2Seq+R	120.127	0.5515	0.955	1.130	160.551	0.7322	0.913	1.125
HRED+R	188.483	0.6147	1.117	2.274	191.247	0.8091	1.056	1.281
VHRED+R	155.229	0.6213	1.255	2.288	171.128	0.7963	1.257	1.276
ReCoSa+R	101.553	1.2961	1.525	3.558	125.988	1.3253	1.418	3.201
STAR-BTM+R	105.284	1.3398	1.775	4.556	134.365	1.3461	1.428	3.257
CHG+R	98.232	1.2925	1.704	4.462	111.745	1.3002	1.451	3.285
HSAN+R	100.165	1.2926	1.770	4.447	138.851	1.3873	1.501	3.306
HRG	**89.735**	**1.3401**	**1.781**	**4.558**	**77.243**	**1.4353**	**1.512**	**3.361**

Table 5. Ablation experiment on daily dialogue dataset.

Model	Hierarchical Encoder	Retrieval	KL Gate	PPL↑	BLEU↓	Dist-1↓	Dist-2 ↓
HRG				19.876	0.0935	0.051	0.188
HRG	✓	✓		11.439	0.0322	0.026	0.115
HRG	✓	✓	✓	0	0	0	0

5 Analysis

The different test results are shown in the table. We use two evaluation criteria, one is human evaluation, and the other is mainstream evaluation algorithms based on machine translation. Under the same dataset, the hierarchical fusion mechanism through KL gate improves the fusion degree of other information, which also shows that the performance of the model can be improved by fusing more information, and the semantic-based information retrieval can recall the most relevant candidates. This shows that the model can help improve the quality of the generated dialogue when more information is considered. The following is a detailed analysis.

The human evaluation focuses more on areas that are not covered by the automatic assessment. In the human evaluation in Table 2, our method has the highest average score in terms of appropriateness, informativeness, and grammaticality. This shows that after integrating the retrieved information, the model can capture more background context and generate richer responses. Due to the use of contextual turn level attention, the model has also achieved good results in maintaining the consistency of the context of the generated response. From Table 2, we find that the improvement of appropriateness is not large. When the model is generated, it has the limitation of retrieval candidate background context direction, which leads to the lack of syntax flexibility. It can be seen from the human evaluation table that our model performs very well at +2 on the index of informativeness, which shows that it is very effective to consider richer information. Compared with the CHG and STAR-BTM model, which merge additional information, we can find that using hierarchical fusion to retrieve information is more effective than historical information and topic information. Overall, there has been an improvement in all indicators.

In automatic evaluation, in order to prevent the influence of retrieval information on the performance of the model, we also add retrieval information to the other baseline. As shown in Table 3, our model has achieved good results on the PPL index and Dist of both datasets, we think this has a great relationship with the basis encoder and decoder architecture and the integration of the retrieved information. Besides, the quality of the sentences retrieved by semantic similarity is much better. Thanks to the powerful RoBERTa model, these sentences are more in line with the context, which is also helpful to the later experimental results. We conduct several experiments with ReCoSa method, but we can't get the performance of the BLEU score like the ReCoSa paper, but other indicators we follow the paper. Dist index can measure the richness of response. On the richness index of response model, our model performs well, due to the fusion of more information, which makes the sentences generated by the model very rich. In Table 3, our HRG model performs well in all indicators. Compared with the HSAN, which uses hierarchical self-attention, our method uses turns attention is much better. In terms of the degree of additional information fusion, our hierarchical fusion model can dynamically measure the information difference according to the data distribution by KL divergence, and fuse according to the importance of different turns on two datasets.

As shown in Table 4, we also add semantic retrieval information to the baseline model, but only simply integrate the retrieval information. A simple way to introduce retrieval information is to use the same semantic retrieval mechanism to find the most relevant retrieval sentence of the current turn of information, and directly splice the vector obtained with the vector of the current turn. The results show that after the simple introduction of the retrieval module, all baseline models improve the performance of BLEU, which also proves that considering more information can improve the richness of the model. The performance of STAR-BTM+R model is improved after integrating the retrieval distribution, this shows that the semantic retrieval features roughly contain the implicit information of the latitude of the topic, and the retrieved sentence information is richer. What's more, in the baseline models, using our separate method to split context makes the baseline models' performance improve, this also shows that our split method is very effective. Maybe because we simply fuse the retrieval information, the PPL index has increased a little compared with other baseline models. Overall, all response's quality of baseline model has a certain improvement, which proves the effectiveness of integrating retrieval background information. Besides, using our hierarchical fusion encoder is more effective than simply integrating method. Both automatic evaluation and human evaluation show that the context consistency of the conversation and the richness of the answers are improved after the retrieval information is added.

In our ablation experiment on daily dialogue dataset as shown in Table 5, we remove retrieval information and hierarchical encoder simultaneously, the results of different indicators are similar to ReCoSa method, but due to we use the turn of sentence embedding (TOS), our HRG model performs a little better, which shows the necessity of TOS embedding. After removing the KL gate and replacing it with a simple attention, we can see the experiment results improve a lot compared with the former ablation experiment, which shows that using KL gate can measure the difference between two distributions effectively.

6   Conclusion and Future Work

In this paper, we propose a novel hierarchical fusion encoder that combines the retrieved information in multi-turn of dialogue generation. We encode current turn and context information respectively and add turn of sentence embedding (TOS) to enhance the sentence level attention. Using hierarchical fusion encoder can effectively utilize retrieved historical information according to the different weights of context sentences. Experiments show the effectiveness of this method. In future research, we will explore how to introduce knowledge graph information or other external forms of knowledge into multi-turn dialogue generation.

References

1. Azzalini, F., Jin, S., Renzi, M., Tanca, L.: Blocking techniques for entity linkage: a semantics-based approach. Data Sci. Eng. **6**(1), 20–38 (2021)
2. Devlin, J., Chang, M.W., Lee, K., Toutanova, K.: Bert: pre-training of deep bidirectional transformers for language understanding. In: Proceedings of the 2019 Conference of the North American Chapter of the Association for Computational Linguistics: Human Language Technologies, vol. 1 (Long and Short Papers), pp. 4171–4186 (2019)

3. Hu, B., Lu, Z., Li, H., Chen, Q.: Convolutional neural network architectures for matching natural language sentences. In: Advances in Neural Information Processing Systems 27: Annual Conference on Neural Information Processing Systems 2014, 8–13 December 13 2014, Montreal, Quebec, Canada, pp. 2042–2050 (2014)
4. Ke, P., Guan, J., Huang, M., Zhu, X.: Generating informative responses with controlled sentence function. In: Proceedings of the 56th Annual Meeting of the Association for Computational Linguistics, pp. 1499–1508 (2018)
5. Kong, Y., Zhang, L., Ma, C., Cao, C.: Hsan: A hierarchical self-attention network for multi-turn dialogue generation. In: ICASSP 2021–2021 IEEE International Conference on Acoustics, Speech and Signal Processing (ICASSP), pp. 7433–7437. IEEE (2021)
6. Li, C., Yang, C., Liu, B., Yuan, Y., Wang, G.: LRSC: learning representations for subspace clustering. In: Proceedings of the AAAI Conference on Artificial Intelligence, vol. 35, pp. 8340–8348 (2021)
7. Li, J., Galley, M., Brockett, C., Gao, J., Dolan, B.: A diversity-promoting objective function for neural conversation models. In: NAACL HLT 2016, The 2016 Conference of the North American Chapter of the Association for Computational Linguistics, pp. 110–119 (2016)
8. Li, Y., Su, H., Shen, X., Li, W., Cao, Z., Niu, S.: DailyDialog: a manually labelled multi-turn dialogue dataset. In: Proceedings of the Eighth International Joint Conference on Natural Language Processing, IJCNLP 2017, Taipei, Taiwan, November 27– December 1, 2017 - Volume 1: Long Papers, pp. 986–995 (2017)
9. Liang, Y., Meng, F., Zhang, Y., Chen, Y., Xu, J., Zhou, J.: Infusing multi-source knowledge with heterogeneous graph neural network for emotional conversation generation. In: Thirty-Fifth AAAI Conference on Artificial Intelligence, AAAI 2021, pp. 13343–13352 (2021)
10. Liu, Y., et al.: A robustly optimized bert pretraining approach. arXiv preprint arXiv:1907.11692 (2019)
11. Lowe, R., Pow, N., Serban, I., Pineau, J.: The ubuntu dialogue corpus: a large dataset for research in unstructured multi-turn dialogue systems. In: Proceedings of the SIGDIAL 2015 Conference, The 16th Annual Meeting of the Special Interest Group on Discourse and Dialogue, pp. 285–294 (2015)
12. Oluwatobi, O., Mueller, E.: DLGNet,: a transformer-based model for dialogue response generation. In: Proceedings of the 2nd Workshop on Natural Language Processing for Conversational AI (2020)
13. Ritter, A., Cherry, C., Dolan, W.B.: Data-driven response generation in social media. In: Proceedings of the 2011 Conference on Empirical Methods in Natural Language Processing, EMNLP 2011, pp. 583–593 (2011)
14. Serban, I., Sordoni, A., Bengio, Y., Courville, A., Pineau, J.: Building end-to-end dialogue systems using generative hierarchical neural network models. In: Proceedings of the AAAI Conference on Artificial Intelligence, vol. 30 (2016)
15. Serban, I., et al.: A hierarchical latent variable encoder-decoder model for generating dialogues. In: Proceedings of the AAAI Conference on Artificial Intelligence, vol. 31 (2017)
16. Shang, L., Lu, Z., Li, H.: Neural responding machine for short-text conversation. In: Proceedings of the 53rd Annual Meeting of the Association for Computational Linguistics, ACL 2015, pp. 1577–1586 (2015)
17. Sordoni, A., Bengio, Y., Vahabi, H., Lioma, C., Grue Simonsen, J., Nie, J.Y.: A hierarchical recurrent encoder-decoder for generative context-aware query suggestion. In: Proceedings of the 24th ACM International on Conference on Information and Knowledge Management, pp. 553–562 (2015)

18. Sutskever, I., Vinyals, O., Le, Q.V.: Sequence to sequence learning with neural networks. In: Advances in Neural Information Processing Systems, pp. 3104–3112. The MIT Press, London (2014)
19. Vaswani, A., et al.: Attention is all you need. In: Advances in Neural Information Processing Systems NIPS 2017, pp. 5998–6008 (2017)
20. Wang, H., Lu, Z., Li, H., Chen, E.: A dataset for research on short-text conversations. In: Proceedings of the 2013 Conference on Empirical Methods in Natural Language Processing, pp. 935–945 (2013)
21. Wang, S., Jiang, J.: Learning natural language inference with LSTM. In: NAACL HLT 2016, The 2016 Conference of the North American Chapter of the Association for Computational Linguistics: Human Language Technologies, San Diego, June 12–17, 2016, pp. 1442–1451 (2016)
22. Xing, C., et al.: Topic aware neural response generation. In: Proceedings of the AAAI Conference on Artificial Intelligence, vol. 31 (2017)
23. Yan, R., Song, Y., Wu, H.: Learning to respond with deep neural networks for retrieval-based human-computer conversation system. In: Proceedings of the 39th International ACM SIGIR conference on Research and Development in Information Retrieval, pp. 55–64 (2016)
24. Zhang, H., Lan, Y., Pang, L., Chen, H., Ding, Z., Yin, D.: Modeling topical relevance for multi-turn dialogue generation. In: Proceedings of the Twenty-Ninth International Joint Conference on Artificial Intelligence, IJCAI (2020)
25. Zhang, H., Lan, Y., Pang, L., Guo, J., Cheng, X.: ReCoSa: detecting the relevant contexts with self-attention for multi-turn dialogue generation. In: Proceedings of ACL 2019, vol. 1: Long Papers, pp. 3721–3730 (2019)
26. Zhang, W., et al.: Multi-turn dialogue generation in e-commerce platform with the context of historical dialogue. In: Proceedings of the 2020 Conference on Empirical Methods in Natural Language Processing: Findings, pp. 1981–1990 (2020)
27. Zhao, X., Wu, W., Xu, C., Tao, C., Zhao, D., Yan, R.: Knowledge-grounded dialogue generation with pre-trained language models. In: Proceedings of the 2020 Conference on Empirical Methods in Natural Language Processing, EMNLP 2020, 16–20 November 2020, pp. 3377–3390 (2020)
28. Zhu, Q., Cui, L., Zhang, W., Wei, F., Liu, T.: Retrieval-enhanced adversarial training for neural response generation. In: Proceedings of the 57th Conference of the Association for Computational Linguistics, ACL, pp. 3763–3773 (2019)

FALCON: A Faithful Contrastive Framework for Response Generation in TableQA Systems

Shineng Fang[1], Jiangjie Chen[1], Xinyao Shen[1], Yunwen Chen[3],

and Yanghua Xiao[1,2(✉)]

[1] Shanghai Key Laboratory of Data Science, School of Computer Science,
Fudan University, Shanghai, China
{snfang19,jjchen19,xinyaoshen19,shawyh}@fudan.edu.cn
[2] Fudan-Aishu Cognitive Intelligence Joint Research Center, Shanghai, China
[3] DataGrand Inc., Shanghai, China
chenyunwen@datagrand.com

Abstract. In a practical TableQA system, response generation is a critical module to generate a natural language description of the SQL and the execution result. Due to the complex syntax of SQL and matching issues with table content, this task is prone to produce factual errors. In this paper, we propose FALCON, a FAithfuL CONtrastive generation framework to improve the factual correctness of generated responses. FALCON forces the generation model to identify examples with factual errors in the latent space during training and takes contrastive examples into consideration during inference. We also propose two new automatic metrics to further evaluate faithfulness specialized to this task. Experimental results show FALCON brings a favorable performance improvement on both automatic and human evaluation amongst various baseline methods (The code of FALCON is released at https://github.com/whuFSN/FalCon).

Keywords: Response generation · Factual correctness · Contrastive learning

1 Introduction

With extensive research on the Natural Language Interface of Databases (NLIDB), semantic parsing task has made significant progress in recent years [4,12,27,31]. However, in a practical TableQA system [37], *response generation* (RG) is also a critical module to interact with users' natural language questions. As shown in Fig. 1, a practical TableQA system consists of semantic parsing and response generation module. RG requires generating a natural language description of the SQL and the execution result. In general, we argue that such a response generation module is necessary due to the following two reasons: 1) The generated response could help users verify whether the query result is consistent with the original question; 2) The generated response provides a concise and easy-to-understand summary about the result table.

However, previous SQL-Interface-related studies rarely investigate RG in any systematic way. Previous work [3,13,33] mainly focus on the SQL-to-Question task, which generates natural language descriptions interpreting the meaning of a given SQL. Different from SQL-to-Question, the unique challenges of RG lie in: 1) In addition to the

A. Bhattacharya et al. (Eds.): DASFAA 2022, LNCS 13247, pp. 197–212, 2022.
https://doi.org/10.1007/978-3-031-00129-1_13

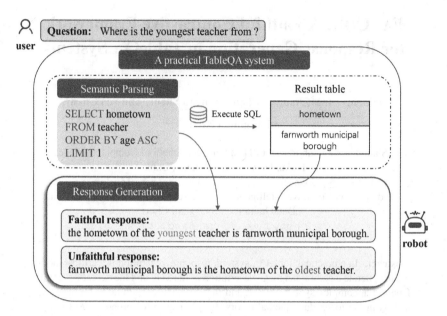

Fig. 1. An illustration of a TableQA system, which consists of semantic parsing and response generation module. RG takes the SQL and result table as input to generate the response. The unfaithful response is highlighted in red, which is not factually consistent with the input. (Color figure online)

SQL, RG also takes the execution result as input, which needs an explanation of table content; 2) The benchmark of RG is CoSQL [35], which includes more complex SQL grammar (e.g., nested queries, multi-table queries).

Besides fluency and grammaticality, factual correctness is also an important factor when evaluating the quality of generated responses. While advanced Natural Language Generation (NLG) techniques have been successful in producing fluent text, these methods still face great challenge caused by factual incorrectness [11,17]. As shown in Fig. 1, an unfaithful response is easily generated by T5 [26], a popular pre-trained generation model, which misunderstands the "*youngest teacher*" as the "*oldest teacher*". This error implies that the generic T5 is unaware of sorting by age (in ascending order) specified in the input SQL. Unfortunately, such an error is not accidentally generated. Our new task has several inherent properties that make models prone to produce factual errors. First, responses contain rich logical information, which is mainly specified in aggregate functions (e.g., MAX), logical operators (e.g., <), and abstract fields (e.g., dept_name) in SQL. Second, content matching between SQL and the table is also a challenge.

In this paper, we propose a FAithfuL CONtrastive generation framework (FALCON) to improve factual correctness of generated responses. Contrastive learning [6,24,30] learns representations by contrasting positive pairs and negative pairs. We intentionally introduce contrastive learning into RG to handle factual errors. Specifically, we utilize a series of heuristic rules to construct samples with factual errors which we call *imposters*. During training, in addition to optimizing naive training objective, our models attempt to distinguish between imposters and ground truths in the latent space. During inference, an imposters-contrastive decoding method is proposed to avoid generating unfaithful responses.

We also proposed two metrics to evaluate the faithfulness of generated responses. Previous research [10] has shown that ROUGE [20], a widely used metric based on n-gram overlap, is not always a valid factuality metric. This motivates us to propose two new faithfulness-aware metrics. The first metric is *Logical Score*, which mainly reflects the logical matching degree between the generated response and input SQL. Specifically, we parse the generated response into SQL components by a trained Response-to-SQL model. Then Logical Score is obtained by calculating the matching degree between parsed SQL and input SQL. The second metric is *Consistency Score*, which is estimated by a consistency classifier trained on positive and negative samples. This metric represents the consistency between the generated response and the input.

In brief, our contributions can be summarized as follows:

- We are the first work to formulate the response generation task in a TableQA system. Focusing on factual correctness, we propose a faithful contrastive generation framework (FALCON) to boost the faithfulness of responses.
- Two metrics, *Logical Score* and *Consistency Score*, are proposed to evaluate the factual correctness of generated responses. Experiments show they have a high correlation with human judgment.
- Extensive evaluations demonstrate the state-of-the-art (SoTA) performance of FALCON on both automatic and human evaluation.

2 Related Work

SQL-to-Question. A natural language generation task that generates a natural language description from the SQL. Iyer *et al.* [14] employ LSTM networks with attention to generate descriptions of the SQL which regards the SQL as a sequential text. Xu *et al.* [33] represent the SQL as a directed graph and use a Graph2seq model to encode this graph-structured SQL. Cai *et al.* [3] leverage the syntactic structure of abstract syntax trees (AST) to encode the SQL with a novel type-associated encoder.

Factual Correctness. Previous work on factual correctness focus on the summarization task. Cao *et al.* [5] propose to improve summarization models by leveraging open information extraction. Falke *et al.* [9] apply natural language inference to evaluate the factual correctness of the generated summaries. Zhang *et al.* [39] utilize reinforcement learning to optimize the faithfulness of radiology summaries, whose reward is obtained from a trained fact extractor. In the Data-to-Text task, Wang *et al.* [32] generate a faithful table description by introducing a new table-text optimal-transport matching loss and a table text embedding similarity loss.

Contrastive Learning. Contrastive learning is to learn representations by contrasting positive pairs and negative pairs. Contrastive representation learning builds a generative model to score positive samples higher than negative samples in computer vision

[6,24]. Contrastive learning also has many applications in natural language processing. To learn a sentence's representation, Lajanugen et al. [23] take its consecutive sentences as positive samples and sentences from another document as negative samples. By assigning a higher probability to a ground-truth translation than erroneous translation, Yang et al. [34] apply contrastive learning to reducing word omission errors in machine translation. The gradient-based method [18], which constructs positive and negative pairs automatically, is proposed to tackle the exposure bias problem.

3   Task Definition and Preliminary

In this section, we formally define the response generation task, and then we describe the backbone model under Sequence-to-Sequence (Seq2Seq) architecture to tackle the task.

Given the input X including a SQL S and a result table T, our goal is to generate the corresponding response Y. The SQL S is represented as a token sequence $S = s_1, s_2, ..., s_{|S|}$. The table $T = \{T_{i,j} | 1 \le i \le R_T, 1 \le j \le C_T\}$ has R_T rows and C_T columns with each $T_{i,j}$ being the content in the (i, j) cell. $T_{i,j}$ could be a word, a phrase or a number. The annotated response is $Y = y_1, y_2, ...y_{|Y|}$. We aim to train a response generator $p(Y|X)$ to generate response from X. The generated response is expected to be both *fluent* and *faithful*.

Modeling the response generation task under Seq2Seq paradigm, we choose T5 [26], a strong pre-trained generation model as our backbone model. In fact, FALCON can also be applied to other Data-to-Text methods, as shown in our experiments. Following previous work on linearizing table as natural language [7], we use template to flatten the table T as a paragraph $T_L = t_1, ..., t_{|T_L|}$. If the result table is empty, we use "No, there is no result" to represent the result. Then we concatenate SQL S and linearized table T_L as a complete sequence input $X = x_1, ..., x_{|X|}$.

A typical approach to train T5 is to minimize the negative log-likelihood (NLL) of the target sequence Y, which we refer to as $\mathcal{L}_{nll}$.

$$\mathcal{L}_{nll} = -\sum_{t=1}^{|Y|} \log p(y_t|X, y_{<t})$$
$$p(y_t|X, y_{<t}) = \texttt{softmax}(\mathbf{W}\mathbf{h}_t^D + \mathbf{b}) \tag{1}$$
$$\mathbf{h}_t^D = \text{Decoder}(y_{<t}, \mathbf{H}^E)$$
$$\mathbf{H}^E = \text{Encoder}(X)$$

where $\mathbf{H}^E = [\mathbf{h}_1^E, ..., \mathbf{h}_{|X|}^E] \in \mathbb{R}^{d \times |X|}$ is the concatenation of hidden state of input tokens and $\mathbf{H}^D = [\mathbf{h}_1^D, ..., \mathbf{h}_{|Y|}^D] \in \mathbb{R}^{d \times |Y|}$ is the concatenation of hidden state of output tokens. d is the hidden size of T5.

4   Proposed Approach

First, we describe the method of constructing imposters, which are crucial for the performance of FALCON. Then we introduce our faithful contrastive generation framework

Table 1. Examples of constructing imposters through text transformations. Blue and red text highlight the changes made by the transformation.

Transformation	Original sentence	Transformered sentence
Construct $\widetilde{X}$	SELECT f_id FROM files WHERE formats = "mp4" UNION SELECT f_id FROM song WHERE resolution > 720	SELECT f_id FROM files WHERE formats = "mp4" EXCEPT SELECT f_id FROM song WHERE resolution > 720
	SELECT address FROM member WHERE age < 30	SELECT address FROM member WHERE age > 30
	SELECT MAX (Length) FROM roller_coaster	SELECT MIN (Length) FROM roller_coaster
	SELECT name FROM country ORDER BY surfacearea DESC LIMIT 1	SELECT name FROM country ORDER BY surfacearea ASC LIMIT 1
	SELECT AVG (salary) FROM instructor	SELECT AVG (salary) FROM dept_name
Construct $\widetilde{Y}$	The number of drivers who are from Hartford City or younger than 40 is 11	The number of drivers who are from Hartford City or older than 40 is 11
	The average damage for all storms is 11.0629 million USD	The maximum damage for all storms is 11.0629 million USD
	The average damage for all storms is 11.0629 million USD	The average storm for all storms is 11.0629 million USD

(FALCON) as shown in Fig. 2, which consists of two major components: contrastive training and contrastive inference. Finally, we elaborate on two metrics to evaluate faithfulness.

4.1 Imposters Construction

To make the model subject to factual constraints when learning to generate target sequence, we construct negative samples that are syntactically identical but have significant differences in factual correctness, which we call imposters.

Given a pair (X, Y), we construct imposters through a series of predefined transformations. To be specific, source imposter $\widetilde{X}$ is obtained by modifying original input X, and original output Y is modified to yield target imposter $\widetilde{Y}$. Consequently, a tuple $(X, Y, \widetilde{X}, \widetilde{Y})$ is constructed to be applied to FALCON. As shown in Table 1, we utilize following heuristic rules to construct imposters.

Construct Source Imposter $\widetilde{X}$: Some keywords play the same role in SQL syntax. When the original keyword is replaced by other keywords, the query intention or condition of SQL will change. Apparently, the modified SQL is factually inconsistent with the original output. Specifically, 1) replace the UNION, INTERSECT, and EXCEPT with another one of the above keywords. e.g., UNION to EXCEPT; 2) replace comparison

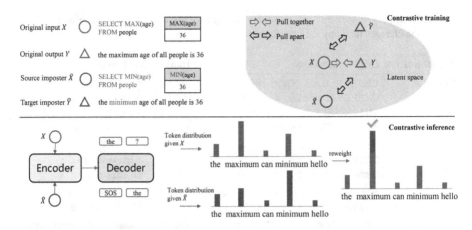

Fig. 2. An illustration of FALCON, which consists of contrastive training and contrastive inference. In contrastive training, the backbone model is enforced to pull factually consistent pairs (X, Y) together and push the factually inconsistent pairs $(\widetilde{X}, Y), (\widetilde{X}, Y)$ apart in the latent space. In contrastive inference, FALCON generates a more faithful response by referring to the token probability distribution of source imposter $\widetilde{X}$.

operators ($>, <, \geq, \leq, ! =$) with another comparison operators. e.g., $>$ to $<$; 3) replace aggregation keywords (MAX, AVG, MIN, COUNT, SUM) with another aggregation keywords. e.g., MAX to MIN; 4) replace order keywords (ASC, DESC) with another order keywords. e.g., DESC to ASC; 5) use string-match based method to recognize columns and tables mentioned in the response, then replace it with another column or table[1].

Construct Target Imposter $\widetilde{Y}$: We modify the logical expression involved in the response so that the modified response is inconsistent with the original input. Specifically, 1) use string-based methods to locate comparative and superlative and replace them with antonyms. e.g., younger to older; 2) replace the tokens that represent aggregate keywords. e.g., average to maximum; 3) replace the span mentioned in SQL or table with another randomly sampled span.

4.2 Contrastive Training

We argue that the generation model should be restricted by factual consistency in the latent space. To achieve this, FALCON encourages the latent semantics of the ground truth response to be consistent with the SQL and table. Specifically, given a pair of input and output (X, Y), we first project the original input and output to the latent space:

$$
\begin{aligned}
\mathbf{h}^E &= \text{MeanPool}(\text{Relu}(\mathbf{W}^E \mathbf{H}^E + \mathbf{b}^E)) \\
\mathbf{h}^D &= \text{MeanPool}(\text{Relu}(\mathbf{W}^D \mathbf{H}^D + \mathbf{b}^D))
\end{aligned}
\tag{2}
$$

---

[1] Synchronously modify headers and values in the result table.

where $\mathbf{h}^E, \mathbf{h}^D \in \mathbb{R}^d$ represent the latent representation of input and output. $\mathbf{W}^E$ and $\mathbf{W}^D$ are the source and target projection matrix, respectively. MeanPool denotes average pooling operation. Then we calculate their consistency $\mathcal{C}(X, Y)$ in the latent space:

$$\mathcal{C}(X, Y) = \sigma(\mathbf{W}^C[\mathbf{h}^E; \mathbf{h}^D] + b^C)) \tag{3}$$

where $\mathbf{W}^C \in \mathbb{R}^{2*d}$ and $b^C \in \mathbb{R}$ are learnable parameters. σ is the sigmoid function and $[\cdot; \cdot]$ denotes the operator of vector concatenation.

Models not only learn from faithful examples but also benefit from identifying imposters. Hence, given a tuple $(X, Y, \widetilde{X}, \widetilde{Y})$ where $\widetilde{X}$ and $\widetilde{Y}$ are the source and target imposter, FALCON enforces the model to provide a lower consistency score for imposters than ground truth.

For the target imposter $\widetilde{Y}$, we first compute its representation $\widetilde{\mathbf{h}}^D$ in the latent space given original input X:

$$\begin{aligned} \widetilde{\mathbf{h}}^D &= \text{MeanPool}(\text{Relu}(\mathbf{W}^D \widetilde{\mathbf{H}}^D + \mathbf{b}^D)) \\ \widetilde{\mathbf{h}}^D_t &= \text{Decoder}(\widetilde{y}_{<t}, \mathbf{H}^E). \end{aligned} \tag{4}$$

Then the consistency $\mathcal{C}(X, \widehat{Y})$ between original input X and target imposter $\widetilde{Y}$ is obtained. According to the principle that a consistent pair should be assigned a higher score, we adopt margin ranking loss [1] to separate original output and target imposter.

$$\begin{aligned} \mathcal{L}_t &= \max(0, \delta - \mathcal{C}(X, Y) + \mathcal{C}(X, \widetilde{Y})) \\ \mathcal{C}(X, \widetilde{Y}) &= \sigma(\mathbf{W}^C[\mathbf{h}^E; \widetilde{\mathbf{h}}^D] + b^C)) \end{aligned} \tag{5}$$

where margin δ is the hyper-parameter.

For the source imposter $\widetilde{X}$, we obtain its latent representation $\widetilde{\mathbf{h}}^E$ through the encoder. Then FALCON requires the model to distinguish between original input and source imposter in the latent space.

$$\begin{aligned} \mathcal{L}_s &= \max(0, \delta - \mathcal{C}(X, Y) + \mathcal{C}(\widetilde{X}, Y)) \\ \mathcal{C}(\widetilde{X}, Y) &= \sigma(\mathbf{W}^C[\widetilde{\mathbf{h}}^E; \mathbf{h}^D] + b^C)) \\ \widetilde{\mathbf{h}}^E &= \text{MeanPool}(\text{Relu}(\mathbf{W}^E \widetilde{\mathbf{H}}^E + \mathbf{b}^E)) \\ \widetilde{\mathbf{H}}^E &= \text{Encoder}(\widetilde{X}). \end{aligned} \tag{6}$$

The overall optimization object is as follows:

$$\mathcal{L} = \mathcal{L}_{nll} + \alpha(\mathcal{L}_s + \mathcal{L}_t) \tag{7}$$

where the contrastive weight α controls the relative importance of contrastive learning.

4.3   Contrastive Inference

A typical decoding approach is that when decoding the t-th step, we compute the probability distribution of each token on target vocabulary $\mathcal{V}$. $p(y_t^k|X, y_{<t})$ represents the probability of the k-th token $\mathcal{V}_k$ on vocabulary at step t:

$$p(y_t^k|X, y_{<t}) = p(y_t = \mathcal{V}_k|X, y_{<t}). \tag{8}$$

We argue that referring to the token probability distribution of source imposter $\widetilde{X}$ can help improve faithfulness. Therefore, we propose an imposters-contrastive decoding method to generate more faithful responses. We first obtain the imposter $\widetilde{X}$ of the original input X through the construction method[2] mentioned in Subsect. 4.1. Intuitively, we should encourage tokens to be generated based on the original input, while suppressing the token probability generated by the source imposter.

Then we compute the proportion of the token $\mathcal{V}_k$ probability generated by the original input X compared to the source imposter $\widetilde{X}$:

$$\alpha_t^k = \frac{\exp p(y_t^k | X, y_{<t})}{\exp p(y_t^k | X, y_{<t}) + \exp p(y_t^k | \widetilde{X}, y_{<t})}. \tag{9}$$

By setting a threshold of $\rho > 0.5$, for token $\mathcal{V}_k$ with α_t^k greater than ρ, it means that it's generated specifically based on the original input compared to imposter, which should be encouraged. Therefore, we increase the probability of these tokens to generate words that are more consistent with the original input:

$$\delta_t^k = \begin{cases} \lambda \alpha_t^k p(y_t | X, y_{<t}), & \alpha_t^k \geq \rho \\ 0, & \alpha_t^k < \rho \end{cases} \tag{10}$$

where λ is the hyper-parameter to control the impact of imposter on responses generation. δ_t^k is the probability increment.

The final probability of the k-th token is computed as:

$$p'(y_t^k | X, \widetilde{X}, y_{<t}) = p(y_t^k | X, y_{<t}) + \delta_t^k. \tag{11}$$

4.4 Metrics

In this subsection, we introduce two metrics inspired by two principles to evaluate faithfulness of generated responses.

Principle 1. *The generated response should contain the logical information of the input SQL, including the query intent and condition.*

Logical Score. According to *principle 1*, we train a model to parse the response into SQL. Through this model, the generated response is translated into various components of SQL. Then we conduct a fine-grained matching between the parsed SQL and the input SQL to compute its logical accuracy. Specifically, we adopt BRIDGE [21], the SoTA Text-to-SQL semantic parser on Spider [36], as our Response-to-SQL model. Following previous Text-to-SQL evaluation methods [40], we employ Exact Match as the logical score to measure the logical accuracy between the generated response and the original SQL.

Principle 2. *A model that can distinguish between ground truth and imposter should be able to evaluate the correctness of the generated response.*

---

[2] We only use the first three rules for constructing source imposters when inference.

Consistency Score. According to *principle 2*, we build a consistency classifier to score the generated response. Given a tuple $(X, Y, \widetilde{X}, \widetilde{Y})$ constructed from Sect. 4.1, (X, Y) is a consistent pair as our positive example. $(\widetilde{X}, Y)$ and $(X, \widetilde{Y})$ are inconsistent pairs as negative examples. A RoBERTa-based [22] consistency classifier is trained to distinguish whether the response is factually consistent with the given SQL and result. Specifically, taking the concatenation of SQL, linearized table, and response as input, this classifier predicts whether the generated output is a faithful statement about the input. We employ the average predicted score as the consistency score.

5   Experiments

In this section, we conduct several experiments to demonstrate the effectiveness of our method.

5.1   Dataset

CoSQL. We evaluate FALCON on CoSQL [35], consisting of 7845/1074 examples for train/development. Each example includes the SQL, the result table, and the corresponding response. Different from WikiSQL [40], the SQL in CoSQL have following complex syntax: 1) multi-table queries; 2) nested queries; 3) advanced keywords (HAVING, ORDER BY, UNION, DISTINCT, LIMIT, etc.)

5.2   Baselines

We apply FALCON to several competitive generation models. Due to the lack of relevant baselines, we extend some Data-to-Text methods to adapt to this task. These methods can be divided into two categories, one is structure-based methods, and the other is textual-based methods. Structure-based methods can explicitly encode the internal structure of SQL and table to generate high-quality text.

- **GraphWriter** [16]: A Graph-to-Text generation model that utilizes a graph network to encode the knowledge graph (KG) and RNN to encode its title. We parse the SQL into an Abstract Syntax Tree (AST) as the input KG and take the linearized table as its input title.
- **GTN** [2]: A Graph Transformer that uses explicit relation encoding and allows direct communication between two distant nodes. Following [41], we convert the result table into a graph. Then we use a global node to connect it to the AST as the entire input graph.

Based on the Seq2Seq architecture, textual-based methods take the sequential SQL and the linearized table as input to generate the response.

- **Transformer** [29]. A representative model of neural machine translation. We add the copy mechanism to enable the model to copy words from the input.
- **BART** [19]: A pre-trained language model (PLM) which is pre-trained as a text-to-text denoising autoencoder. We use BART-base in this paper.
- **T5** [26]: A PLM which converts every language problem into a Text-to-Text format. We use T5-small in this paper.

5.3   Implementation Details

First, we try our best to adjust the hyper-parameters so that the backbone model has a relatively optimal performance. Then according to the same hyper-parameter settings, we apply FALCON to improve the factual correctness. Taking backbone model T5 as an example, we optimize models by Adam [15] with a learning rate of $5e-5$ and the batch size of 32. The margin δ is set to 0.6 and α is set to 5.0. During inference, the greedy search is adopted to decode the response and the maximum decoding length is 25. λ and ρ are set to 0.2 and 0.8, respectively.

To train the Response-to-SQL model, we follow the same hyper-parameter settings as [21] except that the batch size is changed to 12. In order to filter out trivial responses like "yes", we discard the top 20% of the samples with the lowest edit distance between its SQL and response in the dataset. We also add Question-to-SQL examples from spider [36] to train set because we found it can significantly improve its performance. Finally, its exact match accuracy on the development set is 61.7%.

We train the consistency classifier with RoBERTa-base [22]. The synthetic train set includes 7,845 positive samples and 13,250 negative samples. 1,074 positive samples and 1,778 negative samples are included in the synthetic development set. On the synthetic development set, the classifier has an 86.5% F1-score.

5.4   Evaluation Metrics

Following the common practice, we illustrate the n-gram based BLEU [25] and ROUGE-L [20] evaluations to evaluate the quality of our generated response. We also evaluate the results using BERTScore [38] and BLEURT metric [28], which employ contextual embeddings to incorporate semantic information. Logical Score and Consistency Score are also used to evaluate the faithfulness of generated responses.

We invited three graduate students majoring in CS as experiment participants. Each of them is familiar with SQL. To ensure accurate human evaluation, the raters are trained with word instructions and text examples of the grading standard beforehand. Specifically, Two annotators are first asked to evaluate fluency and faithfulness for 600 samples separately. Then they re-check each other's results. If two annotators have conflicts, we will send these samples to the third annotator for final judgment. The following are detailed criteria for evaluating fluency and faithfulness.

Fluency

- **0**: The response is smooth and natural.
- **1**: The response does not flow smoothly but people can understand its meaning.
- **2**: The response is not fluent at all.

Faithfulness

- **0**: The response has no factual errors.
- **1**: The response has a factual error.
- **2**: The response has more than one factual error.

Table 2. Evaluation results of the different models on automatic metrics. "+FALCON" means the model is trained using FALCON. L-Score and C-Score denote Logical Score and Consistency Score, respectively. BLEURT-T and BLEURT-B denote BLEURT-Tiny and BLEURT-Base. The best results in each group are highlighted in **bold**.

	BLEU	ROUGE	BERTScore	BLEURT-T	BLEURT-B	L-Score	C-Score
GraphWriter	16.86	47.44	37.42	−27.68	−58.43	28.70	55.04
Graphwriter+FALCON	**17.06**	**48.39**	**38.30**	**−24.28**	**−57.30**	**30.48**	**55.26**
GTN	18.31	**51.55**	46.68	0.20	−27.95	35.46	66.49
GTN+FALCON	**18.70**	51.48	**47.34**	**1.53**	**−25.84**	**37.19**	**67.81**
Transformer	12.88	43.42	32.05	−23.52	−66.50	17.61	43.52
Transformer+FALCON	**13.13**	**43.62**	**32.60**	**−21.98**	**−63.96**	**17.89**	**43.76**
BART	24.60	57.39	58.72	22.38	9.28	54.10	83.96
BART+FALCON	**24.73**	**57.51**	**59.12**	**23.36**	**9.60**	**54.79**	**84.52**
T5	25.25	57.54	57.89	22.40	6.74	53.79	85.46
T5+FALCON	**25.65**	**57.89**	**58.41**	**23.92**	**7.76**	**54.32**	**85.58**

6 Result and Analysis

6.1 Performance on Automatic Metrics

Table 2 presents the automatic evaluation results of different backbone models with conventional NLL training and FALCON[3]. On the whole, applying FALCON to different models consistently improves the factual correctness of generated responses.

Specifically, FALCON slightly improves the performance of models on BLEU and ROUGE. This is reasonable because these metrics measure the n-gram overlap between a reference response and candidate response, which are not always valid factuality metrics [10]. Even so, applying FALCON to GraphWriter improve 0.20 on BLEU and 0.95 on ROUGE.

FALCON has achieved noticeable improvements on BERTScore and BLEURT. Such PLM-based metrics reflect the factual correctness by matching semantic information in the embedding space [10]. FALCON has achieved an average improvement of 1.42 points on BLEURT-Base, which is a favorable performance boost. These improvements demonstrate that by encouraging the model to distinguish well-designed contrastive pairs, the model learns a more robust representation and produces responses that are more faithful to the input.

Applying FALCON has improved each model on Logical Score and Consistency Score. FALCON has achieved an average improvement of 1.00 points on Logical Score and 0.49 points on Consistency Score. The improvement of Logical Score shows that FALCON has produced a response that is more logically consistent with the input SQL. The enhancement of Consistency Score indicates that the response generated by FAL-CON is more consistent with the input.

---

[3] We report the average best performance observed in 3 runs on the development set of CoSQL since its test set are not public. All improvements of FALCON are significant with $p < 0.01$ compared to backbone models.

Table 3. Human evaluation on fluency and faithfulness. Both metrics are the smaller the better.

	Fluency	Faithfulness	Kappa
GraphWriter	1.23	1.32	0.69
Graphwriter+FALCON	**1.02**	**0.98**	0.72
GTN	0.86	0.84	0.64
GTN+FALCON	**0.71**	**0.65**	0.53
T5	0.45	0.52	0.61
T5+FALCON	**0.35**	**0.27**	0.71

Table 4. Ablation study of our framework components using T5. "C-inference" means using contrastive inference, "S-imposter" means using source imposters, "T-imposter" means using target imposters, and "C-training" means training T5 with both source and target imposters.

	BLUERT-T	BLEURT-B	L-Score	C-Score
(a) T5	22.40	6.74	53.79	85.46
(b) *w/ C-inference*	22.78	6.90	54.23	85.55
(c) *w/ S-imposter*	22.97	**7.78**	54.03	**85.61**
(d) *w/ T-imposter*	22.59	7.48	53.91	85.27
(e) *w/ C-training*	23.41	7.66	54.12	85.46
(f) Full version	**23.92**	7.76	**54.32**	85.58

6.2   Performance on Human Evaluation

FALCON *Performance* We further conduct extensive human evaluations of generated responses. Generation results of six models on 100 samples are provided to three students for blind testing to evaluate the fluency and faithfulness.

Table 3 presents evaluation results and Cohen's kappa scores [8] to measure the intra-rater reliability. With FALCON, T5 has reduced the average number of factual errors per response from 0.52 to 0.27. In general, applying FALCON to GTN and T5 has achieved significant improvement in the faithful aspect, which demonstrates our framework's effectiveness to boost factual correctness.

Metrics Performance. We also measure the correlation (Pearson's r) between the faithfulness and two new metrics. The correlation coefficient between Logical Score and faithfulness is -0.68 (with $p < 0.05$), and -0.71 (with $p < 0.05$) is the correlation coefficient of Consistency Score. This result reveals they have a relatively high correlation with human judgment on faithfulness, which illustrates their effectiveness to evaluate the faithfulness of responses.

6.3   Model Analysis

Effect of Contrastive Training. To constrain the model with factual correctness, we require the model to distinguish ground truths from imposters in the latent space when

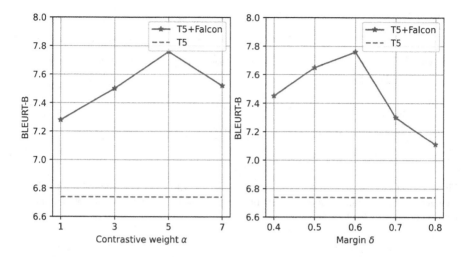

Fig. 3. Evaluation results with different contrastive weight α and margin δ using T5.

training. (a) and (e) in Table 4 demonstrates the impacts of contrastive training. It is observed that with contrastive training, the model improves the performance on all of the automatic metrics, including two new metrics to evaluate faithfulness. We also explore the effects of source imposters and target imposters on the performance of the model. As shown in Table 4 (c) and (d), training with source imposters or target imposters leads to a consistent performance improvement, which suggests that the model can benefit from perceiving factual differences during contrastive training.

Effect of Contrastive Inference. We propose contrastive inference to generate more faithful responses from contrasting source imposters. Comparing (a) and (b) demonstrates that using contrastive inference leads to a consistent performance increase. Comparing (e) and (f) also shows that contrastive inference contributes to improve the factual correctness of the generated responses.

Impact of Hyper-parameter α and δ. As shown in Fig. 3, we explore the impact of using different contrastive weight α and margin δ. We observe that as α or δ increase, the performance of T5 with FALCON presents an earlier increase and later decrease trend. The increase of the two hyper-parameters means that the importance of contrast learning boosts. We conjecture that a higher contrastive learning loss allows the model better distinguish imposters from ground truths, meanwhile it also risks failing to fit the original response.

6.4 Case Study

As shown in Fig. 4, we further demonstrate several typical examples for better understanding how our framework boosts faithfulness. Each example includes an input including SQL and table, a reference response (**Ref**), the T5 output (**T5**), and the generated response by our approach (**T5+FALCON**).

SQL:	SELECT t2.best_finish FROM people t1 JOIN poker_player t2 ON t1.people_id = t2.people_id ORDER BY t2.earnings LIMIT 1	**Ref:** the best finish of the poker player with the lowest earning is 1.0. **T5:** the best finish of the player with the highest earning amount is 1.0
Table:	best finish 1.0	**T5+Falcon:** the best finish of the player with the lowest amount of money is 1.0
SQL:	SELECT country FROM singer WHERE age > 40 INTERSECT SELECT country FROM singer WHERE age < 30	**Ref:** the country where singers above age 40 and below age 30 are from is france. **T5:** france is the country that has singers younger than 40 and younger than 30
Table:	country france	**T5+Falcon:** france is the country of singers with age above 40 and younger than 30.

Fig. 4. Examples of generated responses from T5 and T5 with FALCON. The red color indicates factual errors in generated responses. Text in blue emphasizes the role of FALCON in improving factual correctness. (Color figure online)

In the first example, T5 fails to understand the condition that sort players by earning in ascending order and incorrectly generates a response with "*highest earning*". The reason can be that the "*highest*" occurs frequently in the training corpus. On the other hand, with the contrastive examples' constraint, FALCON guides T5 to generate a faithful response with the key fact "*lowest*".

The second example shows a similar situation, where T5 generates a grammatically correct response but with factual errors. "*above age 40*" is incorrectly translated as "*younger than 40*". The possible reason is that "*40*" is incorrectly associated with the operator < in the latter part of SQL. FALCON helps recover the correct meaning: *with age above 40*".

7   Conclusion

We propose a faithful contrastive generation framework FALCON to improve responses' factual correctness in a TableQA system. Our framework includes contrastive training and contrastive inference. To better evaluate faithfulness, we propose two metrics. Extensive evaluations show FALCON brings a favorable performance improvement amongst various baseline methods.

Acknowledgements. We thank anonymous reviewers for their comments and suggestions. This work was supported by National Key Research and Development Project (No. 2020AAA0109302), Shanghai Science and Technology Innovation Action Plan (No. 19511120400) and Shanghai Municipal Science and Technology Major Project (No. 2021SHZDZX0103).

References

1. Burges, C.J.C., et al.: Learning to rank using gradient descent. In: Proceedings of ICML. ACM International Conference Proceeding Series (2005)
2. Cai, D., Lam, W.: Graph transformer for graph-to-sequence learning. In: The Thirty-Fourth AAAI Conference on Artificial Intelligence, AAAI 2020 (2020)

3. Cai, R., Liang, Z., Xu, B., Li, Z., Hao, Y., Chen, Y.: TAG : type auxiliary guiding for code comment generation. In: Proceedings of ACL (2020)
4. Cao, R., Chen, L., Chen, Z., Zhao, Y., Zhu, S., Yu, K.: LGESQL: line graph enhanced text-to-SQL model with mixed local and non-local relations. In: Proceedings of ACL (2021)
5. Cao, Z., Wei, F., Li, W., Li, S.: Faithful to the original: fact aware neural abstractive summarization. In: Proceedings of AAAI (2018)
6. Chen, T., Kornblith, S., Norouzi, M., Hinton, G.E.: A simple framework for contrastive learning of visual representations. In: Proceedings of ICML. Proceedings of Machine Learning Research (2020)
7. Chen, W., Chen, J., Su, Y., Chen, Z., Wang, W.Y.: Logical natural language generation from open-domain tables. In: Proceedings of ACL (2020)
8. Cohen, J.: A coefficient of agreement for nominal scales. Educ. Psychol. Meas. **20**(1), 37–46 (1960)
9. Falke, T., Ribeiro, L.F.R., Utama, P.A., Dagan, I., Gurevych, I.: Ranking generated summaries by correctness: an interesting but challenging application for natural language inference. In: Proceedings of ACL (2019)
10. Gabriel, S., Celikyilmaz, A., Jha, R., Choi, Y., Gao, J.: GO FIGURE: a meta evaluation of factuality in summarization. In: Findings of the Association for Computational Linguistics: ACL-IJCNLP 2021 (2021)
11. Goodrich, B., Rao, V., Liu, P.J., Saleh, M.: Assessing the factual accuracy of generated text. In: Proceedings of the 25th ACM SIGKDD International Conference on Knowledge Discovery and Data Mining, KDD 2019, 4–8 August 2019, Anchorage (2019)
12. Guo, J., et al.: Towards complex text-to-SQL in cross-domain database with intermediate representation. In: Proceedings of ACL (2019)
13. Hu, X., Li, G., Xia, X., Lo, D., Jin, Z.: Deep code comment generation. In: 2018 IEEE/ACM 26th International Conference on Program Comprehension (ICPC). IEEE (2018)
14. Iyer, S., Konstas, I., Cheung, A., Zettlemoyer, L.: Summarizing source code using a neural attention model. In: Proceedings of ACL (2016)
15. Kingma, D.P., Ba, J.: Adam: a method for stochastic optimization. In: Proceedings of ICLR (2015)
16. Koncel-Kedziorski, R., Bekal, D., Luan, Y., Lapata, M., Hajishirzi, H.: Text generation from knowledge graphs with graph transformers. In: Proceedings of NAACL-HLT (2019)
17. Kryscinski, W., Keskar, N.S., McCann, B., Xiong, C., Socher, R.: Neural text summarization: a critical evaluation. In: Proceedings of EMNLP (2019)
18. Lee, S., Lee, D.B., Hwang, S.J.: Contrastive learning with adversarial perturbations for conditional text generation. In: Proceedings of ICLR (2021)
19. Lewis, M., et al.: BART: denoising sequence-to-sequence pre-training for natural language generation, translation, and comprehension. In: Proceedings of ACL (2020)
20. Lin, C.Y.: ROUGE: a package for automatic evaluation of summaries. In: Text Summarization Branches Out (2004)
21. Lin, X.V., Socher, R., Xiong, C.: Bridging textual and tabular data for cross-domain text-to-SQL semantic parsing. In: Findings of the Association for Computational Linguistics: EMNLP 2020 (2020)
22. Liu, Y., et al.: RoBERTa: A robustly optimized BERT pretraining approach. arXiv preprint (2019)
23. Logeswaran, L., Lee, H.: An efficient framework for learning sentence representations. In: Proceedings of ICLR (2018)
24. Oord, A., Li, Y., Vinyals, O.: Representation learning with contrastive predictive coding. arXiv preprint (2018)
25. Papineni, K., Roukos, S., Ward, T., Zhu, W.J.: BLEU: a method for automatic evaluation of machine translation. In: Proceedings of ACL (2002)

26. Raffel, C., et al.: Exploring the limits of transfer learning with a unified text-to-text transformer. arXiv preprint (2019)

27. Scholak, T., Schucher, N., Bahdanau, D.: PICARD: parsing incrementally for constrained auto-regressive decoding from language models. arXiv preprint (2021)

28. Sellam, T., Das, D., Parikh, A.: BLEURT: learning robust metrics for text generation. In: Proceedings of ACL (2020)

29. Vaswani, A., et al.: Attention is all you need. In: Advances in Neural Information Processing Systems 30: Annual Conference on Neural Information Processing Systems 2017, 4–9 December 2017, Long Beach (2017)

30. Verma, V., Luong, T., Kawaguchi, K., Pham, H., Le, Q.V.: Towards domain-agnostic contrastive learning. In: Proceedings of ICML. Proceedings of Machine Learning Research (2021)

31. Wang, B., Shin, R., Liu, X., Polozov, O., Richardson, M.: RAT-SQL: relation-aware schema encoding and linking for text-to-SQL parsers. In: Proceedings of ACL (2020)

32. Wang, Z., Wang, X., An, B., Yu, D., Chen, C.: Towards faithful neural table-to-text generation with content-matching constraints. In: Proceedings of ACL (2020)

33. Xu, K., Wu, L., Wang, Z., Feng, Y., Sheinin, V.: SQL-to-text generation with graph-to-sequence model. In: Proceedings of EMNLP (2018)

34. Yang, Z., Cheng, Y., Liu, Y., Sun, M.: Reducing word omission errors in neural machine translation: a contrastive learning approach. In: Proceedings of ACL (2019)

35. Yu, T., et al.: CoSQL: a conversational text-to-SQL challenge towards cross-domain natural language interfaces to databases. In: Proceedings of EMNLP (2019)

36. Yu, T., et al.: Spider: a large-scale human-labeled dataset for complex and cross-domain semantic parsing and text-to-SQL task. In: Proceedings of EMNLP (2018)

37. Zeng, J., et al.: Photon: a robust cross-domain text-to-SQL system. In: Proceedings of ACL (2020)

38. Zhang, T., Kishore, V., Wu, F., Weinberger, K.Q., Artzi, Y.: BERTScore: evaluating text generation with BERT. In: Proceedings of ICLR (2020)

39. Zhang, Y., Merck, D., Tsai, E., Manning, C.D., Langlotz, C.: Optimizing the factual correctness of a summary: a study of summarizing radiology reports. In: Proceedings of ACL (2020)

40. Zhong, V., Xiong, C., Socher, R.: Seq2sql: Generating structured queries from natural language using reinforcement learning. arXiv preprint (2017)

41. Zhong, W., et al.: LogicalFactChecker: leveraging logical operations for fact checking with graph module network. In: Proceedings of ACL (2020)

Tipster: A Topic-Guided Language Model for Topic-Aware Text Segmentation

Zheng Gong, Shiwei Tong, Han Wu, Qi Liu$^{(\boxtimes)}$, Hanqing Tao, Wei Huang, and Runlong Yu

Anhui Province Key Laboratory of Big Data Analysis and Application, University of Science and Technology of China, Hefei 230026, China
{gz70229,tongsw,wuhanhan,hqtao,ustc0411,yrunl}@mail.ustc.edu.cn,
qiliuql@ustc.edu.cn

Abstract. The accurate segmentation and structural topics of plain documents not only meet people's reading habit, but also facilitate various downstream tasks. Recently, some works have consistently given positive hints that text segmentation and segment topic labeling could be regarded as a mutual task, and cooperating with word distributions has the potential to model latent topics in a certain document better. To this end, we present a novel model namely *Tipster* to solve text segmentation and segment topic labeling collaboratively. We first utilize a neural topic model to infer latent topic distributions of sentences considering word distributions. Then, our model divides the document into topically coherent segments based on the topic-guided contextual sentence representations of the pre-trained language model and assign relevant topic labels to each segment. Finally, we conduct extensive experiments which demonstrate that Tipster achieves the state-of-the-art performance in both text segmentation and segment topic labeling tasks.

Keywords: Text segmentation · Neural topic model · Language model

1 Introduction

Text segmentation and segment topic labeling tasks are two coupled tasks denoted henceforth **topic-aware text segmentation** (**TATS**) task, which aim to provide accurate text segmentation and structural topics of unlabeled documents. Figure 1 describes a toy example of TATS about *Paris* city. Based on topical coherence, the sentences in the document (*i.e.*, s1, s2, s3, s4) are divided into three segments, which portray various topics of *Paris*, *i.e.*, History, Geography and Culture. Accurate segmentation and structural topics can not only help understand the unlabeled documents better, but also be applied to many downstream tasks, such as passage retrieval and intent detection.

The vast majority of studies on TATS concentrate on supervised methods. SECTOR [1] and S-LSTM [2] exploit pre-trained word embeddings to predict segment boundary and assign related topics. Besides, some BERT-based

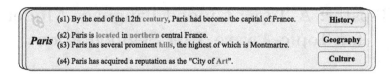

Fig. 1. A raw document includes detailed description of *Paris* covering multiple topics (*e.g.*, History). The colored words are keywords related to the corresponding topic. (Color figure online)

works [10] recently have achieved a high performance on the single text segmentation task by directly representing sentence with pre-trained BERT embedding [4]. However, the native sentence representations generated from BERT are proved to collapse into a small space and produce high similarity between most sentences [9], which hinder the sentence representations of diverse topics.

Actually, *word distributions can be regarded as evident representations of topics.* For instance, as shown in Fig. 1, the word *"century"* appears more frequently in the text related to history topic (in Sentence 1), whereas *"located"* and *"northern"* appear more frequently in geography topic (in Sentence 2). Therefore, it comes to the conclusion that word distributions vary in different topics. We could utilize the word distributions to pay more attention to the words relevant to topics, leading to the distinguishable sentence representation of topics.

However, there are many technical challenges in aligning the pre-trained sentence representation and topic-word distributions with the TATS. First, the mainstream methods of modeling word distributions adopt probabilistic topic models [13], which have a slow speed in estimating parameters. However, due to the complexity of our task, traditional inference methods will be limited by the cost of computation. Second, since the pre-trained sentence representations have some limitations [9] in capturing semantic changes in sentences, how to allocate rational attention on each word has not been explored much.

With this in mind, we bring in a neural variational topic model to capture the relation between distributions of words and topics and propose the **T**opic-guided **p**re-trained **s**entence **r**epresentation of language models (**Tipster**) for TATS. Specifically, our model has a three-stage process for this task. First, we infer the contextual representation of words in the sentence with a pre-trained language model. Second, with the help of neural topic model, we exploit the word distributions to discover informative words in the sentence and obtain topic-guided sentence representations via rational attention weights. Third, we capture the change of semantics via sentence representations of the document and predict topic labels of each segment. Our extensive experimental results show that Tipster achieves the state-of-the-art (SOTA) performance on the TATS. We also show that these improvements are in line with out-of-domain datasets.

2   Problem Formulation

In this section, we formulate the TATS formally. Given the pre-defined topic categories C and a document containing N consecutive sentences $S = [s_1, ..., s_N]$, the goals of TATS are to split S into a sequence of segments $B = [b_1, ..., b_M]$ and

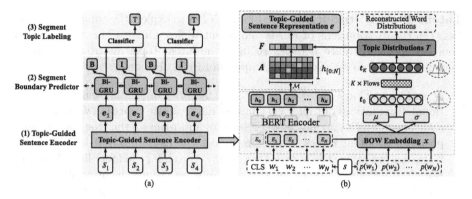

Fig. 2. (a) The Tipster model takes a given sequence of sentences as inputs, and outputs the segment boundaries and each segment topic label. (b) The topic-guided sentence encoder consists of a pre-trained language model and a neural topic model.

assign one topic t_i (single-label task) or multiple topics T_i (multi-label task) to each segment b_i. The process of topic classification is formulated as follows:

$$t_i = f_{\text{single}}(b_i) \quad \text{and} \quad T_i = f_{\text{multiple}}(b_i), \tag{1}$$

where f_{single} is the function that how we assign a single topic $t_i \in C$ to the segment b_i, while f_{multiple} is the function that how we assign multiple topics $T_i \subseteq C$ to the segment b_i. Our model handles both two circumstances.

3 Tipster for Topic-Aware Text Segmentation

3.1 Topic-Guided Sentence Encoder

Contextual and BOW Representations. The topic-guided sentence encoder is designed as a cooperative architecture with the pre-trained BERT [4] and a neural topic model (NTM), as shown in Fig. 2(b). Considering a word sequence of sentence $s = \{w_i\}_{i=1}^{N}$ as input, where N is the number of words, we obtain contextual representations $h_{[0:N]}$ of each word from the last hidden layer of BERT, where $h_0, h_j \in \mathbb{R}^d$ is the contextual representation of CLS token and word w_j respectively, and d is the hidden layer size of BERT.

Inspired by Variational Encoder [7,12], we build a NTM to learn non-linear topic-word distributions from $h_{[1:N]}$. We process the sentence s into a bag-of-words (BOW) input $s_{bow} \in R^{|\mathcal{V}|}$, where $\mathcal{V}$ denotes the word vocabulary with the stopwords removed. Then we derive the BOW embedding x of sentence s with the distribution $p(w)$ and pre-trained word embedding matrix E of BERT:

$$p(w_i) = \frac{(s_{bow})_i}{\sum_{w \in s} (s_{bow})_w} \quad \text{and} \quad x = \sum_{w \in s} (p(w)E_w). \tag{2}$$

Latent Topic Distributions. We build a NTM to exploit the complex relation between the distributions of words and latent topics. Our NTM consists of two components: a inference network to infer the latent topic distributions and a decoder network to reconstruct the input word distributions.

1) *Normalizing Flows-Based Posterior.* The majority of NTMs design the variational distributions of topics via simplest multivariate Gaussian with diagonal covariance, which may not be expressive to approximate the true complicated posterior of latent topics. Therefore, under the assumption of keeping prior Gaussian, we utilize the normalizing flows [11] to infer the more flexible posterior distributions of latent topics. Specifically, we assume the prior of topic distribution is a multivariate Gaussian with diagonal covariance $p_\lambda(t) \sim \mathcal{N}(0, I)$. We formulate the inference process as follows:

$$g = f_{\text{MLP}}(x), \; \mu = l_1(g), \; \log \sigma = l_2(g),$$
$$q_\phi(t \mid x) = \mathcal{N}(t \mid \mu(x), \text{diag}(\sigma^2(x))), \tag{3}$$

where f_{MLP} denotes the multi-layer perceptron (MLP), l_1 and l_2 are linear transformation functions with bias, μ and σ denote the mean and the standard deviation of the Gaussian posterior. To reduce the variance in stochastic estimation, we use the reparameterize trick [7,12] by sampling $\epsilon \sim \mathcal{N}(0, I)$, and reparameterizing $t = \mu + \epsilon \times \sigma$.

Since the Gaussian posterior $q_\phi(t \mid x)$ may not be sufficiently flexible with the true posterior, we utilize the normalizing flows to transform the variational posterior into more complex density via a sequence of invertible transformation $t_K = f_K(f_{K-1}(...f_1(t_0)))$, where we suppose that the initial sample $t_0 = t$ from $q_\phi(t \mid x)$, and apply K parameterized invertible transformation f to obtain the final sample t_K. The probability density of the final sample t_K is defined through the variational method as:

$$q_K(t_K \mid x) = q_\phi(t_0 \mid x) \prod_{k=1}^{K} \left| \det \frac{\partial f_k(t_{k-1}; \phi_k)}{\partial t_{k-1}} \right|^{-1}, \tag{4}$$

where the last term $|\circ|$ denotes the Jacobian determinant and ϕ_k is the parameter of the k-th transformation. As for our model, we consider the Planar flows [11] as the transformation:

$$f_k(t_{k-1}; \phi_k) = t_{k-1} + u_k \cdot \tanh(w_k^T t_{k-1} + b_k), \tag{5}$$

where $\phi_k = \{u_k, w_k \in \mathbb{R}^m, b_k \in \mathbb{R}\}$, $\left| \det \frac{\partial f_k}{\partial t_{k-1}} \right| = \left| 1 + u_k^\top \psi_k(t_{k-1}) \right|$. The planar flows apply contractions and expansions in the perpendicular direction to the hyperplane $w_k^T t_{k-1} + b_k = 0$.

2) *Word Distribution Reconstruction.* After we obtain the posterior distribution of topics t_K, we explicitly reconstruct the sentence s by independently generating the word $e_{\text{NVTM}} \rightarrow w_i$ via a dense layer θ:

$$p_\theta(w_i \mid t_K) = \text{Softmax}(E(w_i; t_K, \theta)), \tag{6}$$

where $E\left(w_i; t_K, \theta\right) = t_K^T W w_i + b_{w_i}$, $\theta = \{W \in \mathbb{R}^{m \times |\mathcal{V}|}, b \in \mathbb{R}^{|\mathcal{V}|}\}$. This reconstruction process encourages the latent topic distributions to contain the most important words in the sentence.

3) *Evidence Lower Bound.* To maximize the marginal likelihood $\log p(x)$ of sentence s, we derive the loss function of NTM from *evidence lower bound* [11]:

$$\mathcal{L}_{\mathrm{NTM}} = -\sum_{i=1}^{|\mathcal{V}|} p\left(w_i\right) \log p_\theta\left(w_i \mid t_K\right) + \mathrm{KL}\left(q_\phi\left(t_0 \mid x\right) \| p_\lambda\left(t_0\right)\right) \\ - \sum_{k=1}^{K} \log \left| \det \frac{\partial f_k\left(t_{k-1}\right)}{\partial t_{k-1}} \right|. \tag{7}$$

where the first term of $\mathcal{L}_{\mathrm{NTM}}$ is the reconstruction loss of word distributions, and the second term of $\mathcal{L}_{\mathrm{NTM}}$ is the Kullback-Leibler divergence between posterior and prior of the first layer topic distributions. Based on the sample $t_0 \sim q_\phi\left(t_0 \mid x\right)$, the parameters of our NTM can be optimized by back propagating the stochastic gradients. Afterwards, we obtain the latent topic distributions $T \in (0, 1)^m$ of sentence s via a Softmax function on the output topic t_K.

Topic-Guided Sentence Representation. Considering that the topics T in the NTM have different meanings, we build a memory bank $\mathcal{M} = \langle m_1, m_2, ..., m_{|T|} \rangle$ to represent each topic. Moreover, such a memory bank facilitates reasonable attention on the meaningful words by projecting the space of NTM to the contextual representations $h_{[1:N]}$. We hypothesise that each word has a different contribution to each topic. The memory bank of topics is used to determine the attention matrix $A \in \mathbb{R}^{(N+1) \times m}$ between contextual representations of words $h_{[0:N]} \in \mathbb{R}^{(N+1) \times d}$ and topic embeddings $\mathcal{M} \in \mathbb{R}^{m \times d}$, and then we obtain multiple topic facets $F \in \mathbb{R}^{m \times d}$ through A and $h_{[0:N]}$:

$$A_{jk} = \frac{\exp\left(\cos\left(m_k, h_j\right)\right)}{\sum_{j'=0}^{N} \exp\left(\cos\left(m_k, h_{j'}\right)\right)} \quad and \quad F = A^T \cdot h_{[0:N]}, \tag{8}$$

where cos denotes the cosine similarity. The topic guided sentence representation $e \in \mathbb{R}^d$ is obtained by aggregating F weighted by topic distributions T:

$$e = \sum_{i=1}^{m} T_i F_i. \tag{9}$$

3.2 Segment Boundary Predictor

After getting the sentence embedding e, we use a two-layer bidirectional Gated Recurrent Units (Bi-GRU) to predict segment boundaries in the document. This sub-network takes a sequence of sentence embedding $e_{1,2,...,M} \in R^d$ as input, feeds them into Bi-GRU and predicts two classes, B or I, representing whether the sentence t is (B)eginning or (I)nside of a segment. We formulate the segment boundary predictor as:

$$\overleftrightarrow{h_t} = f_{\mathrm{Bi\text{-}GRU}}\left(e_t, \overrightarrow{h_{t-1}}, \overleftarrow{h_{t+1}}\right), \quad p_t(\mathrm{B; I}) = \mathrm{Softmax}\left(W_s \overleftrightarrow{h_t} + b_s\right). \tag{10}$$

3.3 Segment Topic Labeling

The last step of our model is to assign related topic labels to each segment. We obtain the embedding H_i of segment i by concatenating the mean pooling, max pooling and attention pooling [14] of contained sentences embedding $H_i = [\text{Mean}(\overleftrightarrow{h_j}); \text{Max}(\overleftrightarrow{h_j}); \text{Attn}(\overleftrightarrow{h_j})]$. Then we assign each segment one topic t with a Softmax activation output layer for single topic classification. While for the multi-topic classification, we assign each segment multiple topics T by replacing the Softmax with Sigmoid function. We formulate segment labeling process as:

$$t_i = \text{Softmax}(W_y H_i + b_y) \quad \text{and} \quad T_i = \text{Sigmoid}(W_z H_i + b_z). \tag{11}$$

3.4 Joint Learning

We jointly train Tipster with multi-task loss function:

$$\mathcal{L} = (1 - \alpha - \beta)\mathcal{L}_{\text{Seg}} + \alpha\mathcal{L}_{\text{Topic}} + \beta\mathcal{L}_{\text{NTM}}, \tag{12}$$

where $\mathcal{L}_{\text{Seg}}$ and $\mathcal{L}_{\text{Topic}}$ are segmentation loss and topic labeling loss, which are cross entropy calculated at sentence level and segment level respectively.

4 Experiments

Experimental Setup. We conduct extensive experiments to demonstrate the effectiveness of Tipster on five public datasets: *Wiki-Section* [1], *Wiki-50* [8], *Cities* [3], *Elements* [3] and *Clinical* [5]. *Wiki-Section* contains four Wikipedia datasets across various languages and domains: English Language City(**en_city**), English Language Disease (**en_disease**), German Language City (**de_city**) and German Language Disease (**de_disease**). Table 1 shows their Statistics. We compare our model with seven popular methods: TopicTiling [13], Textseg [8], SEC>T+emb [1], CATS [6], S-LSTM [2], Cross-Segment BERT (CS-BERT) and BERT-LSTM [10]. For Tipster, we use pre-trained model *bert-base-uncased* for English Datasets as bert embedding, *bert-base-german-cased* for German Datasets. For topic-guided sentence embedding, we set both the number of topics

Table 1. Statistics of all datasets used in our experiments.

Dataset	Wiki-Section				Wiki-50	Cities	Elements	Clinical
	en_disease	de_disease	en_city	de_city				
# Docs	3.6k	2.3k	19.5k	12.5k	50	100	118	227
# Seg/Doc	7.5	7.2	8.3	7.6	3.5	12.2	6.8	5.0
# Sent/Seg	7.8	6.3	6.8	5.3	13.6	5.2	3.3	28.0
Topic label	✓	✓	✓	✓	✓	✓	✓	✗

Table 2. Results for text segmentation, single-label and multi-label classification on *WikiSection* datasets. "n/a" denotes the model is inapplicable to the subtask.

WikiSection single-label model	en_disease 27 topics			de_disease 25 topics			en_city 30 topics			de_city 27 topics		
	↓P_k	↑F_1	↑MAP	↓P_k	↑F_1	↑MAP	↓P_k	↑F_1	↑MAP	↓P_k	↑F_1	↑MAP
TopicTiling	43.4	n/a	n/a	45.4	n/a	n/a	30.5	n/a	n/a	41.3	n/a	n/a
Textseg	24.3	n/a	n/a	35.7	n/a	n/a	19.3	n/a	n/a	27.5	n/a	n/a
SEC>T+emb	26.3	55.8	69.4	27.5	48.9	65.1	15.5	71.6	81.0	16.2	71.0	81.1
S-LSTM	21.2	57.5	70.9	19.7	52.3	67.1	10.5	74.5	82.2	10.2	75.8	83.6
CS-BERT	18.7	n/a	n/a	20.5	n/a	n/a	11.7	n/a	n/a	11.6	n/a	n/a
BERT-LSTM	16.8	n/a	n/a	15.3	n/a	n/a	9.3	n/a	n/a	9.8	n/a	n/a
Tipster	**14.2**	**62.2**	**74.6**	**13.7**	**57.0**	**70.8**	**8.3**	**79.8**	**86.2**	**7.9**	**78.8**	**86.3**
Tipster-NTM	15.5	59.3	72.1	15.3	55.8	69.3	9.0	77.4	84.7	10.7	78.4	85.9

multi-label model	179 topics			115 topics			603 topics			318 topics		
	↓P_k	↑P@1	↑MAP	↓P_k	↑P@1	↑MAP	↓P_k	↑P@1	↑MAP	↓P_k	↑P@1	↑MAP
SEC>T+emb	30.7	50.5	57.3	32.9	26.6	36.7	17.9	72.3	71.1	19.3	68.4	70.2
S-LSTM	22.1	52.7	59.8	19.5	35.4	45.2	10.2	73.1	71.4	10.7	73.7	74.5
Tipster	**13.7**	**60.8**	**66.2**	**15.9**	**46.4**	**56.9**	**8.4**	**79.0**	**76.2**	**7.5**	**78.6**	**79.3**
Tipster-NTM	14.8	58.4	63.5	17.6	45.9	54.8	9.1	77.5	74.3	8.9	76.9	78.1

Table 3. Results for transferring evaluation. Models marked with △ are trained on the big corpus *Wiki-727K*, while the models marked with □ are trained on en_city for *Wiki-50* and *Cities*, en_disease for *Elements* and *Clinical*.

Segmentation multi-label	Wiki-50		Cities		Elements		Clinical
	↓P_k	↑MAP	↓P_k	↑MAP	↓P_k	↑MAP	↓P_k
TextSeg△	18.2	n/a	19.7	n/a	41.6	n/a	**30.8**
SEC>H+emb□	40.5	13.4	33.3	53.6	43.3	9.5	36.5
S-LSTM□	22.7	16.6	21.2	54.2	30.2	19.1	36.1
Tipster□	19.2	**20.8**	18.2	**59.4**	27.3	**22.5**	32.9
CATS△	**16.5**	n/a	**16.9**	n/a	**18.4**	n/a	-

and normalized flow as 5. We set the GRU layer size to 128. We train our model with ADAM optimizer, 5e-4 learning rate. The tradeoff α and β are both set to 0.1 in all experiments. Following [1,2], we adopt P_k as the segmentation measure. As for segment topic labeling, we report *F1* and *Mean Average Precision* (MAP) for single-label task, *Precision@1* and MAP for multi-label task.

Performance in Single-Label and Multi-label Task. As shown in Table 2, we evaluate Tipster on *WikiSection* for both the single-label and multi-label TATS. Regarding segmentation error, our model reduced P_k significantly by 1.5 points on average compared to BERT-LSTM. As for the single-topic classification, our model reached 69.5% on F_1 and 79.5% on MAP on average, which improved over 4.5 and 3.5 points compared to S-LSTM, respectively. While for the multi-topic classification, which is more closely to real-world scenarios,

Tipster has a consistent improvement for the single-topic classification, which demonstrates the favorable robustness of our model under complex scenarios. From the overview, our model achieves the SOTA performance, which suggests that incorporating word distributions and assign rational attention weight to meaningful words contribute to the task. To validate that the proposed module of topic model embedding incorporating word distributions contributes to the TATS, we remove the NTM and obtain the sentence representations via average pooling, which we denote as "Tipster-NTM". Table 2 shows that Tipster without the NTM reduces the performance in segmentation and topic classification. Therefore, performing reasonable attention on the meaningful words assists with TATS.

Transferring Evaluation. We evaluate Tipster transferability on *Wiki-50*, *Cities*, *Elements* and *Clinical* by training it on the *WikiSection* datasets of corresponding domains. Table 3 shows the results for transferring evaluation on four existing datasets. Owing to the large-scale training corpus *Wiki-727k* [8], CATS outperforms other models on *Wiki-50*, *Cities* and *Elements* datasets. Our model, Tipster outperforms the other supervised models trained on the tiny *WikiSection* and other mainstream unsupervised models. This result illustrates that our model is equipped with a favorable transferability.

5   Conclusion

In this paper, we introduced Tipster, a topic-guided model explicitly incorporating word distributions for topic-aware segmentation task. Our model not only achieved the SOTA performance on TATS, but also showed a novel pooling attention mechanism for TATS. Finally, we conduct extensive experiments to demonstrate the effectiveness and transferability of Tipster.

Acknowledgement. This research was partially supported by grants from the National Natural Science Foundation of China (Grants No. 61922073 and U20A20229), and the Foundation of State Key Laboratory of Cognitive Intelligence, iFLYTEK, P. R. China (No. CIOS-2020SC05).

References

1. Arnold, S., Schneider, R., Cudré-Mauroux, P., Gers, F.A., Löser, A.: SECTOR: a neural model for coherent topic segmentation and classification. Trans. Assoc. Comput. Linguist. **7**, 169–184 (2019)
2. Barrow, J., Jain, R., Morariu, V., Manjunatha, V., Oard, D.W., Resnik, P.: A joint model for document segmentation and segment labeling. In: Proceedings of ACL (2020)
3. Chen, H., Branavan, S., Barzilay, R., Karger, D.R.: Content modeling using latent permutations. J. Artif. Intell. Res. **36**, 129–163 (2009)
4. Devlin, J., Chang, M.W., Lee, K., Toutanova, K.: BERT: pre-training of deep bidirectional transformers for language understanding. In: Proceedings of ACL (2019)

5. Eisenstein, J., Barzilay, R.: Bayesian unsupervised topic segmentation. In: Proceedings of EMNLP (2008)
6. Glavaš, G., Somasundaran, S.: Two-level transformer and auxiliary coherence modeling for improved text segmentation. In: Proceedings of AAAI (2020)
7. Kingma, D.P., Welling, M.: Auto-encoding variational Bayes. In: Proceedings of ICLR (2014)
8. Koshorek, O., Cohen, A., Mor, N., Rotman, M., Berant, J.: Text segmentation as a supervised learning task. In: Proceedings of ACL (2018)
9. Li, B., Zhou, H., He, J., Wang, M., Yang, Y., Li, L.: On the sentence embeddings from pre-trained language models. In: Proceedings of EMNLP (2020)
10. Lukasik, M., Dadachev, B., Papineni, K., Simões, G.: Text segmentation by cross segment attention. In: Proceedings of EMNLP (2020)
11. Rezende, D., Mohamed, S.: Variational inference with normalizing flows. In: Proceedings of ICML (2015)
12. Rezende, D.J., Mohamed, S., Wierstra, D.: Stochastic backpropagation and approximate inference in deep generative models. In: Proceedings of ICML (2014)
13. Riedl, M., Biemann, C.: TopicTiling: a text segmentation algorithm based on LDA. In: Proceedings of ACL 2012 Student Research Workshop (2012)
14. Yang, Z., Yang, D., Dyer, C., He, X., Smola, A., Hovy, E.: Hierarchical attention networks for document classification. In: Proceedings of ACL (2016)

SimEmotion: A Simple Knowledgeable Prompt Tuning Method for Image Emotion Classification

Sinuo Deng, Ge Shi, Lifang Wu$^{(\boxtimes)}$, Lehao Xing, Wenjin Hu, Heng Zhang, and Ye Xiang

Faculty of Information Technology, Beijing University of Technology, Beijing, China
lfwu@bjut.edu.cn

Abstract. Image emotion classification is an important computer vision task to extract emotions from images. The state-of-the-art methods for image emotion classification are primarily based on proposing new architectures and fine-tuning them on pre-trained Convolutional Neural Networks. Recently, learning transferable visual models from natural language supervision has shown great success in zero-shot settings due to the easily accessible web-scale training data, i.e., CLIP. In this paper, we present a conceptually simple while empirically powerful framework for supervised image emotion classification, SimEmotion, to effectively leverage the rich image and text semantics entailed in CLIP. Specifically, we propose a prompt-based fine-tuning strategy to learn task-specific representations while preserving knowledge contained in CLIP. As image emotion classification tasks lack text descriptions, *sentiment-level concept* and *entity-level information* are introduced to enrich text semantics, forming knowledgeable prompts and avoiding considerable bias introduced by fixed designed prompts, further improving the model's ability to distinguish emotion categories. Evaluations on four widely-used affective datasets, namely, Flickr and Instagram (FI), EmotionROI, Twitter I, and Twitter II, demonstrate that the proposed algorithm outperforms the state-of-the-art methods to a large margin (i.e., 5.27% absolute accuracy gain on FI) on image emotion classification tasks.

Keywords: Prompt tuning · Image emotion classification · Fine-tuning

1 Introduction

Image emotion classification aims to extract emotions evoked in images [16]. Previous methods tackle this challenging but essential task by employing CNN models as encoders to learn concrete image representations. In most cases, these

---

S. Deng and G. Shi—Equal contribution.

© The Author(s), under exclusive license to Springer Nature Switzerland AG 2022
A. Bhattacharya et al. (Eds.): DASFAA 2022, LNCS 13247, pp. 222–229, 2022.
https://doi.org/10.1007/978-3-031-00129-1_15

methods are initialized by using models pre-trained on manually-labeled datasets like the ImageNet dataset. Pre-training methods can constrain the huge parameter search space brought by learning from scratch and transfer knowledge entailed in the dataset to the target tasks, thus achieving great success [7]. However, these pre-training methods, also depicted as *label-supervised pre-training*, are learned from supervised "gold labels". These restricted forms of supervision greatly confine the model's generality since the scale of training data and label categories are limited [6].

Natural language, by contrast, is of high variation and can be easily accessed. It can provide more diversified and abundant information as supervised signals, so as to express and supervise a wider range of visual concepts through its generality. Many recent studies [5,9] have validated that the generalization ability of language-supervised methods is much stronger than that of label-supervised methods, benefiting from which many zero/few-shot vision-language related tasks such as OCR and image captioning, have made great progress. Similar to these tasks, learning representations of strong generalization ability is the prerequisite to guarantee the high performance of emotion classification, and it is also non-trivial to study how to adopt language-supervised pre-training models to the emotion classification task. While language-supervised pre-training methods can better meet the needs of emotion classification compared to label-supervised methods, most of them are tailored for few/zero-shot problems, and how to effectively adapt them to supervised scenarios is still under-explored [9]. A straightforward method is sequentially fine-tuning to transfer knowledge entailed in pre-training models. However, with the model fine-tuning, some previously learned knowledge will be forgotten [2], leading to catastrophic forgetting. Besides, the language-supervised pre-training methods use a large number of image caption pairs in the training phase, but the image classification task has access to images only. The lack of text descriptions forms a gap between pre-training and fine-tuning, hindering the knowledge transfer and resulting in sub-optimal results.

To tackle the above mentioned problems, we propose an embarrassingly simple while effective prompt tuning framework for image emotion classification. We adapt the CLIP-ViT-B/32 [5] model as the encoder, which is able to encode text-image pairs and obtain their representations. Our emotion classification model is built on top of this encoder by placing a fully connected (FC) layer to map visual and text features to a common emotional space. We then propose a new training loss that combines the traditional cross-entropy loss with cosine similarity between the image and its corresponding text. The newly introduced loss can ensure the model learning task-specific information while avoiding the knowledge catastrophic caused by adjusting parameters too quickly. Furthermore, we propose two knowledge enhancement methods to complement limited language-supervised signals and compose knowledgeable prompts. Specifically, we introduce emotional knowledge information, which contains sentiment-level and entity-level. To introduce the emotion concept, we extend the label information by using the Plunchik emotion wheel model [4], which was proposed by psychologists to explain human emotion categories. Then we use the Detetron2 toolkit [11], a famous ready-made detection module, to detect entity information to enrich the entity level information. After enhancing these two parts, the SimEmotion with enriched text input can better exert its performance on image emotion analysis tasks. Finally, we validate the framework on different emotional image datasets and show that it outperforms the state-of-the-art methods to a large margin.

2   Related Work

Our work is closely related to image emotion classification and large-scale language-supervised pre-training methods.

Image emotion classification is usually formulated as an emotion feature extraction problem. Learning discriminative emotion features will facilitate classification performance. To approach this task, previous methods utilized hand-crafted features [15]. Later on, CNNs or other alternative architectures were proposed to boost classification performance. Based on it, researchers tried to incorporate additional information such as salient regions [12]. Besides these, some research managed to enriching feature representations by incorporating external knowledge, such as using a sentiment dictionary [13] or some statistic information to achieve better performance [8].

In recent years, with the rise of pre-training technical in the Natural Language Processing (NLP) field, many popular works are trained by large-scale language data get amazing performance on several text analysis tasks [6,7]. All these researches make use of the comprehensive information of language to dig out more critical feature, which brings more incredible results on their own scenario. However, no one has made similar efforts in the task of image emotion classification in the image modal up to now, which has a strong semantic relevance with language. In this paper, we study how to adapt the model from the different target-designed scenario to the image emotion classification, and we first propose a method to enrich the corresponding text to realize the effective use of large-scale language model by forming the knowledgeable prompts.

3   SimEmotion

Figure 1 shows the overview of our SimEmotion framework. It consists of three components: Knowledgeable Prompt Generation, Semantic Emotion Feature Extraction, and Similarity Constrains Strategy.

Given an image set $\mathcal{M}$ with its corresponding label $\mathcal{L}$, where $\mathcal{M} = \{m_i\}_{i=1}^{n}$ and $\mathcal{L} = \{l_i\}_{i=1}^{n}$, we first extract all three words of the most relative category of the l_i in emotion wheel model, marked as $e_{m_i}^1$, $e_{m_i}^2$ and $e_{m_i}^3$. And we extract and get the entity word of the image by detector toolkit, marked as $e_{n_i}^1$, $e_{n_i}^2$, and $e_{n_i}^3$. Then we put them together into a template to generate a prompt $\mathcal{P} = \{p_i\}_{i=1}^{n}$. After obtaining the prompt, we get image-prompt pairs m, p. Then we apply a visual encoder and a text encoder to embedding them into emotional space to be sentiment vector v_{img} and v_{txt} respectively, further mapping the visual feature to emotion categories $\mathcal{Y}^{\mathcal{M}} = \{y^{M_j}\}_{i=1}^{n}$ by a FC layer. Then the vision-language sentiment features v_{img} and v_{txt} are fed to calculate the similarity loss while the classification loss get by predict category y^{M_i} and the label l_i.

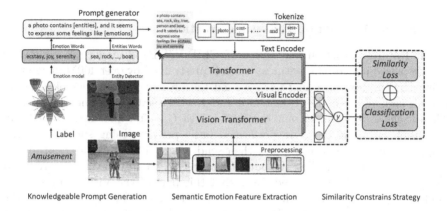

Fig. 1. Illustration of the SimEmotion framework.

3.1 Knowledgeable Prompt Generation

In order to compose a knowledgeable prompt, we added sentiment-level concept information and entity-level information into our designed prompt.

Regarding the description of emotions, psychologists have constructed many classic emotion models to express the emotional state of human beings, such as Mikels emotion model, Ekman emotion model, and "positive-negative" emotion polarity model. Among them, Plutchik's Wheel of Emotions contains 8 basic emotion categories with three different intensities of adjectives to describe each category, which demonstrates a strong ability to describe emotions [4]. However, as a model that also describes human emotions, the models of these emotion datasets can be transformed into Plutchik's emotion model under certain conditions, or the tag words can be generalized in the same style.

Since human have a wealth of associative capabilities for established entity information, which plays a vital role in awakening emotions [12]. It is reasonable to consider extracting entity information in the graph as a supplement to knowledge expansion. Therefore, for the consideration of versatility and convenience, we choose a panoptic segmentation model with relatively better accuracy from a well-known target detection library Detectron2 to detect and extract all entities from the input image $x^{(i)}$ in the image part. Then we extracted the entity words with a confidence level greater than 0.5 in the image and record it as $e_n^i \in W_{en}$ which means entity words.

Through a sentence template similar to "a photo of [*label word*]" in CLIP work, we constructed a sentence template of "a photo contains [*entitywords*], it seems like to express some feelings like [*emotion words*]," denoted as s_p^i which means prompt sentence. Then entered it into the feature extraction part as a pair with the image $\{x^{(i)}, s_p^{(i)}\}_{i=1}^N$.

3.2 Semantic Emotion Feature Extraction

In the feature extraction part, we used CLIP-ViT-B/32 model [5] as LPM to encode paired image-text features, which is pre-trained by over 400M image-text pairs. For the text encoding part, we fed the prompt s_p^i into the tokenization module and get the input of the text encoder, whose setting is same as the CLIP. Then it gets a 512-dimensional output features f_{txt}^i after encoding by LPM. In the vision encoding part, we performed basic pre-processing such as scaling, cropping, and normalization, and then send image to the Vision Transformer model $\mathcal{M}_{img}$. After sending the output feature f_{txt}^i to a FC layer, the predicted category f_{img}^i will be obtained. Note that this part of the Vision Transformer model $\mathcal{M}_{img}$ is different from the model of text branch. The parameters in the training phase are not fixed. We want to learn this part of the visual encoder through training so that the visual branch can capture more useful information from text.

3.3 Similarity Constrains Strategy

Given the feature f_{img}^i and f_{txt}^i from outputs of the text and vision encoder model, we calculated the difference $D(f_{img}^i, f_{txt}^i)$ between the output of text branch and vision branch by the cosine similarity distance.

Therefore, we set a cosine similarity distance D to ensure that the output features of the visual branch can be closer to the processing result of the language branch so that the visual part can extract more emotional features with language information by using the guidance of the language part of the knowledge. The formula is as follows:

$$L_C = \sum_{i=1}^N [1 - D(f_{img}^{(i)}, f_{txt}^{(i)})] = N - \sum_{i=1}^N \frac{f_{img}^{(i)} \cdot f_{txt}^{(i)}}{\|f_{img}^{(i)}\| \|f_{txt}^{(i)}\|}, \tag{1}$$

where i is the order of the sample in batch, $f_{img}^{(i)}$ is the output feature of Vision Transformer, $f_{txt}^{(i)}$ is the output feature of text branch, and N is the scale of the batch.

Recent studies have pointed out that for a pre-trained model with a large number of parameters such as CLIP [5], low-quality fine-tuning data will cause the problem of catastrophic forgetting, which makes it difficult for the model to use language knowledge effectively. Hence, we only use the artificially generated text as an auxiliary supervision signal for vision encoder training to avoid catastrophic forgetting. During training, we fix the parameters of the text encoder part and only train the vision encoder and classifier parts to make it an emotion classification module that obtains semantic information supervision to complete the image emotion classification task during testing independently.

Apart from this, we employ a cross-entropy loss L_S as classification loss to ensure the image will be classified in the right way. The similarity loss allows the visual model to learn more discriminative emotional features extracted by the powerful language processing capabilities of the text LPM, and the classification loss strengthens the mapping of features in the emotional space to different emotional categories, making the distinction between classes more clear. In order to combine the two parts of the loss, we adopted an additive method. The sum of these two losses is used as the total loss constraint.

$$L = \lambda * L_C + (1 - \lambda) * L_S, \tag{2}$$

where λ is the weight to control the tradeoff between the two losses.

4   Experiments and Analysis

We perform our experiment on four datasets, including Flickr and Instagram (FI) [14], EmotionROI [3], Twitter I [13] and Twitter II [1].

4.1   Classification Performance

As can be seen from the Table 1, the content-based methods which require fewer resources rely on more powerful and more targeted feature extractor design. Also, the deep learning methods show it advanced on emotional features extraction by improving the accuracy rate. The methods incorporating external knowledge are shown in the next four lines of the Table 1. [10] built richer semantic relations through GCN and other methods after task-specific design, which get much better performance on the binary polarity classification task. Although [8] combined more complex models to obtain relatively better results, the introduction of the external information of voting probability makes it difficult to fully implement it when judging more constrained datasets or wild images. Our method surpasses this method on majority datasets by a large margin.

Besides, we only utilize the image and its own label information. The emotion model and entity word extraction modules are all existing plug-and-play parts. The main source of knowledge is a large-scale corpus LPM, which is relatively universal. It can be seen from the results in the Table 1 that the best performance can be obtained on most datasets of different emotion categories and scales.

Our full model, combined with task-specific prompt design, achieves SOTA performance on five datasets except for the Twitter II, whose result is fluctuating and hard-tuning due to its small data scale of only 603 images.

Table 1. The results of SimEmotion on different datasets comparing with baseline methods

Method	FI_8	FI_2	EmotionROI_6	EmotionROI_2	Twitter I	Twitter II
Zhao et al.	0.4613	0.5842	0.3484	0.7345	0.6792	0.6751
ResNet101	0.6616	0.7576	0.5160	0.7392	0.7813	0.7823
WSCNet	0.7007	-	**0.5825**	-	0.8425	0.8135
SentiBank	0.4923	0.5647	0.3524	0.6618	0.6663	0.6593
DeepSentiBank	0.5154	0.6154	0.4253	0.7011	0.7125	0.7023
Rao et al.	**0.7546**	0.8751	-	0.8294	-	-
Wu et al.	-	**0.8871**	-	0.8429	**0.8965**	**0.8268**
SimEmotion	**0.7962**	**0.9398**	**0.7003**	**0.8906**	**0.9094**	0.8250

Table 2. The results of ablation study with different prompt type on different datasets.

Lw	Sw	Ew	FI_8	FI_2	EmotionROI_6	EmotionROI_2	Twitter I	Twitter II
✓			0.7924	0.9363	0.6919	0.8889	0.9055	0.7750
	✓		0.7954	0.9392	0.6936	**0.8939**	0.9016	0.8167
✓		✓	0.7959	0.9366	0.6953	0.8838	**0.9094**	**0.8250**
	✓	✓	**0.7962**	**0.9398**	**0.7003**	0.8906	**0.9094**	**0.8250**

4.2 Ablation Study

We did the ablation study by four parts, *"label word"*, *"sentiment words"* and *"entity words"*, which are abbreviated as *"Lw"*, *"Sw"* and *"Ew"* in the Table 2. As the supervised information, the label will always be used. So there are two statuses about label information ablation study, only use the label word or three generated emotion words instead. When chose the *"label word"* part, the prompt type is only a label word just like *"amusement"*, *"anger"*. And the related word of the label word in the Plunthick wheel emotion model will be set to prompt while the *"Sw"* is selected. When choosing the *"Ew"*, we can get better results than the label word. The lines which marked *"Ew"* selected means the part *"contains..."* is added to the first half of the sentence, just like "a photo contains $Ew_1, Ew_2, \ldots, Ew_{M-1}$ and Ew_M...". Since we have confidence limits on the words in the entity word extraction part, the words in this part may not exist in all the data. In the sample without corresponding entity words, we use the most basic label sentence to replace it. The proportion of this part is less than 10% of the total sample, and it is evenly distributed in the training set and test set. The results in the Table 2 show that in most cases, adopting the prompt design of the entity word part can make the classification accuracy of emotion images have better results.

5   Conclusion

In this paper, we proposed a conceptually simple while empirically powerful framework for supervised image emotion classification, SimEmotion, to effectively leverage the rich image and text semantics entailed in CLIP. Especially, a proposed method is experimentally verified to apply image emotion classification tasks on the CLIP model effectively, which is introduced in the way of fusion sentiment-level concept and entity-level information. Compared with the state-of-the-art method, SimEmotion performs better in several widely-used datasets, including binary categories and multi-categories.

Acknowledgements. This work was supported in part by the National Natural Science Foundation of China under Grant NO. 62106010, 61976010, 62176011, 62106011.

References

1. Borth, D., Ji, R., Chen, T., Breuel, T., Chang, S.F.: Large-scale visual sentiment ontology and detectors using adjective noun pairs. In: Proceedings of the 21st ACM International Conference on Multimedia, pp. 223–232 (2013)
2. Chen, S., Hou, Y., Cui, Y., Che, W., Liu, T., Yu, X.: Recall and learn: Fine-tuning deep pretrained language models with less forgetting. arXiv preprint arXiv:2004.12651 (2020)
3. Peng, K.C., Sadovnik, A., Gallagher, A., Chen, T.: Where do emotions come from? predicting the emotion stimuli map. In: 2016 IEEE International Conference on Image Processing (ICIP), pp. 614–618. IEEE (2016)
4. Plutchik, R.: Emotions: a general psychoevolutionary theory. Approaches Emot. **1984**, 197–219 (1984)
5. Radford, A., et al.: Learning transferable visual models from natural language supervision. arXiv preprint arXiv:2103.00020 (2021)
6. Radford, A., Narasimhan, K., Salimans, T., Sutskever, I.: Improving language understanding by generative pre-training (2018)
7. Radford, A., Wu, J., Child, R., Luan, D., Amodei, D., Sutskever, I., et al.: Language models are unsupervised multitask learners. OpenAI Blog **1**(8), 9 (2019)
8. Rao, T., Li, X., Zhang, H., Xu, M.: Multi-level region-based convolutional neural network for image emotion classification. Neurocomputing **333**, 429–439 (2019)
9. Shen, S., et al.: How much can clip benefit vision-and-language tasks? arXiv preprint arXiv:2107.06383 (2021)
10. Wu, L., Zhang, H., Deng, S., Shi, G., Liu, X.: Discovering sentimental interaction via graph convolutional network for visual sentiment prediction. Appl. Sci. **11**(4), 1404 (2021)
11. Wu, Y., Kirillov, A., Massa, F., Lo, W.Y., Girshick, R.: Detectron2 (2019). https://github.com/facebookresearch/detectron2
12. Yang, J., She, D., Lai, Y.K., Rosin, P.L., Yang, M.H.: Weakly supervised coupled networks for visual sentiment analysis. In: Proceedings of the IEEE Conference on Computer Vision and Pattern Recognition, pp. 7584–7592 (2018)
13. You, Q., Luo, J., Jin, H., Yang, J.: Robust image sentiment analysis using progressively trained and domain transferred deep networks. In: Twenty-Ninth AAAI Conference on Artificial Intelligence (2015)
14. You, Q., Luo, J., Jin, H., Yang, J.: Building a large scale dataset for image emotion recognition: the fine print and the benchmark. In: Proceedings of the AAAI Conference On Artificial Intelligence, vol. 30 (2016)
15. Zhao, S., Gao, Y., Jiang, X., Yao, H., Chua, T.S., Sun, X.: Exploring principles-of-art features for image emotion recognition. In: Proceedings of the 22nd ACM International Conference on Multimedia, pp. 47–56 (2014)
16. Zhao, S., et al.: Affective image content analysis: two decades review and new perspectives. IEEE Trans. Pattern Anal. Mach. Intell. (2021)

Predicting Rumor Veracity on Social Media with Graph Structured Multi-task Learning

Yudong Liu[1,2], Xiaoyu Yang[1], Xi Zhang[1(✉)], Zhihao Tang[1], Zongyi Chen[1], and Zheng Liwen[1]

[1] Key Laboratory of Trustworthy Distributed Computing and Service (MoE), Beijing University of Posts and Telecommunications, Beijing, China
{yudong.liu,littlehaes,zhangx,innerone,zongyi_chen,zhenglw}@bupt.edu.cn
[2] Beijing Electronic and Science Technology Institute, Beijing, China

Abstract. Previous studies have shown that the multi-task learning paradigm with the stance classification could facilitate the successful detection of rumours, but the shared layers in multi-task learning tend to yield a compromise between the general and the task-specific representation of structural information. To address this issue, we propose a novel **Multi-Task Learning** framework with **Shared Multi-channel Interactions** (MTL-SMI), which is composed of two shared channels and two task-specific graph channels. The shared channels extract task-invariant text features and structural features, and the task-specific graph channels, by interacting with the shared channels, extract the task-enhanced structural features. Experiments on two realworld datasets show the superiority of MTL-SMI against strong baselines.

Keywords: Rumor veracity · Stance classification · Multi-task learning · Graph neural network

1 Introduction

A rumor is an item of circulating information whose veracity status has not been verified at the time of posting. Recently, related research has proved the high pertinence between the rumor verification and stance classification tasks. Most of these studies aim to use various deep neural networks, such as LSTM [2], GRU [8] and Transformer [18], to extract textual features. However, they failed to make full use of the network structure existed in social media, which has been proven effective for rumor detection tasks [5–7].

As shown in Fig. 1 (a), a conversation thread consists of a source post with a veracity label, comments with stance labels, and the users. Figure 1 (b) is the constructed conversation network. Comment 4 and 2 come from the same user and have similar stance toward the source post, hence having less impact on

---

Y. Liu and X. Yang—Equal contribution.

A. Bhattacharya et al. (Eds.): DASFAA 2022, LNCS 13247, pp. 230–237, 2022.
https://doi.org/10.1007/978-3-031-00129-1_16

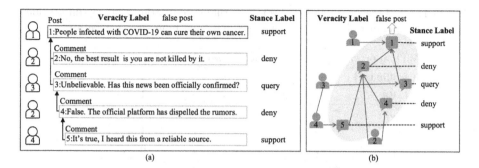

Fig. 1. (a) A conversation thread with four different users. The leading numbers are the timestamp order. (b) The conversation network based on the left thread.

rumor verification task than two independent comments, which motivates us to employ extra task-specific channels in MTL-SMI. Besides, we use combination of different neural networks in the shared channels and interactions between channels to enhance the ability of representation learning.

Our contributions can be summarized as follows:

- To improve the accuracy of rumor verification, we propose a multi-task learning framework MTL-SMI, which utilizes a shared text channel combining different neural networks, three graph channels, and multi-channel interactions to enhance the ability of task-specific representation learning.
- We conduct extensive experiments on two benchmark datasets and the results demonstrate the effectiveness of our key components.

2   Related Work

Rumor verification task focuses on the truthfulness of a rumor post. Previous studies [12–14] commonly rely on traditional hand-crafted features. Nowadays, deep learning methods are more popular. Ma et al. [15] propose two recursive neural models respectively for rumor representation learning and classification. Dungs et al. [16] model the temporal changes in stance information with multi-spaced Hidden Markov Model. Li et al. [17] generates the representation of posts through an attention-based LSTM network and use an ensemble of the traditional classification algorithms and neural network models for rumor verification.

Stance classification has gradually matured in recent years. Kochkina et al. [1] build a multi-task learning framework with hard parameter sharing. Ma et al. [3] propose a model with both shared layers and task-specific layers to enhance each task. Li et al. [4] take the user credibility into consideration. Wei et al. [7] consider both the text feature, and structural features of the conversation network. Yu et al. [10] propose a Coupled Transformer Module to capture the interactions among tasks and promote the accuracy of rumor verification by using the predicted stance labels. Table 1 shows the differences among these multi-task learning frameworks.

Table 1. Multi-task learning frameworks for rumor verification task. "N-Interaction" refers to the number of interaction channels to extract representative features.

MTL approch	Post	Comment	User	Network	#N-Interaction
Kochkina [1]	✓	✓	✗	✗	1
Yu [10]	✓	✓	✗	✗	1
Ma [3]	✓	✓	✗	✗	1
Li [4]	✓	✓	✓	✗	1
Wei [7]	✓	✓	✗	✓	0
MTL-SMI (Ours)	✓	✓	✓	✓	2

3   Problem Statement

We follow the problem setting as previous studies [7, 10] and denote the dataset as a set of conversations $\mathbb{D} = \{C_1, C_2, \ldots, C_{|\mathbb{D}|}\}$, where C_i is composed of a source post and corresponding comments. For conversation $C_i = \{S_0^i, R_1^i, R_2^i, \ldots, R_n^i\}$, the source post is denoted as S_0^i, and the attached n comments are denoted as $R_1^i, R_2^i, \ldots, R_n^i$. The goal of stance classification is to learn a classifier g : $(S_0^i, R_j^i) \rightarrow s_i$, and s_i takes one of the four possible stance labels: support, deny, query and comment. The goal of rumor verification is to learn a classifier $f : S_0^i \rightarrow y_i$, where y_i is of three possible labels: true, false and unverified.

Given a conversation network $G = (V, E, A)$, V is the node set including user, post, and comment nodes, E is the edge set and $\mathbf{A} \in \{0, 1\}^{|V| \times |V|}$ is the adjacency matrix. And an edge is established between the following node pair: (1) a user and the comment or post that he or she published; (2) two users according to the following or followed relationship; (3) two comments if one comment is commented by the other. Our conversation network G is an undigraph.

4   Methodology

In this section, we propose the MTL-SMI framework as Fig. 2, which consists of four channels: RV Graph Channel, Shared Graph Channel, Shared Text Channel, SC Graph Channel; and two MLP classfiers: RV and SC.

SGC (Shared Graph Channel) is a two-layer GCN [9]. Initial nodes, H^0, are represented by average of word vectors of text content or user profile.

STC (Shared Text Channel). Each text sequence is truncated or filled to fixed-length L, to which three steps are applied: (1) the pre-trained BERT [11] to get the initial word vectors; (2) SIGNED-TRANS, which is a signed co-attention layer [19] with the "-softmax" channel to capture both the positive and negative correlation among words; (3) text-cnn to aggregate the words vector into the final feature vector.

RVGC (RV Graph Channel). The difference with the SGC is in the input of the 2nd layer of the GCN. To interact with two shared channels, three vectors

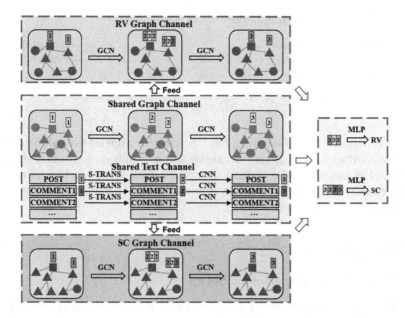

Fig. 2. Multi-task learning framework MTL-SMI. The circle, square, and triangle stand for user node, source post node, and comment node respectively (Best viewed in color).

of the same node from the 1st layer of SGC, STC, and RVGC, are concatenated as input of the 2nd layer of RVGC. In this way, we extend the input information for the second layer and thus extract efficient features with GCN.

SCGC (SC Graph Channel). The SCGC has the same structure as the RVGC, but dedicated to the SC task.

RV (Rumor Verification). We concatenate post node embedding from the RVGC, SGC, and STC to predicate the veracity category with a fully connected layer.

SC (Stance Classification). We concatenate comment features from the SGC, STC, SCGC, and the post feature from the STC, to predicate the stance category with a fully connected layer.

Overall Loss. A cross-entropy based loss function is defined as the overall loss of MTL-SMI:

$$\mathcal{L}(\theta_{rv}, \theta_{sc}) = \sum_{a=1}^{N_{rv}} \sum_{c_{rv} \in \{0,1,2\}} -\mathbf{y}_a^{rv} \log \hat{\mathbf{y}}_a^{rv} + \lambda \sum_{b=1}^{N_{sc}} \sum_{c_{sc} \in \{0,1,2,3\}} -\mathbf{y}_b^{sc} \log \hat{\mathbf{y}}_b^{sc}$$
(1)

where N_{rv} and N_{sc} represent the data volume. c_{rv} and c_{sc} are the categories. $\mathbf{y}_a^{rv}$ and $\mathbf{y}_b^{sc}$ represent the one-hot labels. λ is used to control the proportion of the stance classification task loss in the overall loss.

Table 2. Statistics of the datasets.

Dataset	Threads	Tweets	Stance labels				Rumor veracity labels		
			#Support	#Deny	#Query	#Comment	#True	#False	#Unverified
SemEval	325	5,568	1,004	415	464	3,685	145	74	106
PHEME	2,402	105,534	–	–	–	–	1,067	638	697

STL-GT (Single Task Learning Setting on the Shared Graph and Text Channel). After removing the two task-specific graph channels in Fig. 2, the left shared channels can be trained in a single task setting, which is used to illustrate the efficiency of the combined neural networks and serve as one of the base models in the ablation test.

5   Experiments

5.1   Datasets and Experiment Setup

According to previous studies [1,7,10], we experiment on two datasets: (1) The SemEval [20], whose training and validation sets are related to eight events and the test set covers another two events, with stance and veracity labels. (2) The PHEME [1] dataset contains nine events with only verification labels. We perform leave-one-event-out cross-validation on this dataset for rumor verification task. The statistical information of the two datasets is shown in Table 2.

We use Macro-F_1 as the main evaluation metric and Accuracy as the secondary evaluation metric, aiming to improve the performance of rumor verification without considering the metrics of stance classification. GCN in different channels all have two layers. λ in formula 1 is set to 1. We use the Adam optimizer with $\beta_1 = 0.9$, $\beta_2 = 0.999$, and learning rate is 0.00005.

5.2   Results of Rumor Verification with STL-GT and MTL-SMI

The following baselines are choosen to compair with STL-GT and MTL-SMI respectively.

- **Single Task Baselines**
    - **BranchLSTM** [?] decomposes the tree conversation structure into linear structure, and uses a sequence model based on LSTM to incorporate structural information for classification.
    - **TD-RvNN** [15] models the top-down conversation tree structure to capture complex propagation patterns and classifies rumors with RNN.
    - **Hierarchical GCN-RNN** [7] is a variant of Conversational-GCN for single tasks, which does not use the stance labels in the training process.
- **Multi-Task Baselines**
    - **BranchLSTM + NileTMRG** [21] extracts the textual features with BranchLSTM, and uses SVM for classification.

Table 3. Rumor verification results with STL-GT and MTL-SMI.

Setting	Method	SemEval		PHEME	
		#Macro-F_1	#Acc	#Macro-F_1	#Acc
Single-Task	BranchLSTM	0.491	0.500	0.259	0.314
	TD-RvNN	0.509	0.536	0.264	0.341
	Hierarchical GCN-RNN	0.540	0.536	0.317	0.356
	STL-GT	**0.618**	**0.607**	**0.359**	**0.430**
Multi-Task	BranchLSTM+NileTMRG	0.539	0.570	0.297	0.360
	MTL2(Veracity+Stance)	0.558	0.571	0.318	0.357
	Hierarchical PSV	0.588	0.643	0.333	0.361
	MTL-SMI	**0.685**	**0.679**	**0.409**	**0.468**

- **MTL2 (Veracity+Stance)** [1] adopts a multi-task learning framework with hard parameter sharing at the bottom layers, and use the output of task-specific layer for classification.
- **Hierarchical PSV** [7] uses Conversational-GCN to obtain features of the source post and comments. Then it extracts the temporal features of the conversation with GRU for rumor verification.

Performance Analysis. Table 3 shows the results of STL-GT and MTL-SMI in rumor verification task.

STL-GT achieves the best results on both datasets in the single-task setting. (1) Sequence models like BranchLSTM and TD-RvNN, perform worse than Hierarchical GCN-RNN and STL-GT, which demonstrates the importance of the structural features of conversation graph. (2) STL-GT, using BERT, SIGNED-TRANS in text processing, perform better than Hierarchical GCN-RNN with the traditional text processing technique.

MTL-SMI achieves the best results on both datasets in multi-task settings. (1) BranchLSTM+NileTMRG performs the worst in that it is not trained in an end-to-end manner. (2) Hierarchical PSV and MTL-SMI consider the structural information of conversation networks, hence performing better. (3) MTL-SMI with shared multi-channel interactions could effectively capture the relationship between two tasks and generates more powerful features for rumor verification.

5.3 Ablation Tests

In order to test the effectiveness of different components in MTL-SMI, we set up the following ablation experiments. (1) **M/INTER** is MTL-SMI without considering the interactions between task-specific channels and shared channels. (2) **M/GRAPH** only considers the interaction between Shared Text Channel. (3) **M/TEXT** only considers the interaction between Shared Graph Channel. (4) **M/SIGNED** is MTL-SMI without adding the "-softmax" channel in the multi-head self-attention mechanism. (5) **STL-GT** is the single task model.

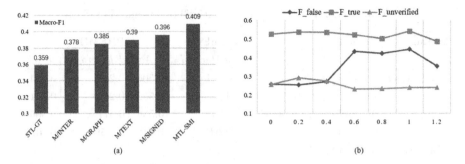

Fig. 3. (a) Results of ablation tests on SemEval dataset. (b) F_1 scores of each category under different λ values on SemEval dataset.

Performance Analysis. It can be seen from Fig. 3 (a) that: (1) Compared with the STL-GT, the MTL-SMI performs better, proving the effectiveness of multi-task learning. (2) M/GRAPH and M/TEXT, considering only one kind of interaction, still perform better than M/INTER, showing that the multi-channel interaction could benefit the performance. (3) M/TEXT and M/GRAPH perform similarly, showing that textual and network structural information are of equal importance. (4) MTL-SMI perform better than M/SIGNED, indicating that negative correlation may extract complementary semantics to help the task.

Relationship of Two Tasks. We adjust the proportion of stance classification loss by changing λ values in formula 1 to investigate the influence of SC on RV task. From Fig. 3(b), we can see that: (1) the SC task helps the RV task when $\lambda < 1$ since the F_1 scores are increasing with the increase of λ; (2) the F_1 scores show a downward trend when $\lambda > 1$. The possible reason is that the model pay more attention to the stance classification. Therefore, adjusting $\lambda = 1$ is crucial for the multi-task learning.

6   Conclusion

In this paper, we construct the conversation networks with three types of nodes and employ three graph channels to best utilize the structure information. MTL-SMI is designed to improve the accuracy of rumor verification. It has two shared channels, two task-specific graph channels and use feature interaction among channels to improve the representation of features. Results of the experiments on two datasets demonstrate the effectiveness of the MTL-SMI. Further investigation analyse the potency of key components and correlation of the two tasks.

Acknowledgements. This work was supported by the Natural Science Foundation of China (No. 61976026, No. 61902394) and 111 Project (B18008).

References

1. Kochkina, E., Liakata, M., Zubiaga, A.: All-in-one: multi-task learning for rumour verification. In: COLING, pp. 3402–3413 (2018)
2. Hochreiter, S., Schmidhuber, J.: Long short-term memory. Neural Comput. **9**(8), 1735–1780 (1997)
3. Ma, J., Gao, W., Wong, K.: Detect rumor and stance jointly by neural multi-task learning. In: Companion of the the Web Conference 2018, pp. 585–593 (2018)
4. Li, Q., Zhang, Q., Si, L.: Rumor detection by exploiting user credibility information, attention and multi-task learning. In: ACL, pp. 1173–1179 (2019)
5. Yang, X., et al.: Rumor detection on social media with graph structured adversarial learning. In: IJCAI, pp. 1417–1423 (2020)
6. Bian, T., Xiao, X., Xu, T.: Rumor detection on social media with bi-directional graph convolutional networks. In: AAAI, pp. 549–556 (2020)
7. Wei, P., Xu, N., Mao, W.: Modeling conversation structure and temporal dynamics for jointly predicting rumor stance and veracity. In: EMNLP, pp. 4786–4797 (2019)
8. Cho, K., van Merrienboer, B.: Learning phrase representations using RNN encoder-decoder for statistical machine translation. In: EMNLP, pp. 1724–1734 (2014)
9. Kipf, T.N., Welling, M.: Semi-supervised classification with graph convolutional networks. In: ICLR, Conference Track Proceedings, OpenReview.net (2017)
10. Yu, J., et al.: Coupled hierarchical transformer for stance-aware rumor verification in social media conversations. In: EMNLP (2020)
11. Devlin, J., Chang, M., Lee, K., Toutanova, K.: BERT: pre-training of deep bidirectional transformers for language understanding. In: NAACL-HLT, pp. 4171–4186 (2019)
12. Castillo, C., Mendoza, M.: Information credibility on Twitter. In: Proceedings of the 20th International Conference on World Wide Web, pp. 675–684 (2011)
13. Ma, J., Gao, W., Wei, Z.: Detect rumors using time series of social context information on microblogging websites. In: CIKM, pp. 1751–1754 (2015)
14. Wu, K., Yang, S.: False rumors detection on Sina Weibo by propagation structures. In: 31st IEEE International Conference on Data Engineering, pp. 651–662 (2015)
15. Ma, J., Gao, W., Wong, K.: Rumor detection on Twitter with tree-structured recursive neural networks. In: ACL, Volume 1: Long Papers, pp. 1980–1989 (2018)
16. Dungs, S., Aker, A., Fuhr, N., Bontcheva, K.: Can rumour stance alone predict veracity? In: COLING, pp. 3360–3370 (2018)
17. Li, Q., Zhang, Q., Si, L.: eventAI at SemEval-2019 task 7: rumor detection on social media by exploiting content, user credibility and propagation information. In: Proceedings of the 13th International Workshop on Semantic Evaluation, pp. 855–859 (2019)
18. Vaswani, A., Shazeer, N., Parmar, N.: Attention is all you need. In: Annual Conference on Neural Information Processing Systems 2017, pp. 5998–6008 (2017)
19. Tian, T., et al.: QSAN: a quantum-probability based signed attention network for explainable false information detection. In: CIKM, pp. 1445–1454 (2020)
20. Derczynski, L., Bontcheva, K., Liakata, M.: SemEval-2017 task 8: rumoureval: determining rumour veracity and support for rumours. In: Proceedings of the 11th International Workshop on Semantic Evaluation, pp. 69–76 (2017)
21. Liu, X., Nourbakhsh, A., Li, Q.: Real-time rumor debunking on Twitter. In: CIKM, pp. 1867–1870 (2015)

Knowing What I Don't Know: A Generation Assisted Rejection Framework in Knowledge Base Question Answering

Junyang Huang[1], Xuantao Lu[1], Jiaqing Liang[1], Qiaoben Bao[1], Chen Huang[3], Yanghua Xiao[1,2(✉)], Bang Liu[4], and Yunwen Chen[5]

[1] Shanghai Key Laboratory of Data Science, School of Computer Science, Fudan University, Shanghai, China
{jyhuang19,xtlu20,qbbao19,shawyh}@fudan.edu.cn
[2] Fudan-Aishu Cognitive Intelligence Joint Research Center, Shanghai, China
[3] CES Finance Co., Ltd., Shanghai, China
huangchen@kiiik.com
[4] Mila & DIRO, Université de Montréal, Montréal, Québec, Canada
bang.liu@umontreal.ca
[5] DataGrand Inc., Shanghai, China
chenyunwen@datagrand.com

Abstract. Existing Knowledge Base Question Answering (KBQA) systems suffer from the sparsity issue of Knowledge Graphs (KG). To alleviate the issue of KG sparsity, some recent research works introduce external text for KBQA. However, such external information is not always readily available. We argue that it is critical for a KBQA system to know whether it lacks the knowledge to answer a given question. In this paper, we present a novel **G**eneration **A**ssisted **R**ejection (GEAR) framework that identifies unanswerable questions well. GEAR can be applied to almost all KBQA systems as an add-on component. Specifically, the backbone of GEAR is a sequence-to-sequence model that generates candidate predicates to rerank the original results of a KBQA system. Furthermore, we devise a Probability Distribution Reranking algorithm to ensemble GEAR and KBQA since the architectural distinctions between GEAR and KBQA are vast. Empirical results and case study demonstrates the effectiveness of our framework in improving the performance of KBQA, particularly in identifying unanswerable questions.

Keywords: Knowledge graph · Knowledge base question answering · Natural language generation

1 Introduction

Knowledge Base Question Answering (KBQA) aims to answer questions by using a large collection of triples in a Knowledge Graph (KG). Nowadays, KBQA

A. Bhattacharya et al. (Eds.): DASFAA 2022, LNCS 13247, pp. 238–246, 2022.
https://doi.org/10.1007/978-3-031-00129-1_17

systems are widely leveraged in lots of web applications due to the fact that a KG is in general highly accurate and contains few noisy data. Unlike many question answering tasks where the answer of a question can be retrieved from the input data, the questions for KBQA tend to be more diverse [12] and some of them are unanswerable because the corresponding facts might be absent [7] in the KG. However, constructing a well-designed and complete KG with high precision requires unaffordable human efforts [11], which limits the coverage of a KG. Thus, more often than not, KG is incomplete to answer an open-domain question.

Building a KBQA system with the ability to identify unanswerable questions is an important but difficult task. However, there has been few research studying this problem. Introducing extra information to supply missing evidence in KG [11] and employing ensemble models with different sources are straightforward ideas to alleviate the problem of unanswerable questions. However, the gold answer may not always appear in external materials. Besides, these methods enlarge noises when retrieving relevant data. Another problem is that predicate linking in KBQA is often modeled as a binary classification task, which requires both positive and negative samples. However, the KG in KBQA can not provide all negative samples because KG is incomplete which incurs overfitting on negative samples and finally negatively impacts on the model's performance. REINFORCE algorithm based KBQA models are able to actively identify unanswerable questions, but they are limited to low accuracy on answerable questions. Godin et al. [6] proposed a ternary reward function that rewards an agent for not answering a question. However, this method realizes this idea by adding a virtual predicate "No Answer" to every entity, which leads to low recall on answerable questions.

To overcome the problems mentioned above, we propose a **G**eneration **A**ssisted **R**ejection (GEAR) framework, which helps KBQA better identify unanswerable questions. The framework is an add-on component for KBQA systems that takes the result of a KBQA system into consideration through the following steps: **(1) Predicate Generation:** GEAR directly generates the predicate to the question through a sequence-to-sequence (seq2seq) model, i.e., "What is the highest point in Canada?" → "highest point". The seq2seq model is trained only on positive samples that overcomes the problem of overfitting on negative samples. **(2) Result Re-ranking:** GEAR re-ranks the answers of the KBQA system through the latent space of the generated predicate from the seq2seq model. Specifically, we apply a Probability Distribution Re-ranking (PDR) algorithm to assemble seq2seq model and KBQA model. The ensemble of a typical seq2seq model and a KBQA model is difficult without the PDR algorithm due to the vast distinction in the structure of seq2seq and KBQA models.

Extensive evaluations and case studies demonstrate that the performance of KBQA models improved significantly when GEAR was incorporated especially for unanswerable questions. In summary, we have made the following contributions:

– We propose a novel framework GEAR for KBQA model that improves the performance of KBQA especially on identifying unanswerable questions. To the best of our knowledge, this is the first attempt to adopt generation methods in identifying unanswerable questions.

- We devise an algorithm PDR to overcome the difficulty in implementing the ensemble of seq2seq models and typical KBQA models. The ensemble of KBQA model and GEAR strengthens robustness and generalization capability of processing a question.
- GEAR has a broad scope of application. The framework can be easily applied to almost every KBQA model, which improves the performance of off-the-shelf KBQA systems.

2   Task Definition

The input of KBQA task is a natural language question, and the output is an answer object from KG or "No Answer" indicating that KBQA can't answer the question. More specifically, a KG is denoted as $\mathcal{K}$, which is a set of triples in the form of (s, p, o), where s, p, o denote subject, predicate and object respectively. Here, the object entity is also called answer entity. The task is, given a natural language question denoted as X containing M tokens $X = [x_1, x_2, \ldots, x_M]$ and the entity of the question, the KBQA system should retrieve the gold answer if KG $\mathcal{K}$ contains the answer. Otherwise, the KBQA system should return "No Answer" indicating no relevant information exists in KG $\mathcal{K}$.

3   Methodology

3.1   Overview of GEAR

There are two core components in GEAR: the Generation Module and the Re-ranking Module. The interaction between GEAR and KBQA model is shown in Fig. 1. First, the question will be sent to a KBQA model. The KBQA model will score every candidate predicate-answer pair retrieved from KG. The score represents the confidence in each candidate predicate-answer pair. Then, the Generation Module generates a sequence probability matrix $\mathbf{C}$ and a predicate sequence in line with the question. Finally, the Re-ranking Module re-ranks the result from KBQA model based on the sequence probability matrix $\mathbf{C}$ through PDR. The output of Re-ranking Module is the final re-ranked score of every candidate predicate-answer pair. If the candidate predicate-answer pair with the highest confidence is lower than hyper-parameter τ, the system will reject answering the question.

3.2   Generation Module

The Generation Module, in this paper, aims to generate the predicate for a given input question sequence via a transformer-based encoder-decoder architecture, which consists of two components: (a) an encoder that computes a representation for each source sentence and (b) a decoder that generates one target word at each timestep since the conditional probability is decomposed into several timesteps.

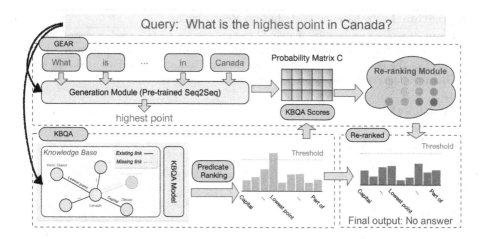

Fig. 1. The illustration of the interaction between GEAR and KBQA. GEAR is composed of a Generation Module (Sect. 3.2) and a Re-ranking Module (Sect. 3.3). GEAR reranks the candidate answers retrieved by the KBQA model.

Given the question token sequence $X = [x_1, x_2, \ldots, x_M]$ as input, GEAR outputs: 1) a sequence represents the predicate of the input question sequence X, which is denoted as $Y = \{y_1, y_2, \ldots, y_L\}$, where y_i is the token-id of the i-th word in Y. M and L are the lengths of input sequence and output sequence respectively. 2) a sequence probability matrix $\mathbf{C}$. The sequence probability matrix is the probability distribution of each candidate character in every step of the generated sequence, which is denoted as $\mathbf{C} = \{\mathbf{c}_1, \ldots, \mathbf{c}_L\}$, L is the length of the sequence. $\mathbf{c}_i = \{c_{i1}, \ldots, c_{iW}\}$ is the character probability distribution of the i-th decoding step, c_{ij} is the j-th token, W is the length of the vocabulary list $\mathcal{V}$. The sequence probability matrix $\mathbf{C}$ will be used as part of the input of the Re-ranking Module.

We adopt the pre-trained language model BART [8] as our transformer-based seq2seq backbone. In this way, the general semantic knowledge of natural language and text generation knowledge can be directly reused. The encoder of the Generation Module first learns the hidden states $\mathbf{H} = \mathbf{h}_i, \ldots, \mathbf{h}_{|M|}$ of the given input question sequence through a multi-layer transformer encoder:

$$\mathbf{H} = \text{Encoder}(x_1, x_2, \ldots, x_M), \tag{1}$$

where each layer of $\text{Encoder}(\cdot)$ is a transformer block with the multi-head attention mechanism. After the input question token sequence is encoded, the decoder predicts the output predicate token-by-token with the sequential input tokens' hidden vectors. At the timestep i of generation, the self-attention decoder predicts the i-th token p_i in the linearized form and decoder hidden state $\mathbf{h}_i^d$ as:

$$y_i, \mathbf{h}_i^d = \text{Decoder}([\mathbf{H}; \mathbf{h}_1^d, \ldots, \mathbf{h}_{i-1}^d], y_{i-1}), \tag{2}$$

Algorithm 1. Probability Distribution Re-ranking

Input: Sequence probability matrix $\mathbf{C}$ from GEAR, candidate predicate sequence $\mathcal{P}$, the score of the candidate predicate $\mathcal{S}_\mathcal{P}$ from KBQA, the weight parameter η, the length of the generated predicate L and the length of the candidate predicate s.

Output: The score $\mathcal{S}$ of the candidate predicate sequence $\mathcal{P}$.

1: $k \leftarrow 0, \mathcal{S} \leftarrow 0$
2: **if** $L \leq s$ **then**
3:    **while** $k \leq s - L$ **do**
4:        $\mathcal{S} \leftarrow \max(\sum_{i=1}^{l} c_{i+k,p_i}, \mathcal{S})$
5:        $k \leftarrow k + 1$
6:    **end while**
7: **else**
8:    **while** $k \leq L - s$ **do**
9:        $\mathcal{S} \leftarrow \max(\sum_{i=1}^{s} c_{i,p_{i+k}}, \mathcal{S})$
10:        $k \leftarrow k + 1$
11:    **end while**
12: **end if**
13: $\mathcal{S} \leftarrow \frac{\eta\mathcal{S}}{max(s,L)} + (1 - \eta)\mathcal{S}_\mathcal{P}$
14: **return** $\mathcal{S}$

---

where each layer of Decoder$(\cdot)$ is a transformer block that includes self-attention with decoder state $\mathbf{h}_i^d$ and cross-attention with encoder state $\mathbf{H}$. The generated output predicate begins with "$<s>$" and ends with the end token "$</s>$". The probability of each step $p(y_i|y_{<i}, x)$ is gradually added to the conditional probability of the entire output sequence $p(y|x)$:

$$p(y|x) = \prod_{i}^{|y|} p(y_i|y_{<i}, X), \tag{3}$$

where $y_{<i} = y_1, \ldots, y_{i-1}$, and $p(y_i|y_{<i}, x)$ is the probability over the target vocabulary list $\mathcal{V}$ normalized by softmax function. At step i, $\mathbf{c}_i$ represents the normalized probability distribution over $\mathcal{V}$. The objective function of GEAR during training is to minimize the negative log-likelihood loss, θ represents the parameters of the model:

$$\mathcal{L} = -\sum_{t=0}^{L} \log P_\theta(y_t|y_{<t}, X), \tag{4}$$

3.3 Re-ranking Module

The goal of the Re-ranking Module is to assign new scores for matched predicates in KG retrieved by KBQA model. Specifically, the PDR algorithm re-ranks the score with the generated results from the Generation Module. PDR is

Table 1. Evaluation of different KBQA models on SimpleQuestions and NLPCCKBQA datasets with a varying fraction of KG-size. "+GEAR" means the result is re-ranked by GEAR. The best results in each group are highlighted in bold.

SimpleQuestions	100%	70%	50%	30%	10%
BiLSTM	80.3/–	79.9/65.0	77.9/68.5	79.7/67.5	**82.1**/66.8
BiLSTM+GEAR	**80.5**/–	**80.2/76.3**	77.9/**77.2**	79.7/**76.2**	79.2/**75.4**
MCCNN	**79.8**/–	**79.4**/71.4	**78.7**/71.9	79.5/71.6	76.8/71.5
MCCNN+GEAR	79.5/–	79.2/**77.2**	77.6/**76.1**	**79.7/76.6**	**77.4/76.6**
BAMnet	**82.1**/–	81.4 / 62.0	80.1/66.4	**81.5**/65.0	**83.9**/64.3
BAMnet+GEAR	81.9/–	80.8/**73.1**	**80.8/74.4**	81.2/**73.5**	83.3/**72.8**
BBKBQA	88.3/–	**88.4**/68.0	88.8/67.0	87.4/66.4	**90.5**/66.2
BBKBQA+GEAR	**88.4**/–	88.2/**72.3**	**89.3/71.2**	**87.8/71.3**	89.2/**71.1**
NLPCCKBQA	100%	70%	50%	30%	10%
BiLSTM	66.8/–	66.7/65.0	65.5/65.4	66.3/64.7	65.3/64.6
BiLSTM+GEAR	**75.7**/–	**75.6/70.2**	**75.1/70.6**	**73.6/69.9**	**75.9/70.1**
MCCNN	77.8/–	77.9/83.2	76.3/84.1	76.9/83.8	77.8/83.6
MCCNN+GEAR	**80.8**/–	**81.0/87.9**	**79.3/88.7**	**80.0/88.3**	**81.1/88.0**
BAMnet	71.6/–	71.9/65.2	70.5/64.4	70.4/64.0	69.6/64.4
BAMnet+GEAR	**79.3**/–	**79.2/73.5**	**78.6/72.3**	**78.7/72.4**	**78.3/72.3**
BBKBQA	90.3/–	90.6/81.3	89.5/81.9	89.7/81.6	89.7/81.1
BBKBQA+GEAR	**91.5**/–	**91.9/83.2**	**91.1/83.5**	**90.8/83.3**	**91.1/82.7**

a re-ranking algorithm that re-ranks the result of KBQA with the help of the sequence probability matrix $\mathbf{C}$ generated by the Generation Module. The input to PDR is a sequence probability matrix $\mathbf{C}$, a candidate predicate sequence $\mathcal{P} = \{p_1, p_2, \ldots, p_s\}$, the score $\mathcal{S}_{\mathcal{P}}$ of the candidate predicate $\mathcal{P}$ calculated by KBQA model and the weight parameter η. The output is the re-ranked final score $\mathcal{S}$ that denotes the confidence of the predicate sequence $\mathcal{P}$. The overall re-ranking procedure is described in Algorithm 1. We set η as the weight parameter that indicates the proportion of GEAR in the final scoring. In other words, the proportion of KBQA in the final scoring is $1 - \eta$.

4 Experiments

4.1 Datasets, Baselines and Training Details

We evaluate our framework on two popular KBQA datasets SimpleQuestions[1] and NLPCCKBQA [4]. The two datasets are modified with 5 different KG-size setups: 100%, 70%, 50%, 30%, 10%. If KG-size parameter is set to 70%, we drop triples related to 30% random selected questions in test sets. Thus, 70% questions are answerable while the other 30% questions are unanswerable. We

Table 2. Evaluation of the predicates generated by the Generation Module on SimpleQuestions and NLPCCKBQA.

Metrics	SimpleQuestions	NLPCCKBQA
EM	86.7	70.9
BLEU-1	90.7	83.5
ROUGE-L	91.1	85.4
HCI	95.8	88.4

take four widely used KBQA models as our baselines: BiLSTM [10], MCCNN [3], BAMnet [2], and BBKBQA [9]. We do the best to adjust the hyperparameters of the baseline KBQA models so that the baseline models have relatively optimal performance. We implement the Generation Module in GEAR based on the public HuggingFace implementation of the BART seq2seq model and optimize it by Adam. The initial learning rate is set to 5e−5 and we set the batch size to 32. We set $\tau = 0.5$ because re-ranking is essentially implementing binary classification for each predicate-answer pair. We set η to 0.3 based on the overall accuracy of the model for all questions on the devset.

4.2 Performance and Analysis

Evaluation of KBQA Model After GEAR is Applied. The main results are shown in Table 1. Each grid has two results separated by "/" indicating accuracy of answerable questions and the rejection rate of unanswerable questions respectively. "-" is a placeholder since no unanswerable question exists when KG-size is 100%. The rate of rejection on unanswerable questions achieves significant improvement after GEAR is applied to baseline KBQA models. However, in SimpleQuestions, the improvement in accuracy on answerable questions is inapparent. This is reasonable because overfitting on negative samples mainly presents in unanswerable questions. This improvement demonstrates that GEAR improves the performance of KBQA particularly on identifying unanswerable questions.

Evaluation of seq2seq in Generation Module. We further discuss the performance of the seq2seq model in Generation Module. The quality of the generated sequence reflects the general text generation and semantic knowledge contained in the seq2seq model. As shown in Table 2, the performance of generated predicate is evaluated by four metrics. EM, BLEU-1 and ROUGE-L are metrics that measure the n-gram overlap between a reference predicate and a candidate predicate, which are not always valid factuality metrics. Metric HCI [5] is a human evaluation method to test whether the predicate generated by GEAR is a reliable predicate from the perspective of humans and how effective the perception of the end-user towards is accepting the "No Answer". Ten volunteers were asked to rate the accuracy of the generated predicates. We select 1000 questions at random from the two datasets. Each of these volunteers is assigned

500 questions to rate the answers given by the KBQA system. We guarantee that each question will be asked five times. For each question, we use hard voting to calculate the final result. The results demonstrate that our seq2seq model is capable of generating a proper predicate in response to a given question.

5   Conclusion

In this paper, we present GEAR, a generation based framework to improve the performance of KBQA task. We also propose PDR, an algorithm to implement the ensemble of generation based model and traditional retrieval based KBQA model. We illustrate that the proposed framework can successfully reduce the noise of negative samples from an incomplete KG. Moreover, the framework improves the ability of a KBQA system to identify unanswerable questions. Empirical results on two popular KBQA datasets under different degrees of KG incompleteness and case study demonstrate the effectiveness of our model.

Acknowledgements. We thank anonymous reviewers for their comments and suggestions. This work was supported by National Key Research and Development Project (No. 2020AAA0109302), Shanghai Science and Technology Innovation Action Plan (No. 19511120400) and Shanghai Municipal Science and Technology Major Project (No. 2021SHZDZX0103).

References

1. Bordes, A., Usunier, N., Chopra, S., Weston, J.: Large-scale simple question answering with memory networks. CoRR (2015)
2. Chen, Y., Wu, L., Zaki, M.J.: Bidirectional attentive memory networks for question answering over knowledge bases. In: Proceedings of ACL (2019)
3. Dong, L., Wei, F., Zhou, M., Xu, K.: Question answering over freebase with multi-column convolutional neural networks. In: Proceedings of ACL (2015)
4. Duan, N.: Overview of the NLPCC-ICCPOL 2016 shared task: open domain Chinese question answering. In: Lin, C.-Y., Xue, N., Zhao, D., Huang, X., Feng, Y. (eds.) ICCPOL/NLPCC -2016. LNCS (LNAI), vol. 10102, pp. 942–948. Springer, Cham (2016). https://doi.org/10.1007/978-3-319-50496-4_89
5. Ehsan, U.: On design and evaluation of human-centered explainable AI systems. In: Human-Centered Machine Learning Perspectives Workshop at CHI (2019)
6. Godin, F., Kumar, A., Mittal, A.: Learning when not to answer: a ternary reward structure for reinforcement learning based question answering. In: Proceedings of ACL (2019)
7. Kim, N., Pavlick, E., Karagol Ayan, B., Ramachandran, D.: Which linguist invented the lightbulb? presupposition verification for question-answering. In: Proceedings of ACL (2021)
8. Lewis, M., et al.: BART: denoising sequence-to-sequence pre-training for natural language generation, translation, and comprehension. In: Proceedings of ACL (2020)

9. Liu, A., Huang, Z., Lu, H., Wang, X., Yuan, C.: BB-KBQA: BERT-based knowledge base question answering. In: Sun, M., Huang, X., Ji, H., Liu, Z., Liu, Y. (eds.) CCL 2019. LNCS (LNAI), vol. 11856, pp. 81–92. Springer, Cham (2019). https://doi.org/10.1007/978-3-030-32381-3_7

10. Petrochuk, M., Zettlemoyer, L.: SimpleQuestions nearly solved: a new upperbound and baseline approach. In: Proceedings of EMNLP (2018)

11. Xiong, W., Yu, M., Chang, S., Guo, X., Wang, W.Y.: Improving question answering over incomplete KBs with knowledge-aware reader. In: Proceedings of ACL (2019)

12. Zhu, S., Cheng, X., Su, S.: Knowledge-based question answering by tree-to-sequence learning. Neurocomputing **372**, 64–72 (2020)

Medical Image Fusion Based on Pixel-Level Nonlocal Self-similarity Prior and Optimization

Rui Zhu[1,2], Xiongfei Li[1,2], Yu Wang[1,2], and Xiaoli Zhang[1,2(✉)]

[1] Key Laboratory of Symbolic Computation and Knowledge Engineering, Ministry of Education, Jilin University, Changchun 130012, China
[2] College of Computer Science and Technology, Jilin University, Changchun 130012, China
zhangxiaoli@jlu.edu.cn

Abstract. "Self-similarity" is a common characteristic of medical images. That is, small-scale features often appear in multiple locations in the image frequently. Therefore, the global search for similar pixels helps to infer the pixel value of a certain location, which can be used for the extraction of image details. In this paper, a two-stage image decomposition framework (TS-PLNSS) is proposed. It combines the pixel-level nonlocal self-similarity prior and pixel intensity attributes of the image. First, the source image is adaptively decomposed into three scales using different thresholds: texture layer, structure layer, and base layer. Then, different fusion strategies are adopted according to the characteristics of different layers. Among them, an optimization function that integrates structural information is designed to preserve the salient information of the source image to the maximum extent. Finally, the fused medical image is obtained through image reconstruction. The proposed method is compared with seven state-of-the-art multimodal medical image fusion methods to verify its effectiveness. The dataset consists of 110 image pairs, including CT/SPECT, MRI/PET, MRI/SPECT, MRI/CT, and GFP/PC. The subjective and objective evaluation results show that the TS-PLNSS method has better ability to distinguish and retain important information and texture details without distortion.

Keywords: Image fusion · Nonlocal self-similarity · Optimization

1 Introduction

Medical image fusion technology can break through the limitations of hardware equipment imaging. The fused single multimodal image can provide a more comprehensive description of the organ than any input to assist in clinical diagnosis [1]. The fusion of magnetic resonance imaging (MRI) and computed tomography (CT) images can describe the adhesion relationship between the lesion and the bone or soft tissues, which benefits the treatment of skull base tumors and

A. Bhattacharya et al. (Eds.): DASFAA 2022, LNCS 13247, pp. 247–254, 2022.
https://doi.org/10.1007/978-3-031-00129-1_18

nasopharyngeal carcinoma [2]. The combination of the anatomical image and the functional image can integrate structural and metabolic information of organs and express them at the same level, such as MRI/positron emission tomography (PET), MRI/single photon emission computed tomography (SPECT), and CT/SPECT. The fused green fluorescent protein (GFP)/phase contrast (PC) images can facilitate the study of protein functional analysis [3].

In recent years, a variety of medical fusion methods have been proposed. Among the fusion methods, the fusion effect based on multi-scale analysis is better [4]. Commonly used image decomposition tools include discrete wavelet transform [5], ridgelet transform [6], non-subsampled contourlet transform [7] and non-subsampled shearlet transform [8]. The medical images are decomposed into different layers according to the imaging characteristics, which is conducive to designing specific fusion rules to integrate features at the same level.

In addition to the methods based on the transform domain, the methods based on edge-preserving filters achieve better fusion performance than the methods using linear filters [9] and maintain low computational cost. However, they determine whether the current pixel is filtered based on the neighborhood intensity characteristics of the pixel rather than the intensity distribution of the entire image. When the intensity of a small-scale structure is lower than its neighboring pixels, the above methods may treat it as insignificant information and smooth it, but it ought to be preserved as details when considering the intensity distribution characteristics of the entire image.

With the development of the deep learning, there are many methods based on convolutional neural networks (CNNs) for image fusion tasks [10,11]. They all train a large amount of labeled data to extract the features of the input image. Moreover, different network frameworks have different sensitivity to details. In order to improve the quality of the fused image, researchers focus on designing a more effective framework to make full use of the intensity distribution characteristics of the entire source image.

1.1   Motivation and Contributions

From the above analysis, it can be found that the quality of the fusion result depends on the degree of preservation of significant information at the same level. It is not ideal to rely on the sensitivity of different tools or network structures to texture information to achieve image decomposition. Combining the imaging characteristics of the source image itself with the above tools can extract salient features more effectively and reasonably, for example, some small-scale but very important soft tissue information. Therefore, we introduce the nonlocal self-similarity (NSS) prior, which can effectively improve the ability to preserve image details [12]. In this paper, a two-stage image decomposition method is proposed. The pixel-level NSS is combined with the intensity attribute to decompose the input image into a texture layer, a structure layer, and a base layer. According to the similarity between the fusion result and the different features of different source images, an optimization function is established. It is expected that the fused image contains the structural details of the MRI/PC image, the intensity information of the CT image, and the pseudo-color information of the PET/SPECT/GFP image. The main contributions can be summarized as follows: (1) Image decomposition introduces pixel-level nonlocal NSS prior to ensure that the intensity information and structural details of the source medical image are completely separated. (2) Combining the mean and median of the pixel intensity of the source image, the thresholds are adaptively set to extract the texture layer and structure layer of the image. (3) An optimization function is designed to retain important structural information of the source image according to the characteristics of the ideal fused structure layer.

2 Related Work

2.1 Nonlocal Self-similarity

As mentioned above, the nonlocal attention mechanism is conducive to fully extract the significant details of the source image as it enables the designed method to consider the image content. It makes the methods find similar image patches that can be matched in the entire image. For a gray-scale image $I \in R^{h \times w}$, it assumes that there are N patches in size of $\sqrt{n} \times \sqrt{n}$. These patches can be stretched to a vector $I_{l,1}$ $(l = 1, ..., N)$. For each $I_{l,1}$, the method searches its $m - 1$ most similar patches by comparing the Euclidean distances. These m vectors (including $I_{l,1}$) form the similar matrix $Y_l = [I_{l,1}, ..., I_{l,m}] \in R^{n \times m}$. In this way, the pixels in the same row are similar [13]. But the details obtained through patch-level NSS prior are not clear enough.

2.2 Nonlocal Similar Pixels

Based on the patch-level NSS, the pixel-level NSS aims to search similar pixels in the similar matrix. That is, it needs to calculate Euclidean distances among n rows in Y_l. Specifically, the distance between the i-th row and j-th row are calculated as $d_l^{ij} = \left\| I_l^i - I_l^j \right\|_2$. In this way, select q rows with the smallest distances from the i-th row to form the matrix $Y_l^{iq} \in R^{q \times m}$. Traverse the entire image to construct the detail matrices $\left\{ Y_l^{iq} \right\}$ where $i = 1, ..., n$, $l = 1, ..., N$. They can be used to estimate texture information. Thus, the source image can be decomposed into different scales.

3 Methodology

As shown in Fig. 1, the input medical image $I(I = MRI, CT)$ is first decomposed into the texture layer T_I, the structure layer S_I, and the base layer B_I according to different adaptive intensity thresholds ξ_{T_I} and ξ_{S_I}. Then, in order to obtain a fused image full of salient and important information, an optimization function is designed for the structure layer fusion. For the texture layer and the base layer, simple but effective fusion strategies are selected. Finally, the fused image can be obtained through image reconstruction.

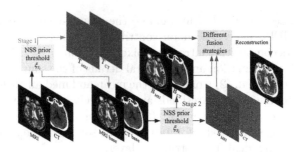

Fig. 1. The framework of two-stage image decomposition and image fusion.

3.1 Detailed Information Estimation

To better retain the detailed information of different scales of the source image, the TS-PLNSS method classifies them into large-scale structure features and small-scale texture features. Thus, the accurate estimation for detailed information is important. In the matrix Y_l^{iq}, the pixels are sufficiently like each other. So, the standard deviation (σ) of the matrix Y_l^{iq} can be used to estimate the texture details. Thus, the global texture estimation is calculated through

$$\sigma_g = \frac{1}{N}\sum_{l=1}^{N}\sigma_l = \frac{1}{N}\sum_{l=1}^{N}\frac{1}{n(q-1)}\sum_{t=2}^{q}\sum_{i=1}^{n}\sqrt{\frac{1}{m}\left(d_l^{ii_t}\right)^2} \tag{1}$$

3.2 Two-Stage Decomposition Framework

The proposed medical image decomposition framework can be regarded as a two-stage image smoothing work in the wavelet domain. Transform the obtained matrices $\{Y_l^q\}_{l=1}^{N}$ composed of Y_l^{iq} to C_l^q. Then, the pixel located at (x, y) can be refined according to the intensity characteristics of medical images through

$$\widehat{C}_l^q(x,y) = \begin{cases} C_l^q(x,y), & \text{if } |C_l^q(x,y)| \geqslant \xi_{T_I} \\ 0, & \text{otherwise} \end{cases} \tag{2}$$

where μ_I and m_I represent the mean and median pixel intensity value of the source image I, $\xi_{T_I} = |\mu_I - m_I| \cdot \sigma_g$. Then, the low frequency $\widetilde{C}_l^q$ of the $\widehat{C}_l^q$ can be regarded as the first-level base layer $\widetilde{I}$ after inverse transformation. The texture layer can be obtained through $T_I = I - \widetilde{I}$. In the second stage, the input image I is updated to $\widetilde{I}$. The threshold ξ_{T_I} is replaced with $\xi_{S_I} = \min(\mu_I, m_I)$. Repeat the first stage of smoothing operation to get the final base layer B_I and the structure layer S_I via $S_I = \widetilde{I} - B_I$.

3.3 Image Fusion and Reconstruction

Structure Layer Fusion. MRI expresses more structural information of organs and tissues and CT describes the bone information. It is desired that the fused medical image can retain the structure of MRI image and the intensity change boundaries of CT image. For fusing the structure layers of multimodal medical images, an optimization function is designed as:

$$\min_{S_F} \|S_F - S_{MRI}\|_2^2 + \beta \|\nabla S_F - \nabla S_{CT}\|_2^2 + \lambda \|\nabla S_F\|_1 \tag{3}$$

where S_F represents the fused structure layer and ∇ means the gradient operator. The β is the balance coefficient and λ is the penalty coefficient. The function can be solved using the Half Quadratic Splitting, coordinate descent, and interior-point methods.

Texture Layer Fusion. As mentioned above, if the features of the source image are extracted enough, simple fusion rules can also ensure a good fusion effect. So, the sum of modified Laplacian algorithm is selected for fusing texture layers, which is calculated by

$$SML(x,y) = \sum_{a=-N}^{N} \sum_{b=-N}^{N} W(a,b) \cdot |ML(x+a, y+b)| \tag{4}$$

where ML calculates the intensity difference between the center pixel and the eight adjacent pixels. The size of sliding window is set to $(2N+1)$ and $N=1$. The weight matrix is defined as $W = \frac{1}{16} \begin{bmatrix} 1 & 2 & 1 \\ 2 & 4 & 2 \\ 1 & 2 & 1 \end{bmatrix}$. It ensures that the weights add up to 1. Thus, the fused texture layer T_F can be obtained by

$$T_F(x,y) = \begin{cases} T_{MRI}(x,y), & \text{if } SML_{T_{MRI}}(x,y) \geqslant SML_{T_{CT}}(x,y) \\ T_{CT}(x,y), & \text{otherwise} \end{cases} \tag{5}$$

Base Layer Fusion. The base layer B_I is smoother than the source image, which is fused using the "choosing maximum" method to obtain the fused base layer B_F.

Image Reconstruction. Obviously, the image reconstruction is a simple linear addition operation, which is the inverse process of image decomposition. Thus, the final fused multimodal medical image F is obtained through

$$F = T_F + S_F + B_F \tag{6}$$

3.4 Fusion of Pseudo-Color Image

In addition to anatomical images, the functional medical images also play a vital role in clinical diagnosis. The difference is that the gray-scale anatomical images describe the exact positional relationship between the lesions and the organs and tissues. The lower-resolution pseudo-color images reflect information about cell metabolism, blood flow, and molecular distribution information. Here, the RGB-YUV color space transformation method is selected to extend the proposed TS-PLNSS algorithm to process color images with three channels.

4   Experiments and Analysis

4.1   Experimental Setup

Dataset. The registered multimodal medical image pairs are from the Whole Brain Atlas database[1] which contain 20 groups of CT/SPECT images, 20 groups of MRI/PET images, 20 groups of MRI/SPECT images, and 20 groups of MRI/CT images. Besides, there are 30 groups of GFP/PC images obtained from http://data.jic.ac.uk/Gfp/.

Comparison Methods. In order to verify the fusion effect of the TS-PLNSS method, seven most advanced medical image fusion methods are selected for comparison, including Zerol [14], U2Fusion [10], LLE [7], JBFLGE [9], EMFusion [11], DPCN [3], and MGFF [4]. All the codes are provided by the authors. In the TS-PLNSS method, β is set to 0.05 and λ is set to 0.01.

Evaluation Metrics Six evaluation metrics are selected to objectively compare the fusion effect, including Q_{NCIE} [15], Q_P [16], Q_S [17], Q_{CB} [18], $Q_{AB/F}$ [19], and Q_{VIFF} [20]. They evaluate the quality of the result from information theory, image feature, structural similarity, human perception [21], and information fidelity. The higher the evaluation score, the more ideal the fusion result.

4.2   Experimental Results and Discussions

There are five types of image pairs in the dataset. In order to comprehensively compare the fusion effect of different methods, one group of images are selected for detailed display considering the length limitation. And the average objective evaluation scores on the entire dataset are calculated in Table 1.

MRI/PET Fusion. Figure 2 shows the fused MRI-PET images in detail. The fusion results 2(d), 2(g), and 2(h) look darker. The MGFF method also has the same problem, but it is not serious. The results obtained by LLE, JBFLGE, and TS-PLNSS methods have better visual quality. The first two methods pay more attention to the white tissue expressed in MRI images. It reduces the color fidelity of the PET image, especially in the marked red box. The proposed method better preserves the color information of PET image.

In addition to the subjective evaluation, Table 1 calculates the average scores of all methods on the six assessment metrics. The bold indicates the highest one. Obviously, the TS-PLNSS method achieves two highest scores and three second highest scores. Two of them are slightly lower than JBFLGE. DPCN achieves the highest Q_P, but the fusion effect is not excellent in visual. The method MGFF is similar. The average result of objective evaluation shows that the TS-PLNSS method can obtain stable fusion effects in all aspects. Combined with the perceived quality of the fused images, the performance of TS-PLNSS is better. It can retain salient and useful information of the source images without distortion.

---

[1] http://www.med.harvard.edu/aanlib/home.html.

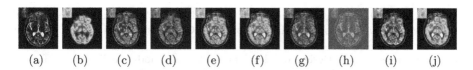

(a) (b) (c) (d) (e) (f) (g) (h) (i) (j)

Fig. 2. The fused MRI-PET images obtained by different fusion methods. (a) MRI (b) PET (c) Zerol (d) U2Fusion (e) LLE (f) JBFLGE (g) EMFusion (h) DPCN (i) MGFF (j) TS-PLNSS.

Table 1. The average objective evaluation scores.

Methods	Zerol	U2Fusion	LLE	JBFLGE	EMFusion	DPCN	MGFF	TS-PLNSS
Q_{NCIE}	0.8059	0.8062	0.8071	0.8089	0.8059	0.8082	0.8057	**0.8092**
Q_S	0.7552	0.4158	0.8305	0.8509	0.8154	0.3800	0.8287	**0.8526**
$Q_{AB/F}$	0.5347	0.4833	0.5931	**0.6154**	0.5323	0.4374	0.5758	0.6053(2)
Q_{CB}	0.6244	0.3331	0.6170	**0.6462**	0.6360	0.2980	0.6388	0.6419(2)
Q_P	0.3017	0.3548	0.4100	0.4338	0.4580(2)	**0.7250**	0.3739	0.4392(3)
Q_{VIFF}	0.4187	0.4787	0.5334	0.5192	0.4427	0.2257	**0.5844**	0.5338(2)

5 Conclusion

Medical image fusion can integrate supplementary information of lesions to assist clinical diagnosis. In this paper, pixel-level NSS attention enables the method to extract texture details based on the imaging characteristics of the source image itself. Therefore, the multimodal medical image can be adaptively decomposed into three scales. In order to retain more useful information of the anatomical image, an optimization function is designed according to the characteristics of the ideal fused image. In addition, the fusion of anatomical images and functional images is achieved through the method of color space transformation. The experiments are conducted on the dataset containing 110 groups of images. Compared with the seven state-of-the-art medical image fusion methods, the proposed TS-PLNSS method performs better in subjective and objective evaluations, and has an excellent ability to retain significant information of the source images without distortion. In the future, a more efficient nonlocal similar pixel search method will be studied to improve the computational efficiency of the method.

References

1. Li, T., Wang, Y.: Biological image fusion using a NSCT based variable-weight method. Inf. Fusion **12**(2), 85–92 (2011)
2. Zhu, R., Li, X., Zhang, X., Wang, J.: HID: the hybrid image decomposition model for MRI and CT fusion. IEEE J. Biomed. Health Inform. **26**(2), 727–739 (2022)
3. Tang, W., Liu, Y., Cheng, J., Li, C., Chen, X.: Green fluorescent protein and phase contrast image fusion via detail preserving cross network. IEEE Trans. Comput. Imag. **7**, 584–597 (2021)

4. Bavirisetti, D., Xiao, G., Zhao, J., Dhuli, R., Liu, G.: Multi-scale guided image and video fusion: a fast and efficient approach. Circ. Syst. Signal Process. **38**(12), 5576–5605 (2019)
5. Abdulkareem, M.B.: Design and development of multimodal medical image fusion using discrete wavelet transform. In: 2nd International Conference on Inventive Communication and Computational Technologies (ICICCT), pp. 1629–1633. IEEE, Coimbatore, India (2018)
6. Arif, M., Wang, G.: Fast curvelet transform through genetic algorithm for multi-modal medical image fusion. Soft Comput. **24**(3), 1815–1836 (2019). https://doi.org/10.1007/s00500-019-04011-5
7. Zhu, Z., Zheng, M., Qi, G., Wang, D., Xiang, Y.: A phase congruency and local Laplacian energy based multi-modality medical image fusion method in NSCT domain. IEEE Access **7**, 20811–20824 (2019)
8. Jose, J., et al.: An image quality enhancement scheme employing adolescent identity search algorithm in the NSST domain for multimodal medical image fusion. Biomed. Signal Process. Control **66**, 102480 (2021)
9. Li, X., Zhou, F., Tan, H., et al.: Multimodal medical image fusion based on joint bilateral filter and local gradient energy. Inf. Sci. **569**, 302–325 (2021)
10. Xu, H., Ma, J., Jiang, J., Guo, X., Ling, H.: U2Fusion: a unified unsupervised image fusion network. IEEE Trans. Pattern Anal. Mach. Intell. **44**(1), 502–518 (2020). https://doi.org/10.1109/TPAMI.2020.3012548
11. Xu, H., Ma, J.: EMFusion: an unsupervised enhanced medical image fusion network. Inf. Fusion **76**, 177–186 (2021)
12. Fan, L., Li, X., Fan, H., Feng, Y., Zhang, C.: Adaptive texture-preserving denoising method using gradient histogram and nonlocal self-similarity priors. IEEE Trans. Circ. Syst. Video Technol. **29**(11), 3222–3235 (2018)
13. Hou, Y., Xu, J., Liu, M., Liu, G., Liu, L., Zhu, F.: NLH: a blind pixel-level non-local method for real-world image denoising. IEEE Trans. Image Process. **29**, 5121–5135 (2020)
14. Lahoud, F., Süsstrunk, S.: Zero-learning fast medical image fusion. In: 22th International Conference on Information Fusion (FUSION), pp. 1–8. IEEE, Ottawa, Canada (2019)
15. Wang, Q., Shen, Y., Jin, J.: Performance evaluation of image fusion techniques. Image Fusion Algorithms Appl. **19**, 469–492 (2008)
16. Zhao, J., Laganiere, R., Liu, Z.: Performance assessment of combinative pixel-level image fusion based on an absolute feature measurement. Int. J. Innov. Comput. Inf. Control **3**(6), 1433–1447 (2007)
17. Piella, G., Heijmans, H.: A new quality metric for image fusion. In: 2003 International Conference on Image Processing (Cat. No. 03CH37429), p. III-173. IEEE, Barcelona, Spain (2003)
18. Chen, Y., Blum, R.: A new automated quality assessment algorithm for image fusion. Image Vis. Comput. **27**(10), 1421–1432 (2009)
19. Xydeas, C., Petrovic, V.: Objective image fusion performance measure. Electron. Lett. **36**(4), 308–309 (2000)
20. Han, Y., Cai, Y., Cao, Y., Xu, X.: A new image fusion performance metric based on visual information fidelity. Inf. Fusion **14**(2), 127–135 (2013)
21. Liu, Z., Blasch, E., Xue, Z., Zhao, J., Laganiere, R., Wu, W.: Objective assessment of multiresolution image fusion algorithms for context enhancement in night vision: a comparative study. IEEE Trans. Pattern Anal. Mach. Intell. **34**(1), 94–109 (2011)

Knowledge-Enhanced Interactive Matching Network for Multi-turn Response Selection in Medical Dialogue Systems

Ying Zhu, Shi Feng$^{(\boxtimes)}$, Daling Wang, Yifei Zhang, and Donghong Han

School of Computer Science and Engineering, Northeastern University, Shenyang, China
{fengshi,wangdaling,zhangyifei,handonghong}@cse.neu.edu.cn

Abstract. Recently, the response selection for retrieval-based dialogue systems has gained enormous attention from both academic and industrial communities. Although the previous methods achieve promising results for intelligent customer service systems and open-domain chatbots, the response selection in medical dialogues suffers from lower performance because of the strong dependency on the domain knowledge. In this paper, we construct two specialized medical knowledge bases and propose a **K**nowledge-enhanced **I**nteractive **M**atching **N**etwork (KIMN) for multi-turn response selection in medical dialogue systems. Compared with previous response selection approaches, the KIMN adopts pre-trained language model to alleviate the limited training data problem, and incorporates internal and external medical domain knowledge to perform interactive matching between responses and contexts. The experiments on a real-world medical dialogue dataset show that our proposed model consistently outperforms the strong baseline methods by large margins, which shows that our proposed model can retrieve more accurate responses for medical dialogue systems.

Keywords: Multi-turn conversation · Knowledge-enhanced matching · Medical dialogue systems

1 Introduction

In recent years, we have witnessed substantial progress in dialogue systems, thanks to the large-scale datasets and seq2seq training models. The dialogue system has attracted extensive attention in medical applications.

Retrieval-based dialogue systems have wide applications including medical dialogue systems. However, the existing medical response matching networks have the following problems that need to be tackled. Firstly, most of the previous medical response matching networks are designed for single-turn conversation, which only consider the last utterance in the dialogue and lead to great limitations in their applications and make them lack of information. However proposed models on multi-turn response selection are not well-generalized in medical dialogue systems due to the limited data of the medical conversations. Secondly, the previous medical response selection methods neglect the domain knowledge that are crucial for this task. In open-domain chatbots, some progress

A. Bhattacharya et al. (Eds.): DASFAA 2022, LNCS 13247, pp. 255–262, 2022.
https://doi.org/10.1007/978-3-031-00129-1_19

has been made for knowledge-grounded response selection [1, 2], but these methods only consider the unstructured external knowledge (such as the user persona sentences), and ignore the internal knowledge of the dialogues (such as the disease entities extracted from the dialogue dataset) that also has great impact on this task. How to incorporate knowledge into the multi-turn response selection models in medical domain is still a considerable challenge.

To tackle the challenges above, in this paper, we propose a **K**nowledge-enhanced **I**nteractive **M**atching **N**etwork (KIMN) for multi-turn response selection in medical dialogue. The pre-trained language model BERT [3] is employed to alleviate the problem caused by limited training data. Moreover, we construct internal and external medical knowledge bases, and fuse them into the encoding layers to make the selected response more factual and accurate. More specifically, the context, response and knowledge are fed into BERT to get knowledge-enhanced representations, and the matching features are calculated via different attention mechanisms. Note that the problem we study in this paper is slightly different from the traditional knowledge-grounded response selection that has pre-defined or off-the-shelf knowledge set. For the medical dialogue, we need to build specialized knowledge bases to boost the matching performance, which we call as knowledge-enhanced response selection.

2 Related Work

2.1 Knowledge-Grounded Dialogue System

Recently, some studies indicate that the responses can be more diverse and engaging by fusing background knowledge into response selection. A few studies have considered grounding open domain dialogues with external knowledge, such as facts or personas. Ghazvininejad et al. [4] conditioned responses on both context history and external "facts" to generalize the vanilla Seq2seq model. For retrieval-based systems, Zhao et al. and Gu et al. [1, 2, 5] have made some achievements. As the medical dialogues have strong dependency with the domain knowledge, we conjecture that incorporating knowledge can further improve the matching performance for the task.

2.2 Medical Dialogue System

Compared with the systems for general tasks, the study on response selection for medical dialogue system is still in its infancy. Ye et al. [6] proposed a convolutional neural network framework which used a multi-level structure that could extract high-level dimensional semantic features. Tian et al. [7] released a Chinese medical Q&A corpus called ChiMed for Chinese medical question–answer matching.

3 Knowledge Construction

3.1 External Knowledge Construction

We crawl the related information of the 12 types of gastrointestinal diseases in MedDG [8] dataset from *Chunyu-Doctor*[1] as external knowledge of the model. We take sentences

---

[1] https://www.chunyuyisheng.com/.

as the format of the knowledge and take the names of diseases as labels corresponding to the knowledge. We calculate the cosine similarity between sentences and utterances in order to select the most similar external knowledge to a dialogue. We add the selected external knowledge sentences and the corresponding name of the disease behind the corresponding dialogue.

3.2 Internal Knowledge Construction

We construct the internal knowledge by extracting diseases and symptoms from the whole MedDG dataset. We first set up a set with 218 symptoms and diseases tuples. As Yuan et al. [9] has mentioned that most of the low-related symptoms to a disease are noise, we calculate a likelihood matrix between diseases and symptoms and we only reserve the top-5 tuples of each symptom in the frequencies of diseases. We then take the disease in external knowledge as the key to select the tuples related to the dialogue as internal knowledge from the set we have established. We finally add the tuples behind the corresponding dialogue and external knowledge.

4 Methodology

4.1 Problem Formalization

Suppose that we have a dataset $D = \{(c_i, r_i, ke_i, ki_i, y)\}_{i=1}^{N}$, and (c, r, ke, ki, y) can be represented as an example of D. $\{c = \{u_1, u_2, \ldots u_{n_c}\}$ represents a context with u_i as the i-th utterance and n_c as the number of the utterances in the context c. r represents a response candidate of c. $ke = \{d_1, d_2, \ldots, d_{n_{ke}}\}$ represents the external knowledge of c. $ki = \{e_1, e_2, \ldots, e_{n_{ki}}\}$ represents the internal knowledge of c. y $\{0, 1\}$ denotes a binary label. Our goal is to learn a matching model g (from D, g (c, r, ke, ki) measures the matching degree between (c, ke, ki) and r. The framework of our proposed model is shown in Fig. 1.

4.2 Representation Phase

An utterance u_i in context c, response candidate r, a sentence d_i in external knowledge and an entry e_i in internal knowledge can be represented as $E_{u_i} = [e_{u_i,1}, e_{u_i,2}, \ldots, e_{u_i,n_{u_i}}]$, $[E_r = e_{r,1}, e_{r,2}, \ldots, e_{r,n_r}]$, $[E_{d_i} = e_{d_i,1}, e_{d_i,2}, \ldots, e_{d_i,n_{d_i}}]$, $[E_{e_i} = e_{e_i,1}, e_{e_i,2}, \ldots, e_{e_i,n_{e_i}}]$ respectively, where $e_{u_i,k}$, $e_{r,k}$, $e_{d_i,k}$, $e_{e_i,k}$ represent the k-th word in u_i, r, d_i, e_i, and $n_{u_i}, n_r, n_{d_i}, n_{e_i}$ denote the numbers of words in u_i, r, d_i, $e_i l_c = \sum_{i=1}^{n_c} n_{u_i}$, $l_r = n_r$, $l_{ke} = \sum_{i=1}^{n_{ke}} n_{d_i}$, $l_{ki} = \sum_{i=1}^{n_{ki}} n_{e_i}$, are the numbers of words.

The embeddings $E_{u_i}, E_r, E_{ke_i}, E_{ki_i}$ are then encoded by BERT (as shown in Fig. 2). The process of BERT encoding can be defined by follows:

$$H = Pool_{mean}[BERT(e_1, e_2, \ldots, e_n)] \qquad (1)$$

We can get encoded representations $H_{u_i} = \left[h_{u_i,1}, h_{u_i,2}, \ldots, h_{u_i,n_{u_i}}\right]$, $H_r = [h_{r,1}, h_{r,2}, \ldots, h_{r,n_r}]$ $H_{d_i} = [h_{d_i,1}, h_{d_i,2}, \ldots, h_{d_i,n_{d_i}}]$, $H_{e_i} = [h_{e_i,1}, h_{e_i,2}, \ldots, h_{e_i,n_{e_i}}]$.

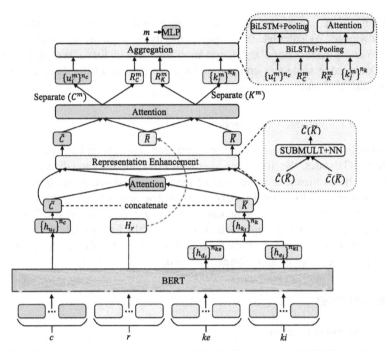

Fig. 1. The framework of our proposed model KIMN. The process of BERT encoding and the format of input embeddings are shown in Fig. 2

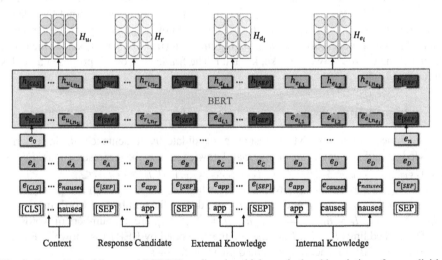

Fig. 2. Input Embeddings and BERT Encoding, in which app is the abbreviation of appendicitis

Specifically, we concatenate the encoded representations of external and internal knowledge and get H_{k_i}.

We enhance knowledge and context representation with each other. We first concatenate H_{u_i} and H_{k_j}:

$$\overline{C} = Concatenate(\{H_{u_i}\}_{i=1}^{n_c}), \overline{K} = Concatenate(\{H_{k_j}\}_{j=1}^{n_k}) \tag{2}$$

where $n_k = sum(n_{ke}, n_{ki})$. The encoded representations of context and knowledge can be denoted as $\overline{C} = \{h_{c_i}\}_{i=1}^{l_c}, \overline{K} = \{h_{k_j}\}_{j=1}^{l_{ke}+l_{ki}}$. We then get knowledge-related context representation and context-related knowledge representation as follows:

$$e_{ij}^k = h_{c_i}^T \cdot h_{k_j}, e_{ij}^c = h_{c_i} \cdot h_{k_j}^T \tag{3}$$

$$a_{ij}^k = \frac{\exp\left(e_{ij}^k\right)}{\sum_{m=1}^{l_k}\exp(e_{im})}, a_{ij}^c = \frac{\exp\left(e_{ij}^c\right)}{\sum_{m=1}^{l_c}\exp\left(e_{mj}\right)} \tag{4}$$

$$\widehat{c_i} = \sum_{j=1}^{l_k} a_{ij}^k \cdot h_{k_j}, \widehat{k_j} = \sum_{i=1}^{l_c} a_{ij}^c \cdot h_{c_i} \tag{5}$$

So far, we have obtained each knowledge-related context word representation $\hat{c}_i$ and each context-related knowledge word representation $\hat{k}_j$. We further define $\hat{C} = \{\hat{c}_i\}_{i=1}^{l_c}$ and $\hat{K} = \{\hat{k}_j\}_{j=1}^{l_k}$ as knowledge-related context and context-related knowledge representation, where $l_k = l_{ke} + l_{ki}$. Here, we take $\hat{C}$ as an example to define an attention function in order to summarize the above progress as

$$\hat{C} = Att(\overline{C}, \overline{K}) \tag{6}$$

To enhance context and knowledge representation, we feed $(\overline{C}, \hat{C})$ and $(\overline{K}, \hat{K})$ into an interactive function called SUBMULT + NN function [10] to get enhanced context and knowledge representation:

$$\tilde{c}_i = \mathrm{ReLU}\left(W_c \begin{bmatrix} (h_{c_i} - \hat{c}_i) \odot (h_{c_i} - \hat{c}_i) \\ h_{c_i} \odot \hat{c}_i \end{bmatrix} + b_c\right), \tilde{k}_j = \mathrm{ReLU}\left(W_k \begin{bmatrix} (h_{k_j} - \hat{k}_j) \odot (h_{k_j} - \hat{k}_j) \\ h_{k_j} \odot \hat{k}_j \end{bmatrix} + b_k\right) \tag{7}$$

W_c, b_c, W_k, b_k are all parameters, $\odot$ refers to element-wise multiplication, and we eed $\tilde{C} = \{\tilde{c}_i\}_{i=1}^{l_c}$ and $\tilde{K} = \{\tilde{k}_j\}_{j=1}^{l_k}$ as knowledge-enhanced context and context-enhanced knowledge representation into the following matching layers.

4.3 Matching Phase

We employ the whole context and knowledge as a sequence respectively to match with response candidate. As for the matching method, the global bidirectional interactions between two sequences are adopted. The context-response matching and knowledge-response matching are in the same way, and we take context-response matching as an example to introduce the matching process.

Response candidate representation is $H_r = \{h_{rj}\}_{j=1}^{l_r}$, and for the sake of better understanding, it is rewritten as $\overline{R} = \{h_{rj}\}_{j=1}^{l_r}$. We calculate the context-related response representation and response-related context representation via different attention mechanism in the same way as which in the representation phase,

$$\dot{C} = Att\left(\tilde{C}, \overline{R}\right), \dot{R} = Att\left(\overline{R}, \tilde{C}\right) \tag{8}$$

So far, we have obtained the whole context representation $\dot{C} = \{\dot{c}_i\}_{i=1}^{l_c}$ and the whole response representation $\dot{R}_c = \{\dot{r}_j\}_{j=1}^{l_r}$

To enhance the information we have obtained, we adopt the method used in the model of Gu et al. [2],

$$C^m = \left[\tilde{C}; \dot{C}; \tilde{C} - \dot{C}; \tilde{C} \odot \dot{C}\right], R_c^m = \left[\overline{R_c}; \dot{R}_c; \overline{R_c} - \dot{R}_c; \overline{R_c} \odot \dot{R}_c\right] \tag{9}$$

Knowledge and response candidate implement matching in the same way as the method above to get K^m and R_k^m. After that we have got context representation C^m, knowledge representation K^m and two response representation R_c^m and R_k^m after matching.

4.4 Aggregation Phase

In the aggregation phase, we separate C^m and K^m into matching matrices of each utterance, and get $\{u_i^m\}_{i=1}^{n_c}$ and $\{k_j^m\}_{j=1}^{n_k}$.

We employ BiLSTM and pooling operation to process matching matrices $u_i^m, R_c^m, k_j^m, R_k^m$, and then we obtain embeddings after aggregation as $u_i^a, R_c^a, k_j^a, R_k^a$. Specifically, the BiLSTMs we use in the aggregation phase share the same parameters.

Moreover, u_i^a and k_j^a implement a further aggregation to get the embedding vectors of context utterances and knowledge entries.

$$\widehat{u_i^a} = BiLSTM\left(U^a, i\right), i \in \{1, \ldots, n_c\} \tag{10}$$

$$c^a = \left[\widehat{u_{max}^a}, \widehat{u_{n_c}^a}\right] \tag{11}$$

$$k^a = Softmax\left[ReLU\left(W_a^T \cdot k_j^a\right) + b_a\right] \tag{12}$$

Here, W_a and b_a are parameters optimized during training. Up till now, we have obtained context, knowledge and two response candidate embedding vectors, c^a, k^a, R_c^a and R_k^a. We further concatenate the four kinds of embedding vectors to get the matching feature vector m that we need. Finally, we feed the matching feature vector into a multi-layer perceptron (MLP) with softmax output to calculate the matching score, which is used to weight the probability of correct response r when (c, k) is given. Specifically, k represents the fusion of external and internal knowledge.

Table 1. Evaluation results of KIMN and other baselines on MedDG-kn dataset

Model	$R_{10}@1$	$R_{10}@2$	$R_{10}@5$
MSN	58.2	74.3	93.9
DAM	54.3	73.5	92.7
DUA	51.3	68.9	89.1
SMN	47.9	61.2	85.6
Starspace	46.3	63.0	77.5
Profile memory	49.0	64.3	77.2
KV profile memory	51.6	65.5	79.1
Transformer	56.3	68.9	85.0
DGMN	62.6	78.1	90.5
DIM	69.5	87.0	95.1
KIMN	**74.2**	**89.6**	**98.7**

5 Experiments

5.1 Dataset

The publicly available medical dialogue dataset, MedDG dataset [8]. We construct a new dataset with external and internal medical knowledge base on MedDG dataset, which is called MedDG-kn, and we evaluate our proposed model on MedDG-kn dataset. We take the correct responses in dialogues as positive responses, and randomly sample responses in MedDG dialogues as negative responses. The ratio between positive responses and negative ones is 1:1 in training set and validation set while 1:9 in testing set.

5.2 Experimental Results

Table 1 shows the evaluation results of KIMN and its baselines. The results are divided into three groups: baselines without knowledge, baselines with knowledge, and our proposed KIMN model.

We choose $R_n@1$, $R_n@2$ and $R_n@5$ to judge the performance of our model. In the first group, the baselines are traditional matching models that do not consider the pre-trained language model (BERT) and knowledge. We can find from Table 1 that the results of these matching models are within reasonable limits, which indicates the defects of the traditional models in the medical domain. All the evaluation metrics of these matching models are worse than our KIMN model, which validates the effectiveness of BERT and knowledge enhancement on the matching network.

The baselines in the second group are matching models with external knowledge. Our KIMN model outperforms them by a large margin especially in terms of $R_n@1$. KIMN takes advantage of the matching features between response candidate representation and fused representation of external and internal knowledge. On the other hand, KIMN adopts

BERT to encode input embeddings, and the large-scale pre-trained corpus alleviates the data limitation of medical dialogue datasets. These make KIMN outperform the other models in Table 1.

We also design several ablation experiments to prove the effectiveness of each module in our model and find out that the performance of the model with all modules is the best. Specially, without BERT, KIMN can still outperform all the baseline models in Table 1.

6   Conclusion

In this paper, we propose a Knowledge-enhanced Interactive Matching Network for multi-turn response selection in medical dialogue systems, which fuses external and internal knowledge into the response selection process and utilizes pre-trained language model BERT to alleviate the limited data problem. Extensive experiments on a medical dialogue dataset with external and internal knowledge have confirmed the higher performance of KIMN than the other strong baseline models.

Acknowledgements. The work was supported by National Natural Science Foundation of China (61872074, 62172086, 62106039).

References

1. Xueliang, Z., Chongyang, T., Wei, W., Can, X., Dongyan, Z., Rui, Y.: A document-grounded matching network for response selection in retrieval-based chatbots. In: IJCAI, pp. 5443–5449 (2019)
2. Gu. J.-C., Ling, Z.-H., Znu, X., Liu, Q.: Dually interactive matching network for personalized response selection in retrieval-based chatbots. In: EMNLP-IJCNLP 2019, pp. 1845–1854 (2019)
3. Devlin, J., Chang, M.-W., Lee, K., Toutanova, K.: BERT: pre-training of deep bidirectional transformers for language understanding. In: NAACL-HLT, pp. 4171–4186 (2019)
4. Ghazvininejad, M., et al.: A knowledge-grounded neural conversation model. In: AAAI, pp. 5110–5117 (2018)
5. Gu, J.-C., Ling, Z.-H., Liu, Q., Chen, Z., Zhu. X.: Filtering before Iteratively referring for knowledge-grounded response selection in retrieval-based chatbots. In: EMNLP (Findings), pp. 1412–1422 (2020)
6. Ye, D., et al.: Multi-level composite neural networks for medical question answer matching. In: DSC 2018, pp. 139–145 (2018)
7. Tian, Y., Ma, W., Xia, F., Song, F.: ChiMed: a Chinese medical corpus for question answering. In: BioNLP@ACL 2019, pp. 250–260 (2019)
8. Liu, W., Tang, J., Qin, J., Xu, L., Li, Z., Liang, X.: MedDG: a large-scale medical consultation dataset for building medical dialogue system. arXiv preprint arXiv: 2010.07497
9. Yuan, Q., Chen, J., Lu, C., Huang, H.: The graph-based mutual attentive network for automatic diagnosis. In: IJCAI 2020, pp. 3393–3399 (2020)
10. Wang, S., Jiang, J.: Compare-aggregate model for matching text sequences. In: ICLR (2017)

KAAS: A Keyword-Aware Attention Abstractive Summarization Model for Scientific Articles

Shuaimin Li[iD] and Jungang Xu[(⊠)][iD]

School of Computer Science and Technology, University of Chinese Academy of Sciences, Beijing, China
lishuaimin17@mails.ucas.ac.cn, xujg@ucas.ac.cn

Abstract. In this work, we focus on abstractive summarization methods for assisting medical researchers in effectively managing information. Particularly, we introduce a COVID-19-related summarization dataset (COVID-SUM) and propose a novel **K**eyword-aware **A**ttention **A**bstractive **S**ummarization (KAAS) model. The KAAS model consists of two encoders and one decoder. As for the encoders, one is a standard article encoder built on transformer layers, while the other one is a hierarchical keyword encoder that first encodes the words in a keyword using BiLSTM, and then passes the keyword representations to a transformer layer to connect the keywords in an example. Additionally, a decoder with keyword-focused attention is utilized to further direct the decoding process to generate comprehensive summaries of the scientific articles. We benchmark several summarization methods on the new COVID-SUM dataset and release this dataset in the hope to promote advances to summarization in the COVID-19 medical area (https://github.com/ccip-author/COVID-SUM/releases). Furthermore, we evaluate the KAAS on COVID-SUM, ArXiv, and PubMed datasets. Experimental results demonstrate that KAAS outperforms several state-of-the-art models on these datasets.

Keywords: Abstractive summarization · COVID-19 · Scientific articles · Keyword-aware model

1 Introduction

COVID-19 has created a worldwide pandemic in recent years, and a huge number of scientific studies related to COVID-19 have been published, posing a challenge for medical researchers in terms of managing large-scale publications and determining critical information. In this work, we focus on improving the work efficiency of researchers with automatic summarization. Automatic summarization is a popular NLP task that seeks to provide a concise summary of a document. Obviously, an effective summarization system can aid readers and

© The Author(s), under exclusive license to Springer Nature Switzerland AG 2022
A. Bhattacharya et al. (Eds.): DASFAA 2022, LNCS 13247, pp. 263–271, 2022.
https://doi.org/10.1007/978-3-031-00129-1_20

Source article: The optimal anesthesia management for patients with acute ischemic stroke undergoing mechanical thrombectomy (MT) during the COVID-19 pandemic...... 4 Endotracheal intubation is a high risk procedure for the transmission of SARS-CoV-2 infection due to the excessive generation of aerosols. ... 5 6 In order to avoid the added risks associated with unplanned intraprocedural intubation, the Society for Neuroscience in Anesthesiology and Critical Care (SNACC) recommended the upfront use of general anesthesia (GA) for all stroke patients undergoing MT who are perceived as having high risk for intraprocedural conversion from sedation to GA, including those with dominant hemispheric occlusions/aphasia, posterior circulation occlusions, or severe neurological deficits (baseline National Institutes of Health Stroke Scale (NIHSS) score >15), while favoring monitored anesthesia care (MAC) for those presenting with non-dominant hemispheric strokes and a baseline NIHSS score <15. 1. ...

Keywords: patients ; mac ; acute ischemic stroke ; endotracheal intubation ; intraprocedural conversion ; nihss ; severe neurological deficits ; default anesthesia modality ; dominant hemispheric occlusions ; volume center ; high risk ; primary analysis ; posterior circulation occlusions ; Neuroscience in Anesthesiology and Critical Care; covid-19 pandemic ; anesthesia care

Fig. 1. A COVID-SUM example displaying the original article and its corresponding keywords.

scientific researchers in quickly skimming long materials. In this paper, we concentrate on abstractive summarization methods.

The majority of the efforts in abstractive summarization [5,7,15,17] focus on short document summaries, such as news stories, but the primary inputs for scientific summarization are substantially longer. Considering this, the researchers attempt to explore the discourse structure of scientific articles [6], pre-trained models [13] or diverse attention mechanisms [10] to deal with the long inputs. Although these efforts obtain promising results, they overlook the fact that scientific publications are more challenging than original texts since they include more professional terms that are difficult to understand. For instance, as shown in Fig. 1, it is obvious that the keywords are almost sophisticated medical terms on COVID-SUM dataset, such as *"acute ischemic stroke"*, and *"endotracheal intubation"*. Furthermore, the research towards abstractive summarization of COVID-19-related scientific articles is still in the early stage. One work [11] takes the original article as input, and the other [19] just takes the scientific summarization of COVID-19 related articles as the downstream tasks of the language model without training on the COVID-19 related datasets.

To address the aforementioned issues, we first constructed a new COVID-19-related summarization dataset (COVID-SUM) including 37,000 document-summary pairs by cleaning and filtering the COVID-19 Open Research Dataset (CORD-19) [21]. Secondly, we propose a novel keyword-aware attention abstractive summarization (KAAS) model for scientific articles, which contains two encoders and a decoder. Specifically, one encoder encodes the long original scientific articles with Transformer [20] architecture, which is used for processing large inputs with self-attention mechanism. The other encoder is a hierarchical keyword encoder, which first encodes words in each keyword with a BiLSTM model [8] and then encodes the keywords with a Transformer model to aid in communication among the keywords in an example. Additionally, the decoder in KAAS is equipped with focused attention to keyword representations, which leads the decoding process with key information of the input and helps to generate comprehensive summaries for scientific articles.

We benchmark several models on COVID-SUM dataset, including traditional extractive summarization models and several state-of-the-art abstractive models. Additionally, we evaluate our KAAS model on COVID-SUM dataset and

Table 1. Statistics of Arxiv, PubMed, and COVID-SUM datasets.

Dataset	Pairs	% novel unigrams	% novel bigrams	% novel trigrams
ArXiv	202,914/6,436/6,440	27.58	61.46	81.28
PubMed	117,103/6,631/6,658	28.67	59.83	76.45
COVID-SUM	30,000/5,000/2,000	23.87	54.30	71.96

the other two public summarization datasets of scientific articles, ArXiv and PubMed datasets [6]. Experimental results demonstrate that our KAAS model outperforms the compared state-of-the-art summarization models on these three datasets.

2 Related Works

Most neural network-based abstractive summarization models [4–7,15–17] are often used to short source data (e.g. news articles), while scientific article summarization models are applied to long publications. To address this issue, a discourse-aware model is first proposed for abstractive summarization of long scientific documents [6], they exploit a hierarchical encoder to encode the structure of a document and then generate the summary with an attentive discourse-aware decoder. PEGASUS [13] is a transformer-based pre-trained model with a gap sentence optimization target. Additionally, several works concentrate on optimizing attention processes in order to deal with long document input [10].

In this paper, we propose a keyword-aware attention model to summarize the scientific articles. Additionally, there have been several studies conducted in the past that examined the usage of keywords or entities in abstractive summarization. Specifically, a keyword-aware summarization model [9] is designed to encode and merge the keyword information using four encoders, while an entity-aware abstractive summarization model [18] first identifies salient sentences using entity information and then generates the final summaries based on the selected contexts. There are two main distinctions between KAAS and these two models: (1) Both of these two models are constructed on LSTM, while KAAS is built on Transformer architecture. (2) These two models only employ the keyword or entity information in the encoder part, but our approach further extends the exploration of the keyword information through a focused-attention decoder.

3 COVID-SUM Dataset

In this paper, we clean and filter CORD-19 to produce a new dataset, COVID-SUM. Specifically, first of all, we get the CORD-19 dataset, which was published on August 30th, 2021. The abstract of the article is then used as the ground-truth summary, and the entire text is used as the source text. A canonical scientific article, we believe, should comprise *"introduction"* and *"conclusion"* sections.

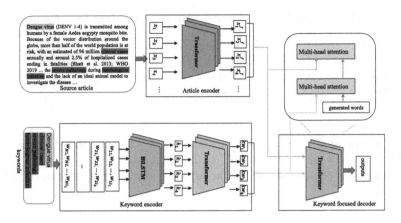

Fig. 2. The framework of KAAS.

Therefore, we remove the articles that lack *"introduction"* or *"conclusion"* sections in CORD-19 dataset, as well as the instances without *"abstract"*. Finally, we randomly choose 37,000 article-summary pairs among examples that satisfy the aforementioned standards to create our COVID-SUM dataset. Additionally, we compute the fraction of n-grams in ground-truth summaries that do not occur in the respective source articles. The results are shown in Table 1, it indicates the abstractiveness of COVID-SUM is similar to the popular summarization datasets.

4   Model Description

A summarization model aims to generate a sequence $y = \{y_1, y_2, ..., y_m\}$ as the summary of an article $A = \{x_1, x_2, ..., x_n\}$, where n specifies the length of the input article and m denotes the length of the summary. Figure 2 shows the framework of KAAS. As is shown, Transformer is employed to encode the input tokens as article encoder and obtain the source document representation D^{final}.

4.1   Keyword Encoder

In this work, the keyword encoder is a hierarchical encoder composed of a bidirectional LSTM for encoding the words associated with each keyword and a transformer layer for connecting all the keywords in an example.

Keyword Extraction. The keywords are extracted using the open-source key extraction toolkit *pke* [1]. Specifically, we chose the Multipartite [2] model to extract keywords. We analyze four possible keyword sources for an example. (1) We begin with attempting to extract the guiding keywords from the target summaries. And in this situation, in the testing time, the keywords are extracted from the source articles. (2) A robust extractive approach, Lexrank, is utilized to extract key sentences from the source articles, and then keywords are taken from these sentences and used during both training and testing. (3) Considering that the title is a short but informative text, we also try to extract the keywords from the title of the examples and used them during training and testing. (4) Keywords are directly taken from the source articles in this situation.

Keyword Encoding. After obtaining the keywords of the examples, we encode them with a hierarchical keyword encoder. Considering that each keyword often contains several words, the keyword encoder first generates the corresponding embedding for each keyword using a two-layer bidirectional LSTM. Specifically, giving a keyword $k_i = \{w_{i1}, w_{i2}, ..., w_{il}\}$, the representation kr_i of the i_{th} keyword k_i can be defined as: $kr_i = BiLSTM(w_{i1}, w_{i2}, .., w_{il})$, where l denotes the number of the words containing in k_i.

Since the keywords in an article are related rather than distinct, a transformer layer is applied to the article's keyword representations to connect one keyword to the others.

$$kw_1, kw_2, ..., kw_N = Transformer(kr_1, kr_2, .., kr_N) \tag{1}$$

where N denotes the keyword number of an article, and kw_i denotes the keyword representation of i_{th} keyword after transformer layers. Thus, the keyword sequence of an article can be expressed as $K = \{kw_1, kw_2, ..., kw_N\}$.

4.2 Keyword-Aware Focused Attention Decoder

After obtaining the representations of the article and the keyword sequence, we employ a keyword-aware focused attention decoder to generate the summaries for the article. In the decoder, generated word representations (target word representations in training time) are first attended to themselves with a self-attention layer $SelfAttn$.

$$y = Layernorm(y + SelfAttn(y)) \tag{2}$$

Then the decoder attends to the keyword representations via multi-head self-attention mechanism $MultiAttn$, which informs the decoder with the keyword information.

$$y = Layernorm(y + MultiAttn(K, K, y)) \tag{3}$$

In the final step, based on the keyword-aware representations, the decoder decodes the summary of an article by attending to the source input article.

$$y = Layernorm(y + MultiAttn(D^{final}, D^{final}, y)) \tag{4}$$

Table 2. Rouge scores on COVID-SUM, ArXiv and PubMed dataset.

Model	COVID-SUM dataset			ArXiv dataset			PubMed dataset		
	R-1	R-2	R-L	R-1	R-2	R-L	R-1	R-2	R-L
MMR	33.62	8.09	15.34	–	–	–	–	–	–
Textrank	40.78	14.80	15.33	–	–	–	–	–	–
PG-BRNN	42.16	16.00	22.34	32.06	9.04	25.16	35.85	10.22	29.69
Discourse-Aware Model	–	–	–	35.80	11.05	31.80	38.93	15.37	35.21
PEGASUS-base	39.15	14.89	**23.46**	34.81	10.16	30.14	39.98	15.15	**35.89**
CopyTrans	41.73	15.41	22.34	42.23	14.62	**30.55**	40.97	15.49	35.42
KAAS+tgt+ke	**43.35**	**16.34**	22.60	**42.36**	**14.91**	30.39	**42.14**	**16.01**	35.71
KAAS+src+ke	43.18	16.26	22.54	–	–	–	–	–	–
KAAS+title+ke	42.63	16.03	22.56	–	–	–	–	–	–
KAAS+lexr+kw	41.50	15.24	22.17	–	–	–	–	–	–

5   Experiments

Evaluation Metrics. We evaluate with ROUGE F1 scores [12], and report ROUGE-1, ROUGE-2, and ROUGE-L F1 scores in this paper, which are the most widely used automated metrics for summarization tasks.

Experimental Setting. Hidden size of transformers is set to 512. Keyword encoder is built using a two-layer BiLSTM with a hidden size of 512. The embedding size is set to 512. The keyword number of the KAAS model is assigned to 20. In the decoding process, beam size is set to 5. Furthermore, for each of the three datasets, the maximum and minimum decoding lengths are set to 200 and 300, respectively.

Baselines. We compare our KAAS model with several state-of-the-art summarization models as follows: **Textrank** [14] calculates sentence significance ratings based on eigenvector centrality in a global network and then ranks the sentences using the acquired scores; **MMR** [3] (Maximal Marginal Relevance) is an extractive summarization model that ranks sentences according to their relevance and redundancy to a given query; **PG-BRNN** [17] is the pointer generator model that consists of a one-layer bi-LSTM encoder and a single-layer LSTM decoder; **CopyTrans** [7] consists of a 4-layer transformer encoder and a 4-layer transformer decoder; **PEGASUS** [13] is a pre-trained Transformer-based encoder-decoder model; **Discourse-Aware Model** [6] encodes the structure of one document with a hierarchical encoder.

5.1   Results

Table 2 shows the ROUGE F1 scores on COVID-SUM, Arxiv, and PubMed datasets. As shown in Table 2, nearly all state-of-the-art abstractive summarization models outperform extractive approaches. We think this is because that the extractive models just extract from the source article, while the original extraction sentences usually do not appear in the target human-written summaries.

Reference: BACKGROUND: The current COVID-19 pandemic has challenged the infrastructure of the healthcare systems. To cope with the pandemic, substantial changes were introduced to surgical practice and education all over the world. METHODS: A scoping search in PubMed and Google scholar was done using the search terms: "Coronavirus," "COVID-19", "SARS-CoV-2", "nCoV-2019", and "surgery." They were either searched individually or in combination. All relevant articles of any study design (published within December 15, 2019, till the mid of June 2020), was included and narratively discussed in this review. RESULTS: Sixty-six articles were reviewed in this article. Through these articles, we provide guidance and recommendations on the preoperative preparation and safety precautions, intraoperative precautions, postoperative precautions, ..., scheduling, and staffing. Special precautions were taken before, during, or after surgeries. New treatment and teaching modalities emerged in response to the pandemic. Psychological support and training platforms are necessary for the surgical team.

PEGASUS: Thevirus disease (COVID-19) is a global pandemic affecting over 3 million people and has vastly impacted healthcare systems worldwide. The surgical field has been influenced as a result of the massive redirection of medical attention and priority towards caring for the COVID-19 infected patients. The economic demands of this global issue are increasing, and they are being exacerbated by the needs of surgical procedures. This narrative review provides a global view of the impact of the COVID-19 pandemic on the surgical field. We summarized the safety measures required in the different operative and postoperative phases. We also highlighted the impact of COVID-19 on elective and cancer surgery scheduling and the evolvement of the process of taking surgical decisions. We discussed the impact of the pandemic on the surgical staff mental health and education.

KAAS: BACKGROUND: Coronavirus disease (COVID-19) is a global pandemic affecting over 3 million people worldwide. The World Health Organization (WHO) recognized this public health issue as an international emergency on March 11, 2020. OBJECTIVE: The current COVID-19 pandemic has challenged the infrastructure of the healthcare systems. To cope with the impact of the pandemic on the surgical field. METHODS: A scoping review was done using the search terms: "Coronavirus," and "surgery." They were searched individually or in combination. All relevant articles of any study design (published within December 15, 2019, till the mid of June 2020), was included and narratively discussed in this scoping review. A computer literature search of PubMed and Google scholar was performed. RESULTS: Of 66 articles and reports were eligible for inclusion in this review. The safety measures required in the surgical field, and the evolvement of the process of taking surgical procedures that need a rapid intervention (4,5). ....

Fig. 3. System summaries of different methods on COVID-SUM.

We study four variants of KAAS based on the various sources of the retrieved keywords on COVID-SUM dataset. In Table 2, KAAS+tgt+ke refers to the KAAS model in which the keywords are derived from target summaries during training and from source articles during testing. Similarly, KAAS+src+ke denotes a KAAS variant that employs keywords from source articles during both training and testing. KAAS+title+ke represents a KAAS model which extracts keywords from the titles of the articles. KAAS+lexr+kw is one KAAS variant in which the keywords are extracted from the extractive summaries generated by the Lexrank method.

As shown in Table 2, KAAS+tgt+ke achieves the best scores on ROUGE-1 and ROUGE-2 metrics on three datasets, which demonstrates the effectiveness of KAAS. This result shows that although pre-trained models achieve great successes benefiting from a huge number of parameters, there is still a gap between the generic datasets used to train the models and the specific scientific papers. Additionally, among all the KAAS variants, KAAS+tgt+ke performs best, we believe this is because that the keywords of the target summaries are more suitable for guiding the summarization. We further evaluate KAAS on ArXiv and PubMed datasets, which are commonly used in previous relevant publications, and the results are shown in Table 2. KAAS+tgt+ke outperforms all the other state-of-the-art abstractive models with ROUGE-1 and ROUGE2 scores.

Case Study. As shown in Fig. 3, we display several summaries generated by PEGASUS, KAAS, and the corresponding reference. In this case, the context highlighted in blue represents the important details. Specifically, the presented case shows that all the models can generate topic-related summaries, while the generated summary of our model is more comprehensive and able to give the details. For instance, only the summary of our model presents that *"a scoping search in PubMed and Goole scholar was done using search terms"*, and the timeline *"a scoping search in PubMed and Goole scholar was done using search terms"* December 15, 2019, till the mid of June 2020".

6   Conclusion

In this paper, we introduce a new scientific summarization dataset, COVID-SUM, which is the first COVID-19-related summarization dataset. We hope that this dataset can promote the advancement of the abstractive summarization of scientific articles. Additionally, we propose a novel keyword-aware attention model for abstractive summarization, which outperforms some state-of-the-art summarization models on COVID-SUM, ArXiv, and PubMed datasets.

References

1. Boudin, F.: PKE: an open source python-based keyphrase extraction toolkit. In: COLING, pp. 69–73 (2016)
2. Boudin, F.: Unsupervised keyphrase extraction with multipartite graphs. In: NAACL-HLT, pp. 667–672 (2018)
3. Carbonell, J.G., Goldstein, J.: The use of MMR, diversity-based reranking for reordering documents and producing summaries. In: SIGIR, pp. 335–336 (1998)
4. Celikyilmaz, A., Bosselut, A., He, X., Choi, Y.: Deep communicating agents for abstractive summarization. In: NAACL-HLT, pp. 1662–1675 (2018)
5. Chopra, S., Auli, M., Rush, A.M.: Abstractive sentence summarization with attentive recurrent neural networks. In: NAACL-HLT, pp. 93–98 (2016)
6. Cohan, A., et al.: A discourse-aware attention model for abstractive summarization of long documents. In: NAACL-HLT, pp. 615–621 (2018)
7. Gehrmann, S., Deng, Y., Rush, A.M.: Bottom-up abstractive summarization. In: EMNLP, pp. 4098–4109 (2018)
8. Hochreiter, S., Schmidhuber, J.: Long short-term memory. Neural Comput. 9(8), 1735–1780 (1997)
9. Hu, T., Liang, J., Ye, W., Zhang, S.: Keyword-aware encoder for abstractive text summarization. In: Jensen, C.S., et al. (eds.) DASFAA 2021. LNCS, vol. 12682, pp. 37–52. Springer, Cham (2021). https://doi.org/10.1007/978-3-030-73197-7_3
10. Huang, L., Cao, S., Parulian, N.N., Ji, H., Wang, L.: Efficient attentions for long document summarization. In: NAACL-HLT, pp. 1419–1436 (2021)
11. Kieuvongngam, V., Tan, B., Niu, Y.: Automatic text summarization of COVID-19 medical research articles using BERT and GPT-2. CoRR abs/2006.01997 (2020)
12. Lin, C.Y.: ROUGE: A package for automatic evaluation of summaries. In: ACL-04 Workshop, pp. 74–81 (2004)
13. Lu, W., Huang, Z., Hong, C., Ma, Y., Qu, H.: PEGASUS: bridging polynomial and non-polynomial evaluations in homomorphic encryption. In: IEEE SP, pp. 1057–1073 (2021)
14. Mihalcea, R., Tarau, P.: TextRank: bringing order into text. In: EMNLP, pp. 404–411 (2004)
15. Nallapati, R., Zhou, B., dos Santos, C.N., Gülçehre, Ç., Xiang, B.: Abstractive text summarization using sequence-to-sequence RNNs and beyond. In: SIGNLL, pp. 280–290 (2016)
16. Paulus, R., Xiong, C., Socher, R.: A deep reinforced model for abstractive summarization. ArXiv abs/1705.04304 (2018)
17. See, A., Liu, P.J., Manning, C.D.: Get to the point: summarization with pointer-generator networks. In: ACL, pp. 1073–1083 (2017)

18. Sharma, E., Huang, L., Hu, Z., Wang, L.: An entity-driven framework for abstractive summarization. In: EMNLP-IJCNLP, pp. 3278–3289 (2019)
19. Su, D., Xu, Y., Yu, T., Siddique, F.B., Barezi, E.J., Fung, P.: CAiRE-COVID: a question answering and query-focused multi-document summarization system for COVID-19 scholarly information management. In: EMNLP (2020)
20. Vaswani, A., et al.: Attention is all you need. In: NIPS, pp. 5998–6008 (2017)
21. Wang, L.L., et al.: CORD-19: the COVID-19 open research dataset. CoRR abs/2004.10706 (2020)

E-Commerce Knowledge Extraction via Multi-modal Machine Reading Comprehension

Chaoyu Bai[✉]

School of Cyber Science and Engineering, Southeast University, Nanjing, China
baichaoyu@seu.edu.cn

Abstract. Extracting commodity attributes from unstructured data is an essential information extraction task in the e-commerce domain. It plays an important role in tasks such as commodity recommendation and commodity knowledge base extension. Traditional models for these tasks often only used text modal information, which is not enough to describe the commodity. In recent years, we increasingly see using multi-modal data to describe commodities such as text, images, and videos, which provides the possibility for better product attribute extraction. To this end, we propose a novel model named E-commerce Knowledge Extraction via Multi-modal Machine Reading Comprehension (EKE-MMRC). Specifically, it finds the missing attributes from the existing knowledge base and generates problems. Then packages them with multi-modal descriptions and encode as fusion vector. And then, decode the correlation of description with the attribute and generate an answer from the fusion vector. Finally, use the correlation as the weight of voting for the answer. At the same time, we contributed a dataset for this task based on public e-commerce data: E-commence Multi-modal Commodity Attributes Extraction Dataset (E-MCAE). We also conducted experiments on a public dataset. The experimental results show that our proposed method is effective, and it achieves more than 15% improvement compared with the SOTA single-modal extraction method.

Keywords: Multi-modal · Machine reading comprehension · Information extraction

1 Introduction

A commodity attribute is the pair of an attribute name and an attribute value. For example, (*color, blue*) is an attribute which means the commodity's color is blue. Extracting commodity attributes from unstructured data is an essential information extraction task in the e-commerce domain. These extracted commodity attributes can constitute a structured commodity knowledge base. These structured data are widely used in commodity recommendation, which greatly improves the transaction efficiency between consumers and merchants.

© The Author(s), under exclusive license to Springer Nature Switzerland AG 2022
A. Bhattacharya et al. (Eds.): DASFAA 2022, LNCS 13247, pp. 272–280, 2022.
https://doi.org/10.1007/978-3-031-00129-1_21

In recent years, breakthroughs have been made in extracting information from semi-structured and unstructured text [3,11]. In the e-commerce domain, these methods have also been successfully used to extract valuable information from commodity text descriptions and user reviews.

In the real world, text and visual and auditory information also play important roles in the e-commerce platform. Multi-modal information, the information in different forms or obtained from different sources, helps us better understand the world. It is the same for the machine. For example, in machine translation [16], machine dialog [12], co-reference resolution [14], after combining multi-modal data, the model can handle more complex and realistic tasks. It also shows obvious advantages compared with the single-modal model. The phenomenon of multi-modal data explosion is particularly obvious in the field of e-commerce.

So we defined a new task. The task is to extract commodity attributes from multi-modal commodity descriptions. For this task, we construct a dataset based on public e-commerce data: E-commence Multi-modal Commodity Attributes Extraction Dataset (E-MCAE). It contains structured information of 5000 commodities and their multi-modal descriptions.

Facing the above challenges, we propose a novel model named E-commerce Knowledge Extraction via Multi-modal Machine Reading Comprehension (EKE-MMRC). Specifically, it finds the missing attributes from the existing knowledge base and generates problems. Then packages them with multi-modal descriptions and encode as fusion vector. And then, decode the correlation of description with the attribute and generate an answer from the fusion vector. Finally, use the correlation as the weight of voting for the answer.

The contributions of this paper are summarized as: We are the first to consider extracting attributes from multi-modal descriptions with existing knowledge base and have constructed a new dataset for the challenging task; We conduct extensive experiments to evaluate our model against popular methods. Results on the constructed dataset show that our proposed method is effective as it significantly improves the F_1 score from 51.52% to 63.67% compared with baselines.

2   Related Work

2.1   Machine Reading Comprehension

Teaching machines to read and understand large-scale text descriptions is a promising long-term goal of natural language understanding. Machine reading comprehension (MRC) model is aimed at accomplishing this task. In this regard, there have been several benchmark datasets proposed in recent years, which pushes forward the development of MRC, including SQuAD and Natural Questions.

In the past one or two years, there has been a trend of transforming information extraction tasks to MRC question answering. Levy et al. [6] formulate relation extraction as QA tasks. For example, the relation MARRY-WITH can

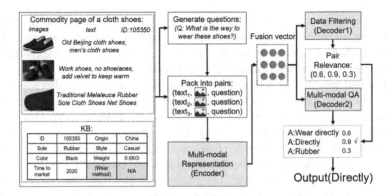

Fig. 1. The framework finds the missing attributes from the existing knowledge base and generates problems. Then packages them with multi-modal descriptions and encode as fusion vector. Decode the correlation of description with the attribute and generate an answer from the fusion vector. Finally, use the correlation as the weight of voting for the answer.

be mapped to *"Who is X's husband/wife?"*. Inspired by this [6], our work formalizes attribute value extraction as a multi-passage machine reading comprehension task. Unlike this work, we use a generative model to generate more diverse attribute values instead of classifying on a predefined set of relationships.

2.2   Multi-modal Information Extraction

Information in the real world generally appears in the form of multi-modality, but due to technical problems, multi-modal research progress is slow. In recent years, due to single-mode research progress, multi-modal research has a more solid foundation. Multi-modal information extraction is a research direction that combines multi-modal learning and information extraction technology [4].

In the task of entity linking, Moon et al. [10] combines pictures and text for entity disambiguation and uses the attention mechanism to fuse pictures, text, and knowledge base information. Finally, the entity link result is obtained by calculating the similarity between the entity mention and the entity in the knowledge base. In the task of link prediction, IKRL [15] adopts the method of expanding the energy function of TransE [1] and adding the energy function between the original representation of the entity and the representation of the entity picture to fuse multi-modal information. Our work also uses multi-modal information to extract knowledge better. Zhu et al. [17] tried to extract product attributes through pictures and text information, compared with this work, our model performs better multi-modal feature fusion, and can extract attributes that are not in the predefined dictionary.

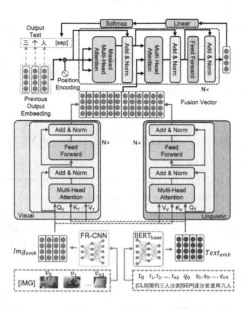

Fig. 2. The text description and the generated questions are spliced together and sent to the multi-modal reading comprehension model together with the image sequence.

3   Framework

3.1   Description-Question Pair Preparation

This step converts the missing attributes into questions. For the missing attributes found in the previous step, the model will pack it into a triple. Then, according to the simple template, the question will be generated from the triples, and the IE task is transformed into a QA task. Based on these query triples, we can use simple templates to generate questions. Unlike traditional QA tasks, this step is not designed to generate real-world questions but to guide the model to find the correct answer in the descriptions.

3.2   Multi-modal Encoder

The Multi-modal Encoder is used to extract and fuse the features of multiple modalities of descriptions and questions. The overview of this component is shown in the bottom part of Fig. 2.

Transformer Module for Modal Fusion. In recent years, Transformer-based models have shown good performance on various tasks. Specifically, the Transformer is formed by connecting multiple identical layers, and each layer includes two sub-layers. The first sub-layer is a self-attention layer, and the second sub-layer is a fully connected layer. Residual connection and layer normalization are added between layers. We introduce a special multi-modal fusion model for the framework based on the Transformer, as shown in Fig. 2. We denote the l-th hidden layer features of visual and linguistic as $\mathbf{H}_V^l$ and $\mathbf{H}_T^l$. The module calculates query, key, and value as in the standard Transformer encoder. However, the Transformer of each modal will send key and value into the Transformer of another modal. The attention layer of the l-th visual transformer can be formalized as follows:

$$\text{Co-Att}_V^l = \text{softmax}\left(\frac{\mathbf{Q}_V^{l-1}(\mathbf{K}_T^{l-1})^T}{\sqrt{d_k}}\right)\mathbf{V}_T^{l-1}, \tag{1}$$

where $\mathbf{Q}_V^{l-1}$ generated by the $(l-1)$-th visual transformer, while $\mathbf{K}_T^{l-1}$ and $\mathbf{V}_T^{l-1}$ generated by the $(l-1)$-th linguistic transformer.

By using this structure, the attention block will produce attention-pooled features of another modality. In the above formula, the attention block performs language attention in visual data. The other block structures are the same as the standard Transformer, including residual connection and layer normalization. This visual and language common attention block has been used in computer vision (first proposed in [13]), and its effectiveness has been verified through experiments. In the dual-stream multi-modal representation learning model [8], by stacking more common attention blocks, a good representation is obtained, and good performance is achieved in multiple downstream tasks.

3.3 Data Filter

This step is used to filter out multi-modal descriptions, which more likely to containing answers. With the development of network technology, more and more information appears every day. While generating valuable information, it also generates a lot of irrelevant data, considered noise in the task. It is difficult for the reading comprehension module to have the ability to understand questions, search for answers, and filter irrelevant information at the same time. In the actual operation process, the text-image pair and the problems generated in the previous step will be encoded through a multi-modal representation model. The representation model will be introduced in detail in the next section. This step can be regarded as a decoder. The model will learn a linear layer, which performs binary classification on the feature vector.

3.4 Attribute Extraction by Machine Reading Comprehension

This component is the second decoder used to extract attribute values from multi-modal data. We cannot guarantee that the attribute value appears in the text description in this task, so the traditional extractive question and answer method cannot be used here. Hence, we adopt the auto-regressive model is used in the process of generating answers. Auto-regressive generation means that the decoding stage is recursively generated word by word, and its formal probability distribution is as follows:

$$p(y_1,\ldots,y_n|x) = p(y_1|x)\ldots p(y_n|x,y_1,\ldots,y_{n-1}). \tag{2}$$

The generation process of y_i will be affected by y_1 to y_{i-1}. In each step, the model will select y_i with the highest probability. Since multiple multi-modal descriptions describe the same entity, each description with a question can decode an answer, so we use voting to determine the final answer. The weight of the vote uses the probability generated in Data Filter. Descriptions that are more relevant to the problem should be given higher weight.

4   Experiments

4.1   Dataset

We get some commodities and their introduction pages on JingDong. It is a popular e-commerce platform with rich pictures and text descriptions for commodities.

<table>
<tr><td colspan="2">Table 1. Details of data</td><td colspan="2">Table 2. Some attributes</td></tr>
<tr><td>Key</td><td>Value</td><td>Attributes name</td><td>Number</td></tr>
<tr><td>Number of commodities</td><td>5000</td><td>Commodity Title</td><td>5000</td></tr>
<tr><td>Number of image</td><td>101,789</td><td>Commodity Brand</td><td>4336</td></tr>
<tr><td>Number of triples</td><td>113,467</td><td>Commodity Color</td><td>3918</td></tr>
<tr><td>Average length of text</td><td>17.77</td><td>Commodity Origin</td><td>3590</td></tr>
<tr><td>Average number of attr</td><td>22.69</td><td>Total</td><td>28,334</td></tr>
</table>

The data we collected contains information about 5000 commodities, and details of the dataset are shown in Table 1. We collect five types of commodities: shoes, cabinets, tables, chairs, and sofas. Table 2 shows some statistical information about the collected multi-modal descriptions and knowledge base.

We construct E-commence Multi-modal Commodity Attributes Extraction Dataset (E-MCAE) based on the knowledge base and entity descriptions construction. Given entity and attribute names, the task is to extract attribute values from the descriptions.

We also conducted experiments on a public dataset MEPAVE [17], which ensures that the attributes that need to be extracted are in the text description.

4.2   Baselines

Since this task is relatively novel and there exist few related models, we report the popular methods using different modal information as baselines. We compare the following methods:

- Single-LSTM [5] is a classic model, widely used in various encoders and decoders.

- BERT QA [2] uses Whole Word Masking, which mitigates the drawbacks of masking partial WordPiece tokens in pre-training BERT.
- Neural Image QA [9] is a classic VQA model, which uses CNN and LSTM to encode images and questions, and finally outputs the answer.
- Multi-modal Attribute Extraction [7] connects text and image features, sends them to the fully connected layer, and then uses LSTM for decoding.
- M-JAVE [17] is a multimodal input model used to extract attributes of e-commerce commodities.
- **(proposed)** Our method: is the proposed method as described in Fig. 2.

Table 3. Main results

Model	E-MCAE(F1)	MEPAVE(F1)
Single-LSTM	36.50	82.08
BERT-QA	51.52	87.43
Neural Image QA	22.94	–
Multi-modal Extraction	38.62	–
M-JAVE	49.32	87.17
EKE-MMRC	63.67	93.52

4.3   Results and Analysis

Main Result. Table 3 shows the F_1 score of each model on E-MCAE and MEPAVE. Comparing the first two experiments indicates that the model with BERT has greatly improved the performance due to the introduction of background knowledge. The multi-modal extraction method over-perform the Single-LSTM and Neural Image QA, which prove the effectiveness of combining multi-modal descriptions. All the methods with background knowledge over-perform others illustrate that it is necessary to combine background knowledge and understand the description to answer the correct attribute value. The method of directly connecting two modal features cannot make good use of image features.

On MEPAVE, since all the attributes are mentioned in the text, the model using only the text modal (LSTM/BERT-QA) can also achieve good performance. Compared with M-JAVE, our model uses a more complex multi-modal feature fusion layer, and its extraction results are better.

5   Conclusion

In this paper, we propose a novel attribute extraction framework named EKE-MMRC to extract entity attributes from multi-modal descriptions. In our framework (Fig. 1), we divide the complicated multi-modal attribute extraction problem into four steps and show that the first steps can be solved well enough by existing work. For the most difficult extraction, we transform it to a machine reading comprehension problem and solve it with our encoder-decoder model. We also construct a large-scale E-commence Multi-modal Commodity Attributes Extraction Dataset (E-MCAE). Our multi-modal encoder is pre-trained on a large-scale general caption dataset, and continues to be pre-trained after collecting domain data, and finally fine-tuned on the E-MCAE. We compare the model with the popular information extraction model on E-MCAE, and the results show that our model has been significantly improved. In future work, we will consider the relationship between commodities and commodities and expand from the knowledge base to graphs, which is excluded from problem scope in this paper. Besides, better fusion of multi-modal information is also a challenging problem as we mentioned previously. We hope this work can provide new thoughts for the area of multi-modal information extraction.

References

1. Bordes, A., Usunier, N., Garcia-Duran, A., Weston, J., Yakhnenko, O.: Translating embeddings for modeling multi-relational data. In: Advances in Neural Information Processing Systems, pp. 2787–2795 (2013)
2. Cui, Y., et al.: Pre-training with whole word masking for Chinese BERT. arXiv preprint arXiv:1906.08101 (2019)
3. Gu, J., Meng, G., Da, C., Xiang, S., Pan, C.: No-reference image quality assessment with reinforcement recursive list-wise ranking. In: Proceedings of the AAAI Conference on Artificial Intelligence, vol. 33, pp. 8336–8343 (2019)
4. Gu, J., et al.: Exploiting behavioral consistence for universal user representation. arXiv preprint arXiv:2012.06146 (2020)
5. Hochreiter, S., Schmidhuber, J.: Long short-term memory. Neural Comput. 9(8), 1735–1780 (1997)
6. Levy, O., Seo, M., Choi, E., Zettlemoyer, L.: Zero-shot relation extraction via reading comprehension. arXiv preprint arXiv:1706.04115 (2017)
7. Logan IV, R.L., Humeau, S., Singh, S.: Multimodal attribute extraction. arXiv preprint arXiv:1711.11118 (2017)
8. Lu, J., Batra, D., Parikh, D., Lee, S.: ViLBERT: Pretraining task-agnostic visiolinguistic representations for vision-and-language tasks. In: Advances in Neural Information Processing Systems, pp. 13–23 (2019)
9. Malinowski, M., Rohrbach, M., Fritz, M.: Ask your neurons: a neural-based approach to answering questions about images. In: Proceedings of the IEEE International Conference on Computer Vision, pp. 1–9 (2015)
10. Moon, S., Neves, L., Carvalho, V.: Multimodal named entity disambiguation for noisy social media posts. In: Proceedings of the 56th Annual Meeting of the Association for Computational Linguistics (Volume 1: Long Papers), pp. 2000–2008 (2018)
11. Niklaus, C., Cetto, M., Freitas, A., Handschuh, S.: A survey on open information extraction. arXiv preprint arXiv:1806.05599 (2018)
12. Poria, S., Hazarika, D., Majumder, N., Naik, G., Cambria, E., Mihalcea, R.: MELD: a multimodal multi-party dataset for emotion recognition in conversations. arXiv preprint arXiv:1810.02508 (2018)

13. Ren, S., He, K., Girshick, R., Sun, J.: Faster R-CNN: towards real-time object detection with region proposal networks. In: Advances in Neural Information Processing Systems, pp. 91–99 (2015)
14. Sigurdsson, G.A., et al.: Visual grounding in video for unsupervised word translation. In: Proceedings of the IEEE/CVF Conference on Computer Vision and Pattern Recognition, pp. 10850–10859 (2020)
15. Xie, R., Liu, Z., Luan, H., Sun, M.: Image-embodied knowledge representation learning. arXiv preprint arXiv:1609.07028 (2016)
16. Zhang, Z., et al.: Neural machine translation with universal visual representation. In: International Conference on Learning Representations (2019)
17. Zhu, T., Wang, Y., Li, H., Wu, Y., He, X., Zhou, B.: Multimodal joint attribute prediction and value extraction for e-commerce product. arXiv preprint arXiv:2009.07162 (2020)

PERM: Pre-training Question Embeddings via Relation Map for Improving Knowledge Tracing

Wentao Wang[1], Huifang Ma[1(✉)], Yan Zhao[1], Fanyi Yang[1], and Liang Chang[2]

[1] College of Computer Science and Engineering, Northwest Normal University, Lanzhou 730070, Gansu, China
mahuifang@yeah.net

[2] School of Computer Science and Information Security, Guilin University of Electronic Technology, Guilin 541004, Guangxi, China

Abstract. Learning informative embedding for educational question (aka. representations) lies at the core of online learning systems. Recent solutions mainly focus on learning question embedding via the question-concept bipartite graph. However, the student-question-concept global relation is inadequately exploited. Moreover, finer-grained semantic information from student-question and student-concept interactions should also be further revealed. To this end, in this paper, we propose to **P**re-train question **E**mbeddings via **R**elation **M**ap for knowledge tracing, namely **PERM**. Extensive experiments conducted on two real-world datasets show that PERM has higher expressive power which enables knowledge tracing methods to effectively predict students' performance.

Keywords: Question embedding · Relation map · Attention aggregation mechanism · Knowledge tracing

1 Introduction

Knowledge tracing (KT), an essential task in CAE systems, aims at modeling students' knowledge proficiency based on their learning history. To be specific, the objective of KT is to predict whether a student can answer the next question correctly or not according to all his/her previous response logs [9].

To solve the KT problem, many great efforts have been made, and basically they can be divided into three broad strands of works: memory-based, attention-based and graph-based methods. First, in memory-based methods, DKT [8] and DKVMN [11] are two typical approaches that try to learn sequential patterns from individual student response logs. Second, some attention-based methods including SAKT [6], RKT [7], and AKT [3] are proposed and improve performances of those memory-based methods. Third, graph-based methods can excavate complex interactions between questions and concepts on certain graphs, such as GKT [5] and GIKT [10].

A. Bhattacharya et al. (Eds.): DASFAA 2022, LNCS 13247, pp. 281–288, 2022.
https://doi.org/10.1007/978-3-031-00129-1_22

Although existing methods have utilized question embeddings for KT, these question embeddings' comprehensiveness is insufficient yet. To be more specific, there exist two problems along this line. First, the student-question-concept global relation has remained inadequately explored. Second, it is also necessary to move one step forward to distill finer-grained semantic information from the student-question interaction and the student-concept interaction. Thus, we strive to take a further step towards maximally extracting and mining plentiful underlying semantic information in the student-question-concept interaction to address those aforementioned two issues. However, challenges lie ahead owing to the complexity of the student-question-concept interaction. On one hand, due to the high heterogeneity of different types of nodes and edges by viewing the student-question-concept interaction as a graph, it is difficult to further exploit semantic information on this graph. On the other hand, the structure of the interaction graph is pretty complex and its hierarchy is also directly explicit, which indicates that modeling question embeddings by exploiting the student-question-concept interaction graph is very challenging.

Taking everything into consideration, in this paper, we propose a pre-training method, namely Pre-training question Embeddings via Relation Map (PERM) for improving knowledge tracing, to learn a general embedding for each question. Concretely, from two semantic perspectives (co-occurrence and answer-agreement), we first exploit relation maps between questions (and concepts) as prior and initialize embeddings of question as well as concept. Afterwards, by extracting the question-concept bipartite graph from the student-question-concept interaction, we utilize a two-level attention aggregation mechanism with the prior to attain the updated embeddings of question and concept. Furthermore, for obtaining the final question embeddings, the difficulty information of question is modeled depending on the updated question embedding, the updated concept embedding and the feature vector of question. Finally, extensive experiments conducted on two real-world datasets clearly demonstrate that PERM has higher expressive power that enables KT methods to effectively predict students' performance. In summary, our key contributions are listed as follows: (1) From two perspectives of co-occurrence and answer-agreement, we exploit two relation maps between questions (and concepts) as the prior; (2) To comprehensively model the question embedding with difficulty, we leverage a two-level attention mechanism with the prior to organically fuse information from the question, related concepts and features of the question; (3) Extensive experiments on real-world datasets validate the effectiveness and superiority of PERM, including quantitative comparison, qualitative analysis and case study.

2   Problem Formulation

Let $\mathcal{S} = \{s_1, s_2, ..., s_N\}$ be the set of N students, $\mathcal{Q} = \{q_1, q_2, ...q_M\}$ be the set of M questions and $\mathcal{C} = \{c_1, c_2, ...c_K\}$ be the set of K concepts. In KT task, given a student's history interaction records $\mathcal{X} = \{(q_1, a_1), ..., (q_{t-1}, a_{t-1})\}$ where $a_i \in \{0, 1\}$ is the answer that the student responds to the question q_i correctly

$(a_i = 1)$ or incorrectly $(a_i = 0)$ at the time step i. Let $\mathcal{L}$ be the set of all students' response records. The goal of KT is to predict the probability that the student will answer next question correctly, that is, $p(a_t = 1|q_t, \mathcal{X})$. For clarity, here we introduce some definitions we will leverage to train question embeddings in our model.

Definition 1 (question-concept bipartite graph). The question-concept bipartite graph denoted as $\mathcal{G}_{qc} = (\mathcal{Q} \cup \mathcal{C}, \mathcal{R}_{qc})$ is an undirected graph, where $\mathcal{R}_{qc} = \{r_{q_i \leftrightarrow c_k}|q_i \in \mathcal{Q}, c_k \in \mathcal{C}, r_{q_i \leftrightarrow c_k} \in \{0,1\}\}$. In addition, $r_{q_i \leftrightarrow c_k} = 1$ refers that the question q_i contains the concept c_k, otherwise $r_{q_i \leftrightarrow c_k} = 0$.

Definition 2 (question difficulty). Following the previous work [4], the question difficulty d_i for a certain question q_i is defined as the ratio of correctly answered by computing from the training dataset.

3   Method

3.1   Relation Map Exploitation

In terms of the co-occurrence perspective, we focus on the question-concept interaction and dedicate ourselves to selecting homogeneous neighbors for all questions. Specifically, the question transaction matrix can be specified as $\mathbf{T}^q = [t_{ij}] \in \{r_{q_j \leftrightarrow c_i}\}^{K \times M}$. Therefore, the question relation matrix under the co-occurrence perspective is $\mathbf{S}_q^{co}$, and its element $S_q^{co}[ij]$ indicates the relation strength between q_i and q_j. To be more specific, the relation strength between any two questions can be calculated as:

$$S_q^{co}[ij] = \frac{\frac{\left\|\mathbf{T}_{:,i}^q \circ \mathbf{T}_{:,j}^q\right\|_1}{K}}{\frac{\left\|\mathbf{T}_{:,i}^q\right\|_1}{K} \times \frac{\left\|\mathbf{T}_{:,j}^q\right\|_1}{K}} = \frac{\left\|\mathbf{T}_{:,i}^q \circ \mathbf{T}_{:,j}^q\right\|_1}{\left\|\mathbf{T}_{:,i}^q\right\|_1 \times \left\|\mathbf{T}_{:,j}^q\right\|_1/K} \tag{1}$$

where $\mathbf{T}_{:,i}^q$ and $\mathbf{T}_{:,j}^q$ are the i-th and the j-th column of the matrix $\mathbf{T}^q$, $\circ$ is the element-wise product, and $\|\cdot\|_1$ is the 1-norm of a vector.

Afterwards, to obtain homogeneous neighbors of the target question q_i, we define the normalized question relation matrix under the co-occurrence perspective as $\hat{\mathbf{S}}_q^{co} \in \{0,1\}^{M \times M}$:

$$\hat{S}_q^{co}[ij] = \begin{cases} 1, \text{if } \left(S_q^{co}[ij] > 1\right) \wedge \left(\left(S_q^{co}[ij]/\max\left(\mathbf{S}_q^{co}\right)\right) \geq \lambda_q\right) \\ 0, \text{otherwise} \end{cases} \tag{2}$$

where without generality, $\sigma(\cdot)$ is the sigmoid function in this paper, and λ_q is a hyperparameter to control the sparsity of $\hat{\mathbf{S}}_q^{co}$. Hence, the set of homogeneous neighbors $N_{q_i}^{co} = \{q_j|\hat{S}_q^{co}[ij] = 1, q_j \in \mathcal{Q}\}$ is readily available for the target question q_i. Along this line, the set of homogeneous neighbors $N_{c_k}^{co}$ for a certain concept c_k can be also obtained by the hyperparameter λ_c.

We now describe how students' performance data can be used to obtain relations between questions. First, by solely considering the pairs of q_i and q_j, we can find all the question pairs (q_i, q_j) in each student's interaction sequence. Then, we compute the *NMI* (Normalized Mutual Information) to represent the relation strength $S_q^{ag}[ij]$ between any two questions. We view the response of q_i and q_j (i.e., a_i and a_j) as two random variables respectively, mathematically $S_q^{ag}[ij]$ can be calculated as:

$$S_q^{ag}[ij] = \frac{-2 \sum\limits_{a\in\{0,1\}} \sum\limits_{b\in\{0,1\}} p(a_i = a, a_j = b) \times \log_2 \frac{p(a_i=a,a_j=b)}{p(a_i=a)p(a_j=b)}}{\sum\limits_{a\in\{0,1\}} p(a_i = a)\log_2 p(a_i = a) + \sum\limits_{b\in\{0,1\}} p(a_j = b)\log_2 p(a_j = b)} \quad (3)$$

Concretely, probabilities in the Eq. (3) can be derived by counting frequencies within all the satisfied interaction pairs. Actually, $S_q^{ag}[ij]$ reflects the degree of dependency between q_i and q_j. Similarly, we denote the normalized question relation matrix under the answer-agreement perspective like Eq. (2). Consequently, the question relation map $\mathcal{G}_q$ (and the concept relation map $\mathcal{G}_c$) can be pinpointed by $\hat{\mathbf{S}}_q^{co}$ and $\hat{\mathbf{S}}_q^{ag}$ ($\hat{\mathbf{S}}_c^{co}$ and $\hat{\mathbf{S}}_c^{ag}$).

3.2 Embedding Aggregation

Before embedding aggregation, we need to extract the bipartite graph $\mathcal{G}_{qc}$ from the student-question-concept interaction and initialize all node embeddings. Thus, we encode questions and concepts with d-dimensional trainable matrices $\mathbf{Q} \in \mathbb{R}^{M\times d}$ and $\mathbf{C} \in \mathbb{R}^{K\times d}$. For question q_i, it is initialized by multiplying an one-hot representation vector $\mathbf{x}_i \in \mathbb{R}^{d\times 1}$ with the trainable matrix $\mathbf{Q}$ (i.e., the transpose of i-th row from matrix $\mathbf{Q}$). Similarly, concept c_k's initialized embedding is the transpose of k-th row from matrix $\mathbf{C}$.

For the target question q_i, we utilize self-attention to learn the weight of the node pair (i, j) [2], where the node j is one of the homogeneous neighbors of the node i, and the weight of the homogeneous node pair (i, j) can be directly represented as $\alpha_{ij}^{\psi} = \gamma \alpha_{ij}^a + (1 - \gamma)\alpha_{ij}^p$. This equation models how important homogeneous neighbor j will be for the target question node i under a certain aforementioned perspective $\psi \in \{co, ag\}$, and γ is a tunable hyperparameter to keep a trade-off between the attention score α_{ij}^a learned adaptively and the attention score α_{ij}^p computed from the prior. Furthermore, Eq. (4) and Eq. (5) model the adaptive weight coefficient and the prior weight coefficient of (i, j), respectively:

$$\alpha_{ij}^a = \frac{\exp\left[\sigma(\mathbf{W}_\psi^T(\mathbf{q}_i \| \mathbf{q}_j))\right]}{\sum_{j' \in N_i^\psi} \exp\left[\sigma(\mathbf{W}_\psi^T(\mathbf{q}_i \| \mathbf{q}_{j'}))\right]} \quad (4)$$

$$\alpha_{ij}^p = \frac{\exp(S_q^\psi[ij])}{\sum_{j'=1}^M \exp(S_q^\psi[ij'])} \quad (5)$$

where $||$ denotes the concatenate operation and $\mathbf{W}_\psi \in \mathbb{R}^{2d \times 1}$ is the node-level attention vector for a certain perspective ψ.

Then, under a certain perspective ψ, the node-level embedding $\mathbf{e}_{q_i}^\psi$ of the target question q_i can be aggregated by homogeneous neighbors' mapped features with the corresponding attention weights as follows: $\mathbf{e}_{q_i}^\psi = \sigma\left(\sum_{j \in N_{q_i}^\psi} \alpha_{ij}^\psi \cdot \mathbf{q}_j\right)$. After injecting node features and prior information into node-level attention, we can aggregate two groups of specific both question node embeddings and concept node embeddings, denoted as $\{\mathbf{e}_{q_i}^{co}, \mathbf{e}_{q_i}^{ag}\}$ and $\{\mathbf{e}_{c_k}^{co}, \mathbf{e}_{c_k}^{ag}\}$.

Although the target question node embeddings under two semantic perspectives are obtained after node-level attention aggregation, to learn more comprehensive node embeddings, it is necessary to fuse multiple semantics. Specifically, to learn the weight of each semantic perspective of the target question node i, we first transform semantic-specific embedding through a nonlinear transformation. Then, we average all contributions from each type of neighbors of the node i. The normalized importance of each semantics for the node i, denoted as $\beta_i^\phi = \frac{\exp(w_i^\phi)}{\sum_{\phi' \in \{co, ag, dir\}} w_i^{\phi'}}$, and w_i^ϕ is shown as follows:

$$w_i^\phi = \begin{cases} \frac{1}{|N_{q_i}^\phi|} \sum_{j \in N_{q_i}^\phi} \rho^T \cdot \tanh(\mathbf{W} \cdot \mathbf{e}_{q_j}^\phi + \mathbf{b}), \text{if } \phi \in \{co, ag\} \\ \frac{1}{|N_{q_i}^\phi|} \sum_{j \in N_{q_i}^\phi} \rho^T \cdot \tanh(\mathbf{W} \cdot \mathbf{c}_j + \mathbf{b}), \text{ if } \phi \in \{dir\} \end{cases} \quad (6)$$

where $\mathbf{W} \in \mathbb{R}^{d \times d}$ is a weight matrix, $\mathbf{b} \in \mathbb{R}^{d \times 1}$ is a bias vector and $\rho \in \mathbb{R}^{d \times 1}$ is an attention vector. Note that ϕ is the element of the semantic set $\{co, ag, dir\}$. With the help of learned weights, for the target question node i, we can fuse these embeddings to obtain the updated question embedding $\tilde{\mathbf{q}}_i = \beta_i^{co} \cdot \mathbf{e}_{q_i}^{co} + \beta_i^{ag} \cdot \mathbf{e}_{q_i}^{ag} + \frac{\beta_i^{dir}}{|N_{q_i}^{dir}|} \sum_{c_k \in N_{q_i}^{dir}} \mathbf{c}_k$. Similarly, for a certain concept c_k, we can also obtain its updated concept embedding $\tilde{\mathbf{c}}_k$.

3.3 Difficulty Prediction

First, question q_i's feature vector is $\tilde{\mathbf{f}}_i = \sigma\left(\mathbf{W}_f [\mathbf{f}_{i1}||\mathbf{f}_{i2}||...||\mathbf{f}_{iF}] + \mathbf{b}_f\right)$, where $\mathbf{W}_f$ is a weight matrix, $\mathbf{b}_f$ is a bias vector, F is the number of features and $\tilde{\mathbf{f}}_i \in \mathbb{R}^d$ is the feature vector of q_i. Note that $\mathbf{f}_{ij}$ is a one-hot vector if j-th feature is categorical (e.g., question type); if j-th feature is numerical (e.g., average response time of students), then $\mathbf{f}_{ij}$ is a scalar value. Furthermore, for the target question q_i, we assume that the set of concepts related to q_i is $N_{q_i}^{dir}$ and we use the average representation of all concept embeddings in $N_{q_i}^{dir}$ as the related concept embedding of q_i, denoted as $\tilde{\mathbf{c}}_{N_i} = \frac{1}{|N_{q_i}^{dir}|} \sum_{c_k \in N_{q_i}^{dir}} \tilde{\mathbf{c}}_k$. Then, to recover the difficulty information from the heterogeneous graph effectively, we project the concatenated vector $[\tilde{\mathbf{q}}_i||\tilde{\mathbf{c}}_{N_i}||\tilde{\mathbf{f}}_i]$ into the final target question embedding $\hat{\mathbf{q}}_i \in \mathbb{R}^d$ by using a MLP (Multilayer Perceptron) with three layers, and further leverage a non-linear layer to map $\hat{\mathbf{q}}_i$ to a difficulty approximation: $\hat{d}_i = \sigma(\mathbf{w}_d^T \cdot \hat{\mathbf{q}}_i + b_d)$, where $\mathbf{w}_d$ and b_d are trainable parameters. Finally, following the Definition 2, we design $L = \sum_i^M (d_i - \hat{d}_i)^2$ to measure the difficulty approximation error.

4   Experiments

4.1   Datasets

To evaluate our method, the experiments are conducted on two widely-used datasets in KT. One is ASSISTment09[1], the other dataset is Junyi[2]. Specifically, for the ASSISTment09 dataset, we remove students who attempted fewer than 15 response logs similar to [1] and simultaneously remove those questions whose related concepts less than 2. Moreover, for the Junyi dataset, we filter out students with less than 500 response logs for reducing the cost of computation, and we further fuse some similar concepts (e.g., add_1, add_2, add_3) into one concept (e.g., add) to guarantee that each concept has enough related questions for embedding.

4.2   Experimental Setup

To show the improvement of our method to the mainstream KT models and verify the effectiveness of our proposed method, we compare it with some baselines. Specifically, we first applied it on several kinds of KT models like memory-based models (DKT, DKVMN), attention-based models (SAKT, AKT) and graph-based models (GIKT), and then we compare it with the state-of-the-art question embedding method without text information PEBG.

4.3   Performance Comparison

Table 1 reports the overall performance of all methods on the student performance prediction task on ASSISTment09 and the best scores are denoted in bold. From the result, we summarize several important observations. Firstly, almost all of these PERM+ methods perform better than baselines at all data sparsities. Secondly, PERM+DKT and PERM+DKVMN consistently perform better than PEBG+DKT and PEBG+DKVMN. Lastly, almost all PERM+ methods are superior to baselines.

4.4   Ablation Study

We further set three comparative settings, and their performances have been shown in Fig. 1. The details of three settings are listed below: (1) PERM+ GIKT/P does not consider the prior weights in the node-level aggregation with

---

[1] https://sites.google.com/site/assistmentsdata/home/assistment-2009-2010-data/skill-builder-data-2009-2010.

[2] https://pslcdatashop.web.cmu.edu/DatasetInfo?datasetId=1198.

$\gamma = 1$; (2) PERM+GIKT/R does not consider the relation map exploitation, where select homogeneous neighbors for questions and concepts like PEBG; (3) PERM+GIKT/A does not consider the embedding aggregation, where we choose another aggregation strategy following the literature [9].

Table 1. Comparison results on student performance prediction on ASSISTment09.

Train/Test ratio	50%/50%			60%/40%			70%/30%			80%/20%		
Methods	RMSE	ACC	AUC	RMSE	ACC	AUC	RMSE	ACC	AUC	RMSE	ACC	AUC
DKT	0.471	0.631	0.683	0.467	0.638	0.687	0.464	0.714	0.737	0.458	0.731	0.742
DKVMN	0.468	0.648	0.692	0.453	0.647	0.694	0.457	0.721	0.746	0.441	0.747	0.753
SAKT	0.463	0.651	0.708	0.449	0.659	0.702	0.441	0.734	0.751	0.435	0.758	0.762
AKT-NR	0.458	0.667	0.718	0.442	0.664	0.716	0.438	0.742	0.768	0.428	0.764	0.778
GIKT	0.441	0.672	0.721	0.436	0.678	0.722	0.429	0.759	0.776	0.411	0.771	0.794
PEBG+DKT	0.452	0.658	0.709	0.441	0.673	0.711	0.435	0.747	0.764	0.424	0.768	0.781
PEBG+DKVMN	0.446	0.662	0.715	0.438	0.679	0.719	0.427	0.758	0.778	0.412	0.774	0.797
PERM+DKT	0.432	0.673	0.724	0.431	0.687	0.721	0.422	0.764	0.784	0.408	0.783	0.803
PERM+DKVMN	0.421	0.681	0.728	0.427	0.694	0.729	0.413	0.771	0.799	0.401	0.794	0.814
PERM+SAKT	0.419	0.687	0.731	0.414	0.709	0.739	0.407	0.779	0.808	0.394	0.807	0.827
PERM+AKT-NR	0.408	0.698	0.736	0.412	0.718	0.748	0.398	0.788	0.814	0.388	0.818	0.839
PERM+GIKT	**0.401**	**0.712**	**0.741**	**0.405**	**0.727**	**0.754**	**0.392**	**0.792**	**0.821**	**0.383**	**0.834**	**0.842**

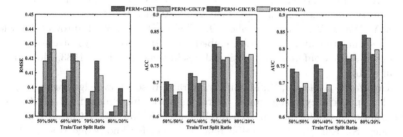

Fig. 1. Performance comparison of ablation study on ASSISTment09.

From the Fig. 1 we can find that (1) PERM+GIKT performs the best indicating the effectiveness of different components of our PERM method; (2) the performances decline slightly when removing the prior in the node-level aggregation; (3) removing the relation map exploitation hurts the performances badly and utilizing the folded bipartite graph also has lower performances; (4) replacing the embedding aggregation with a light aggregation layer, the performances also show a shrinking trend.

5   Conclusion

In this paper, we present a novel PERM method, concentrating on exploiting semantic relation maps and further fusing the embeddings of question and concept under two perspective semantics to obtain the final question embedding for improving performances of KT methods. Extensive experiments conducted on two real-world datasets show PERM has higher expressive power that enables KT methods to effectively predict students' performance.

Acknowledgments. This work is supported by the National Natural Science Foundation of China (U1811264, 61762078, 61363058, 61966004), Gansu Natural Science Foundation Project (21JR7RA114), Research Fund of Guangxi Key Lab of Multisource Information Mining and Security (MIMS18-08), Northwest Normal University Young Teachers Research Capacity Promotion Plan (NWNU-LKQN2019-2) and Research Fund of Guangxi Key Laboratory of Trusted Software (kx202003).

References

1. Wang, W., Ma, H., Zhao, Y., Li, Z., He, X.: Relevance-aware Q-matrix calibration for knowledge tracing. In: Farkaš, I., Masulli, P., Otte, S., Wermter, S. (eds.) ICANN 2021. LNCS, vol. 12893, pp. 101–112. Springer, Cham (2021). https://doi.org/10.1007/978-3-030-86365-4_9
2. Fu, X., Zhang, J., Meng, Z., King, I.: MAGNN: Metapath aggregated graph neural network for heterogeneous graph embedding. In: Proceedings of the 29th International Conference on World Wide Web, pp. 2331–2341. ACM (2020)
3. Ghosh, A., Heffernan, N., Lan, A.S.: Context-aware attentive knowledge tracing. In: Proceedings of the 26th ACM SIGKDD International Conference on Knowledge Discovery and Data Mining, pp. 2330–2339. ACM (2020)
4. Liu, Y., Yang, Y., Chen, X., Shen, J., Zhang, H., Yu, Y.: Improving knowledge tracing via pre-training question embeddings. In: Proceedings of the 29th International Joint Conference on Artificial Intelligence, pp. 1577–1583. Morgan Kaufmann (2020)
5. Nakagawa, H., Iwasawa, Y., Matsuo, Y.: Graph-based knowledge tracing: modeling student proficiency using graph neural network. In: Proceedings of 2019 IEEE International Conference on Web Intelligence, pp. 156–163. ACM (2019)
6. Pandey, S., Karypis, G.: A self-attentive model for knowledge tracing. In: Proceedings of the 12th International Conference on Educational Data Mining, pp. 384–389. EDM (2019)
7. Pandey, S., Srivastava, J.: RKT: relation-aware self-attention for knowledge tracing. In: Proceeding of the 29th ACM International Conference on Information and Knowledge Management, pp. 1205–1214. ACM (2020)
8. Piech, C., et al.: Deep knowledge tracing. In: Proceedings of the 28th International Conference on Neural Information Processing Systems, pp. 505–513. MIT Press (2015)
9. Wang, W., Ma, H., Zhao, Y., Li, Z., He, X.: Tracking knowledge proficiency of students with calibrated Q-matrix. Expert Syst. Appl. **192**, 116454 (2022)
10. Yang, Y.: GIKT: a graph-based interaction model for knowledge tracing. In: Hutter, F., Kersting, K., Lijffijt, J., Valera, I. (eds.) ECML PKDD 2020. LNCS (LNAI), vol. 12457, pp. 299–315. Springer, Cham (2021). https://doi.org/10.1007/978-3-030-67658-2_18
11. Zhang, J., Shi, X., King, I., Yeung, D.Y.: Dynamic key-value memory networks for knowledge tracing. In: Proceedings of the 26th International Conference on World Wide Web, pp. 765–774. ACM (2017)

A Three-Stage Curriculum Learning Framework with Hierarchical Label Smoothing for Fine-Grained Entity Typing

Bo Xu[1], Zhengqi Zhang[1], Chaofeng Sha[2,3]([✉]), Ming Du[1], Hui Song[1], and Hongya Wang[1]

[1] School of Computer Science and Technology, Donghua University, Shanghai, China
{xubo,duming,songhui,hywang}@dhu.edu.cn, 2202405@mail.dhu.edu.cn
[2] School of Computer Science, Fudan University, Shanghai, China
cfsha@fudan.edu.cn
[3] Shanghai Key Laboratory of Intelligence Processing, Shanghai, China

Abstract. In this paper, we study the *noisy labeling* problem on the fine-grained entity typing (FET) task. Most existing methods propose to divide the training data into "clean" and "noisy" sets and use different strategies to deal with them during the training process. However, the "clean" samples used in these methods are not actually clean, some of them also contain noisy labels. To overcome this issue, we propose a three-stage curriculum learning framework with hierarchical label smoothing to train the FET model, which can use relatively clean data to train the model and prevent the model from overfitting to noisy labels. Experiments conducted on three widely used FET datasets show that our method achieves the new state-of-the-art performance. Our code is publicly available at https://github.com/xubodhu/NFETC-CLHLS.

1 Introduction

Fine-grained entity typing (FET) is a task that classifies an entity mention in a sentence into a predefined set of fine-grained types. Typically, each fine-grained entity typing task requires its own annotated data for training the model, which is expensive and time-consuming. To address this problem, distant supervision has been adopted to automatically annotate a large number of unlabeled mentions in the training corpus. Specifically, an unlabeled entity mention will be linked to an existing entity in the knowledge base (KB), and then all types of that entity will be assigned to the entity mention.

Despite its efficiency, distant supervision often suffers from the *noisy labeling* problem. To address the problem, most existing FET methods propose to divide the training data into "clean" and "noisy" sets and use different strategies to deal with them during the training process. A training sample is considered "clean" if the candidate types of its mention form a single path in the type hierarchy.

A. Bhattacharya et al. (Eds.): DASFAA 2022, LNCS 13247, pp. 289–296, 2022.
https://doi.org/10.1007/978-3-031-00129-1_23

Otherwise, the training sample is considered "noisy" if the candidate types of its entity include multiple type paths. Based on this assumption, AFET [6] used an adaptive-margin rank loss to model the "clean" set and used a partial-label loss to model the "noisy" set. NFETC [11] trained on "clean" set with standard cross-entropy loss and on "noisy" set with a variant of cross-entropy loss respectively. NFETC-CLSC [2] calculated classification loss only on the "clean" set while regularizing the model using both "clean" and "noisy" sets.

However, the "clean" samples used in these methods are not actually clean, some of them also contain noisy labels [12]. To solve the problem, we propose a three-stage curriculum learning framework with hierarchical label smoothing to train the FET model, which can use relatively clean data to train the model and prevent the model from overfitting to noisy labels.

To the best of our knowledge, this is the first attempt to use curriculum learning with label smoothing to solve the noisy labeling problem on the FET task, which can also be applied to many other denoising tasks.

2   Overview

2.1   Problem Definition

We adopt the same setting adopted by [12]. The input is a training corpus $\mathcal{D}$ labeled with type hierarchy $\mathcal{Y}$ of the knowledge base by using distant supervision. The output is a *type-path* in $\mathcal{Y}$ for each entity mention from the test set.

Specifically, we define $\Gamma = \{t_1, t_2, \cdots, t_K\}$ as all candidate *type-path* labels, where K is the total number of types in $\mathcal{Y}$. Each label is a *type-path* from root node to a terminal node, which could be either a leaf node or a non-leaf node (e.g., /person, /person/artist and /person/artist/actor). The training corpus $\mathcal{D}$ consists of triples with form $\{(m_i, c_i, \boldsymbol{y_i})\}_{i=1}^{N}$. For each training sample, we denote the context sentence as a word sequence $c_i = \{w_1, w_2, \cdots, w_n\}$, and entity mention $m_i = \{w_j, \cdots, w_k\}$ as a continuous sub-sequence from the context sentence. The type-path label $\boldsymbol{y_i} = \{y_{i1}, y_{i2}, \cdots, y_{iK}\}$ is a binary vector, only the type-path from the root node to the most specific nodes are labeled as 1.

2.2   Framework

Our framework is shown in Fig. 1, which consists of a basic FET model and a three-stage training process.

The fine-grained entity typing model is used to classifies an entity mention in a sentence into a single *type-path* label. For a fair comparison with previous works [2,11,12], we adopt the same structure proposed by NFETC [11], which consists of a feature encoder and an MLP + softmax decoder. The feature encoder is used to learn representations for entity mentions and their context jointly, and the MLP + softmax decoder is used to output the label probability distribution.

The hierarchical label smoothing (HLS) method is used to smooth the original labels to prevent the model from overfitting the noisy labels. The entire training process based on curriculum learning is divided into three stages:

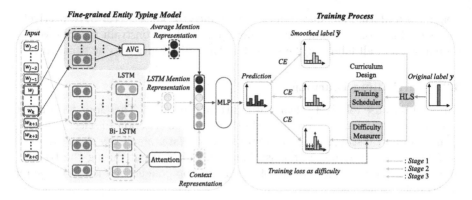

Fig. 1. Our Curriculum Learning Framework with Hierarchical Label Smoothing (**CLHLS**) for the fine-grained entity typing task.

- In the first stage, we use the hierarchical label smoothing (HLS) method to generate smoothed label distributions of all training data and use the smoothed labels to train the FET model.
- In the second stage, we train the FET model with a curriculum that focuses on relatively clean data with smoothed labels to reduce the interference of noisy data.
- In the third stage, we train the FET model with a curriculum that gradually sharpens the smoothed labels of relatively clean data to their original labels to achieve a better fit.

3   Method

3.1   Hierarchical Label Smoothing

Label smoothing has been used to prevent the model from overfitting the noisy labels [5]. In our FET task, considering the type hierarchy, predicting an ancestor type of the true type is better than some other unrelated types. For example, if one sample is labeled as /person/athlete, it is reasonable to predict its type as /person. However, predicting other high-level types like /location or /organization would be inappropriate.

Therefore, we propose a novel hierarchical label smoothing method to generate smoothed label distribution $\widetilde{y}$ from the original label vector y considering the type hierarchy. The smoothing process can be formulated as follows:

$$\widetilde{y} = \frac{1-\alpha-\beta}{sum(y)} \cdot y + \alpha \cdot u + \frac{\beta}{sum(\widehat{y})} \cdot \widehat{y}, \qquad (1)$$

where $u = \frac{1}{K}$ is a uniform label distribution and $\widehat{y}$ is the ancestor label vector. The ancestor label vector $\widehat{y} = \{\widehat{y}_{i1}, \widehat{y}_{i2}, \cdots, \widehat{y}_{iK}\}$ is obtained based on its original label and the type hierarchy, which is also a binary vector and the ancestor type-paths of y are also labeled as 1. α and β are hyperparameters used to control the weights of uniform label distribution $\boldsymbol{u}$ and ancestor label vector $\widehat{\boldsymbol{y}}$, respectively. The *sum* function is a summation operation used to constrain the sum of the probabilities of all labels in the smoothed label distribution to be equal to 1.

3.2 Curriculum Learning

Our goal is to use relatively clean data to train the model. Therefore, we propose a curriculum learning framework to design a reliable criterion to select/reweight clean data and distinguish them from the noisy data, so all the clean data can be fully exploited while most noisy labels are filtered out during the training process. The framework consists of two main components: difficulty measurer and training scheduler.

A widely used difficulty measurer is to treat the samples with small *instantaneous loss* as "easy" (clean) data, which is the loss evaluated at the current step. The reason is that clean labels are consistent with each other when generating gradient updates, so the model can fit them better and faster. On the other hand, noisy labels may contain mutually inconsistent information, thus forming a lasting "tug of war" between them [13].

However, due to the randomness of deep neural network training (e.g., random initialization), the instantaneous loss of the sample fluctuates rapidly, which will lead to unstable model training. To solve this problem, we use the exponential moving average (EMA) of instantaneous loss [13] (EMA loss) to evaluate each sample. The hypothesis is that if the loss of a sample continues to maintain a low value during the training step, the label of the sample is more likely to be correct.

Specifically, the EMA loss of the i-th sample at the $t + 1$ training step is defined as follows:

$$l_i^{t+1} = \begin{cases} \gamma \cdot H(\widetilde{\boldsymbol{y}}_i, \boldsymbol{p}_i^t) + (1 - \gamma) \cdot l_i^t, & if \ i \in D_t \\ l_i^t, & else \end{cases}, \quad (2)$$

where $\gamma \in [0, 1]$ is the discounting factor, $\boldsymbol{p}_i^t$ is the predicted label distribution of the i-th sample at the t-th training step, and $\mathcal{D}_t$ represents the batch data selected at t-th step.

The training scheduler is used to arrange the sequence of data subsets throughout the training process based on the judgment from the difficulty measurer. In our task, the samples with a low EMA loss are often selected with a high probability. Specifically, at each training step t, the training scheduler samples a batch of training data from the samples with low EMA losses to train the FET model.

In order to avoid getting trapped in a local minimum dominated by a small set of clean data, we reweight the sampling probability distribution of the training data at each training step. The probability of the i-th sample being sampled at t-th training step is calculated as follows:

$$P_i^t = \frac{\exp(-l_i^t)}{\sum_{j=1}^{N} \exp(-l_j^t)}, \quad (3)$$

where N is the number of samples in the training data.

3.3 Training Process

Algorithm 1. The Training Process of Three-Stage Curriculum Learning

---

Inputs: The training data $\mathcal{D} = \{(m_i, c_i, y_i)\}_{i=1}^{N}$, hyperparameters of HLS α and β, the number of epochs in the three stages E_1, E_2, and E_3, the batch size B.

Output: the fine-grained entity typing model f_θ.

1: **Initialize** training step $t \leftarrow 1$ and parameters θ_t for the model f_θ
2: **for** $epoch \leftarrow 1$ to $E_1 + E_2 + E_3$ **do**
3:     Shuffle the training data $\mathcal{D}$
4:     **for** $iteration \leftarrow 1$ to $\lceil N/B \rceil$ **do**
5:         **if** $epoch \leq E_1$ **then**
6:             Select a batch data $\mathcal{D}_t = \{(m_j, c_j, y_j)\}_{j=1}^{B}$ from $\mathcal{D}$ in traversal order at the t-th training step
7:         **else if** $epoch \leq E_1 + E_2$ **then**
8:             Sample $\mathcal{D}_t$ from $\mathcal{D}$ according to their probabilities calculated by Eq. 3
9:         **else if** $epoch \leq E_1 + E_2 + E_3$ **then**
10:             Sample $\mathcal{D}_t$ from $\mathcal{D}$ according to their probabilities calculated by Eq. 3
11:             Decay α gradually to sharpen the smoothed labels by Eq. 4
12:         **end if**
13:         Obtain smoothed label distributions $\{\widetilde{\boldsymbol{y}}_j\}_{j=1}^{B}$ by Eq. 1 according to α and β
14:         Update θ_{t+1} w.r.t minimize $\sum_{j=1}^{B} H(\widetilde{\boldsymbol{y}}_j, f(m_j, c_j; \theta))$
15:         Update the EMA loss of the training data $\mathcal{D}$ by Eq. 2
16:         Update $t \leftarrow t + 1$
17:     **end for**
18: **end for**

The training process is shown in Algorithm 1. In the first stage, since the model is trained from scratch and cannot distinguish between clean data and noisy data, we use HLS to smooth the labels of all training data, and use the smoothed labels to train the model to avoid overfitting noisy labels. The trained model can have a preliminary estimate of the difficulty of the training data. In the second stage, based on the evaluation of the difficulty of the training samples in the previous stage, we use a curriculum training model to train the FET model. This curriculum focuses on selecting relatively clean samples with smooth labels to train the model to reduce the impact of noisy data. After this stage, our model can fit the smoothed labels of the clean samples. In the third stage, we gradually decay the weight of the uniform distribution in the smoothed label distribution, because the uniform distribution will interfere with the model fitting the original labels that are correct.

The main differences between these three stages are the selection of each batch of samples and the smoothing method used. In the first stage, we select the batch of samples $\mathcal{D}_t$ from the training data $\mathcal{D}$ in traversal order (Step 6) and obtain the smoothed label distributions according to the fixed hyperparameters α and β (Step 13). In the second stage, we select the batch of samples $\mathcal{D}_t$ from the training data $\mathcal{D}$ based on their probabilities calculated by Eq. 3 (Step 8) and obtain the smoothed label distributions according to the fixed hyperparameters α and β (Step 13). In the third stage, we select the batch of samples $\mathcal{D}_t$ from the training data $\mathcal{D}$ based on their probabilities calculated by Eq. 3 (Step 10), and obtain the smoothed label distributions according to the dynamic hyperparameter α and fixed hyperparameter β (Step 13). The dynamic hyperparameter α_λ is gradually annealed from α to 0 to sharpen the smoothed labels (Step 11) as follows:

$$\alpha_\lambda = \alpha \cdot (1 - \frac{\lambda}{E_3}), \tag{4}$$

where λ is the training epochs in the third stage.

4   Experiment

4.1   Datasets and Metrics

We conduct experiments on three widely used FET datasets, namely Wiki [4], OntoNotes [9] and BBN [8]. For a fair comparison, we use the same datasets provided by [11][1] and use the same metrics for evaluation [2,6,11,12].

4.2   Baselines

We compare our method *NFETC-CLHLS* with several state-of-the-art FET methods, including *AFET* [6], *Attentive* [7], *AAA* [1], *NDP* [10], *HET* [3], *NFETC* [11], *NFETC-CLSC* [2] and *NFETC-AR* [12].

4.3   Performance Comparison

We report the metrics of strict accuracy (ACC), loose macro-averaged F1 score (Ma-F1) and loose micro-averaged F1 score (Mi-F1) on three benchmarks. Table 1 shows the overall performance. We find that our method achieves the new state-of-the-art performance on three benchmarks, indicating that our curriculum learning framework with hierarchical label smoothing can indeed effectively train the FET model. More detailed comparisons are as follows.

Firstly, compared with the basic NFETC method, our method improves the strict accuracy on the three benchmarks from 68.9 to 71.3 (Wiki), 60.2 to 65.1 (OntoNotes) and 73.9 to 77.2 (BBN), respectively. That indicates the necessity of noise reduction in the distantly supervised FET task and the effectiveness of our proposed method.

Secondly, compared with other NFETC-based methods, our method *NFETC-CLHLS* performs better than *NFETC* and *NFETC-CLSC*, indicating that our clean data detection method is better than their heuristic methods. Our method *NFETC-CLHLS* performs better than the state-of-the-art method *NFETC-AR*, indicating that filtering out the noisy data and only using the clean data for training is better than relabeling all the data.

---

[1] https://github.com/billy-inn/NFETC.

Table 1. Performance comparison on the three benchmarks.

Methods	Wiki			OntoNotes			BBN		
	ACC	Ma-F1	Mi-F1	ACC	Ma-F1	Mi-F1	ACC	Ma-F1	Mi-F1
Attentive [7]	59.7	80.0	75.4	51.7	71.0	64.9	48.4	73.2	72.4
AFET [6]	53.3	69.3	66.4	55.3	71.2	64.6	68.3	74.4	74.7
AAA [1]	65.8	81.2	77.4	52.2	68.5	63.3	65.5	73.6	75.2
NDP [10]	67.7	81.8	78.0	58.0	71.2	64.8	72.7	76.4	77.7
HET [3]	69.1	82.6	80.8	58.7	73.0	68.1	75.2	79.7	80.5
NFETC [11]	68.9	81.9	79.0	60.2	76.4	70.2	73.9	78.8	79.4
NFETC-CLSC [2]	–	–	–	62.8	77.8	72.0	74.7	80.7	80.5
NFETC-AR [12]	70.1	83.2	80.1	64.0	78.8	73.0	76.7	81.4	81.5
NFETC-CLHLS	**71.3**	**85.1**	**80.9**	**65.1**	**80.1**	**74.6**	**77.2**	**81.7**	**81.8**

4.4 Ablation Study

To investigate the effectiveness of each component proposed in our framework, we conduct an ablation study to compare the performance between the full model and its ablation methods on the OntoNotes dataset.

Table 2. Ablation Study of *NFETC-CLHLS* on the OntoNotes Dataset.

Methods	Stric-Acc	Macro-F1	Micro-F1
NFETC-CLHLS	**65.1**	**80.1**	**74.6**
w/o CL	63.9	79.3	73.5
– w/o u	62.6	78.3	72.9
– w/o $\widehat{y}$	62.1	77.1	71.1
– w/o HLS	60.5	75.9	69.6

As shown in Table 2, without the CL module (**w/o CL**), i.e., only the first stage is used to train the FET model, the performance (i.e., Strict-Acc score, Macro-F1 score and Micro-F1 score) drop 1.2, 0.8 and 1.1, respectively. This shows that curriculum learning module is beneficial for training in FET model.

We also investigate the effectiveness of our HLS module on the first stage. Without considering the uniform distribution u (i.e., **w/o u**), the performance drop 1.3, 1.0 and 0.6, respectively. It shows that using the uniform distribution on our HLS module can significantly reduce confidence in the noisy label. Without considering the ancestor label vector $\widehat{y}$ (i.e., **w/o $\widehat{y}$**), the performance drop 1.8, 2.2 and 2.4, respectively. It shows that using the ancestor label vector on our HLS module can significantly alleviate the overly specific problem. And without using the entire hierarchical label smoothing module (i.e., **w/o HLS**), the performance drop 3.4, 3.4 and 3.9, respectively. That demonstrates the effectiveness of our HLS module.

5   Conclusion

In this paper, we propose a three-stage curriculum learning framework with hierarchical label smoothing to train the fine-grained entity typing model. Experimental results show the effectiveness of our method.

Acknowledgement. This paper was supported by the National Natural Science Foundation of China (61906035), Shanghai Sailing Program (19YF1402300), Shanghai Municipal Commission of Economy and Information (202002009) and the Fundamental Research Funds for the Central Universities (2232021A-08).

References

1. Abhishek, A., Anand, A., Awekar, A.: Fine-grained entity type classification by jointly learning representations and label embeddings. In: EACL, pp. 797–807 (2017)
2. Chen, B., et al.: Improving distantly-supervised entity typing with compact latent space clustering. In: NAACL, pp. 2862–2872 (2019)
3. Chen, T., Chen, Y., Van Durme, B.: Hierarchical entity typing via multi-level learning to rank. In: ACL, pp. 8465–8475 (2020)
4. Ling, X., Weld, D.S.: Fine-grained entity recognition. In: AAAI, pp. 94–100 (2012)
5. Lukasik, M., Bhojanapalli, S., Menon, A., Kumar, S.: Does label smoothing mitigate label noise? In: ICML, pp. 6448–6458. PMLR (2020)
6. Ren, X., He, W., Qu, M., Huang, L., Ji, H., Han, J.: AFET: automatic fine-grained entity typing by hierarchical partial-label embedding. In: EMNLP, pp. 1369–1378 (2016)
7. Shimaoka, S., Stenetorp, P., Inui, K., Riedel, S.: An attentive neural architecture for fine-grained entity type classification. In: Proceedings of the 5th Workshop on Automated Knowledge Base Construction, pp. 69–74 (2016)
8. Weischedel, R., Brunstein, A.: BBN pronoun coreference and entity type corpus. Linguistic Data Consortium, Philadelphia 112 (2005)
9. Weischedel, R., et al.: Ontonotes release 5.0 ldc2013t19. Linguistic Data Consortium, Philadelphia, PA 23 (2013)
10. Wu, J., Zhang, R., Mao, Y., Guo, H., Huai, J.: Modeling noisy hierarchical types in fine-grained entity typing: a content-based weighting approach. In: IJCAI, pp. 5264–5270 (2019)
11. Xu, P., Barbosa, D.: Neural fine-grained entity type classification with hierarchy-aware loss. In: NAACL, pp. 16–25 (2018)
12. Zhang, H., et al.: Learning with noise: improving distantly-supervised fine-grained entity typing via automatic relabeling. In: IJCAI, pp. 3808–3815 (2020)
13. Zhou, T., Wang, S., Bilmes, J.A.: Robust curriculum learning: from clean label detection to noisy label self-correction. In: ICLR, pp. 1–18 (2021)

PromptMNER: Prompt-Based Entity-Related Visual Clue Extraction and Integration for Multimodal Named Entity Recognition

Xuwu Wang[1], Junfeng Tian[2], Min Gui[3], Zhixu Li[1(✉)], Jiabo Ye[4], Ming Yan[2], and Yanghua Xiao[1,5(✉)]

[1] Shanghai Key Laboratory of Data Science, School of Computer Science, Fudan University, Shanghai, China
{xwwang18,zhixuli,shawyh}@fudan.edu.cn
[2] Alibaba DAMO Academy, Hangzhou, China
{tjf141457,ym119608}@alibaba-inc.com
[3] Shopee, Singapore, Singapore
min.gui@shopee.com
[4] East China Normal University, Shanghai, China
jiabo.ye@stu.ecnu.edu.cn
[5] Fudan-Aishu Cognitive Intelligence Joint Research Center, Shanghai, China

Abstract. Multimodal named entity recognition (MNER) is an emerging task that incorporates visual and textual inputs to detect named entities and predicts their corresponding entity types. However, existing MNER methods often fail to capture certain entity-related but text-loosely-related visual clues from the image, which may introduce task-irrelevant noises or even errors. To address this problem, we propose to utilize entity-related prompts for extracting proper visual clues with a pre-trained vision-language model. To better integrate different modalities and address the popular semantic gap problem, we further propose a modality-aware attention mechanism for better cross-modal fusion. Experimental results on two benchmarks show that our MNER approach outperforms the state-of-the-art MNER approaches with a large margin.

Keywords: Named entity recognition · Multimodal learning · Knowledge graph

1 Introduction

Named entity recognition (NER) is an indispensable task for information extraction, knowledge graph construction, and question answering, etc. Recently, posts on social media platforms are becoming increasingly multimodal. It becomes more common that the text information can only be understood with correlated

This work was conducted when Min Gui worked at Alibaba.

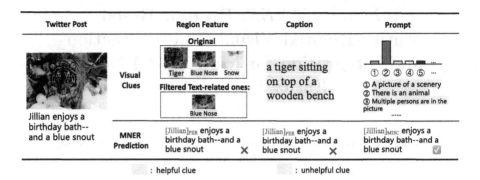

Fig. 1. Comparison of methods used to get the visual clues. The region feature is obtained from an object detector. The caption is generated from an image captioning model. (Color figure online)

images. Therefore, the research of multimodal named entity recognition (MNER) is flourishing, which aims at detecting named entities and classifying them based on a text and the related image.

Previous researches of MNER mainly focus on two issues: extracting visual clues [12] from images and integrating the information of the two modalities. For extracting visual clues, existing works tend to: 1) encode the entire image as a global feature in the form of a visual feature vector or a caption [2,9,11,18], or 2) extract local fine-grained region features from the image [17]. For the integration of multiple modalities, existing efforts focus on: 1) using attention mechanism to fuse the two modalities based on the relevance between them [9,11,18], or 2) transforming the image into a caption or a label set and then directly integrating them with the original sentence [2,14].

However, there are two critical problems that existing works may often neglect. **Problem 1.** It is entity-related visual clues, instead of text-related visual clues or the global image features, that really helps to improve MNER. The potential entities in the sentence are often presented with specific entity names. It is difficult to align these names with the image through image-text relevance (e.g., it is hard to align the entity 'Jillian' to the 'tiger' in the image as shown in Fig. 1). And some text-related visual features may only correspond to non-entity words (e.g., the 'blue nose' in the image corresponds to the word 'blue snout'), which has few benefits to the MNER task. **Problem 2.** Besides, there are semantic gap and structure gap between the two modalities: the visual features of images and the textual features of texts belong to different semantic spaces and have different structures, which hinders the fusion between the two modalities.

In this paper, we propose a novel framework named PROMPTMNER to address the above problems. Firstly, we leverage Prompt Learning [7] to extract entity-related visual clues from images. Prompt learning is an emerging way of releasing the immense power of Pre-trained Language Models (PLMs) to tackle downstream tasks. With the pre-trained vision language models' ability to predict the matching degree between an image and a text, we design multiple

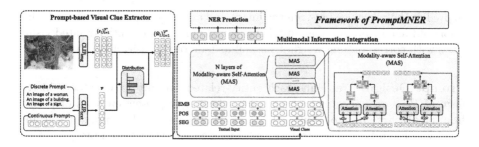

Fig. 2. Framework of the PROMPTMNER

entity-related prompts (see Fig. 1 for example) and use the matching degrees between the image and the prompts to select the entity-related visual clues. Secondly, to bridge the semantic gap in integrating the entity-related visual clues and the textual features, a modality-aware transformer encoder is adopted to achieve better intra-modal homogeneous attention and inter-modal heterogeneous attention. The experiments demonstrate that PROMPTMNER can achieve significant improvement compared with the state-of-the-art (SOTA) methods.

2 Methodology

2.1 Task Definition

Let $T = (w_1, w_2, ..., w_n)$ denote a sentence consisting of multiple words and I denote an image. Given a sentence T and its corresponding image I as input, MNER aims at detecting a set of entities from T and classifying each entity into one of the pre-defined types including person (PER), location (LOC), organization (ORG) and miscellaneous (MISC).

2.2 Overview

As shown in Fig. 2, the proposed PROMPTMNER mainly consists of the following components: Firstly, a Prompt-based Visual Clue Extractor (Sect. 2.3) is used to extract entity-related visual clues with a pre-trained vision-language model (VLM) from the input image. Secondly, a Multimodal Information Integration Module (Sect. 2.4) is designed to fuse the extracted visual clues and the input sentence. Finally, to train the model for NER (Sect. 2.5), we adopt the span-based method [3] that enumerates all candidate spans of the sentence and predicts the corresponding entity types.

2.3 Prompt-Based Visual Clue Extractor

Let $\mathbf{v}$ represent the presentation of the input image $\mathbf{I}$. As discussed in Sect. 1, as a form of expression with huge amount of information, $\mathbf{v}$ contains lots of task-independent noises. By designing a set of entity-related prompts, a VLM is able to predict the relevance between the image and every prompt due to the valuable priors obtained from pre-training. It helps us to extract the entity-related visual clues with corresponding weights, as well as fade out task-irrelevant noises.

Prompt Design. We define each entity-related prompt as a sentence in the form of $P_i = an\ image\ of\ [w_i]$, $w_i \in \mathcal{V}^*$, where w_i represents a word or phrase from a entity-related vocabulary $\mathcal{V}^*$. Following recent work of prompt learning [1,4], we design both discrete and continuous entity-related prompts.

Discrete Prompts. For discrete prompts, the words $w_i \in \mathcal{V}^*$ come from a human-readable vocabulary. We propose three methods to find the $\mathcal{V}^*$:

- **Task Labels.** The labels of MNER: `person`, `location`, `organization` can be used to build $\mathcal{V}^*$.
- **KB-Retrieved Labels.** However, the above task labels have multiple short-comings. 1) The limited number of task labels cannot guarantee the high coverage of visual clues. 2) The task labels may be quite different from the frequently used expressions of the VLM's pre-training corpus. It would severely impair VLM's ability to predict the relevance between the image and the prompt. Thus, we incorporate off-the-shelf Related Words[1] to retrieve related words of the task labels to enrich the label set. Related Words is a KB of related words that incorporates multiple KBs including WordNet [10], Concept Net [6], etc. For example, `person` can be expanded with `people`, `someone`, `individual`, `worker`, `child`, etc.
- **Expert Provided Labels.** Some related words that rely on the complicated relationships such as whole-part and collective-individual are difficult to be retrieved through a KB, because this retrieving procedure often involves complex knowledge such as common sense and cognition. So we also obtain entity-related labels from experts who are familiar with the task as well as VLMs. For example, `person` can be expanded with `player`, `pants`, `hat`, `suit`, `group of people`, `team`, etc.

Continuous Prompts. However, finding as many discrete prompts as possible is a non-trivial task, which requires a large amount of time for word tuning as well as expertise about the task and the VLM. So alternatively, we can use continuous vectors as the prompts and optimize them during the training procedure: $P_i = \mathbf{w}_i$, where $\mathbf{w}_i$ is a learnable embedding. This type of prompt is denoted as the continuous prompt, which is able to automatically learn open-set entity-related visual clues.

Visual Clue Extraction. In this paper, we adopt CLIP as the VLM. Given the input image $\mathbf{I} \in \mathbb{R}^{C \times H \times W}$, its embedding is obtained through: $\mathbf{v} = \text{CLIP}_{\text{img}}(\mathbf{I})$. Here $\text{CLIP}_{\text{img}}(\cdot)$ represents the convolutional neural network-based image encoder of CLIP. For each prompt P_i, we obtain its embedding from the CLIP's textual encoder: $\mathbf{s_i} = \text{CLIP}_{\text{text}}(P_i)$.

---

[1] http://relatedwords.org.

After that, given the image embedding $\mathbf{v}$ and entity-related prompt embeddings $\{\mathbf{s}_i\}_{i=1}^{|\mathcal{V}^*|}$, we are able to get the relevance between the image $\mathbf{I}$ and every prompt P_i:

$$p\left(P_i|\mathbf{I}\right) = \frac{\exp\left(<\mathbf{s_i}, \mathbf{v}>/\tau\right)}{\sum_{j=1}^{|\mathcal{V}^*|} \exp\left(<\mathbf{s_j}, \mathbf{v}>/\tau\right)} \tag{1}$$

where $<\cdot,\cdot>$ represents the cosine similarity between two vectors and τ is a temperature parameter. Both of them are set as the same paradigm of the pre-training of CLIP.

Finally, the weighted entity-related visual clue $\hat{\mathbf{w}}_i$ is represented as the weighted embedding of the prompt: $\hat{\mathbf{w}}_i = p\left(P_i|\mathbf{I}\right) \times \mathbf{s}_i$.

2.4 Multimodal Information Integration

In this subsection, we introduce how to integrate the visual clues $\{\hat{\mathbf{w}}_i\}_{i=1}^{|\mathcal{V}^*|}$ and the sentence $T = (w_1, ..., w_N)$ to capture the interaction between the two modalities.

Embedding Layer. Initially, for each word w_i of the input sentence $\{w_i\}_{i=1}^N$, it is mapped to a distributed vector from the embedding layer of a pre-trained language model (LM): $\mathbf{w}_i = \text{LM}_{\text{WordEmb}}(w_i)$. Given the visual clues $\{\hat{\mathbf{w}}_i\}_{i=1}^{\mathcal{V}^*}$ as well as $\{\mathbf{w}_i\}_{i=1}^N$, we then concatenate the corresponding sequences and add the special tokens of [CLS] and [SEP]:

$$[\text{CLS}], \mathbf{w}_1, ..., \mathbf{w}_N, [\text{SEP}], \hat{\mathbf{w}}_1, ..., \hat{\mathbf{w}}_{|\mathcal{V}^*|}, [\text{SEP}]. \tag{2}$$

For each word $\mathbf{w}_i$ and $\hat{\mathbf{w}}_i$, the position embedding and segmentation embedding are also added to it followed by a LayerNorm layer to capture the positional information and the type information.

Modality-aware Self-Attention (MAS). Then the embeddings sequence of textual and visual tokens are fed into multiple layers of self-attention. Note that the visual clue tokens and textual tokens are heterogeneous, which will influence the information fusion achieved through self-attention. Yamada [15] noticed that it is beneficial to take the target token type into consideration when computing the attention score. Inspired by this discovery, we propose to take the modality type into consideration when computing the attention query. Let $\mathbf{x}_i$ represent the input token that corresponds to $\mathbf{w}_i$ or $\hat{\mathbf{w}}_i$. Its attention weight corresponding to the jth token in the sequence is calculated as:

$$\alpha_{ij} = \text{softmax}\left(\frac{(\mathbf{K}\mathbf{x}_j)^T (\mathbf{Q}_{Q2V}\mathbf{x}_i)}{\sqrt{N + |\mathcal{V}^*|}}\right) \tag{3}$$

where $\mathbf{Q}_{Q2V}$, $\mathbf{K}$ represent the matrices used to projecting queries and keys. $\mathbf{Q}_{Q2V}$ is designed to be modality-aware: it may be $\mathbf{Q}_{\text{w2w}}$, $\mathbf{Q}_{\text{w2}\hat{\text{w}}}$, $\mathbf{Q}_{\hat{\text{w}}\text{2w}}$, or

$Q_{\hat{w}2\hat{w}}$ depending on the modality of the query and the value. The output is then calculated as:

$$\mathbf{x}_i := \sum_{j=1}^{N+|\mathcal{V}^*|} \alpha_{ij}\mathbf{x}_j \qquad (4)$$

We apply K layers of MAS to embed the input sequence and take the hidden states of the textual words $\{\mathbf{h}_i\}_{i=1}^{N}$ from the last layer for the prediction.

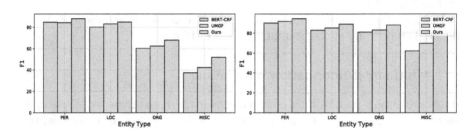

Fig. 3. Model performance in terms of different entity types. Our results come from PROMPTMNER Best.

2.5 Prediction

Given the textual word representations: $\{\mathbf{h}_i\}_{i=1}^{N}$, which have integrated the information of the visual clues, we then reformulate NER as the task of identifying start and end indices of an entity span as well as assigning a category label to the span [5]. So we enumerate all possible spans in the sentence $\{w_1, ..., w_N\}$ and classify whether it is an entity and predict the entity type. For each text span $\{w_i, ..., w_j\}$, we concatenate the embeddings of the first and the last tokens and then feed them into a MLP to predict its entity type:

$$l_c = \text{MLP}\left([\mathbf{h}_i||\mathbf{h}_j]\right) \qquad (5)$$

where l_c represents the entity type from {person, location, organization, misc, not_entity}.

Our objective is assigning a correct entity type to each enumerated span. So the loss function of MNER is formulated as the softmax cross-entropy loss:

$$p\left(l_c\right) = \frac{\exp(l_c)}{\sum_{\hat{c}=1}^{C}\exp(l_{\hat{c}})}, \quad \mathcal{L}_{\text{MNER}} = -\sum_{i=1}^{N}\sum_{c=1}^{C} y_{i_c}\log p(l_c) \qquad (6)$$

where C represents the number of the entity types. y_c is the correct span label.

3   Experiment

3.1   Experimental Setups

Datasets. We test on two benchmarks: Twitter-2015 [18] and Twitter-2017 [9].

Baselines. We compare with a wide range of baselines including both textual and multimodal baselines: CNN-BiLSTM-CRF, BERT-CRF, BERT-BiLSTM-CRF, ACoA [18], IAIK [17], UMT [16], RIVA [13], RpBERT [12], UMGF [17].

Implementation Details. We conduct all the experiments on 8 NVIDIA V100 GPUs using Pytorch 1.7. We use RoBERTa [8] as the LM and CLIP as the VLM. We set the number of self-attention layers K as 12. We set AdamW optimizer with the learning rate of 1e−4 and a warmup linear scheduler to control the learning rate. The batch size is set as 32 and the dropout is set as 0.5.

Table 1. Performance of different methods of the MNER task. 'T' and 'T+V' represent textual methods and multimodal methods respectively. 'CPrompt' represents continuous prompts. PROMPTMNER Best is achieved with 100 CPrompts

	Methods	Twitter-2015			Twitter-2017		
		Prec.	Recall	F1	Prec.	Recall	F1
T	CNN-BiLSTM-CRF	66.24	68.09	67.15	80.00	78.76	79.37
	BERT-CRF	69.22	74.59	71.81	83.32	83.57	83.44
	BERT-BiLSTM-CRF	–	–	71.60	–	–	–
T+V	ACoA	72.75	68.74	70.69	84.16	80.24	82.15
	UMT	71.67	75.23	73.41	85.28	85.34	85.31
	RIVA	–	–	73.80	–	–	–
	IAIK	74.78	71.82	73.27	–	–	–
	RpBERT	–	–	74.80	–	–	85.51
	UMGF	74.49	75.21	74.85	86.54	84.50	85.51
	PROMPTMNER Best	78.03	**79.17**	**78.60**	**89.93**	90.60	**90.27**
	PROMPTMNER w/ Task Labels	77.89	77.44	77.66	88.27	90.82	89.53
	PROMPTMNER w/ KB Prompts	78.15	78.23	78.19	89.78	89.71	89.74
	PROMPTMNER w/ Expert Prompts	77.96	78.65	78.30	89.59	90.45	90.02
	PROMPTMNER w/ 30 CPrompts	77.84	78.90	78.37	89.32	**90.97**	90.14
	PROMPTMNER w/ 50 CPrompts	78.33	78.75	78.54	89.67	90.60	90.13
	PROMPTMNER w/o MAS	**78.47**	77.62	78.04	89.12	90.07	89.59

3.2   Experimental Results

We report the performance of different methods in Table 1. We can see that: 1) Our method greatly outperforms previous methods. So the prompt-based entity-related visual clues can effectively extract helpful information from the image. 2) Both KB-retrieved prompts and expert-provided prompts achieve competitive performance. But the continuous prompts help the model to achieve the best results. And more continuous prompts lead to better model performance.

We also present the performance of each entity category in Fig. 3. In all entity categories, our method shows superior performance compared with BERT-CRF and UMGF, especially in the categories of ORG and MISC. We speculate that the reason may be that existing methods are difficult to extract visual clues helpful to these two types of entities.

4   Conclusion

In this paper, we propose a novel framework PROMPTMNER to extract and integrate entity-related visual clues for multimodal named entity recognition. The model includes a novel prompt-based visual clue extractor to obtain useful task-related image features and a new multimodal integration module to fuse the visual features with the textual features. Experiments show the superiority of our method compared with previous methods on two MNER datasets.

Acknowledgement. This research was supported by the National Key Research and Development Project (No. 2020AAA0109302), National Natural Science Foundation of China (No. 62072323), Shanghai Science and Technology Innovation Action Plan (No. 19511120400), Shanghai Municipal Science an Technology Major Project (No. 2021SHZDZX0103) and Alibaba Research Intern Program.

References

1. Chen, D., Li, Z., Gu, B., Chen, Z.: Multimodal named entity recognition with image attributes and image knowledge. In: Jensen, C.S., et al. (eds.) DASFAA 2021. LNCS, vol. 12682, pp. 186–201. Springer, Cham (2021). https://doi.org/10.1007/978-3-030-73197-7_12
2. Chen, S., Aguilar, G., et al.: Can images help recognize entities? A study of the role of images for multimodal NER (2021)
3. Fu, J., Huang, X., Liu, P.: SpanNER: Named entity re-/recognition as span prediction. arXiv preprint arXiv:2106.00641 (2021)
4. Li, X.L., Liang, P.: Prefix-Tuning: optimizing continuous prompts for generation. In: Proceedings of ACL, pp. 4582–4597 (2021)
5. Liu, C., Fan, H., Liu, J.: Span-based nested named entity recognition with pre-trained language model. In: Jensen, C.S., et al. (eds.) DASFAA 2021. LNCS, vol. 12682, pp. 620–628. Springer, Cham (2021). https://doi.org/10.1007/978-3-030-73197-7_42
6. Liu, H., Singh, P.: ConceptNet-a practical commonsense reasoning tool-kit. BT Technol. J. **22**(4), 211–226 (2004). https://doi.org/10.1023/B:BTTJ.0000047600.45421.6d
7. Liu, P., Yuan, W., et al.: Pre-train, prompt, and predict: a systematic survey of prompting methods in natural language processing. arXiv:2107.13586 (2021)
8. Liu, Y., Ott, M., et al.: RoBERTa: a robustly optimized BERT pretraining approach. arXiv:1907.11692 (2019)

9. Lu, D., Neves, L., et al.: Visual attention model for name tagging in multimodal social media. In: Proceedings of ACL, pp. 1990–1999 (2018)
10. Miller, G.A.: WordNet: a lexical database for English. Commun. ACM **38**(11), 39–41 (1995)
11. Moon, S., Neves, L., et al.: Multimodal named entity recognition for short social media posts. In: Proceedings of NAACL, pp. 852–860 (2018)
12. Sun, L., Wang, J., et al.: RpBERT: a text-image relation propagation-based BERT model for multimodal NER. In: Proceedings of AAAI, vol. 35 (2021)
13. Sun, L., Wang, J., et al.: RIVA: a pre-trained tweet multimodal model based on text-image relation for multimodal NER. In: COLING, pp. 1852–1862 (2020)
14. Wu, Z., Zheng, C., et al.: Multimodal representation with embedded visual guiding objects for named entity recognition in social media posts. In: MM (2020)
15. Yamada, I., Asai, A., et al.: LUKE: deep contextualized entity representations with entity-aware self-attention. arXiv preprint arXiv:2010.01057 (2020)
16. Yu, J., Jiang, J., et al.: Improving multimodal named entity recognition via entity span detection with unified multimodal transformer. In: Proceedings of ACL (2020)
17. Zhang, D., Wei, S., et al.: Multi-modal graph fusion for named entity recognition with targeted visual guidance. In: Proceedings of AAAI, pp. 14347–14355 (2021)
18. Zhang, Q., Fu, J., et al.: Adaptive co-attention network for named entity recognition in tweets. In: Proceedings of AAAI (2018)

TaskSum: Task-Driven Extractive Text Summarization for Long News Documents Based on Reinforcement Learning

Moming Tang[1], Dawei Cheng[3], Cen Chen[1(✉)], Yuqi Liang[2], Yifeng Luo[1], and Weining Qian[1]

[1] East China Normal University, Shanghai, China
mmtang@stu.ecnu.edu.cn, {cenchen,yfluo,wnqian}@dase.ecnu.edu.cn
[2] Seek Data Inc., Shanghai, China
yuqi.liang@seek-data.com
[3] Tongji University, Shanghai, China
dcheng@tongji.edu.cn

Abstract. A popular and state-of-the-art family of extractive summarization is to explore pre-trained language models through reinforcement learning (RL). Despite gaining promising results, existing RL-based methods suffer from three drawbacks. First, they often adopt sparse reward signal schemes, which only give rewards to some of the extracted sentences, and result in *neglecting salient sentences*. Second, they often deem summarization as an independent task and *neglect the latent connections* existing between summarization and other downstream tasks, that could provide insightful hints to guide the upstream summarization task in return. Third, the length of input sequences in most summarization methods is restricted by the utilized pre-trained language models. To address these problems, we propose a novel RL-based Seq2Seq extractive summarization model, namely TaskSum, which combines extractive text summarization with multiple associated tasks via a dense reward signal scheme. Moreover, we implement a BERT-based hierarchical encoder to effectively encode documents of arbitrary length. Empirical results demonstrate that TaskSum can overcome the above-mentioned drawbacks of existing RL-based summarization methods and achieve significantly better results *for long documents*.

Keywords: Extractive summarization · Reinforcement learning

1 Introduction

Extractive summarization builds summaries by selecting a subset of snippets from the source documents, thus can keep summaries faithful to the source documents. In contrast, abstractive summarization generates word sequences based on the documents and thus may result in unfaithful summaries [15]. For many applications where information correctness needs to be guaranteed, extractive summarization is more preferable to abstractive summarization [17]. Most

© The Author(s), under exclusive license to Springer Nature Switzerland AG 2022
A. Bhattacharya et al. (Eds.): DASFAA 2022, LNCS 13247, pp. 306–313, 2022.
https://doi.org/10.1007/978-3-031-00129-1_25

extractive summarization methods are supervised [2]. However, extractive labels are generated heuristically [5], and thus limit the model performance. Besides, there exists a mismatch between maximizing the likelihood function and optimizing the desired evaluation metric. Recent works adopt Reinforcement learning (RL) to directly optimize the ROUGE-based evaluation metrics [3,4].

Existing RL-based summarization methods usually utilize *sparse reward signals* for quick converge, where only evaluation scores for the whole summaries [13,18] or scores for the first few extracted sentences are returned [3,20], which may result in neglecting salient sentences. On the other hand, *dense reward signal* mechanism evaluates each extracted sentence according to its relative gain to the draft summary [13]. However, the generated dense signals are *order-sensitive*, sentences extracted at later decoding steps can hardly improve the draft summary regardless of their semantic quality [19]. Besides, RL algorithms mainly utilize reference summaries-based or human ratings-based reward schemes [1,13,18], which are costly to obtain in real world. Critical tasks associated with text summarization could be utilized to guide summary extraction and generalize the summarization methods to the real-world application.

Recently, pre-trained language models have become the most well-recognized paradigm in text summarization that facilitates the summarization methods to achieve state-of-the-art performance [6,7,11,20]. However, these methods have to truncate documents to a length that pre-trained language models can handle.

In this paper, we propose a novel task-driven RL-based extractive text summarization method, namely TaskSum, with a dense reward signal scheme. We introduce the sentence entailment into the reward scheme, to ensure the extracted sentences are semantic consistent with the core meaning of source document. The proposed dense reward signal mechanism evaluates each sentence according to corresponding candidate summaries, thus alleviating the order-sensitive optimization problem. Finally, we implement a BERT-based hierarchical encoder to generate contextual representations for all sentences in a document of arbitrary length. We conduct extensive experiments to evaluate TaskSum. Empirical results show our model outperforms the baseline models for long documents. And, we will open-source our codes and dataset along with the paper.

2 Related Work

RL-based extractive summarization learns a summarization policy by letting the agent interacts with the reward function [3,4]. They usually utilizes ROUGE-based rewards to guide sentence extraction. However, [18] demonstrate ROUGE metrics are sometimes inconsistent with human preference. Therefore, heuristic reward schemes based on reference summaries or human ratings [1,13,18] are proposed to strengthen certain properties of extractive summaries.

Besides, most RL-based extractive summarization suffer from sparse reward signal mechanism. [1,13,18] only provide evaluation score for the extracted summaries, while [3,20] give rewards to few extracted sentences. These methods mainly rely on cumbersome pre-training or a hybrid objective to converge. Existing dense reward signal mechanisms calculate reward for each extracted sentence based on the its relative gain to the draft summary [14] or its individual evaluation score [13]. However, sentences extracted at early decoding steps always have larger global returns than sentences extracted at late decoding steps.

Pre-trained language models have help extractive summarization to achieve state-of-the-art performance [6,7,11,20]. However, [6,7,11,20] have to truncate the document, due to the length restriction of pre-trained language models, and extract summary sentences from the first few encoded sentences. To encode long documents, [16] pre-train a hierarchical encoder, [21] propose novel sparse attention mechanisms for Transformer architecture. However, [16,21] cannot capture relationships among all tokens which results in a drastic performance drop for text summarization.

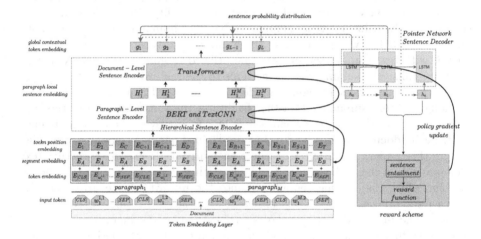

Fig. 1. Overall architecture of the TaskSum text summarization model

3   Model Architecture

Our extractive summarization model follows the encoder-decoder architecture, consisting of the Hierarchical Sentence Encoder (HSE) and the Pointer Network Sentence Decoder, just as illustrated in Fig. 1.

3.1   Hierarchical Sentence Encoder

The HSE consists of a paragraph-level encoder and a document-level encoder. As long documents consist of paragraphs with reasonable lengths, the paragraph-level encoder is capable of encoding each paragraph with BERT to get paragraph-level contextual sentences representations, the document-level encoder can further encode paragraph-level sentences representations of all paragraphs to get document-level sentences representations.

Paragraph-Level Sentence Encoder. We denote document D with M paragraphs as $D = \{P_q | q = 1 \cdots M\}$, paragraph P_q with K sentences as $P_q = \{s_j^q | j = 1 \cdots K\}$, and sentence s_j^q with N tokens as $\mathbf{s}_j^q = \{w_v^{(q_j)} | v = 1 \cdots N\}$. We denote embeddings for tokens, token segments and token positions in P_q as:

$$Tok_{P_q} = \left\{ e_{[CLS]}^{(q_j)} \cdots e_v^{(q_j)} \cdots e_{[SEP]}^{(q_j)} | j = 1 \cdots K \right\} \tag{1}$$

$$Seg_{P_q} = \{E_j \cdots E_j \cdots E_j | j = 1 \cdots K\} \tag{2}$$

$$Pos_{P_q} = \left\{ p_{[CLS]}^{(q_j)} \cdots p_v^{(q_j)} \cdots p_{[SEP]}^{(q_j)} | j = 1 \cdots K \right\} \tag{3}$$

where $e_v^{(q_j)}$, E_j and $p_v^{(q_j)}$ respectively denote token embedding, segment embedding, and position embedding of $w_v^{(q_j)}$. Our paragraph-level sentence encoder encodes each paragraph with BERT to get representations of all tokens, as:

$$B^{(q_j)} = \left\{ B_{[CLS]}^{(q_j)} \cdots B_{[SEP]}^{(q_j)} | j = 1 \cdots K \right\} = \text{BERT}\left(Tok_{P_q}, Seg_{P_q}, Pos_{P_q} \right) \tag{4}$$

Then, we utilize TextCNN [12] to get sentences representations. Representations for sentences in paragraph P_q and in document D could be obtained and respectively denoted as $H^{(P_q)}$ and H^D, just defined as follows:

$$H^{(P_q)} = \left\{ \text{CNN}(B^{(q_j)}) | j = 1 \cdots K \right\} \tag{5}$$

$$H^D = \left\{ H^{(P_q)} | q = 1 \cdots M \right\} = \{h_i^D | i = 1 \cdots L\} \tag{6}$$

where h_i^D denotes the paragraph-level sentence representation of the i-th sentence in document D.

Document-Level Sentence Encoder. To capture long-range dependencies among sentences in different paragraphs, we utilize t Transformer layers [8] to get document-level sentences representations. We deem hidden states output by the last Transformer layer, namely S_t^D, as the document-level sentence representations, defined as:

$$S_l^D = \{g_i^D | i = 1 \cdots L\} = \text{Transformer}^{(l)}(S_{l-1}^D), l = 1 \cdots t \tag{7}$$

where $S_0^D = H^D$ and g_i^D indicates the document-level sentence representation of sentence s_i in D.

3.2 Pointer Network Sentence Decoder

We adopt LSTM Pointer Network as decoder. As the decoder structure is almost the same with the previous work, we refer the interested readers to [3] for more details.

4 RL with Our Reward Scheme

To alleviate the order sensitive optimization and generalize extractive summarization to real world scenarios where human ratings and extractive labels are costly, we design a novel RL reward scheme based on sentence entailment task, and a novel dense reward signal mechanism in TaskSum.

Entailment-Based Reward Scheme: The sentence entailment task aims to determine the entailment relationship between the "premise" sentence and a "hypothesis" sentence as "true", "false" or "undetermined", respectively denoting they are semantically entailed, contradictory or neutral. Since some extracted sentences may deviate from the core meaning of the original document, the entailment task is utilized to rule out trivial sentences. The core meaning of the document is often expressed by its title and reference summaries, either of which can be utilized as the "premise", and extracted sentences are deemed as "hypothesis". We utilize the title for an explanation. We denote document D's title as t_D, the reward of extracted sentence d_i, namely $r_e(t_D, d_i)$, is defined as:

$$\{p_e, p_n, p_c\} = \text{Entailment}(t_D, d_i), \ r_e(t_D, d_i) = p_e, \tag{8}$$

where p_e, p_n, p_c respectively denote the probabilities that d_i and t_D are semantically entailed, contradictory or neutral.

Dense-Reward-signal mechanism: Supposing K sentences are extracted at decoding stage, namely $E_K = \{d_1, \cdots, d_K\}$, K is usually far larger than the required number of a summary [3,13,20]. Only the first N of the K extracted sentences, namely lead-N, are utilized to form the summary. We deem the extracted sentences E_K as a concise version of D that has ruled out trivial sentences, from which candidate summaries could be sampled. Each candidate summary is a length-M sequence of unique sentences induced by sentence indices $i = \{i_1, \cdots, i_M\}$. Its generation is as follows: Firstly, extracted sentences are evaluated individually according to the reward scheme to get their scores S_K. Secondly, we sample from remaining sentences according to their score to obtain a new summary sentence. We repeat the second step for M times to form a candidate summary $E_K^{(i)}$. We generate C candidate summaries. $E_K^{(i)}$ is evaluated according to the proposed heuristic reward scheme, just as:

$$r_{E_K^{(i)}} = \sum_{d \in E_K^{(i)}} r_e(t_D, d) \tag{9}$$

The evaluation score of the lead-N extracted sentences $\{d_1, \cdots, d_N\}$, namely r_K^{lead}, is deemed as baseline score. We utilize top-k candidate summaries with the higher evaluation scores than r_K^{lead}, denoted as $\{E_K^{(i)} | i = 1 \cdots k\}$, to calculate the reward of extracted sentence d_t, namely r_t, just as:

$$r_t = \sum_{E_K^{(i)}, i=1 \cdots k} I(d_t \in E_K^{(i)}) \cdot (r_{E_K^{(i)}} - r_K^{lead}) \tag{10}$$

where $I(\cdots)$ denotes the indicator function. An extracted sentence is ranked high for selection if it often occurs in the top-k high-scoring candidate summaries.

5   Experiment Results

5.1   Experiment Setup

Datasets: CNN/DailyMail Long. Documents in CNN/DailyMail [9] contains 24 sentences on average, and around 75% of oracle sentences appear in lead-10 sentences [11]. Such a position bias problem, may skew the evaluation towards methods that focus on extracting the first few sentences. We construct a subset dataset by filtering documents with more than 1000 tokens whose oracle sentences do not in the lead-6 sentences. **CN-Fin** is a Chinese financial news dataset crawled from professional financial websites, such as East Money etc. It consists of 69,728 news documents with titles and human-validated summaries. We split the dataset into training, validation, and test sets as 62,098/3,000/4,630. The average length of document and reference summaries are 1200 and 116 tokens respectively. The dataset will be *open-sourced* with this paper.

Auxiliary Labels: We use MNLI dataset [10] for CNN/DailyMail Long. For CN-Fin, we built a dataset from it by handcrafted rules and human validation.

Evaluation Metric: The models are automatically evaluated on ROUGE [5].

Baselines: LEAD-3 [2] extracts the first 3 sentences as a document's summary; **rnn-ext+rl** [3] is an RL-based Seq2Seq summarization method with sparse ROUGE-based rewards; **BanditSum** [4] formulates the extractive summarization as contextual bandit; **PNBert** [20] introduced BERT into [3]; **BertSumExt** [7] utilize BERT to encode long input sequences with multiple sentences; **MatchSum** [6] formulates summarization as text-matching based on [7].

5.2   Main Results

Table 1. ROUGE evaluations on all the datasets. Here, R-1/2/L stands for ROUGE-1/2/L. Best results are in bold.

Model	CNN/DailyMail Long			CN-Fin		
	R-1	R-2	R-L	R-1	R-2	R-L
LEAD	28.67	7.46	25.05	39.18	15.36	35.23
rnn-ext+rl	31.84	10.31	29.54	50.30	22.07	44.98
BnditSum	31.40	9.52	27.8	49.35	22.28	43.85
PNBert	32.19	10.01	28.39	45.48	20.92	41.57
BertSumExt	33.21	10.73	29.54	48.97	22.89	44.69
MatchSum	33.91	10.94	30.18	49.30	23.59	46.43
TaskSum	**34.25**	**11.50**	**30.64**	**55.57**	**25.50**	**50.53**

The ROUGE results of TaskSum and baseline methods are presented in Table 1. TaskSum outperforms BertSumExt and MatchSum on both datasets.

Lengths of documents in CNN/DailyMail Long are beyond encoding length of pre-trained language models utilized in PNBert, MatchSum, and Bert-SumExt (up to 512 tokens). They have to truncate the document, thus dismiss salient sentences beyond their encoding length. For news in CN-Fin, the oracle sentences are evenly distributed across the document. BertSumExt utilizes a heuristic-based Trigram Blocking for redundancy elimination which has little effect for documents that do not have the position bias problem [6]. Summaries extracted by MatchSum are heavily restricted by BertSumExt. Both methods cannot effectively rule out redundant sentences from summaries. PNBert and rnn-ext+rl are RL-based auto-regressive summarization methods and with the same sparse reward signal mechanism. PNBert does not outperform rnn-ext+rl, as it neglects salient sentences beyond encoding length, thus are more likely to make mistakes at early decoding steps, thus causing error propagation.

5.3 Ablation Study

We detach task-driven reward scheme from TaskSum, i.e., TaskSum w/o ENT, and only train the model on a supervised objective (i.e., TaskSum w/o RL).

We detach HSE from TaskSum (i.e., TaskSum w/o HSE) and utilize sentence encoder in [7]. We also evaluate TaskSum's variants that utilize sparse reward signal mechanism during training, as [3]. The ROUGE results of TaskSum and its variants on the CN-Fin dataset are presented in Table 2. Detaching HSE from TaskSum, drastically drops ROUGE results, as salient sentences beyond the encoding length cannot be extracted. Detaching entailment reward schemes also drop ROUGE results, which demonstrates that the heuristic reward scheme can effectively guide the model to extract semantic reliable sentences. RL can directly optimize the ROUGE, thus TaskSum outperforms TaskSum w/o RL. TaskSum outperforms TaskSum w/ Sparse, which demonstrates the proposed dense reward signal can indeed alleviate the order-sensitive optimization problem.

Table 2. ROUGE-F scores of TaskSum and its variants on the CN-Fin dataset. Best results are in bold.

Models	R-1	R-2	R-L
TaskSum	55.57	25.50	50.53
w/o ENT	52.37	23.84	47.35
w/o RL	51.02	23.52	46.67
w/o HSE	47.54	22.13	43.38
w/ Sprase	**52.61**	**24.15**	**47.86**

6   Conclusion

In this paper, we propose TaskSum, a novel RL-based Seq2Seq extractive summarization model with dense reward signal and heuristic-task-based reward schemes. TaskSum can help to alleviate the major problems of the existing RL-based extractive summarization methods including sparse rewards, order-sensitive optimization, and lacking of reference summaries. Extensive experiments show that the proposed TaskSum can achieve better performance than the competing methods.

References

1. Arumae, K., Liu, F.: Guiding extractive summarization with question-answering rewards. In: NAACL, pp. 2566–2577 (2019)
2. Nallapati, R., Zhai, F., Zhou, B.: SummaRuNNer: a recurrent neural network based sequence model for extractive summarization of documents. In: AAAI, pp. 3075–3081 (2016)
3. Chen, Y.C., Bansal, M.: Fast abstractive summarization with reinforce-selected sentence rewriting. In: ACL, pp. 675–686 (2018)
4. Dong, Y., Shen, Y., Crawford, E., van Hoof, H., Cheung, J.C.K.: BanditSum: extractive summarization as a contextual bandit. In: EMNLP, pp. 3739–3748 (2018)
5. Lin, C.Y.: Rouge: a package for automatic evaluation of summaries. In: ACL, pp. 74–81 (2004)
6. Zhong, M., Liu, P., Chen, Y., Wang, D., Qiu, X., Huang, X.: Extractive summarization as text matching. In: ACL, pp. 6197–6208 (2020)
7. Liu, Y., Lapata, M.: Text summarization with pretrained encoders. In: EMNLP, pp. 3728–3738 (2019)
8. Vaswani, A., et al.: Attention is all you need. In: NeurIPS, pp. 5998–6008 (2017)
9. Hermann, K.M., et al.: Teaching machines to read and comprehend. In: NeuralIPS, pp. 1693–1701 (2015)
10. Williams, A., Nangia, N., Bowman, S.R.: A broad-coverage challenge corpus for sentence understanding through inference. In: NAACL, vol. 1, pp. 1112–1122 (2018)
11. Xu, J., Gan, Z., Cheng, Y., Liu, J.: Discourse-aware neural extractive text summarization. In: ACL, pp. 5021–5031 (2020)
12. Kim, Y.: Convolutional neural networks for sentence classification. In: EMNLP, pp. 1746–1751 (2014)
13. Wu, Y., Hu, B.: Learning to extract coherent summary via deep reinforcement learning. In: AAAI (2018)
14. Henß, S., Mieskes, M., Gurevych, I.: A reinforcement learning approach for adaptive single- and multi-document summarization
15. Kryscinski, W., McCann, B., Xiong, C., Socher, R.: Evaluating the factual consistency of abstractive text summarization. In: EMNLP, pp. 9332–9346 (2020)
16. Zhang, X., Wei, F., Zhou, M.: HIBERT: document level pre-training of hierarchical bidirectional transformers for document summarization. In: ACL, pp. 5059–5069 (2019)
17. La Quatra, M., Cagliero, L.: End-to-end training for financial report summarization. In: COLING, pp. 118–123 (2020)
18. Böhm, F., Gao, Y., Meyer, C.M., Shapira, O., Dagan, I., Gurevych, I.: Better rewards yield better summaries: learning to summarise without references. In: EMNLP, pp. 3108–3118 (2019)
19. Bi, K., Jha, R., Croft, W.B., Celikyilmaz, A.: AREDSUM: adaptive redundancy-aware iterative sentence ranking for extractive document summarization. In: EACL, pp. 281–291 (2021)
20. Zhong, M., Liu, P., Wang, D., Qiu, X., Huang, X.: Searching for effective neural extractive summarization: what works and what's next. In: ACL, pp. 1049–1058 (2019)
21. Beltagy, I., Peters, M.E., Cohan, A.: Longformer: the long-document transformer arXiv:2004.05150 (2020)

Concurrent Transformer
for Spatial-Temporal Graph Modeling

Yi Xie[1,2], Yun Xiong[1,2(✉)], Yangyong Zhu[1,2], Philip S. Yu[3], Cheng Jin[4],
Qiang Wang[4], and Haihong Li[4]

[1] School of Computer Science, Fudan University, Shanghai, China
{18110240043,yunx,yyzhu}@fudan.edu.cn
[2] Shanghai Key Laboratory of Data Science, Fudan University, Shanghai, China
[3] University of Illinois at Chicago, Chicago, USA
psyu@uic.edu
[4] Shanghai Meteorological Disaster Prevention Technology Center, Shanghai, China

Abstract. Previous studies have shown that concurrently extracting spatial and temporal information is a better way to model spatial-temporal data. However, in these studies, the receptive field has been fixed to construct the carrier of concurrent extraction, resulting in the lack of flexibility in selecting receptive fields, and the loss of scalability for capturing long-range temporal dependencies. Furthermore, these studies will result in static weights that are insufficient to describe the complex spatial and temporal dependencies. In this paper, we propose the **C**oncurrent **S**patial-**T**emporal **T**ransformer (CSTT), which ensures the denseness of the carrier of concurrent extraction, so that messages under different receptive fields can be passed more efficiently, thus making the selection of receptive field more flexible. Therefore, the scalability of capturing long-range temporal dependencies can also be promised. Additionally, a unified self-attention mechanism will be applied on the carrier of concurrent extraction to capture spatial and temporal information, so that the dependencies between both dimensions under different contextual information can be preserved. On these bases, we design an iterative strategy to further tackle long sequences. Experiments on four traffic real-world datasets illustrate that our algorithm can achieve significant improvement on the classical spatial-temporal modeling task.

Keywords: Concurrent extraction · Spatial-temporal graph · Attention

1 Introduction

3D-CNNs [3,7] for concurrently extracting spatial-temporal information in Euclidean space can achieve significant improvements, which illustrated that spatial-temporal information can influence each other, and concurrent extraction for both dimensions is a better way. This property inspires us consider that how to apply the concurrent extraction to spatial-temporal graphs.

A. Bhattacharya et al. (Eds.): DASFAA 2022, LNCS 13247, pp. 314–321, 2022.
https://doi.org/10.1007/978-3-031-00129-1_26

In this paper, we proposed **C**oncurrent **S**patial-**T**emporal **T**ransformer (CSTT) to solve these problems. We implemented Graph Neural Networks by Message Passing Neural Networks [9], The concurrent extraction is accomplished through a unified attention-based message passing that simultaneously happens in the spatial and temporal dimensions. The attention mechanism assigns weights dynamically in different inputs also enabled the modeling more accurately. Furthermore, we designed an iterative strategy to further improve the ability to process long historical sequences.

The main contributions are as follows:

- We designed an improved carrier of concurrent extraction: the Concurrent Spatial-Temporal Graph, which is more flexible in receptive field selections and more scalable in long-range temporal dependencies.
- We proposed an algorithm **C**oncurrent **S**patial-**T**emporal **T**ransformer (CSTT), entirely consists of attention mechanism, which can concurrently capture the complex spatial-temporal information more accurately.
- Experiments on four real-world datasets illustrate our algorithm consistently achieves significant performance.

2 Problem Definition

The snapshot on timestamp t of a graph G is represented as $G^{(t)} = (V, E, W, X^{(t)})$, where V is the set of vertices and E is the set of edges. $W \in \mathbb{R}^{N \times N}$ is the weighted adjacency matrix of G, N is the number of vertices. $X^{(t)} \in \mathbb{R}^{N \times D}$ is the representations of vertices in G on timestamp t, D is the hidden dimension. $\mathbb{G}^{(t_s:t_d)} = [G^{(t_s)}, G^{(t_s+1)}, \cdots, G^{(t_d)}]$ is the sequence of snapshots from the timestamp t_s to the t_d. Our task to find a mapping function $f(.)$, so that we can infer the snapshots of future F timestamps:

$$f(\mathbb{G}^{(t_{p-T+1}:t_p)}) = \mathbb{G}^{(t_{p+1}:t_{p+F})}. \tag{1}$$

3 Methodology

The overall framework of CSTT is given in Fig. 1.

3.1 Concurrent Spatial-Temporal Graph

Firstly, we constructed a Concurrent Spatial-Temporal Graph as the carrier of concurrent extraction for the spatial-temporal information as shown in Fig. 1(a). The core idea of constructing the Concurrent Spatial-Temporal Graph is that **vertices pass message concurrently to adjacent vertices and subsequent snapshots with a consistent strategy.**

Formally, given a sequence of historical observations $\mathbb{G}^{(t_0:t_{T-1})}$ with T snapshots, we generate directed edges to connect with vertices in subsequent snapshots. Therefore, the Concurrent Spatial-Temporal Graph is

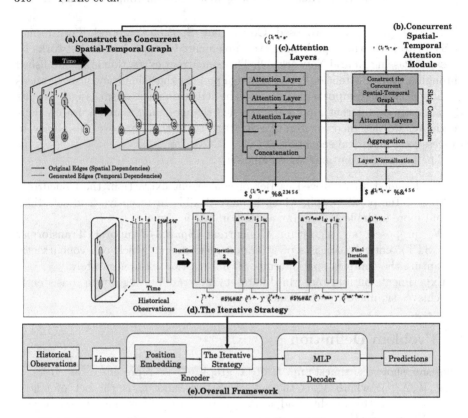

Fig. 1. The overall framework of Concurrent Spatial-Temporal Transformer.

constructed based on the sequence $\mathbb{G}^{(t_0:t_{T-1})}$, denoted as $\mathcal{G}^{(t_0:t_{T-1})} = (\mathcal{V}^{(t_0:t_{T-1})}, \mathcal{E}^{(t_0:t_{T-1})}, \mathcal{W}^{(t_0:t_{T-1})}, \mathcal{X}^{(t_0:t_{T-1})})$. $\mathcal{V}^{(t_0:t_{T-1})}$ denotes the set of vertices in $\mathcal{G}^{(t_0:t_{T-1})}$ with $T*N$ vertices from all snapshots in $\mathbb{G}^{(t_0:t_{T-1})}$. $\mathcal{E}^{(t_0:t_{T-1})}$ is the set of weighted edges in $\mathcal{G}^{(t_0:t_{T-1})}$, which contains two sets of edges: the set of original edges in each snapshot, and the set of generated directed edges connecting vertices in different snapshots. $\mathcal{X}^{(t_0:t_{T-1})} = ||_{\zeta=0}^{T-1} X^{(t_\zeta)} \in \mathbb{R}^{T*N \times D}$ represents the representations of vertices in $\mathcal{G}^{(t_0:t_{T-1})}$, which concatenates the representations of all snapshots in the sequence $\mathbb{G}^{(t_0:t_{T-1})}$. The adjacency matrix of $\mathcal{G}^{(t_0:t_{T-1})}$ is shown as follows:

$$\mathcal{W}^{(t_0:t_{T-1})} = \begin{pmatrix} W & I_N & I_N & \cdots & I_N \\ 0 & W & I_N & \cdots & I_N \\ \vdots & \vdots & \vdots & \ddots & \vdots \\ 0 & 0 & 0 & \cdots & I_N \\ 0 & 0 & 0 & \cdots & W \end{pmatrix}, \tag{2}$$

where $\mathcal{W}^{(t_0:t_{T-1})} \in \mathbb{R}^{T*N \times T*N}$, I_N is the N-order identity matrix. The purpose of constructing the Concurrent Spatial-Temporal Graph is performing a unified process of message passing on where there exists a directed edge $e_{a \to b} \in \mathcal{E}^{(t_0:t_{T-1})}$ from vertex a to vertex b.

There will be a problem if we only connect snapshots on adjacent timestamps: the longer the sequence, the larger the T, and the sparser the Concurrent Spatial-Temporal Graph. According to [10], the optimal number of layers in Message Passing Neural Networks is depends on the largest shortest path distance of graphs, and sparser graphs require deeper layers in Message Passing Neural Networks. Therefore, it is difficult and complicated if not impossible to assign an determined optimal number of layers in Message Passing Neural Networks for different sequences with T snapshots, which limits the flexibility of CSTT under different values of T. However, it can be addressed by keeping the denseness of the Concurrent Spatial-Temporal Graph Specifically, we added dense connections to densify the Concurrent Spatial-Temporal Graph, which can be reflected in the phenomenon that all entries above the diagonal are not zero matrices in $\mathcal{W}^{(t_0:t_{T-1})}$. The operation promises the denseness of the Concurrent Spatial-Temporal Graph, which cannot get sparser any longer as T increases, thus it is not necessary to assign different optimal depth for different values of T, and the optimal number of layers in Message Passing Neural Networks and the value of T can be decoupled. Therefore, CSTT can handle dependencies of sequence with arbitrary length, without considering the performance degradation caused by inappropriate number of layers in Message Passing Neural Networks.

3.2 Concurrent Spatial-Temporal Attention Module

Secondly, we performed a unified attention mechanism on the Concurrent Spatial-Temporal Graph $\mathcal{G}^{(t_0:t_{T-1})}$ by the Concurrent Spatial-Temporal Attention Module, to concurrently extract information from both the spatial and the temporal dimensions, which is shown in Fig. 1(b).

Formally, given a sequence of snapshots $\mathbb{G}^{(t_0:t_{T-1})}$ with T snapshots as inputs, we firstly constructed a Concurrent Spatial-Temporal Graph $\mathcal{G}^{(t_0:t_{T-1})}$. For arbitrary vertex a and vertex b, where $a, b \in \mathcal{V}^{(t_0:t_{T-1})}$, if there is a directed edge $e_{a \to b} \in \mathcal{E}^{(t_0:t_{T-1})}$ from the vertex a to vertex b with weight w_{ab}, the vertex-wise correlation coefficients $c_{a \to b}$ from the vertex a to b are defined as follows:

$$c_{a \to b} = w_{ab} \sigma(\boldsymbol{\theta}^\top (\boldsymbol{x_a} \Theta_1 + \boldsymbol{x_b} \Theta_2)), \tag{3}$$

where $\Theta_1, \Theta_2 \in \mathbb{R}^{D \times D}$ are trainable parameters, $\boldsymbol{x_a}, \boldsymbol{x_b} \in \mathbb{R}^D$ are the representations of vertex a and vertex b, respectively, $\boldsymbol{\theta} \in \mathbb{R}^D$ represents a trainable vector for an inner product, and σ is an activation function (LeakyReLU herein). Then we perform a Softmax function to calculate attention coefficients $\alpha_{a \to b}$:

$$\alpha_{a \to b} = \frac{\exp(c_{a \to b})}{\sum_{n \in N_{(b)}} \exp(c_{n \to b})}, \tag{4}$$

where $N_{(b)}$ denotes the set of vertices with a directed edge to the vertex b, $i.e.$ for all $n \in N_{(b)}$ with edge $e_{n \to b} \in \mathcal{E}^{(t_0:t_{T-1})}$.

Similar to previous works [11,12], we applied multi-head attention mechanism to capture different attention coefficients to extract information in different subspaces:

$$x_n^{'m} = \sigma(\alpha_{n \to b}^m x_n \mathcal{H}_n^m), \tag{5}$$

where $\mathcal{M}$ denotes the number of attention heads, vertex n denotes one of the neighbors in vertex b, $x_n \in \mathbb{R}^D$ is the representation of vertex n, $\alpha_{n \to b}^m$ represents the attention coefficients in the m-th head, while $\mathcal{H}_n^m \in \mathbb{R}^{D \times \frac{D}{\mathcal{M}}}$ is a trainable parameter matrix for the m-th head, $x_n^{'m} \in \mathbb{R}^{\frac{D}{\mathcal{M}}}$ denotes the intermediate representation of the vertex n. It is noteworthy that attention coefficients in different head are independent, $i.e.$ parameters are not shared. And then, intermediate representations of vertex n in each attention head are concatenated to to form the representation x_n' of vertex n:

$$x_n' = ||_{m=0}^{\mathcal{M}}(x_n^{'m}), \tag{6}$$

where $x_n' \in \mathbb{R}^D$, $||$ is the operation of concatenation. The aggregation phase:

$$x_b = \sum_{n \in N_{(b)}} x_n'. \tag{7}$$

where $x_b \in \mathbb{R}^D$. Therefore, a layer of vertex-wise attention-based message passing in the Concurrent Spatial-Temporal Graph can be completed by the calculations from the Eq. (3) to the Eq. (7), the vertex-wise operations shown above will be applied on all vertices in the Concurrent Spatial-Temporal Graph $\mathcal{G}^{(t_0:t_{T-1})}$ parallelly to update representations of all vertices. We stacked multiple layers to capture more information. Outputs of each attention layer are concatenated to capture both local and global information, as shown in Fig. 1(c).

Subsequently, all vertices fully updated information from both dimensions, and the intermediate representations $\mathcal{X}_{int} \in \mathbb{R}^{T*N \times D}$ can be obtained. We had to aggregate the intermediate representations and obtain the compressed final results $\mathcal{X}' \in \mathbb{R}^{N \times D}$ for the sequence of snapshots $\mathbb{G}^{(t_0:t_{T-1})}$. Specifically, we adopted an attention function for aggregation, calculations are as follows:

$$\mathcal{C} = \sigma(\Theta_3 \mathcal{X}_{int} \mathcal{X}_{int}^\top \Theta_4)\Theta_5, \tag{8}$$

$$\mathcal{A}_{qr} = \frac{\exp(\mathcal{C}_{qr})}{\sum_{t=0}^T \exp(\mathcal{C}_{tr})}, \tag{9}$$

$$\mathcal{X}' = \mathcal{A}\mathcal{X}_{int}. \tag{10}$$

where $\Theta_3 \in \mathbb{R}^{T \times T*N}, \Theta_4 \in \mathbb{R}^{T*N \times T}$, and $\Theta_5 \in \mathbb{R}^{T \times T}$ are trainable parameter matrices, $\mathcal{C} \in \mathbb{R}^{T \times T}$ denotes the correlation matrix describing the correlation coefficient between snapshots to be aggregated, where $\mathcal{C}_{qr}$ denotes the correlation coefficient between the snapshots on timestamp q and timestamp r. $\mathcal{A} \in \mathbb{R}^{N \times N}$ denotes the attention matrix after Softmax. Analogously, $\mathcal{A}_{qr}$ denotes the attention coefficient between the snapshots on timestamp q and timestamp r. Therefore, after aggregation by attention mechanism, we obtained a compressed final results $\mathcal{X}' \in \mathbb{R}^{N \times D}$ of aggregation, followed by subsequent operations.

3.3 Iterative Strategy

Longer sequences with T snapshots result in larger Concurrent Spatial-Temporal Graph. Thus, the length of historical observation sequences is limited by memory. We propose Interactive Strategy to handle the issue, shown as Fig. 1(d).

Formally, given a sequence of snapshots $\mathbb{G}^{(t_0:t_{T-1})}$, we divided it into multiple subsequences: $\mathbb{G}^{(t_0:t_{k-1})}$, $\mathbb{G}^{(t_k:t_{2k-2})}$, ..., $\mathbb{G}^{(t_g:t_{T-1})}$. Note that each subsequence contains $k-1$ snapshots except for the first, which contains k snapshots. Specifically, for the last subsequence $\mathbb{G}^{(t_g:t_{T-1})}$, if $t_g > t_{T-k}$, there is no enough snapshots to form a subsequence with $k-1$ snapshots, in order to ensure the consistency of each iteration, we pad with zero tensors to satisfy there are $k-1$ snapshots in the last subsequence. In our paper, k is named as **Receptive Field**.

We firstly adopted the Concurrent Spatial-Temporal Attention Module to handle $\mathbb{G}^{(t_0:t_{k-1})}$ according to Subsect. 3.2, the input representations for the first iteration is denoted as $\mathcal{X}^{(t_0:t_{k-1})} \in \mathbb{R}^{k*N \times D}$. After the Concurrent Spatial-Temporal Attention Module, the output representations are denoted as $\mathcal{X}'^{(t_0:t_{k-1})} \in \mathbb{R}^{N \times D}$. $\mathcal{X}'^{(t_0:t_{k-1})}$ compresses all information of $\mathbb{G}^{(t_0:t_{k-1})}$ which is further concatenated with hidden representations $\mathcal{X}^{(t_k:t_{2k-2})}$ in $\mathbb{G}^{(t_k:t_{2k-2})}$ to form a new subsequence with k snapshots, and then performs the same operations iteratively until the end of the historical sequence $\mathbb{G}^{(t_0:t_{T-1})}$. Finally, we can obtain a final compressed result $\mathcal{X}'^{(t_0:t_{T-1})} \in \mathbb{R}^{N \times D}$ that compresses all information for the input sequence $\mathbb{G}^{(t_0:t_{T-1})}$. Note that the number of iterations is not finite, therefore, we can dispose of a historical sequence with arbitrary length.

3.4 Overall Framework

The overall is an Encoder-Decoder architecture [13], shown as Fig. 1(e).

Firstly, we performed a Multi-Layer Perceptron (MLP) to map the input data $X =\in \mathbb{R}^{T \times N \times D_i}$ into the hidden space with dimension D and to obtain the hidden representations of input data. Considering that our model concurrently extracts information from both dimensions discriminatively, we added another trainable matrix $\mathcal{S} \in \mathbb{R}^{N \times D}$ to mark the spatial information:

$$X \leftarrow X + \mathcal{T} + \mathcal{S} \in \mathbb{R}^{T \times N \times D}. \tag{11}$$

Then, we performed the iterative strategy for encoding to obtain the compressed results $\mathcal{X}'^{(t_0:t_{T-1})}$. The spatial and temporal embeddings form the encoder of CSTT together with the iterative strategy. In order to eliminate the accumulative error caused by RNN-based decoders, we selected a MLP for parallelly decoding. We incorporated MAE as the objective function with L2 regularization.

4 Experiments

4.1 Datasets

Datasets Introduction. We evaluated CSTT on four traffic flow datasets: PeMSD3, PeMSD4, PeMSD7 and PeMSD8. All datasets record the total traffic flow in the past 5 min.

Datasets Pre-processing. We adopted Z-score normalization to the input features of vertices. We constructed the adjacency matrix according to ASTGCN [6]: We splitted the training set, validation set, and test set by the corresponding ratio of 60%, 20% and 20%, respectively.

4.2 Experimental Introduction

Experimental Description. We predicted traffic flow information on future continuous 12 timestamps (60 min) according to information on historical 12 timestamps taking the settings in ASTGCN [6] as the reference.

Evaluation Metrics. Mean Absolute Error (MAE), Root Mean Square Error (RMSE) and Mean Absolute Percentage Errors (MAPE) are evaluation metrics:

Baselines. We compared our model with following baselines: **DCRNN** [2]; **STGCN** [1]; **Graph WaveNet** [5]; **ASTGCN** [6]; **STSGCN** [8].

4.3 Performance Comparison

Table 1. Performance comparison

Datasets	PeMSD3			PeMSD4		
Algorithm	MAE	RMSE	MAPE (%)	MAE	RMSE	MAPE (%)
DCRNN	18.18	30.31	18.91	24.70	38.12	17.12
STGCN	17.49	30.12	17.15	22.70	35.55	14.59
GraphWavenet	19.85	32.94	19.31	25.45	39.70	17.29
ASTGCN	17.69	29.66	19.40	22.93	35.22	16.56
STSGCN	17.48	29.21	16.78	21.19	33.65	13.90
CSTT	16.21 ± 0.11	27.48 ± 0.09	15.40 ± 0.18	19.97 ± 0.09	29.64 ± 0.10	13.86 ± 0.17
Datasets	PeMSD7			PeMSD8		
Algorithm	MAE	RMSE	MAPE (%)	MAE	RMSE	MAPE (%)
DCRNN	25.30	38.58	11.66	17.86	27.83	11.45
STGCN	25.38	38.78	11.08	18.02	27.83	11.40
GraphWavenet	25.45	39.70	17.29	26.85	42.78	12.12
ASTGCN	22.93	35.22	16.56	18.61	28.16	13.08
STSGCN	24.26	39.03	10.21	17.13	26.80	10.96
CSTT	23.63 ± 0.11	36.02 ± 0.13	10.13 ± 0.12	15.75 ± 0.08	23.68 ± 0.07	10.26 ± 0.10

Performance comparison is shown in Table 1. In contrast, CSTT outperformed the other baselines on all tasks.

5 Conclusion

In this paper, we proposed CSTT for spatial-temporal graph modeling. A unified attention mechanism is adopted to model spatial and temporal information concurrently. To this end, we designed an improved carrier of concurrent extraction, which enables our algorithm to capture long-range temporal dependencies. The iterative strategy further improves the ability to handle long sequences. On four public datasets, CSTT achieves the state-of-the-art performance.

Acknowledgements. This work is funded in part by the National Natural Science Foundation of China Projects No. U1936213. This work is also supported in part by NSF under grants III-1763325, III-1909323, III-2106758, and SaTC-1930941.

References

1. Yu, B., Yin, H., Zhu, Z.: Spatio-temporal graph convolutional networks: a deep learning framework for traffic forecasting. In: IJCAI (2018)
2. Li, Y., Yu, R., Shahabi, C., Liu, Y.: Diffusion convolutional recurrent neural network: data-driven traffic forecasting. In: ICLR (2018)
3. Ji, S., Wei, X., Yang, M., Kai, Yu.: 3D convolutional neural networks for human action recognition. T-PAMI **35**, 221–231 (2013)
4. Yan, S., Xiong, Y., Lin, D.: Spatial temporal graph convolutional networks for skeleton-based action recognition. ArXiv abs/1801.07455 (2018)
5. Wu, Z., Pan, S., Long, G., Jiang, J., Zhang, C.: Graph WaveNet for deep spatial-temporal graph modeling. In: IJCAI (2019)
6. Guo, S., Lin, Y., Feng, N., Song, C., Wan, H.: Attention based spatial-temporal graph convolutional networks for traffic flow forecasting. In: AAAI (2019)
7. Tran, D., Bourdev, L., Fergus, R., Torresani, L., Paluri, M.: Learning spatiotemporal features with 3D convolutional networks. In: ICCV (2015)
8. Song, C., Lin, Y., Guo, S., Wan, H.: Spatial-temporal synchronous graph convolutional networks: a new framework for spatial-temporal network data forecasting. In: AAAI (2020)
9. Gilmer, J., Schoenholz, S.S., Riley, P.F., Vinyals, O., Dahl, G.E.: Neural message passing for quantum chemistry. In: ICML (2017)
10. Klicpera, J., Bojchevski, A., Günnemann, S.: Predict then propagate: graph neural networks meet personalized PageRank. In: ICLR (2019)
11. Vaswani, A., et al.: Attention is all you need. In: NIPS (2017)
12. Velickovic, P., Cucurull, G., Casanova, A., Romero, A., Lio, P., Bengio, Y.: Graph attention networks. In: ICLR (2018)
13. Bahdanau, D., Cho, K., Bengio, Y.: Neural machine translation by jointly learning to align and translate. CoRR abs/1409.0473 (2015)
14. Hochreiter, S., Schmidhuber, J.: Long short-term memory. Neural Comput. **9** (1997)
15. Drucker, H., Burges, C.J., Kaufman, L., Smola, A., Vapnik, V.: Support vector regression machines. In: NIPS (1996)

Towards Personalized Review Generation with Gated Multi-source Fusion Network

Hongtao Liu[1,2], Wenjun Wang[1,2,4], Hongyan Xu[1,2], Qiyao Peng[1,3], Pengfei Jiao[5], and Yueheng Sun[1,2(✉)]

[1] State Key Laboratory of Communication Content Cognition, Beijing, China
[2] College of Intelligence and Computing, Tianjin University, Tianjin, China
{htliu,wjwang,hongyanxu,qypeng,yhs}@tju.edu.cn
[3] School of New Media and Communication, Tianjin University, Tianjin, China
[4] College of Information Science and Technology, Shihezi University, Xinjiang, China
[5] School of Cyberspace, Hangzhou Dianzi University, Hangzhou, China
pjiao@hdu.edu.cn

Abstract. Product review generation is an important task in recommender systems, which could provide explanation and persuasiveness for the recommendation. The key challenge of review generation is a lack of semantic information that can be used, resulting in more uncertainty in the generation processing. In this paper we propose a novel encoder-decoder personalized review generation model which aims to effectively and selectively fuse multiple resources including the user/product attributes and their textual information (e.g., historical reviews, short phrases). In the encoder part, we first design an attention layer to accordingly retrieve useful information from the users/product historical reviews since not all reviews are relevant to the target review generation. Note that the attention vector is derived from the semantic features of the short phrases (e.g., review summaries, product titles) which could indicate user preference and product characteristics. In the decoder part, we design a hierarchical gated multi-source network from both intra- and inter-gate perspectives to effectively fuse the multiple context features from encoders. Experimental results on the real-world dataset show that our proposed model could achieve better review generation performance than the baselines, in terms of fluency, readability, and personalization.

Keywords: Review generation · Recommender system · Multi-source gate

1 Introduction

Most online e-commerce platforms (e.g., Amazon.com) allow users to write reviews of products that have already been purchased. The reviews are widely used in many recommender systems to help model users and products, which could improve the performance of the recommendation. However, users usually could not understand why the product is recommended, that is, the recommender

A. Bhattacharya et al. (Eds.): DASFAA 2022, LNCS 13247, pp. 322–330, 2022.
https://doi.org/10.1007/978-3-031-00129-1_27

system has poor interpretability [7]. Considering reviews are the purchase experiences of users towards products, they can provide semantic interpretability for the recommended results. Therefore, review generation task has attracted more and more attentions [3].

Compared with other text generation tasks (such as text summarization), the review generation is more challenging due to the lack of sufficient semantic input information. Many review generation methods are based on the encoder-decoder framework. Give a user and a product, the encoder module is to learn context features from the user/product available information, and the decoder module (always based on Recurrent Neural Network) to generate review according to features from the encoder. However, it is nontrivial to effectively combine these information together in encoder side and decoder side. There remain two main challenges: (1) Many existing works use very little input clues, which would lead to more uncertainty in the following review generation. It is obvious that more useful information about users/products is used, the generated reviews will be more accurate. (2) How to select more relevant parts from the various information such as attributes, historical reviews, etc. Different kinds of encoded information are of different importance for the target review generation. In all, it is necessary to design a unified and selective framework to encode as much useful information as possible to generate more comprehensive reviews.

Hence, we propose a novel History-aware Review generation method with a Gated multi-source Fusion, named HRGF. There are two main components: an encoder to learn various features from input information and a decoder with a gated multi-source fusion to generate reviews. (1) In the encoder, we first use an Attribute Encoder to learn the features of user/product IDs, and a Phrase Encoder to learn the semantic representations from product titles and review summaries. Besides, we propose a personalized attention network to learn user/product representations from their historical reviews. In this way, through the three encoders, the model can learn rich context features for the following decoder. (2) In the decoder, we first initialize the hidden states of GRU-based decoder with embeddings learned in the above encoder. Afterwards, we propose a novel inter- and intra-gate hierarchical network to fully fuse the context features from the encoder, which makes the decoder be able to access to useful information from the various input clues. The experimental results demonstrate our method could generate more readable and accurate reviews that could use for explainable recommendation.

2 Proposed Method

2.1 Problem Formulation

Given a user u and a product p, the available data resources contain the IDs attributes, the product title, the review summary, and their historical reviews. The generated review is denoted as $\hat{Y} = \{\hat{y}_1, \hat{y}_2, \cdots, \hat{y}_L\}$ and L is the generated review length; the reference review is denoted as $Y = \{y_1, y_2, \cdots, y_{L'}\}$ and L' is the reference review length. Review generation aims to automatically generate reviews for the given user towards the target products according to the above information. Figure 1 illustrates the overall architecture of our model.

2.2 History-Aware Personalized Encoder

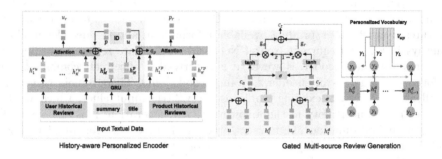

Fig. 1. The framework of our *HRGF* approach.

This section presents the first part of the model, the personalized encoder, which is used to learn the feature of the rich input data. Three types of encoders are designed, which will be introduced below.

Attribute Encoder. The ID attributes of users and products can reflect unique characteristics, which is of great help to the generation of personalized reviews. For the ID attributes, we encode the user and product ID embedding $\mathbf{u}, \mathbf{p} \in \mathcal{R}^{d_a}$ via a multi-layer perceptron: $\mathbf{a} = \tanh(\mathbf{W}_a[\mathbf{u}; \mathbf{p}])$. $[;]$ indicates the concatenation operation, and the feature $\mathbf{a}$ is the final attribution vector.

Phrase Encoder. Phrase information typically includes product titles, review summaries and has been shown to help generate coherent review [4]. A phrase encoder is designed to extract their semantic features via a Bi-GRU encoder. Given the product title $T = \{w_1, w_2, \cdots, w_N\}$, the representation can be computed by:

$$\overrightarrow{\mathbf{h_i}} = \overrightarrow{\mathrm{GRU}}(\mathbf{w_i}) , \overleftarrow{\mathbf{h_i}} = \overleftarrow{\mathrm{GRU}}(\mathbf{w_i}) , \ 1 \leq i \leq N \tag{1}$$

where $\mathbf{w_i}$ is the word embedding of the i-th word in the product title. By combining the two hidden state vectors, the semantic feature of each word in the title can be obtained via $\mathbf{h_i^P} = [\overrightarrow{\mathbf{h_i}}, \overleftarrow{\mathbf{h_i}}]$. After that, the hidden state of the last word is as the final representation of the entire title, which is defined as $\mathbf{h_N^p}$. Likewise, Passing the review summary through the GRU network, its semantic feature can be obtained, defined as $\mathbf{h_M^s}$.

Personalized Historical Reviews Encoder. For users, their historical review information can effectively show their preferences and writing style; for products, historical review information can reflect their general characteristics. For a given user u, we first use the Bi-GRU network in the phrase encoder above to learn features on all historical reviews of the user, and obtain the semantic features of all K comments, denoted as: $\{\mathbf{h}_1^{ru}, \cdots, \mathbf{h}_K^{ru}\}$. Similarly, for the current product p, we can learn features of all the historical reviews, denoted as: $\{\mathbf{h}_1^{rp}, \cdots, \mathbf{h}_K^{rp}\}$.

Considering not all historical reviews are relevant to the target review generation. Hence, we design an attention network to focus on those more useful reviews in terms of the target review. Instead of using general query vectors for all users and products, we integrate the attributes and short phrases into the attention module to indicate the personalization of different users and products. The query vectors for the user u and the product p respectively are defined via a cross manner:

$$\mathbf{q_u} = [\mathbf{p}; \mathbf{h}_M^s; \mathbf{h}_N^p], \mathbf{q_i} = [\mathbf{u}; \mathbf{h}_M^s; \mathbf{h}_N^p]. \qquad (2)$$

The motivation of this "cross-attention query vector" is as follows: the same historical review of user u would be differently relevant towards different products and would be different usefulness for the corresponding review generation. The same is for the product query vector construction. For the i-th history review of the user u, the weight is computed as:

$$\alpha_i = \tanh(\mathbf{W}_q\mathbf{q_u} + \mathbf{W}_k\mathbf{h}_i^{ru}), \ s_i = \frac{\exp(\alpha_i)}{\sum_{j=1}^{K} \exp(\alpha_j)} \ \mathbf{u}_r = \sum_{i=1}^{K} s_i\mathbf{h}_i^r. \qquad (3)$$

where $\mathbf{h}_i^{ru}$ is the feature of the i-th review of the user u, and α_i is the corresponding weight. We can obtain the history-aware representation of the user u, denoted as $\mathbf{u}_r$: Likewise, the history-aware representation of the product p could be learned from its historical reviews, denoted as $\mathbf{p}_r$.

Through the three encoders, we have prepared rich context embeddings, including the ID embeddings (i.e., $\mathbf{u}, \mathbf{p}$), the short phrase features (i.e., $\mathbf{h}_M^s, \mathbf{h}_N^p]$) and the history-aware embeddings (i.e., $\mathbf{u}_r, \mathbf{p}_r$).

2.3 Review Generation with Gated Multi-source Decoder

Our decoder is derived from the GRU network [1] to generate reviews word by word. We first initialize the hidden state of the decoder via combining all of the context features, denoted as $\mathbf{h}_0^d = \mathbf{h}_M^s + \mathbf{h}_N^p + \mathbf{u}_r + \mathbf{p}_r + \mathbf{a}$.

Intuitively, not all encoding information is related to the word generation at the current decoding step. Therefore, in order to control the multi-source information flow, we design elaborately a hierarchical gate mechanism to filter important context features from encoders: 1) a gate over historical reviews and a gate over ID attributes, i.e., an intra-gate mechanism. 2) a multi-source gated network over the different types of features, i.e., an inter-gate mechanism.

Specifically, the first intra-gate network is to align the current GRU state and the encoder-side information including historical review and ID attribute features respectively. The gate mechanism over the historical reviews is denoted as:

$$\beta_a = \tanh(\mathbf{W}_u^{(1)}\mathbf{u}_r + \mathbf{W}_p^{(1)}\mathbf{p}_r), \; g_{a_t} = \sigma(\mathbf{W}_a^{(1)}\mathbf{h}_t^d), \; \mathbf{c}_r = g_{a_t} \odot \beta_a . \quad (4)$$

where $\mathbf{h}_t^d$ is the generation state at time step t. $\mathbf{c}_r$ is the output of the current gate unit. Likewise, the gate mechanism applied over the ID attributes is calculated denoted as $\mathbf{c}_a$.

$\mathbf{c}_r$ and $\mathbf{c}_a$ model the correlation between the two types of features on the encoder side and the current decoder state respectively. However, different types of features are differently important for the decoding process. Hence, we design the second inter-gate network to decide how different sources influence the generation, i.e., user/product id attributes and historical reviews. Given the output feature of the first gate layer, i.e., $\mathbf{c}_r, \mathbf{c}_a$, the multi-source gated unit is computed as:

$$\mathbf{g}_a = \tanh(\mathbf{W}_{ga}\mathbf{c}_a), \mathbf{g}_r = \tanh(\mathbf{W}_{gr}\mathbf{c}_r), \quad (5)$$

$$z = \sigma(\mathbf{W}_z[\mathbf{c}_a; \mathbf{c}_r]), \; \mathbf{c}_t = z \times \mathbf{g}_a + (1 - z) \times \mathbf{g}_r. \quad (6)$$

$z \in (0,1)$ is the parameter that controls the flow of multi-source information. $\mathbf{c}_t$ is the final context vector that selectively incorporates the features from the ID attributes and historical reviews.

Besides, inspired by [5], we propose to establish a personalized vocabulary V_{up} for current user u and product p. V_{up} consists of a small number of personalized words from the historical reviews of the user and product, which can reflect the personalized writing styles and the product features. First, calculate the probability distribution of each word in the personalized vocabulary as follows:

$$\gamma_{t,i} = \texttt{softmax}(\mathbf{W}_v([\mathbf{h}_t^d : \mathbf{c}_t : \mathbf{e}_i^V])) \quad (7)$$

where $\mathbf{e}_i^V$ is the embedding of the i-th word in the personalized vocabulary V_{up}, $\gamma t, i$ is the weight score of this word at time step t. Through combining these two parts word distribution, we can eventually obtain the final word distribution: $P(\hat{y}_t) = P_v(\hat{y}_t) + \mathbb{I}_{v_i = y_t}\gamma_{t,i}$, We adopt the negative log-likelihood as the loss function:

$$\mathcal{L}_\phi(\hat{Y}|A) = \sum_{t=1}^{L'} -logP(\hat{y}_t), \quad (8)$$

where ϕ are the model parameters.

3 Experiments

3.1 Experimental Setting

We use a real-world dataset from Amazon Electronics[1] to validate the performance of our model. Each sample of the dataset contains the user id, product id,

---

[1] http://jmcauley.ucsd.edu/data/amazon/.

review text, summary text and product title. Following the previous works [2,4], we only reserve the reviews given by active users to popular products, where each user and each product has at least 5 historical reviews. The dataset contains 182,850 users, 59,043 products and 992,172 reviews. Then, the dataset is randomly split into (80%, 10%, 10%) as training, validation and test datasets respectively. the dimension of all the learned features (i.e., $\mathbf{h}_M^s, \mathbf{h}_N^p$) is 512. We randomly initialize the ID embeddings of users and products, and the size is 64.

We use four widely used metrics in text generation to evaluate the performance of our model, including Perplexity, Distinct-1 and Distinct-2, BLEU-1/4 and ROUGE-1/2/L.

3.2 Performance Evaluation

In this section, we compare with several competitive review-generation methods to validate our model, including: (1) **HR-p:** randomly choose a review from historical reviews of the product; (2) **HR-u:** randomly choose a review from product historical reviews of the user; (3) **gC2S** [8]; (4) **Attr2Seq** [2]; (5) **Cyclegen** [6]; (6) **ExpansionNet** [4];

Table 1. Performance of different methods for review generation.

Methods	Perplexity	BLEU-1 (%)	BLEU-4 (%)	ROUGE-1	ROUGE-2	ROUGE-L	Distinct-1	Distinct-2
HR-p	–	22.34	0.74	21.40	1.97	12.69	0.80	0.95
HR-u	–	23.41	1.06	22.38	2.98	13.29	0.81	0.95
gC2S	34.70	26.13	1.37	24.98	3.73	15.97	0.60	0.72
Att2Seq	34.29	26.26	1.54	24.90	3.87	15.93	0.61	0.73
Cyclegen	32.42	26.74	1.68	25.12	4.05	16.09	0.65	0.76
ExpansionNet	31.26	28.35	2.12	28.76	4.47	16.83	0.72	0.81
HRGF	**28.65**	**31.29**	**2.44**	**30.17**	**5.43**	**18.57**	**0.82**	0.94

The comparison results are shown in Table 1. Our model HRGF has achieved the best results in almost evaluation metrics, which reflects the effectiveness of our model. Firstly, in terms of encoders, HRGF further integrates historical reviews which can directly reflect the user's writing style and habits, this can help the model to generate more personalized reviews that are closer to real users. In addition, on the decoder, we propose to design a hierarchical multi-source gating mechanism for different types of features, which can characterize the importance of different features in the decoding hidden state, so more accurate reviews can be generated.

3.3 Ablation Study

In order to further analyze the effectiveness of the important modules in our model, we design ablation experiments in this section. Three model variants are designed: (1) **HRGF-E** denotes that removing the history-aware attention, i.e., using the average operation of the user/product historical reviews embeddings to obtain u_r, p_r. (2) **HRGF-G** denotes that directly concatenating context vectors c_a and c_r without the multi-source gate fusion mechanism. (3) **HRGF-P** denotes that we remove the personalized vocabulary in the decoder.

Table 2. Performance of HRGF and its variants.

Methods	BLEU-1 (%)	BLEU-4 (%)	ROUGE-1	ROUGE-2	ROUGE-L
HRGF-E	29.91	2.05	28.01	5.17	18.01
HRGF-G	30.60	2.12	29.57	5.22	18.13
HRGF-P	30.78	2.22	28.18	5.25	17.93
HRGF	**31.29**	**2.44**	**30.17**	**5.43**	**18.57**

HRGF-E and HRGF-G are used to verify the effect of the attention mechanism proposed in the encoder, and the multi-source gating mechanism in the decoder respectively, and HRGF-P is designed to verify the effect of the personalized vocabulary. The results are shown in the Table 2. We can find that no matter which module is removed, compared with the full model HRGF, the three variant models all have a decline in the evaluation metrics of BLEU and ROUGE. This proves that the above three modules are all helpful to improve performance for review generation tasks.

3.4   Case Study

In order to demonstrate the effect of the HRGF on review generation more intuitively, We show some examples of generated and reference reviews in Fig. 2. We can find the reviews generated by the model and the real reviews are relatively similar in terms of keywords and the user sentiment towards products features. For example, the "screen protector" in the first example and "great little mouse" in the third example would indicate that. Furthermore, user writing style of generated reviews is consistent with the reference, e.g., "would recommend this to anyone" in the second case. This reflects the effectiveness of the proposed model in generating personalized reviews. The generated reviews can give a semantic explanation of why a certain product is recommended, which can enhance the interpretability and user-friendliness of the recommendation.

Reference Review: awesome screen protector. install was a breeze and can't even notice it 's on there, and it comes with two! recommend over other screen protectors. fast shipping.

Generated Review: this screen protector is great. easy to install, and very durable. i have n't had any issues with it at all. highly recommend.

Reference Review: pen writes very good and is just like putting pen to paper. i am very pleased with my purchase. would recommend this to anyone that would like to have the feel of a regular size pen when using the note.

Generated Review: i have been using this pen for about a month now and it works great. i am very pleased with the product. i would recommend this product to anyone who needs a pen.

Reference Review: great little mouse for my wife 's laptop. love the way the usb connection fits into the underside of the mouse and automaticly turns off the mouse when you slide it underneath. easily fits in a laptop bag and could easily be used for any computer (desktops included).

Generated Review: this is a great little mouse. it is small and works well with my laptop. i like the fact that it has a usb port on the back of the mouse so you can leave it in your lap while reading. i have n't had any problems with it at all and i am very happy with my purchase.

Fig. 2. Examples of generated and reference reviews.

4 Conclusion

In this paper, we propose a novel review generation method which aims to fully utilize the rich information including attributes, short phrases and historical reviews. The personalized attention encoder is proposed to focus on useful historical reviews to learn user/product representations via a well-designed cross attention mechanism. In order to make full use of the rich features provided by the encode, the gated multi-source decoder is used to fuse the context features selectively for the review generation word by word. The experimental results on real-world recommendation dataset prove that the proposed method could generate high-quality reviews in terms of fluency, readability, and personalization.

Acknowledgments. This work was supported by State Key Laboratory of Communication Content Cognition (A32002) and the Sustainable Development Project of Shenzhen (KCXFZ20201221173013036).

References

1. Cho, K., et al.: Learning phrase representations using rnn encoder–decoder for statistical machine translation. In: EMNLP, pp. 1724–1734 (2014)
2. Dong, L., Huang, S., Wei, F., Lapata, M., Zhou, M., Xu, K.: Learning to generate product reviews from attributes. In: ECAL, pp. 623–632 (2017)
3. Lu, Y., Dong, R., Smyth, B.: Why i like it: multi-task learning for recommendation and explanation. In: Proceedings of the 12th ACM Conference on Recommender Systems, pp. 4–12 (2018)
4. Ni, J., McAuley, J.: Personalized review generation by expanding phrases and attending on aspect-aware representations. In: ACL, pp. 706–711 (2018)

5. See, A., Liu, P.J., Manning, C.D.: Get to the point: summarization with pointer-generator networks. In: ACL, pp. 1073–1083 (2017)
6. Sharma, V., Sharma, H.V., Bishnu, A., Patel, L.: Cyclegen: cyclic consistency based product review generator from attributes. In: Proceedings of the 11th International Conference on Natural Language Generation, pp. 426–430 (2018)
7. Sun, P., Wu, L., Zhang, K., Fu, Y., Hong, R., Wang, M.: Dual learning for explainable recommendation: towards unifying user preference prediction and review generation. In: Proceedings of the Web Conference 2020, pp. 837–847 (2020)
8. Tang, J., Yang, Y., Carton, S., Zhang, M., Mei, Q.: Context-aware natural language generation with recurrent neural networks. arXiv preprint arXiv:1611.09900 (2016)

Definition-Augmented Jointly Training Framework for Intention Phrase Mining

Denghao Ma[1](✉), Yueguo Chen[2], Changyu Wang[3], Hongbin Pei[4], Yitao Zhai[1], Gang Zheng[1], and Qi Chen[1]

[1] Meituan, Beijing, China
madenghao@meituan.com
[2] DEKE Lab, Renmin University of China, Beijing, China
chenyueguo@ruc.edu.cn
[3] JD.com, Inc., Beijing, China
wangchangyu7@jd.com
[4] Xi'an Jiaotong University, Xian, China
peihongbin@xjtu.edu.cn

Abstract. We propose to mine intention phrases from large numbers of queries, for enabling rich query interpretation that identifies both query intentions and associated intention types. We formalize the notion of intention phrase as a sequence of keywords and an intention type, propose its three criteria (relevance, completeness, and clarity), and identify two key challenges. To handle the criterion modeling challenge, we design a jointly training framework with a sequence labeling model and a clarity classification model. To untangle the data sparsity challenge, we are the first to leverage definitions to learn the embeddings of words, discover a new data source (dictionary), and develop the definition-augmented encoder to generate good semantic representations for words and sentences. Our experiments over three large corpora (hotel, tourism, and product domains) verify the advantage of our model over baselines.

Keywords: Jointly training · Intention phrase mining · Definition

1 Introduction

With the rapid developments of e-commerce platforms, e.g., Amazon, Alibaba and Meituan, people have been used to searching specific products by some keywords with clear searching intentions, e.g., the user's queries in Fig. 1. In the queries, some phrases are vital for interpreting query intentions and intention types, and we name them as intention phrases. For example, the intention phrase "double room" can accurately represent the intention of the query "I want to book a double room" and can be used for recognizing the intention type of the query, i.e., room type. Besides, an intention knowledge graph of intention phrases, their intention typessynonyms, and hyponymy can bridge the semantic gap between users' queries and goods. For example in Fig. 1, different queries with the double

A. Bhattacharya et al. (Eds.): DASFAA 2022, LNCS 13247, pp. 331–339, 2022.
https://doi.org/10.1007/978-3-031-00129-1_28

room intention can be mapped to the standard intention phrase "double room", and thus the relevant goods can be retrieved. With such an intention knowledge graph, e-commerce platforms can rich query interpretation, accurately understand users' information needs, and then return satisfactory results to users. The intention knowledge graph can be widely applied to product service in various domains of manufacturing data.

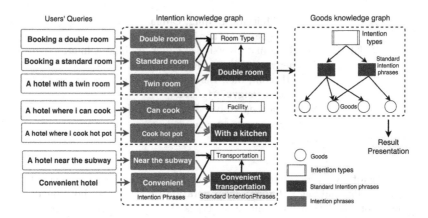

Fig. 1. Query interpretation with an intention knowledge graph.

Problem. To construct the intention knowledge graph, the first step is mining large numbers of intention phrases. Even intention phrases have been exploited in many e-commerce platforms, few studies have introduced the intention phrase mining task. So we define the task as $T = f(I, S, Q_t, Q_p)$: given intention types I of a domain, seed intention phrases S (each $i \in I$ has some seed intention phrases $S_i \in S$), training query corpus Q_t and testing query corpus Q_p, discover new intention phrases for each $i \in I$ from Q_p. The I, S and Q_t are used for constructing training samples; I and Q_p are used for testing the effectiveness of models. Since intention phrases are used for interpreting query intentions and intention types, we define an intention phrase as $t = <W, i>$ where W is a consecutive word sequence and i is the intention type that t belongs to.

Criteria. In a specific domain, an intention phrase t should only belong to one intention type i (*Relevance*). E.g., given the hotel domain, the intention phrase "double room" belongs to the intention type of room type. Secondly, an intention phrase should have correct segmentation boundary (*Completeness*). Next, an intention phrase should have a clear and specific meaning (*Clarity*), which can be easily understood by users. E.g., in a hotel domain, the meanings of "smoking free" and "allow smoking" are clarified, while that of "smoking" is not because it may refer to "smoking free" or "allow smoking". The three criteria provide us with evidence to identify intention phrases from queries.

Addressing the task, we encounter two-fold **challenges**:

- **Criterion modeling challenge.** How to discover intention phrases that satisfy the above three criteria? First, the boundary identification has been very challenging in entity recognition (NER) and phrase segmentation tasks. In our task, it still is challenging and needs to be solved. Second, the relevance of new intention phrases to an intention type is hard to be estimated, because of the semantic gap between them. Third, the clarity estimation is non-trivial because 1) with different domains, the clarity of a phrase may be different; 2) intention phrases are so short texts that they can not provide rich semantic.
- **Data sparsity challenge.** Data sparsity refers to that phrases or words appear rarely in a corpus and have very few contexts. The few contexts can not provide enough semantic to context-based encoders to learn good representations for the sparse words. E.g., BERT [1] takes some sparse words as Out-Of-Vocabulary (OOV) words, and can not learn good semantic representations for them.

To address the above challenges, we propose two **insights**:

Insight: Jointly Training. To address the first challenge, we find an intrinsic relation among the three criteria (insight): phrases with clarity are more likely to be relevant and complete; The relevant phrases with completeness are more likely to be clarified. Based on the insight, we design a jointly-training framework with a domain-specific clarity classification model and sequence labeling model, to simultaneously model the three criteria. In the classification model, for a phrase, we interpret its current query context as a pattern $<bs, as>$ where bs and as are the continuous word sequences before and after the phrase, and develop a pattern-based method to augment the semantic representation of the phrase.

Insight: Definition-Augmented Representation. To address the second challenge, our key idea is learning the definition-based representation from dictionaries, and combining the definition-based representation and the context-based representation to represent each word. Since the definition breaks away from context constraints, it can overcome the data sparsity. A definition is a credible and authoritative explanation, which can represent the semantic of a word. To get the definitions, we discover a type of authoritative and available expert dictionary, such as *Cambridge Dictionary* and *Xinhua Dictionary*. To realize our idea and get good semantic representations of words, phrases, and queries, we design a definition-augmented encoder over the expert dictionary and BERT [1].

Our contributions are presented: 1) We define the intention phrase mining task, discover its three criteria and identify its two key challenges; 2) To address the criterion modeling challenge, we develop the jointly training framework with the clarity classification model and sequence labeling model; 3) To address the data sparsity, we are the first to propose the definition-based representation, and develop the definition-augmented encoder; 4) Extensive experiments over three large corpora show our model significantly outperforms the baselines.

2   Solution

As described in Sect. 1, we develop a jointly-training framework to mine intention phrases. The framework contains two components: 1) the sequence labeling model is to capture the completeness and relevance; 2) the domain-specific clarity classification model is to capture the clarity of a phrase. For both models, one key component is how to effectively represent words and queries. To obtain the high-quality representations, we propose to incorporate definitions and contexts to learn the semantics of words, and develop the definition-augmented encoder.

2.1   Jointly-Training Framework

Training Data Generation. The format of training data is $<q, \hat{y}_q, t, \hat{v}_t>$ where q is a query, $\hat{y}_q$ is the intention type sequence, t is a phrase and $\hat{v}_t$ is the clarity label of t. As described in Sect. 1, the seed intention phrases S and training query corpus Q_t are used for generating the training data. Given $q \in Q_t$ and seed intention phrases S, we use the backward longest common sequence matching to generate the intention type sequence $\hat{y}_q$ for q. We randomly extract some texts T_q from q. If $t \in T_q$ is not in the seed set S, $\hat{v}_t$ of t is 0; otherwise, $\hat{v}_t$ is 1.

Domain-Specific Clarity Classification Model. As motivated in Sect. 1, the clarity estimation encounters two difficulties: 1) with different domains, the clarity of a phrase may be different; 2) intention phrases are very short texts which can not provide rich semantic. Given a specific domain sd and a phrase t, we adopt the classification strategy to estimate the probability $p(t|sd)$ that t is clarified in the domain sd, addressing the first difficulty. Based on the training examples T_t and their probabilities $p(t|sd)$, we design the loss function as follows:

$$L_{clarity} = -\frac{1}{|T_t|} \sum_{t \in T_t} \hat{v}_t \log(p(t|sd)) + (1 - \hat{v}_t) \log(1 - p(t|sd)) \qquad (1)$$

where $\hat{v}_t$ is the ground truth label of t. The $p(t|sd)$ is estimated as follows:

$$p(t|sd) = \phi_1(W_1[e(t) \oplus e(sd)] + b_1) \qquad (2)$$

where $e(t)$ and $e(sd)$ are the embeddings of t and sd. The $\oplus$ is the concatenating operation, and $[\cdot]$ is the matrix transposing operation. The $\phi_1(\cdot)$ is a three-layer MLP where the activation functions of the first two layers are Elu [2], and that of the last layer is Sigmoid [3]. W_1 is a weight matrix and b_1 is a bias matrix.

To address the second difficulty, we use the query context of t to improve the semantic representation of t. Firstly, we interpret the query context as a pattern $<bs, as>$ where bs and as are the continuous word sequences before and after the phrase t. Secondly, we use bs and as to improve the representation of t: $e(t) = e_1(bs) \oplus e_1(t) \oplus e_1(as)$ where $e_1(bs)$ and $e_1(as)$ are the embeddings of bs and as. The $e(sd)$, $e_1(bs)$, $e_1(as)$ and $e_1(t)$ are constructed in Sect. 2.2.

Sequence Labeling Model. Given a query $q = \{x_1, \cdots, x_i, x_n\}$, the semantic representations $[e(x_1), \cdots, e(x_i), e(x_n)]$ and a label sequence $y = \{y_1, \cdots, y_i, y_n\}$, the probability of generating y from q is estimated as follows:

$$p(y|q; \theta) = \frac{\exp(\sum_{i=1}^{n}(W^{y_i}e(x_i) + b^{y_{i-1}, y_i}))}{\sum_{y' \in y^*}\exp(\sum_i(W^{y_i'}e(x_i) + b^{y_{i-1}', y_i'}))} \qquad (3)$$

where y^* is all possible label sequences. The b^{y_{i-1}, y_i} is the transition weight between y_{i-1} and y_i, and W^{y_i} is the probability of generating the label y_i from the representation $e(x_i)$. Both $\theta = \{W, b\}$ are learnt during the training phase. We adopt the negative log-likelihood as the loss function:

$$L_{sl} = -\frac{1}{|Q_t|}\sum_{q \in Q_t}\log(p(\hat{y}_q|q; \theta)) \qquad (4)$$

where $\hat{y}_q$ is the ground truth intention type sequence of q. In this model, one key component is the semantic representations of words, i.e., $[e(x_1), \cdots, e(x_i), e(x_n)]$, which are constructed in Sect. 2.2.

Jointly Training. The final objective loss function is the combination of the sequence labeling loss and the clarity classification loss: $L = L_{clarity} + L_{sl}$ where L_{sl} and $L_{clarity}$ are defined in the Eq. 1 and 4, respectively. In the prediction phase, given a query q, our model firstly gets all intention phrase candidates C by extracting contiguous texts of arbitrary length from q. Secondly, we pair q and each candidate $c \in C$, and input each $<q,c>$ pair into our model. The model outputs an intention type sequence for q and the clarity score for c. According to the outputs, we can collect a set of candidate intention phrases. If c is in the set and its clarity score is greater than a threshold β, c is a true intention phrase.

2.2 Definition-Augmented Representation

To address the sparsity challenge, we are the first to learn the semantic representations from both the definition and context perspectives:

$$R(x_i) = DR(x_i) \oplus CR(x_i) \qquad (5)$$

where $CR(x_i)$ is the context-based representation and $DR(x_i)$ is the definition-based representation. We use BERTLSTM [1,6] to get the context-based representation of a word $x_i \in q$. For a sentence (e.g., a query or a sub-query), we use the embedding of $[CLS]$ in BERT as its context-based semantic representation.

Definition-Based Encoder. We discover a type of authoritative and available expert dictionaries, i.e., *Xinhua dictionary* and *Cambridge dictionary*. In the dictionaries, each word has one or multiple definitions, since the word may be polysemantic. The definitions of words are manually designed by experts so that they are authoritative. Based on the new data source, we design a three-layer model to generate the definition-based representations for words and queries. In the first layer, we collect word-definition pairs from the dictionaries. Taking each definition d as a document, we run BERT to generate the embedding representation for each definition, denoted as V_d. As well, the embedding representation of a query q in data corpus can be constructed in this way, denoted as V_q.

In the second layer, we leverage the definition embeddings to construct the initial definition-based representations. For a word x_i, it may have multiple definitions D_{x_i}. Given a query q including x_i, the definitions of x_i may have different relevance to q. Therefore, the definitions should have different importance to x_i under q. So the initial definition-based representation of x_i is derived as follows:

$$O_{x_i} = \sum_{d_{x_i} \in D_{x_i}} \frac{sim(V_q, V_{d_{x_i}})}{\sum_{d_1 \in D_{x_i}} sim(V_q, V_{d_1})} \cdot V_{d_{x_i}} \tag{6}$$

where V_* is the embedding of a definition or a query. The $sim(V_q, V_{d_{x_i}})$ is the relevance between q and the definition d_{x_i}, estimated as follows:

$$sim(V_q, V_{d_{x_i}}) = \phi_3(W_3[V_q \oplus V_{d_{x_i}} \oplus (V_q - V_{d_{x_i}})] + b_3) \tag{7}$$

where $\phi_3(\cdot)$ is a three-layer deep neural network where the activation functions of the first two layers are Elu [2], and that of the last layer is Sigmoid [3]. The W_3 is a weight matrix, b_3 is a bias matrix, and $-$ is the element-wise subtraction.

In the third layer, we incorporate the sequence dependency relations among words into the definition-based representations. To model the sequence dependency relations, we adopt the bidirectional LSTM to learn the dependency. Bi-LSTM takes the initial definition-based representations of words $([O_{x_1}, \cdots, O_{x_i}])$ as input, and outputs the forward and backward LSTM hidden states. The final definition-based representation is derived by concatenating the backward and forward hidden states: $DR(x_i) = \overrightarrow{LSTM}(x_i) \oplus \overleftarrow{LSTM}(x_i)$ For a sentence, we use the average embedding of words in the sentence as its definition-based semantic representation.

3   Experiments

1. **Comparison Solutions.** Since the task is a new research problem, we adopt state-of-the-art models of the entity mining task as our baselines: 1) LR-CNN [4], incorporating lexicons into CNN by using a rethinking mechanism; 2) LA Lattice LSTM [5], encoding lexicons by BI-LSTM model, and designing a scheme to preserve the lexicon matching results; 3) BERT+CRF, using BERT [1] to encode each word and putting the encoded representation into CRF; 4) BERT+SOFT, using softmax function to replace CRF; 5) BERT+LSTM+CRF, using BERT+LSTM [1,6] to encode each word, and using CRF to predict the label sequence of a sentence. We denote our proposed model as IPM.

2. **Dataset and Evaluation Metrics.** We run all models over three domains, i.e., hotel, tourism and product. The training sample construction has been

Table 1. Performance comparison over three corpora.

Corpus	Model	F1	Precision	Recall
Hotel	LR-CNN	0.2332	0.2199	0.2482
	LA Lattice LSTM	0.2790	0.3672	0.2249
	BERT+SOFT	0.2256	0.3754	0.1613
	BERT+CRF	0.2273	0.6083	0.1398
	BERT+LSTM+CRF	0.2616	0.5800	0.1689
	IPM	**0.3984**	**0.6149**	**0.2970**
Product	LR-CNN	0.2153	0.4412	0.1424
	LA Lattice LSTM	0.3792	0.4439	0.3309
	BERT+SOFT	0.2702	0.5039	0.1846
	BERT+CRF	0.2486	0.4796	0.1678
	BERT+LSTM+CRF	0.2908	0.5458	0.1982
	IPM	**0.5026**	**0.6865**	**0.3964**
Tourism	LR-CNN	0.3075	0.2848	0.3341
	LA Lattice LSTM	0.5144	0.5738	**0.4661**
	BERT+SOFT	0.3632	0.6138	0.2579
	BERT+CRF	0.3847	0.6160	0.2797
	BERT+LSTM+CRF	0.3656	0.7034	0.2470
	IPM	**0.5509**	**0.9388**	0.3898

introduced in Sect. 2.1. As the test set, we firstly perform all trained models over the test question corpus Q_p and collect their mined results into a pool. Three experts identify intention phrases from the pool. A phrase is labeled as a intention phrase, if two or three experts approve it; Otherwise, it is not. We denote intention phrases as T. Each model mines a set of intention phrase candidates C. Based on T and C, we use the precision, recall and F1-measure as metrics.

3. **Performance comparison.** We report the overall performance of our model and baselines on three individual test sets in Table 1. It can be seen that 1) our proposed IPM achieves the best F1 and precision metrics; 2) On the hotel and product domains, IPM still achieves the best recall metric. But in the tourism domain, the LA Lattice LSTM baseline achieves the best recall which verifies LA Lattice LSTM as the state-of-the-art model of entity mining task [5]. According to the above overall comparisons, we can conclude that the jointly-training model better accomplishes the intention phrase mining task than baseline models.

We investigate the effect of incorporating the clarity classification model on the three metrics. In Table 2, we compare the performance of the jointly training model IPM and the sequence labeling model IPM_{SL}. It can be seen that IPM significantly outperforms IPM_{SL} with respect to all metrics on the three test sets. The metric improvements illustrate the sequence labeling model and clarity classification model can benefit each other.

Table 2. The effect of using the clarity classification model.

Corpus	Model	F1	Precision	Recall
Hotel	IPM_{SL}	0.2824	0.3533	0.2352
	IPM	**0.3984**	**0.6149**	**0.2970**
Product	IPM_{SL}	0.3395	0.4813	0.2623
	IPM	**0.5026**	**0.6865**	**0.3964**
Tourism	IPM_{SL}	0.4289	0.7664	0.2978
	IPM	**0.5509**	**0.9388**	**0.3898**

Table 3. The effect of using definition-based representation.

Corpus	Model	F1	Precision	Recall
Hotel	IPM_{SL}	0.2824	0.3533	0.2352
	IPM_{SL}-DR	0.2616	0.5800	0.1689
	IPM_{SL}-DR+WR	0.2499	0.4036	0.1810
	IPM-DR	0.2721	0.4366	0.1976
	IPM-DR+WR	0.2273	0.2782	0.1922
	IPM	**0.3984**	**0.6149**	**0.2970**

We compare the performance of using the definition-based representation and alternative representations in Table 3. The IPM_{SL} denotes the sequence labeling model, -DR denotes that the definition-based representation is not applied, and +WR denotes that the word2vec-based representation is applied. A query WR representation is the average WR representation of all words in the query. In Table 3, IPM_{SL} performs better than IPM_{SL}-DR, and IPM performs better than IPM-DR. This phenomenon illustrates that using the definition-based representation can bring in performance improvements. Besides, it can be seen that IPM_{SL}-DR performs better than IPM_{SL}-DR+WR, and IPM-DR performs better than IPM-DR+WR. This phenomenon illustrates that the definition-based representation is more effective than the word2vec-based representation in our task. The performance comparisons among IPM and its variants verify the effectiveness of our proposed definition-based representation.

4 Conclusion

We propose the intention phrase mining task and its three criteria, i.e., relevance, completeness and clarity. To model the criteria, we develop a jointly training framework with a sequence labeling model (for completeness and relevance) and a clarity classification model (for clarity). By experiments, we find that the jointly training model is better than the single sequence labeling model. To untangle the data sparsity challenge, we propose the definition-based representation which can break away from the context constraint, and find that the combination of the definition-based representation and context-based representation is better than any single one. Dictionaries, e.g., *Cambridge Dictionary* and *Xinhua Dictionary*, are the high quality and available expert knowledge-base, which can provide rich semantic for sparse words and phrases.

Acknowledgements. This work was supported by National Key Research and Development Program of China (2020YFB1710004), and NSFC grant No. U1711261.

References

1. Devlin, J., Chang, M.W., Lee, K., Toutanova, K.: Pre-training of deep bidirectional transformers for language understanding. NAACL-HLT 4171–4186 (2019). https://doi.org/10.18653/v1/n19-1423
2. Clevert, D.A., Unterthiner, T., Hochreiter, S.: Fast and accurate deep network learning by exponential linear units (ELUs). In: ICLR (2016)
3. Sharma, S.K., Chandra, P.: An adaptive sigmoidal activation function cascading neural networks. In: Corchado, E., Snášel, V., Sedano, J., Hassanien, A.E., Calvo, J.L., Ślezak, D. (eds.) Soft Computing Models in Industrial and Environmental Applications, 6th International Conference SOCO 2011. AISC, vol. 87, pp. 105–116 . Springer, Berlin, Heidelberg (2011). https://doi.org/10.1007/978-3-642-19644-7_12
4. Gui, T., Ma, R., Zhang, Q., Zhao, L., Jiang, Y.G., Huang, X.: CNN-based Chinese NER with lexicon rethinking. In: IJCAI, pp. 4982–4988 (2019). https://doi.org/10.24963/ijcai.2019/692
5. Ma, R., Peng, M., Zhang, Q., Huang, X.: Simplify the usage of lexicon in Chinese NER. In: ACL, pp. 5951–5960 (2020). https://doi.org/10.18653/v1/2020.acl-main.528
6. Hochreiter, S., Schmidhuber, J.: Long short-term memory. Neural Comput. **9**(8), 1735–1780 (1997). https://doi.org/10.1162/neco.1997.9.8.1735

Modeling Uncertainty in Neural Relation Extraction

Yu Hong[1], Yanghua Xiao[1,2(✉)], Wei Wang[1], and Yunwen Chen[3]

[1] School of Computer Science, Fudan University, Shanghai, China
{yhong17,shawyh,weiwang1}@fudan.edu.cn
[2] Fudan-Aishu Cognitive Intelligence Joint Research Center, Shanghai, China
[3] DataGrand Inc., Shanghai, China
chenyunwen@datagrand.com

Abstract. Previous work on neural relation extraction mainly focus on point-based methods and ignores uncertainty within the bag, thus making poor predictions when there are insufficient instances of the bag. To solve this problem, in this paper, we propose two density-based methods. Specifically, we assume each bag is a Gaussian distribution and sentences in the bag are drawn from it. We use predicted variance, capturing bag's uncertainty, as well as predicted mean to draw more samples to enrich one-instance bags. We also use predicted variance to vote for good representation and temper the loss. To the best of our knowledge, this is the first paper to model uncertainty in neural relation extraction. Experiment results on NYT-10 show significant improvements over baselines.

Keywords: Neural relation extraction · Uncertainty · Robustness

1 Introduction

Encoding a wealth of structured semantic information, Knowledge Graph (KG) is now widely used to assist various tasks in Natural Language Processing (NLP). Despite hundreds of thousands of nodes and edges, KGs are still far from complete compared with facts in real-world corpus. Relation Extraction (RE), a crucial task in NLP, is proposed to enrich KGs by finding unknown facts from unstructured texts. However, many neural RE methods heavily rely on large amounts of labeled data, which comes at a big cost in both labor and time.

To solve this problem, Distant Supervision [1] (DS) emerges to automatically label large-scaled training data by aligning texts with facts. Specifically, sentences with the same entity pair are all assumed to express the relation of these two entities in KG. Though fast and easy, this strategy obviously introduces noises, since sentences with the same entity pair may semantically vary a lot. To address this problem, recent work turns DS-trained RE into multi-instance learning [2,3], where each piece of the dataset is a bag, composed of multiple sentences with the same entity pair.

However, in real-world corpus, DS-generated data may contain many bags with only *one* instance, which does not perfectly fit the setting of *multi-instance*

A. Bhattacharya et al. (Eds.): DASFAA 2022, LNCS 13247, pp. 340–348, 2022.
https://doi.org/10.1007/978-3-031-00129-1_29

Table 1. Statistics of NYT-10 dataset.

	# Bags	# One-instance bags	One-instance rate
Training set	292,200	244,800	83.78%
Test set	96,867	74,857	77.28%
	# Positive bags	# Positive one-instance bags	Positive one-instance rate
Training set	19,423	9,381	48.30%
Test set	1,950	1,010	51.79%

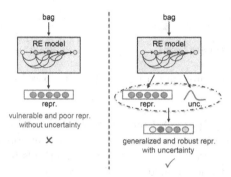

Fig. 1. Point-based method (left) vs. density-based method (right) in neural RE.

learning. Table 1 shows the statistics of NYT-10 [2], a popular DS dataset generated by aligning New York Times corpus with Freebase[1]. In Table 1, one-instance bags in both training and test set significantly overwhelm multi-instance ones. When only considering positive bags (i.e. entity pairs with actual relations), one-instance bags account for around half of the whole bags. This phenomenon makes most bags' representation come from the only instance, which is vulnerable and poor compared with that of the bags with multiple instances.

Many methods have been proposed to tackle noise, but few of them tackles one-instance bags. One important reason is that in conventional neural RE methods, we often make *point-based* estimation about the bag and assume it is always correct, as shown on the left of Fig. 1. When there are insufficient instances about the bag, the representation tends to be vulnerable and poor, since it contains very large *uncertainty* against the golden representation. Due to the nature of point estimation, uncertainty is totally ignored, leading to less robust and generalized predictions which badly hurt the performance.

To tackle this problem, in this paper, we propose to *model uncertainty of the bag* and make *density-based estimations* to learn generalized and robust representations for one-instance bags. A simple illustration is shown on the right of Fig. 1. Specifically, we assume each bag as a Gaussian distribution and sentences in the bag are drawn from it. Instead of learning the parameters in each bag, we make RE model directly output predicted mean and variance. Predicted

---

[1] https://developers.google.com/freebase.

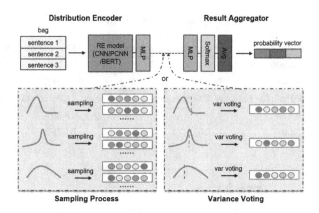

Fig. 2. Architecture of proposed density-based methods.

mean is taken as bag's representation (i.e. repr. on the right of Fig. 1), while predicted variance naturally captures uncertainty (i.e. unc. on the right of Fig. 1). To further leverage uncertainty, we propose two methods to make density-based estimations. The first one (named Sampling Process) is to draw more samples from the estimated distribution, and the second one (named Variance Voting) leverages uncertainty to stretch representations. Intuitively, Sampling Process enriches one-instance bags and Variance Voting votes for robust representations, thus significantly improves performance on one-instance bags.

2   Proposed Methods

There are four modules in our methods, namely *Distribution Encoder, Sampling Process, Variance Voting* and *Result Aggregator*, as shown in the architecture of Fig. 2. In this section, we give details about these four modules one by one.

2.1   Distribution Encoder

We first model the distribution of the bag as a Gaussian distribution and predict its mean and variance through Distribution Encoder. To do so, we first adopt CNN [4], PCNN [5] and BERT [6] to extract features from the sentence to get entity pair's representation c. Then we use a Multi-Layer Perceptron (MLP) to transform c into predicted mean and variance by splitting the output of MLP:

$$\mu = W_2 p + b_2 \tag{1}$$

$$\sigma^2 = \text{Softplus}(W_3 q + b_3) \tag{2}$$

where

$$[p; q] = \text{Softplus}(W_1 c + b_1) \tag{3}$$

In Eq. 1, 2 and 3, $W_1, b_1, W_2, b_2, W_3, b_3$ are parameters, μ and σ^2 are predicted mean and variance, $\mu, \sigma^2 \in \mathbb{R}^t$, t is the dimension of bag's representation. Softplus [7] is the activation function which can also make sure σ^2 is always positive during training and test process.

2.2   Sampling Process

After obtaining predicted mean and variance, bag g is represented by a Gaussian distribution with mean μ and variance σ^2, i.e. $g \sim \mathcal{N}(\mu, \sigma^2)$. To draw samples from bag's distribution, we make

$$f = \mu + \epsilon\sigma \tag{4}$$

where $\epsilon \sim \mathcal{N}(0, 1)$, $f \in \mathbb{R}^t$. By doing Eq. 4 n times, we get samples $\{f_1, \ldots, f_n\}$ for bag g. These samples can be regarded as diverse representations of the bag, which are generalized and robustified by the square root of uncertainty σ^2. This can be very useful since we explicitly leverage uncertainty in bag's representation.

2.3   Variance Voting

Previous method may be expensive since we have to temporarily store intermediate samples. To avoid this limitation, we propose Variance Voting. Recall that σ^2 captures uncertainty about bag's representation, so if our model is not confident about its prediction, σ^2 will be large and $1/\sigma^2 \to 0$. Therefore, $1/\sigma^2$ can be used to adjust features and *vote* for robust and generalized representations. Specifically, bag's representation f adjusted by its uncertainty σ^2 is given by:

$$f = \mu \odot (1/\sigma^2) \tag{5}$$

Here μ and σ^2 are both vectors rather than scalars, since we allow each feature of the representation to possess different uncertainties. Intuitively, features with low uncertainty can be enlarged while others are suppressed. On the contrary, conventional attention mechanism equally scales all features, hence is not robust and generalized compared with Variance Voting.

2.4   Result Aggregator

Finally, we use Result Aggregator to get probability vectors. We adopt another MLP and Softmax to transform representation f into probability vector r, i.e.

$$r = \text{Softmax}(v), \quad \text{where} \quad v = W_4 f + b_4 \tag{6}$$

W_4, b_4 in Eq. 6 are parameters, $r \in \mathbb{R}^k$ where k is the number of relations.

For Sampling Process, we get mn probability vectors by drawing n samples for m sentences in the bag. Bag's probability vector z is defined as the average over its samples:

$$z = \frac{1}{m} \sum_{i=1}^{m} \frac{1}{n} \sum_{j=1}^{n} r_{ij} \tag{7}$$

For Variance Voting, bag's probability vector z is defined as the average over probability vectors of its voted representations:

$$z = \frac{1}{m} \sum_{i=1}^{m} r_i \tag{8}$$

2.5   Loss Functions

For Sampling Process, we define the loss as cross-entropy between bag's label y and bag's probability vector z:

$$L_{SAM}(\theta) = -\log P(y|z,\theta) \tag{9}$$

where θ represents parameters in our methods.

For Variance Voting, voted representation in Eq. 5 can be considered as a Gaussian distribution g_{vote} with mean f and variance $\mathbf{0}$, i.e. $g_{vote} \sim \mathcal{N}(f, \mathbf{0})$. We calculate KL divergence between estimated distribution g and g_{vote}:

$$\mathrm{KL}(g||g_{vote}) = \mathrm{KL}(\mathcal{N}(\mu,\sigma^2)||\mathcal{N}(f,\mathbf{0})) \propto \frac{(\mu - f)^2}{2\sigma^2} + \frac{1}{2}\log(\sigma^2) \tag{10}$$

We minimize Eq. 10 by regressing bag's golden representation f with predicted mean μ and adjusting its weight with uncertainty σ^2. During optimization, Eq. 10 makes RE model learn to *assign uncertainty* for bags and automatically attenuate their effects, thus are more robust for one-instance bags. Total loss of Variance Voting is defined as the sum of KL divergence and cross-entropy:

$$L_{VAR}(\theta) = \mathrm{KL}(g||g_{vote}) - \log P(y|z,\theta) \tag{11}$$

3   Experiments and Analysis

In this section, we introduce dataset and baselines then evaluate performance on one-instance bags and show the effectiveness of proposed methods.

3.1   Dataset

We conduct experiments on NYT-10, a popular RE benchmark which is widely-used in [5,8–11]. It is a large-scale dataset with 522,611 sentences, 292,200 entity pairs and 172,448 sentences, 96,867 entity pairs in training and test sets. There are 53 relations in total, including NA (i.e. no relation).

3.2   Baselines

We denote proposed Sampling Process as SAM and Variance Voting as VAR. We select 5 strong methods to be compared with:

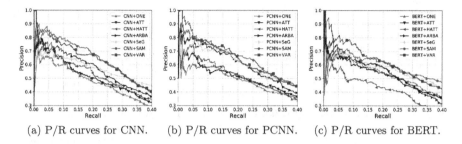

(a) P/R curves for CNN.   (b) P/R curves for PCNN.   (c) P/R curves for BERT.

Fig. 3. P/R curves on three RE models for one-instance bags in NYT-10.

Table 2. Probabilities on relation `location/location/contains` of two one-instance-bag examples. Values before the arrow are estimated by μ in Eq. 1 (i.e. w/o VAR) and values after the arrow are estimated by f in Eq. 5 (i.e. w/VAR).

(a) A simple bag example.

bag1	... campus of *hendrix_college* in central *arkansas* ...
CNN+ONE	0.9241
+ATT	0.9312
+HATT	0.9969
+ARBA	**0.9985**
+SeG	0.9453
+VAR	0.8862 → 0.9797

(b) A hard bag example.

bag2	...; *kenai, alaska*; ...
CNN+ONE	0.6282
+ATT	0.9267
+HATT	0.9742
+ARBA	0.9050
+SeG	0.9660
+VAR	0.8482 → **0.9809**

- ONE[2] [5] is a bag-level RE method which selects the representation of the sentence with maximum probability as the representation of the bag.
- ATT (See footnote 2) [8] uses selective attention over sentences to form bag's representation.
- HATT[3] [9] incorporates hierarchical attention over sentences to compose bag's representation.
- ATT_RA+BAG_ATT[4] [10] considers intra- and inter-bag attention at both sentence and bag level. We denote this method as ARBA.
- Selective Gate[5] [11] leverages gate mechanism to adjust the contribution of global and local features of sentences to form bag's representation. It is the current state-of-the-art method for one-instance bags and is denoted as SeG.

---

[2] https://github.com/thunlp/NRE.
[3] https://github.com/thunlp/HNRE.
[4] https://github.com/ZhixiuYe/Intra-Bag-and-Inter-Bag-Attentions.
[5] https://github.com/tmliang/SeG.

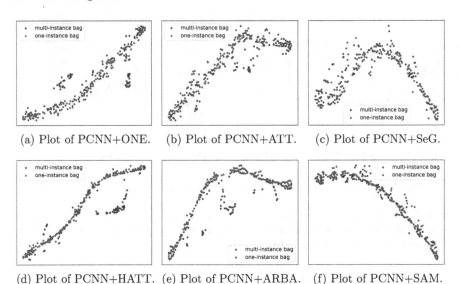

(a) Plot of PCNN+ONE.   (b) Plot of PCNN+ATT.   (c) Plot of PCNN+SeG.

(d) Plot of PCNN+HATT. (e) Plot of PCNN+ARBA.   (f) Plot of PCNN+SAM.

Fig. 4. Scatter plots for bags labeled with `people/person/nationality`.

3.3   Evaluation on One-Instance Bags

In this section, we evaluate the performance of baselines and proposed methods on one-instance bags. Figure 3 shows P/R curves for different methods applied on CNN, PCNN and BERT on NYT-10. In Fig. 3(a), SAM and VAR gain significant improvements on both low and high recalls, which shows great benefits of making density-based estimations. When applied on PCNN, in Fig. 3(b), VAR achieves the highest precision on low recalls and SAM gets the best performance on high recalls. In Fig. 3(c), SAM seems unstable at the beginning, but with the increase of recall, precision becomes steady. VAR is the best on both low and high recalls.

3.4   Effectiveness of Variance Voting

To evaluate the effectiveness of VAR, we select two one-instance bags of relation `location/location/contains` and show predictive probabilities of VAR and baselines in Table 2. In Table 2(a), there is rich context to identify the relation, so the probabilities of VAR and baselines are all quite high. In Table 2(b), there is nearly no context indicating the relation, but VAR still results in a very high probability due to introducing uncertainty into prediction. It improves the probability from 0.8482 to 0.9809, which is much higher than SeG.

3.5   Effectiveness of Sampling Process

To evaluate the effectiveness of SAM, we select all bags labeled with `people/per-son/nationality` to show their probability vectors with t-SNE [12] in Fig. 4. There are 84 multi-instance bags and 191 one-instance bags, in which one-instance bags are more than twice of multi-instance ones. In Fig. 4 (f), SAM makes one-instance bags distribute densely and comparable with multi-instance ones. While in Fig. 4 (a), (b), (d) and (e), one-instance bags and multi-instance bags are not mixed uniformly and bags in Fig. 4 (c) (i.e. SeG) are too scattered, indicating unknown modes and poor representations for one-instance bags.

4 Related Work

In neural RE, [4] first proposes a CNN architecture for sentence encoding, then [5] proposes a PCNN architecture and adopts at-least-one assumption to avoid noise. To leverage more information, [8] adopts setence-level attention over sentences, and [9] extends attention with hierarchy to learn coarse-to-fine representations. [10] makes selective attention not only in sentence level but also in bag level. [11] proposes a pooling-equipped gate mechanism to adjust global dependency on the sentences. However, these methods are all based on point-based estimations without taking uncertainty into consideration.

5 Conclusion

In this paper, we propose two density-based methods, Sampling Process and Variance Voting, to tackle one-instance bags in neural RE. By modeling the bag as a Gaussian distribution, we capture the uncertainty with predicted variance, which is leveraged as well as predicted mean to get robust and generalized bag representations. Experiment results show significant improvements over baselines, indicating the effectiveness of modeling uncertainty in neural RE.

Acknowledgements. This work is supported by National Key Research and Development Project (No.2020AAA0109302), Shanghai Science and Technology Innovation Action Plan (No.19511120400) and Shanghai Municipal Science and Technology Major Project (No.2021SHZDZX0103).

References

1. Mintz, M., Bills, S., Snow, R., Jurafsky, D.: Distant supervision for relation extraction without labeled data. In: ACL (2009)
2. Riedel, S., Yao, L., McCallum, A.: Modeling relations and their mentions without labeled text. In: ECML-KDD (2010)
3. Surdeanu, M., Tibshirani, J., Nallapati, R., Manning, C.: Multi-instance multi-label learning for relation extraction. In: EMNLP (2012)
4. Zeng, D., Liu, K., Lai, S., Zhou, G., Zhao, J.: Relation classification via convolutional deep neural network. In: COLING (2014)
5. Zeng, D., Liu, K., Chen, Y., Zhao, J.: Distant supervision for relation extraction via piecewise convolutional neural networks. In: EMNLP (2015)
6. Soares, L., FitzGerald, N., Ling, J., Kwiatkowski, T.: Matching the blanks: distributional similarity for relation learning. In: ACL (2019)

7. Glorot, X., Bordes, A., Bengio, Y.: Deep sparse rectifier neural networks. In: AIS-TATS (2011)
8. Lin, Y., Shen, S., Liu, Z., Luan, H., Sun, M.: Neural relation extraction with selective attention over instances. In: ACL (2016)
9. Han, X., Yu, P., Liu, Z., Sun, M., Li, P.: Hierarchical relation extraction with coarse-to-fine grained attention. In: EMNLP (2018)
10. Ye, Z., Ling, Z.: Distant supervision relation extraction with intra-bag and inter-bag attentions. In: ACL (2019)
11. Li, Y., et al.: Self-attention enhanced selective gate with entity-aware embedding for distantly supervised relation extraction. In: AAAI (2020)
12. Hinton, G.: Visualizing high-dimensional data using t-SNE. J. Mach. Learn. Res. **9**, 2579–2605 (2008)

Industry Papers

A Joint Framework for Explainable Recommendation with Knowledge Reasoning and Graph Representation

Luhao Zhang[1($\boxtimes$)], Ruiyu Fang[1], Tianchi Yang[1,2], Maodi Hu[1], Tao Li[1], Chuan Shi[2], and Dong Wang[1]

[1] Meituan, Beijing, China
{zhangluhao,fangruiyu,humaodi,litao19,wangdong07}@meituan.com
[2] Beijing University of Posts and Telecommunications, Beijing, China
{yangtianchi,shichuan}@bupt.edu.cn

Abstract. With the development of recommendation systems (RSs), researchers are no longer only satisfied with the recommendation results, but also put forward requirements for the recommendation reasons, which helps improve user experience and discover system defects. Recently, some methods develop knowledge graph reasoning via reinforcement learning for explainable recommendation. Different from traditional RSs, these methods generate corresponding paths reasoned from KG to achieve explicit explainability while providing recommended items. But they suffer from a limitation of the fixed representations that are pre-trained on the KG, which leads to a gap between KG representation and explainable recommendation. To tackle this issue, we propose a joint framework for explainable recommendation with knowledge reasoning and graph representation. A sub-graph is constructed from the paths generated through knowledge reasoning and utilized to optimize the KG representations. In this way, knowledge reasoning and graph representation are optimized jointly and form a positive regulation system. Besides, due to more than one candidate in the step of knowledge reasoning, an attention mechanism is also employed to capture the preference. Extensive experiments are conducted on public real-world datasets to show the superior performance of the proposed method. Moreover, the results of the online A/B test on the large-scale Meituan Waimai (MTWM) KG consistently show our method brings benefits to the industry.

Keywords: Recommendation systems · Knowledge graph · Explainability · Graph representation

1 Introduction

With the rapid development of the Internet, recommender systems (RSs) play a pivotal role in various online services, which have been a basic service of e-commerce platforms. It aims to provide personalized recommendation sticking

© The Author(s), under exclusive license to Springer Nature Switzerland AG 2022
A. Bhattacharya et al. (Eds.): DASFAA 2022, LNCS 13247, pp. 351–363, 2022.
https://doi.org/10.1007/978-3-031-00129-1_30

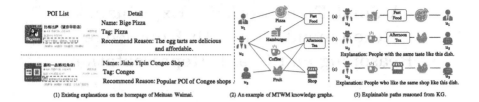

Fig. 1. Existing ways of explainability on the homepage of MTWM App, example of MTWM KGs and semantic paths implicit in the KG.

[5] and help users to make proper choices from the huge amount of products. However, with the development of RSs, not only is knowing what users' need in their mind important but also explaining why the recommended items meet their need is significant [18,21]. Meanwhile, users are often curious about why those products are shown to them. Similar to offline shopping, RSs should provide reasons when recommending products, which will improve the recognition and acceptance of users. Therefore, the interpretability of RSs is of great significance for consumer decision-making and experience [8].

Many efforts have been made to develop an explainable RS in many e-commerce Apps [20]. For example, Fig. 1 (1) shows some different recommended reasons (framed by the red dotted line) for different POIs[1] to interpret the corresponding recommended item on the homepage of Meituan Waimai[2] (MTWM) App. They are either mined from user-generated content (e.g., reviews) or based on statistics. Despite the positive performance, both of these methods provide explainability in a post-hoc way and are irrelevant to the process of RS. An ideal explainable RS should offer interpretation for the decision making procedure simultaneously. Some recent researches have noticed that using KGs as auxiliary information makes RSs have the strong capability of interpretation which is endowed by the multi-hop connections between users and items [2,18,21]. Based on these characteristics, many KG-based recommendation systems have been designed for explainability [2]. For example, path-based methods [15] explore to model the high-order connectivity in KG by graph neural network framework. The explainable paths can be generated through visualizing the attention score calculated among users, recommended items, and other entities in the KG, which relies on manual implementation.

To conduct explicit reasoning over KGs during making decisions automatically, researchers also propose methods based on KG reasoning [10,12,16]. In this way, not only are the candidate products recommended but also the explainable paths between users and items are simultaneously inferred. Specifically, researchers define the problem as a deterministic Markov Decision Process (MDP) over the KG and solve the problem with a Reinforcement Learning (RL) framework [18]. Starting from a given user, the agent aims to find a trajectory

---

[1] POI, Point of Interest, a specific store or restaurant in the MTWM App.

[2] Meituan Waimai, a local business service platform, https://waimai.meituan.com.

that is a connected path from the user for explaining why a potential item is selected. As shown in Fig. 1 (2), an example of WTWM knowledge graphs, when recommending coffee to the user u_3, there could be a semantic path expected to be reasoned through walking on the KG (i.e., path (c) in the Fig. 1 (3), which means people who like the same shop like this dish). Further, [21] introduces some weakly supervised information, such as labeled meta-paths [6], for speeding up the training and imporving the performance.

However, existing methods suffer from the static representation of KGs. First, the used representation can not be updated with the knowledge reasoning, hence resulting in the loss of some explainability-aware information. Second, many methods for learning representations rely on hand designed meta-paths or sampling strategies, which may also lose some valuable structures or paths, especially when the KG scale increases. Both of them lead to a gap between KG representation and explainable recommendation and restricts the reasoning capability.

To bridge the gap, it is assumed that the KG representations can provide guidance for more accurate path reasoning. Simultaneously, the reasoned paths contribute to learning a more reasonable KG representation. For instance, the shown paths in Fig. 1 (3) can benefit to learning the representations of related entities and relations. In turn, if the information of these paths is captured, path reasoning may be more efficient. To this end, we propose a joINt Framework for Explainable Recommendation with knowledge reasoning and representation (INFER). It constructs a sub-graph from the reasoned paths and inputs into a GNN layer for learning KG embeddings fine-tunely. Based on our assumption, the optimized KG representations are then utilized in turn to guide the generation of more meaningful paths. Finally, a joint learning framework is designed to make the two parts enhance each other. In this way, the two modules form a positive regulation system, which optimizes the path finding policy and the KG representation jointly. Additionally, it is observed that there is more than one candidate to select in every step (e.g., both *Afternoon Tea* and *Shop* are candidates of *Coffee*). To select the next step efficiently, our model also employs an attention to learn a probability distribution with preference in the process of reasoning explainable paths. The contributions are summarized as follows:

- We propose a framework for explainable recommendation, which builds the knowledge reasoning and GNN jointly as a positive regulation system.
- A sub-graph is constructed from the reasoned paths and inputs into a GNN layer for learning KG embeddings fine-tunely. Besides, an attention mechanism is also utilized to learn a probability distribution with preference when selecting candidates in the process of reasoning paths.
- Extensive experiments on three benchmark datasets demonstrate the superior performance of the proposed method. Moreover, the results on real industrial scene consistently gain a performance improvement by training the model on MTWM KG.

2   Preliminaries

In this section, we first formalize the problem of explainable recommendation based on the knowledge graph. Then, some related concepts about Markov Decision Process (MDP) for knowledge graph reasoning are introduced.

2.1   Problem Definition

A typical knowledge graph consists of large-scale entity-relation triplets, denoted as $\mathcal{G} = \{(e_{head}, r, e_{tail}) | e_{head}, e_{tail} \in \mathcal{E}, r \in \mathcal{R}\}$, where $\mathcal{E}$ is the entity set, $\mathcal{R}$ is the relation set, e_{head} and e_{tail} represent head entity and tail entity respectively. Each triplet (e_{head}, r, e_{tail}) indicates that there exists a relation r from e_h to e_t, such as (*Titanic, directed_by, James Francis Cameron*). Specially, in the scenario of recommendation, there exist user entities $\mathcal{U} \subset \mathcal{E}$ and item entities $\mathcal{V} \subset \mathcal{E}$ in the knowledge graph $\mathcal{G}$. Different user-item interactions are also viewed as different relations in the knowledge graph, such as *purchase* and *click*. Given the predefined knowledge graph $\mathcal{G}$ and a specific user $u \in \mathcal{U}$, our task is to recommend n items to him/her, denoted as $\mathcal{V}_u = \{v_{u,i} | i = 1, \cdots, n\}$, each of which is associated with an explainable k-hop path $p_i(u, v_{u,i})$. For path $p_i(u, v_{u,i}) = \{u \overset{r_1}{\longleftrightarrow} e_1 \overset{r_2}{\longleftrightarrow} e_2 \cdots \overset{r_k}{\longleftrightarrow} v_{u,i}\}$, $v_{u,i}$ is the recommended items of the user and $e_i \overset{r}{\longleftrightarrow} e_{i+1}$ represents either $e_i \overset{r}{\longleftarrow} e_{i+1}$ or $e_i \overset{r}{\longrightarrow} e_{i+1}$.

2.2   Markov Decision Process for Knowledge Reasoning

Following previous work [11,18], the knowledge graph reasoning can be formulated as a partially observable Markov Decision Process (MDP). Formally, the MDP is usually defined as a quadruplet $(\mathcal{S}, \mathcal{A}, \tau, \gamma)$, where $\mathcal{S}$ denotes the state space, $\mathcal{A}$ is the candidate action space, τ is the state transition probability function, and γ refers to the reward function of the environment. At each time step, the process is some state and the decision maker may choose any action in the space $\mathcal{A}$. The process responds at the next time step by moving into a new state and giving the decision maker a corresponding reward.

State. The state $s_t \in \mathcal{S}$ at step t is defined as a triplet (u, h_t, e_t), where $u \in \mathcal{U}$ is the given user entity, h_t is the historical state, consisting of all entities and relations in the k-step history, i.e., $\{e_{t-k}, r_{t-k+1}, \cdots, e_{t-1}, r_t\}$, and e_t is the currently selected entity based on the historical state.

Action. For state s_t at each time t, the actor selects an action $a_t = (r_{t+1}, e_{t+1})$ from the action space $\mathcal{A}_t$, which is the set of all outgoing edges of the current entity e_t in the KG. Therefore, the size of the action space is affected by the out-degree. Unfortunately, as the dimensionality of the state space increases, the difficulty to solve the MDP grows explosively. Thus, following [18], a score function is used to score all possible actions, and top-k of them are selected as candidates and combined into a new space $\hat{\mathcal{A}}_t$.

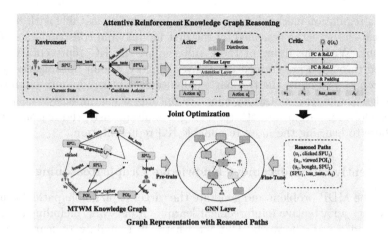

Fig. 2. Illustration of our proposed framework for explainable recommendation on MTWM with knowledge reasoning and graph representation on the MTWM KG.

Transition. The transition function τ is to generate the transition probability distribution of the next state. Given the current state s_t and the candidate action a_t, the next state s_{t+1} is transitioned with the probability $\tau(s_{t+1}|s_t, a_t)$.

Reward. When reached the target entity e_d, the agent will infer a k-hop path $p(u, e_d) = \{u \xleftrightarrow{r_1} e_1 \xleftrightarrow{r_2} \cdots \xleftrightarrow{r_k} e_d\}$ starting from the user u. Naturally, the agent is expected to generate as many high-quality paths as possible and satisfactory recommendation results. Additionally, there should be different rewards for paths of different quality. Based on [18,21], a reward function is designed to evaluate the reasoned path only in the final step T due to lack of supervision in the middle steps. Consequently, the reward function is defined as:

$$R_T = \gamma(p) = \begin{cases} 1, & \text{if } e_d \in \mathcal{V}_u, \\ \sigma(f(u, r_{purchase}, e_d)), & \text{if } e_d \in \mathcal{V} \backslash \mathcal{V}_u, \\ 0, & \text{others,} \end{cases} \quad (1)$$

where $\sigma(\cdot)$ refers to Sigmoid to measure the similarity between the started user and finally reached entity e_d, and f is a score function that is explained later.

3   The Proposed Model

In this section, we will detail our proposed reinforcement knowledge graph reasoning with graph representation for explainable recommendation.

As Fig. 2 shows, INFER is designed for mainly two parts: 1) attentive reinforcement knowledge graph reasoning and 2) graph representation learning with reasoned paths. The former designs an attention to learn a probability distribution with preference when selecting the next action from candidates. The latter employs a regulation mechanism for graph representation learning with reasoned paths. It uses the paths reasoned by the first module and constructs a series of sub-graphs. With the input of these sub-graphs, the KG representations are optimized through a pre-trained GNN layer. Finally, a joint optimization strategy is proposed to make the two parts enhance each other. As our assumption, knowledge graph representations that are optimized continuously can provide guidance for more accurate path reasoning. Simultaneously, the reasoned paths contribute to learning the more reasonable KG representation.

3.1 Attentive Reinforcement Knowledge Graph Reasoning

To solve the MDP problem and generate the target explainable paths for users, we introduce an attentive reinforcement learning framework, including an actor-network and a critic-network. The actor-network is mainly to learn a policy for selecting a next action from the action space. The critic-network aims to effectively maximize the rewards from all external environments. Next, these two sub-networks are introduced in detail.

Actor Network. As mentioned before, in step t, the dimensionality of the action space $\mathcal{A}_t$ may be wide variances due to the out-degrees of the entities. Though some pruned strategies are designed before selecting an action [18], there is still no capability to capture the preference of the current state, when making the decision of the next step. Therefore, we design an attention for selecting preferential action.

First, in the step t, for encoding the current input and previous contexts, the representation $\mathbf{s}_t$ of the current state is constructed by the representations of the started user $\mathbf{u}$, history states $\mathbf{h}_t$, current relation $\mathbf{r}_t$ and the current entity $\mathbf{e}_t$. Following [18], the concatenation operation is applied, formed as,

$$\mathbf{s}_t = \mathbf{u} \oplus \mathbf{h}_t \oplus \mathbf{r}_t \oplus \mathbf{e}_t, \tag{2}$$

$$\mathbf{h}_t = \mathbf{s}_{t-K_h} \oplus \cdots \oplus \mathbf{s}_{t-1}, \tag{3}$$

where K_h is the pre-defined history state length, and $\oplus$ represents the concatenation operator. Specially, $\mathbf{h}_0 = \mathbf{u}$.

Then, we design an attention mechanism to model the interaction between current state and remaining candidate actions after pruning (seeing Sect. 2.2). The state embedding $\mathbf{s}_t$ is treated as the attention query over the candidate actions. Formally, for each state s_t, given the pruned action space $\hat{\mathcal{A}}_t = \{a_t^1, a_t^2, \cdots, a_t^N\}(|\hat{\mathcal{A}}_t| = N)$, the attention vector $\boldsymbol{\alpha}$ is calculated as following,

$$\mathbf{s}_t' = \mathrm{ReLU}(\mathbf{s}_t \mathbf{W}_s), \tag{4}$$

$$\mathbf{a}_t^i = \mathbf{r}_{t+1}^i \oplus \mathbf{e}_{t+1}^i, \tag{5}$$

$$\alpha_i = \frac{\exp\left(\mathbf{s}_t' \mathbf{W}_a \mathbf{a}_t^i\right)}{\sum_{i=1}^{|\hat{\mathcal{A}}_t|} \exp\left(\mathbf{s}_t' \mathbf{W}_a \mathbf{a}_t^i\right)}, \tag{6}$$

where $\mathbf{W}_s$ and $\mathbf{W}_a$ are the parameter matrices, and ReLU is the activation function. $\mathbf{a}_t^i$ is the representation of action $a_t^i \in \hat{\mathcal{A}}_t$, computed by concatenate the embeddings of relation $r_{t+1}^i \in a_t^i$ and entity $e_{t+1}^i \in a_t^i$. The learned attention values are viewed as a probability distribution with preference of remaining actions, and indicate the preference of the next step.

Then, the policy network $\pi(\cdot)$ calculates the probability of each candidate action $a_t \in \hat{\mathcal{A}}_t$ and selects an action based on the distribution at each state. In detail, it takes as input the attention values, which are defined as

$$\pi(a_t|s_t, \hat{\mathcal{A}}_t) = \text{Softmax}(\hat{\mathbf{A}}_t \odot \boldsymbol{\alpha}), \tag{7}$$

where $\hat{\mathbf{A}}_t$ is a masked binarized vector of the action space $\hat{\mathcal{A}}_t$ to avoid overfitting and $\odot$ refers to point-wise product.

Critic Network. The critic network aims to feedback to the actor a temporal difference value for measuring the contribution of the last action that is selected by the actor. If the value is large, it means that the currently selected action is unstable and needs more appearances to ensure convergence. In other words, the critic network is to evaluate the quality of the path reasoning process. Specifically, given the current state s_t , the critic network $Q(\cdot)$ maps its representation to a value with a linear layer

$$Q(s_t) = \text{ReLU}(\mathbf{s}_t'\mathbf{W}_\rho), \tag{8}$$

where $\mathbf{W}_\rho$ is the transformation vector to be learned. Then, the temporal difference method is adopted to optimize the critic network.

Optimization. For maximizing the reward, the policy gradient is defined as

$$\nabla_\Theta \mathbb{L}_A = \mathbb{E}_\pi[\nabla_\Theta \log \pi_\Theta(\mathbf{s}_t', \hat{\mathcal{A}}_t)Q(s_t)]. \tag{9}$$

The critic network is learned through the Temporal Difference method [21], and updated by minimizing the square error as follows,

$$\mathbb{L}_Q = (Q(s_t) - \delta_t)^2, \tag{10}$$

where the target δ_t is calculated as $\delta_t = R_T + \gamma Q(s_{t+1})$ with the decay factor γ.

3.2 Graph Representation Learning with Reasoned Paths

Although the above actor-critic network is already able to reason the path from the user to the item, the representations of the knowledge graph are not updated continuously. With the great success of graph neural network on graph representation learning [3,17], to fine-tune the representation adaptively, we propose a two-step graph convolution network as the graph encoder based on the paths that are inferred by the actor-critic network.

As shown in the Fig. 2, the GNN layer is pre-trained on the whole KG firstly, following [9]. Then, as reinforcement KG reasoning, given the explainable path set, we construct a sub-graph $\mathcal{G}_{sub}(\mathcal{E}_{sub}, \mathcal{V}_{sub})$, $\mathcal{E}_{sub} \subset \mathcal{E}$ and $\mathcal{V}_{sub} \subset \mathcal{V}$ with those paths. Taking the sub-graph as input, the relational graph convolution network is employed to optimize the representation of entities and relations. Formally, the encoder exploits iterative aggregation of neighborhood information to learn the embedding of the source entity,

$$\mathbf{z}_i^{l+1} = \sigma(\sum\nolimits_{r \in \mathcal{R}} \sum\nolimits_{j \in \mathcal{N}_i^r} \frac{1}{|\mathcal{N}_i^r|} \mathbf{W}_r^l \mathbf{z}_j^l + \mathbf{W}_0^l \mathbf{z}_i^l), \tag{11}$$

where $\mathcal{N}_i^r$ denotes the set of neighboring nodes of entity e_i under relation $r \in \mathcal{R}$, $\mathbf{W}_r^l$ and $\mathbf{W}_0^l$ are the corresponding layer-wise weight matrices of the neighbors and itself. Note that as multiple layers are stacked, information from high-order neighbors are integrated, which could be viewed as distantly supervised information when inferring the next-hop entity of the path. The final representation of entity e_i is obtained from the last layer, denoted as $\mathbf{e}_i = \mathbf{z}_i^L$, where L is the number of the graph convolution layers. Then the relation embedding $\mathbf{r}$ of relation r is learnt through the following score function [19],

$$f(e_{head}, r, e_{tail}) = \mathbf{e}_{head}^T \cdot \mathrm{diag}\{\mathbf{r}\} \cdot \mathbf{e}_{tail}, \tag{12}$$

where $\mathbf{e}_{head}$ and $\mathbf{e}_{tail}$ are embeddings of head and tail entities respectively. This function is also used in the reward for measuring the similarity in Eq. 1.

3.3   Joint Learning

Aiming to make the two parts of INFER can enhance each other, we design a joint optimization strategy. First, link prediction is used as the target of the GNN module for optimization in the form of cross entropy,

$$\hat{y}_i = \mathrm{Sigmoid}(f(e_{head}, r, e_{tail})), \tag{13}$$

$$\mathbb{L}_G = -\frac{1}{N} \sum\nolimits_{i=1}^{N} [y_i \log \hat{y}_i + (1 - y_i) \log(1 - \hat{y}_i)], \tag{14}$$

where $\hat{y}_i$ is the predicted probability of whether there exists a relation $r \in \mathcal{R}_{sub}$ between entity $e_h, e_t \in \mathcal{E}_{sub}$, and y_i is the corresponding ground truth.

Note that the GNN module is fine-tuned with the currently reasoned paths by the RL module. Therefore, to prevent the RL module from being limited to these local structures, the link prediction loss is developed as an adversarial term of the reinforcement learning framework, formalized as

$$R_T' = R_T - \eta \mathbb{L}_G, \tag{15}$$

where η is a hyper parameter. Therefore, the original reward R_T is replaced with R_T' when calculating the policy gradient and optimizing the actor-critic network.

In summary, the two parts of INFER can fully leverage the effective information of each other. Specifically, more accurate representations of the KG conduct more reliable path reasoning and generate more valuable paths. In turn, the high-quality paths are exploited to fine-tune and make KG representation more faithful.

Table 1. Statistics of all datasets

Dataset	# Users	# Items	# Interactions	# Entities	# Triplets
Beauty	22,363	12,101	198,502	224,074	3,916,360
Cell Phones	27,879	10,429	194,439	163,249	5,335,545
Clothing	39,387	23,033	278,677	425,528	3,149,749
MTWM	10,000,001	2,683,977	37,295,253	20,414,412	279,119,356

Table 2. Parameter settings.

Parameter	Embedding dimension K_w	State history length K_h	Max path length K_p	Max action number K_a	Graph aggregation layer L	Negative sampling for GNN K_{neg}	Learning rate for RL α_r	Learning rate for GNN α_g	Weight η
Amazon	128	1	3	250	2	5	0.0001	0.001	0.3
MTWM	64	2	4	250	1	10	0.1	0.1	0.2

4   Experiments

In this section, we first show the advance compared with baselines on benchmarks through overall evaluation and ablation study. Further, the results of the online A/B test show INFER brings benefits to the industry.

4.1   Experimental Settings

Datasets. We evaluate the proposed model based on three public datasets from Amazon e-commerce [18,21], and an industry MTWM dataset for online A/B test. All datasets are summarized in Table 1 and detailed in the following.

Amazon E-commerce. This dataset contains three categories: Beauty, Cell Phones, and Clothing. Each of them is considered as a benchmark that consists of six types of entities, including users, products, brands, reviews and categories of products. Besides, there are total eight types of relations in each dataset.

MTWM. This dataset is collected from MTWM mobile application online. The train data is built from the records and attributes of both user clicks and purchases of food. There are six types of entities, including user, POI (Point of Interest, a specific store or restaurant), SPU (Standard Product Unit, such as specific food ordered by user), the category of and labels (including the taste C and the ingredient A) of an SPU. Meanwhile, there are total eight types of relations among these entities.

Detailed Implementation. For INFER, parameters are mainly following [18] in terms of Amazon E-commerce dataset, and apply them to MTWM dataset after some adjustments. Similarly, an action pruning strategy is employed to keep the size of the action space not greater than K_a and the maximum length of path reasoning is K_p. The size of negative sampling is K_{neg} for each entity. We also utilize the dropout of rate 0.5 for alleviating overfitting. The settings of hyper parameters are detailed in Table 2.

Table 3. Comparison of recommendation results on Amazon e-commerce datasets. The results are reported in percentage.

Datasets	Beauty			Cell phones			Clothing		
	P	R	HR	P	R	HR	P	R	HR
DKN	1.030	2.489	8.600	0.465	3.187	4.484	0.106	0.727	1.012
RippletNet	1.133	5.251	9.224	0.688	3.858	5.727	0.201	1.112	1.885
PGPR	1.736	8.448	14.642	1.274	8.416	11.904	0.723	4.827	7.023
ADAC-M	1.824	8.809	14.932	1.308	8.691	12.078	0.745	4.916	7.214
ADAC	1.991	9.424	16.036	1.358	8.943	12.537	0.783	5.152	7.502
INFER	**2.020**	**9.456**	**16.10**	**1.403**	**8.994**	**12.779**	**0.821**	**5.429**	**7.903**

Evaluation Metrics. Following previous studies [18,21], we evaluate the proposed method on Amazon E-commerce datasets with three representative metrics, including Precision (**P**), Recall (**R**), and Hit Ratio (**HR**). All metrics are computed based on top-10 recommendations for every user in the test set.

For MTWM dataset, we report three online indicators concerned by industry: (1) **RPM**, Revenue PerMille, (2) **UVCTR**, Click Through Rate (CTR) of unique visitor, calculated as $\frac{\#click\ users}{\#view\ users}$, (3) **UVCVR**, Click Conversion Rate (CVR) of unique visitor, calculated as $\frac{\#order\ users}{\#click\ users}$.

Baselines. To evaluate the performance on the Amazon E-commerce, we compare the model with the following baselines: **DKN** [14] is a deep knowledge-aware network in news recommendations. **RippleNet** [13] is an end-to-end framework that naturally incorporates the knowledge graph into recommender systems. **PGPR** [18] is a policy-guided path reasoning model that exposes the corresponding reasoning procedure for explainability. **ADAC** [21] is the SOTA method for the explainable recommendation that designs an adversarial actor-critic model for the demonstration-guided (e.g., labeled meta-paths) path finding. To make it more convincing, we also compare INFER with the variant ADAC-M that removes the supervision of pre-defined meta-paths.

In terms of the industry MTWM dataset, an online A/B test is conducted in the recommendation system of MTWM App.

4.2   Experimental Results

Overall Performance on Benchmarks. Table 3 shows the comparison results in terms of the recommendation accuracy. First, it is observed that explainable methods outperform KG-based methods (DKN and RippletNet), indicating they can guarantee interpretability and recommendation performance simultaneously. Additionally, one can observe that in all cases, compared with the most competitive baselines PGPR and ADAC-M on all datasets, INFER shows the best performance. Note that even though ADAC integrates extra supervised demonstration to guide the reasoning, which achieves the most competitive performance,

Table 4. Ablation studies of INFER compared with baselines on Amazon Beauty.

Variants	Precision (%)	Recall (%)	NDCG (%)	HR (%)
PGPR	1.736	8.448	5.511	14.642
INFER$_{transe}$	1.918	9.204	5.943	15.788
INFER$_{w/o\ ATT}$	1.974	9.304	6.017	15.942
INFER	**2.020**	**9.456**	**6.177**	**16.10**

INFER still shows the best performance. It demonstrates that even without the extra supervision, INFER can provide more accurate recommendation results. Due to the static mentioned above, there is a gap between KG representation and explainable recommendation in previous methods, which limits the performance. To bridge the gap, we model the GNN and explainable recommendation jointly in a manner of positive regulation adjustment.

Ablation Studies. In order to verify the effectiveness of the designed modules of INFER, we choose the representative method PGPR [18] as baseline and design two variant models. **INFER$_{w/o\ ATT}$** replaces the attention mechanism with an MLP layer like [18] and only takes the embedding of the current state as input. **INFER$_{transe}$** replaces the GNN layer with a TransE [1] model and still keeps embeddings being optimized while reasoning paths, compared with PGPR that exploits fixed representations pre-trained by Trans-E.

As shown in Table 4, after replacing the GNN layer with a TransE model, the performance of INFER$_{transe}$ decreases obviously, demonstrating that our GNN layer learns more effective representations. However, it still outperforms PGPR, showing that the joint framework can greatly benefit the performance even based on the traditional TransE model. Without the designed attention mechanism, the performance of INFER$_{w/o\ ATT}$ is also reduced a bit, indicating that the necessity of modeling the reasoning process with preference. In summary, all results prove our hypothesis that KG representation can guide more accurate path reasoning. In turn, the reasoned paths contribute to learning the KG representation.

Online A/B Test on MTWM Dataset. To further verify the effectiveness of INFER, we apply it in the recommendation system of MTWM App and focus on the task of SPU recommendation for users. Specifically, the compared baseline is a multi-channel recall model (BASE), including an XGBoost-based model [4], a DSSM-based model [7] and so on. The model INFER is trained on the MTWM dataset and processed as a new channel and added into the BASE for evaluating the online performance. Meanwhile, due to tens of millions of users involved in the offline training and online testing of the model, the open source graph database Nebula[3] is used to build the online graph reasoning service. It is able to host super large-scale graphs with billions of nodes and trillions of edges.

---

[3] https://nebula-graph.io/.

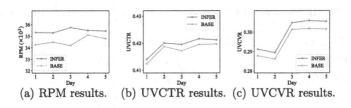

(a) RPM results.   (b) UVCTR results. (c) UVCVR results.

Fig. 3. Online A/B test on the MTWM App. We report the results on 3 indicators for 5 consecutive days, all of which are positive.

As shown in Fig. 3, we report three key indicators observed from our online scene. Compared with the baseline through online testing for five consecutive days, the INFER continues to maintain positive online returns. In detail, experimental results are improved by an average of 2.5%, 0.7%, 1.2% for RMP, UVCTR, UVCVR respectively. The results demonstrate that the proposed INFER is able to improve user satisfaction and performance in the online recommender system significantly.

5   Conclusions

In this paper, we propose a novel attentive reinforcement knowledge graph reasoning method with graph representation for explainable recommendation. First, an attention mechanism is employed to compute a personalized probability distribution when inferring explainable paths from the current state to candidate actions. Then, the reasoned paths are utilized to construct a new sub-graph and optimize the KG embedding in a fine-tuned way. We highlight that the two parts can enhance each other and form a positive regulation system, meaning that knowledge graph representations can guide more accurate policy for reasoning path, and the reasoned paths contribute to learning a more reasonable and robust KG representation in turn. We conduct extensive experiments on public real-world datasets to verify the superior performance of the proposed method. Moreover, the results on large-scale MTWM KG consistently show our method gains benefits to the industry.

Acknowledgements. This research was supported by Meituan and in part by the National Natural Science Foundation of China (No. U20B2045, 62172052, 61772082, 62002029).

References

1. Bordes, A., Usunier, N., Garcia-Duran, A., Weston, J., Yakhnenko, O.: Translating embeddings for modeling multi-relational data. In: NIPS, pp. 1–9 (2013)
2. Cao, Y., Wang, X., He, X., Hu, Z., Chua, T.S.: Unifying knowledge graph learning and recommendation: towards a better understanding of user preferences. In: WWW, pp. 151–161 (2019)

3. Chen, J., Ma, T., Xiao, C.: Fastgcn: fast learning with graph convolutional networks via importance sampling. In: ICLR (2018)
4. Chen, T., Guestrin, C.: Xgboost: a scalable tree boosting system. In: KDD (2016)
5. Fan, S., et al.: Metapath-guided heterogeneous graph neural network for intent recommendation. In: KDD (2019)
6. Hu, B., Shi, C., Zhao, W.X., Yu, P.S.: Leveraging meta-path based context for top-n recommendation with a neural co-attention model. In: KDD (2018)
7. Huang, P.S., He, X., Gao, J., Deng, L., Acero, A., Heck, L.: Learning deep structured semantic models for web search using clickthrough data. In: KDD (2013)
8. Ma, W., et al.: Jointly learning explainable rules for recommendation with knowledge graph. In: WWW, pp. 1210–1221 (2019)
9. Schlichtkrull, M.S., Kipf, T.N., Bloem, P., van den Berg, R., Titov, I., Welling, M.: Modeling relational data with graph convolutional networks. In: ESWC (2018)
10. Song, W., Duan, Z., Yang, Z., Zhu, H., Zhang, M., Tang, J.: Explainable knowledge graph-based recommendation via deep reinforcement learning. arXiv (2019)
11. Wan, G., Du, B., Pan, S., Haffari, G.: Reinforcement learning based meta-path discovery in large-scale heterogeneous information networks. In: AAAI (2020)
12. Wan, G., Pan, S., Gong, C., Zhou, C., Haffari, G.: Reasoning like human: hierarchical reinforcement learning for knowledge graph reasoning. In: IJCAI (2020)
13. Wang, H., et al.: Ripplenet: propagating user preferences on the knowledge graph for recommender systems. In: CIKM, pp. 417–426 (2018)
14. Wang, H., Zhang, F., Xie, X., Guo, M.: DKN: deep knowledge-aware network for news recommendation. In: WWW, pp. 1835–1844 (2018)
15. Wang, X., He, X., Cao, Y., Liu, M., Chua, T.: KGAT: knowledge graph attention network for recommendation. In: KDD, pp. 950–958 (2019)
16. Wang, X., Wang, D., Xu, C., He, X., Cao, Y., Chua, T.S.: Explainable reasoning over knowledge graphs for recommendation. In: AAAI, pp. 5329–5336 (2019)
17. Wang, X., et al.: Heterogeneous graph attention network. In: WWW, pp. 2022–2032 (2019)
18. Xian, Y., Fu, Z., Muthukrishnan, S., de Melo, G., Zhang, Y.: Reinforcement knowledge graph reasoning for explainable recommendation. In: SIGIR (2019)
19. Yang, B., Yih, W., He, X., Gao, J., Deng, L.: Embedding entities and relations for learning and inference in knowledge bases. In: ICLR (2015)
20. Yang, Y., et al.: Query-aware tip generation for vertical search. In: CIKM (2020)
21. Zhao, K., et al.: Leveraging demonstrations for reinforcement recommendation reasoning over knowledge graphs. In: SIGIR, pp. 239–248 (2020)

XDM: Improving Sequential Deep Matching with Unclicked User Behaviors for Recommender System

Fuyu Lv[1(✉)], Mengxue Li[1], Tonglei Guo[1], Changlong Yu[2], Fei Sun[1], Taiwei Jin[1], and Wilfred Ng[2]

[1] Alibaba Group, Hangzhou, China
{fuyu.lfy,lydia.lmx,tonglei.gtl,ofey.sf,taiwei.jtw}@alibaba-inc.com
[2] The Hong Kong University of Science and Technology, Hong Kong, China
{cyuaq,wilfred}@cse.ust.hk

Abstract. Deep learning-based sequential recommender systems have recently attracted increasing attention from both academia and industry. Most of industrial Embedding-Based Retrieval (EBR) systems for recommendation share the similar ideas with sequential recommenders. Among them, how to comprehensively capture sequential user interest is a fundamental problem. However, most existing sequential recommendation models take as input clicked or purchased behavior sequences from user-item interactions. This leads to incomprehensive user representation and suboptimal model performance, since they ignore the complete user behavior exposure data, *i.e.,* items impressed yet unclicked by users. In this work, we attempt to incorporate and model those unclicked item sequences using a new learning approach in order to explore better sequential recommendation technique. An efficient triplet metric learning algorithm is proposed to appropriately learn the representation of unclicked items. Our method can be simply integrated with existing sequential recommendation models by a confidence fusion network and further gain better user representation. The offline experimental results based on real-world E-commerce data demonstrate the effectiveness and verify the importance of unclicked items in sequential recommendation. Moreover we deploy our new model (named XDM) into EBR of recommender system at Taobao, outperforming the previous deployed generation SDM.

Keywords: User behavior modeling · Sequential recommendation · Metric learning · Embedding-based retrieval

1 Introduction

In order to reduce information overload and satisfy customers' diverse online service needs (*e.g.,* E-commerce, music, and movies), personalized recommender systems (RS) have become increasingly important. Traditional recommendation algorithms (collaborative filtering [11] and content-based filtering [10]) only model users' long-term preference, while ignore dynamic interest in users' behavior sequences. Hence sequential recommendation (SR) is introduced to model

© The Author(s), under exclusive license to Springer Nature Switzerland AG 2022
A. Bhattacharya et al. (Eds.): DASFAA 2022, LNCS 13247, pp. 364–376, 2022.
https://doi.org/10.1007/978-3-031-00129-1_31

sequential user behaviors in history to generate user representation by considering time dependency of user-item interactions.

Moreover, SR sheds light on the rapid development of embedding-based retrieval (EBR) system (*a.k.a* deep matching or deep candidate generation) for recommendation in industry (*e.g.,* YouTubeDNN [1], SDM [9], and MIND [6]). The key to EBR system is understanding the evolution of users' preference. However, those models as well as most existing SR models (*e.g.,* GRU4REC [4], NARM [7], and Caser [12]) only take as input sequential clicked or purchased behaviors for user modeling. They pay little attention to model more abundant exposure data in users' complete behavior sequences, *i.e.,* those items that were impressed to users yet not clicked (refer to *unclicked items* in this paper). They are of less interest to users, which influence users' future behaviors and can bring better understandings about users' preference. Those items also contain valuable signal on users' dynamic preference, which can complement the clicked data. Modeling users' preference ignoring the unclicked behavior sequences leads to incomprehensive user representation and limits the capacity and performance of SR.

In this work, we aim to integrate the valuable unclicked item sequences with clicked ones as complete user behaviors into SR models' input to enhance performances of sequential deep matching. Though it is novel in SR, prior works [2,13,15] explore for the general recommendations. Compared with SR, they focus on quite different tasks (*i.e.,* matrix factorization [2], reinforcement recommender [15] or click-through rate prediction [13]) and specially show different settings such as task definition, training/test sample construction and evaluation. Besides, their modelings of unclicked sequences remain at the feature level without complex interactions with clicked ones. It is believable that clicked and unclicked behaviors affect each other. Naturally we start to think about effectively incorporating them together from the model level. Firstly, we derive two important characteristics observed from real-life cases. **1)** As introduced, it is obvious that unclicked items reflect users' dislikes to some extent compared with clicked ones as shown in Fig. 1(A). **2)** On the other hand, this kind of items are not those that users particularly dislike compared with a random recommended item. The skipped unclicked items can be seen as an *intermediate feedback* between clicked and random recommended items. Because a modern RS recommends items in which users are probably interested by personalized algorithms. Users choose to skip items possibly due to many other complex factors, such as price of items displayed nearby, seasonal nature of items or hot consumer trends. Illustrated in Fig. 1(B), all of these items impressed to users at least partly conform to users' preference, but the user only select a few of them to click. Those unclicked items obviously are not random items.

Based on these observations, we propose a new metric learning algorithm to learn the "intermediate" representations of unclicked item sequences in SR. Specifically, we first project sequential clicked and unclicked behaviors as well as labels into the vector space by deep neural networks (*e.g.,* LSTM, self-attention, and MLP), where Euclidean distance is used as metric measurement. The labels

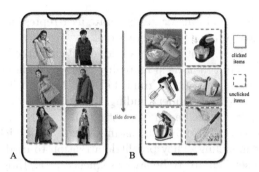

Fig. 1. Previous sequential recommendations only consider clicked sequences while unclicked sequences are also informative. For example, (A) Users choose women's wear rather than men's wear. (B) Users selectively clicked the same type of products.

are next clicked items after the current user sequence, which represent the true vector of user interest. We consider triplet relations among those different vectors: **1)** clicked and unclicked item vectors, and **2)** clicked and label item vectors. The key idea is to regularize the model by enforcing that the representation of clicked sequence should be far away from the unclicked one. Meanwhile the accompanying direction of regularization is applied to clicked and label item representations, which pushes the correct optimization of clicked representation towards the label vector. Moreover, the properties of *intermediate feedback* of unclicked items are ensured by adding a predefined margin, which controls the maximum distance between clicked and unclicked vectors. The clicked and unclicked vectors are combined by a confidence fusion network, which dynamically learns the fusion weight of unclicked items, to get the final user representation.

The offline experimental results based on two real-world E-commerce datasets demonstrate the effectiveness. Further experiments have been conducted to understand the importance of unclicked items in the sequential recommendation. We successfully deploy our new model, named XDM, into EBR of recommender system at Taobao, replacing the previous generation SDM [9]. Online experiments demonstrate that XDM leads to improved engagement metrics over SDM. The main contributions of this paper are summarized below:

- We identify the importance of unclicked items in SR and integrate them into models for complete sequential user behavior modeling.
- We propose XDM based on triplet metric learning and a confidence fusion network to model users' unclicked together with clicked item sequences. It dynamically controls relationships between different representations to achieve accurate recommendation.
- We demonstrate the effectiveness of XDM on real-world E-commerce data for this topic, which would shed light on more research of incorporating unclicked item sequences. Our model has also been successfully deployed on the production environment of recommender system at Taobao.

2   Our Approach

2.1   Problem Formulation

Let $\mathcal{U} = \{u_1, \ldots, u_m\}$ denote the set of users, and $\mathcal{I} = \{i_1, \ldots, i_n\}$ denote the set of items. Our task focuses on implicit recommender systems. For a user $u \in \mathcal{U}$, we record the user's clicking interactions in the ascending order with time t and get the clicked sequence, namely $\mathcal{S}_u^+ = \{i_1^+, \ldots, i_t^+, \ldots, i_{n_p}^+\}$. The unclicked sequence (items impressed to u yet without clicking interactions) is formed by the same way, namely $\mathcal{S}_u^- = \{i_1^-, \ldots, i_t^-, \ldots, i_{n_n}^-\}$. The two sequences make up the complete sequential user behaviors $\mathcal{S}_u = \mathcal{S}_u^+ \cup \mathcal{S}_u^-$. In fact, clicked and unclicked items appear alternately in the same user sequence. We partition them into individual sequences to simplify problem definition in our work.

Given $\mathcal{S}_{u,t}$, we would like to predict the items set $\mathcal{I}_{u,t}^{pre} \subset \mathcal{I}$ that the user will interact after t. In the process of modeling, all types of user behaviors are encoded into vectors of the same dimension L_e. Following [9], we take next k clicked items after $\mathcal{S}_{u,t}$ as target items (labels) denoted as $\mathcal{C}_{u,t} = \{c_1, \ldots, c_k\}$. In practice, due to the strict requirement of latency, industrial recommender systems usually consist of two stages, matching and ranking. The matching, also called EBR if embedding techniques used, corresponds to retrieving Top-k candidates. Our paper mainly focuses on improving the effectiveness in EBR.

2.2   Base Sequential Recommendation

Given $\mathcal{S}_{u,t}^+ = \{i_1^+, i_2^+, \ldots, i_t^+\}$, a deep sequential recommender computes the user representation vector $\boldsymbol{h}_{u,t} \in \mathbb{R}^{L_e}$ as:

$$\boldsymbol{h}_{u,t} = \text{DSR}(\mathcal{S}_{u,t}^+, \boldsymbol{e_u}; \Theta) \tag{1}$$

where $\boldsymbol{e}_u \in \mathbb{R}^{L_e}$ is the user profile (gender, sex, etc.) embedding. DSR means **D**eep **S**equential **R**ecommenders for short. Θ denotes all the model parameters. Each item $i \in \mathcal{S}_{u,t}^+$ is mapped into an *item embedding* vector $\boldsymbol{q}_i \in \mathbb{R}^{L_e}$.

To generate sequential recommendations for user u at time t, we rank a candidate item i by computing the recommendation score $\hat{y}_{u,i,t}$ according to:

$$\hat{y}_{u,i,t} = g(u, i, t) = \boldsymbol{h}_{u,t}^{\text{T}} \cdot \boldsymbol{q}_i \tag{2}$$

where $g(\cdot)$ is the score function, implemented as the inner product between $\boldsymbol{h}_{u,t}$ and $\boldsymbol{q}_i$. After obtaining scores of all items, we can select Top-k items for recommendation. As the item candidates of industrial recommender systems are from a very large corpus, the online process of scoring all items is generally replaced with fast K nearest neighbors (KNN) algorithm.

2.3   Sequential Recommendation with Unclicked User Behaviors

The base DSR only take as input $\mathcal{S}_{u,t}^+$ and recommends items according to $\boldsymbol{h}_{u,t}$. They ignore the influence of $\mathcal{S}_{u,t}^-$. We propose to model $\mathcal{S}_{u,t}^- = \{i_1^-, i_2^-, \ldots, i_t^-\}$ as a plug-in module on basis of DSR. Here we choose *SDM* [9] as the base DSR due to its capacity of handling with large-scale data for efficient deployed industry applications.

Metric Learning for Unclicked Items. Compared with clicked ones, unclicked items reflect users' dislikes to some extent, but they are not those users particularly dislike compared to a random recommended item. Because a modern RS recommends items which at least partly conform to users' preference by personalized algorithms. Thus unclicked items can be intuitively treated as the *intermediate feedback* between clicked and random recommended items. Therefore, for user u's $\mathcal{S}_{u,t}^-$, it should have an intermediate representation of vector $\boldsymbol{n}_{u,t}$ between $\mathcal{S}_{u,t}^+$ and random recommended items.

To solve this problem, we introduce metric learning to control the representation of $\mathcal{S}_{u,t}^-$. Specifically, the first step is to project the sequence $\mathcal{S}_{u,t}^-$ into vector space. We encode item $i \in \mathcal{S}_{u,t}^-$ denoted as $\boldsymbol{q}_i$, which is the same as $i \in \mathcal{S}_{u,t}^+$. On account of huge volume of unclicked items, we simply average all the $\boldsymbol{q}_i$ in $\mathcal{S}_{u,t}^-$ and then use feed-forward network to generate the embedding of unclicked items $\boldsymbol{n}_{u,t} \in \mathbb{R}^{L_e}$, described as:

$$\boldsymbol{n}_{u,t} = f\left(\frac{1}{|\mathcal{S}_{u,t}^-|} \sum_{i=1}^{|\mathcal{S}_{u,t}^-|} \boldsymbol{q}_i\right) \tag{3}$$

where $f(\cdot)$ represents non-linear function implemented by feed-forward network with tanh activation. More complex neural structures *e.g.*, Transformer, remain for future work and are not the major points in this paper.

Given a user u, now we have $\boldsymbol{h}_{u,t}$, $\boldsymbol{n}_{u,t}$, and label representation $\boldsymbol{c}_{u,t}$. Here $\boldsymbol{c}_{u,t}$ generated from $\mathcal{C}_{u,t}$ is embedded in the same way of $\boldsymbol{n}_{u,t}$. Then we use triplet metric learning to construct triple structures among $\boldsymbol{h}_{u,t}$, $\boldsymbol{n}_{u,t}$, and $\boldsymbol{c}_{u,t}$. The optimization goal is to make $\boldsymbol{h}_{u,t}$ and $\boldsymbol{c}_{u,t}$ closer while to make $\boldsymbol{n}_{u,t}$ and $\boldsymbol{h}_{u,t}$ far away from each other. The overall triplet optimization is to minimize:

$$\mathcal{L}_{tri} = \sum_{u \in \mathcal{U}} \left[\|\boldsymbol{h}_{u,t} - \boldsymbol{c}_{u,t}\|_2^2 - \|\boldsymbol{h}_{u,t} - \boldsymbol{n}_{u,t}\|_2^2 + m\right]_+ \tag{4}$$

where $\|\boldsymbol{x}\|_2^2 = \sum_{i=1}^n x_i^2$ denotes the squared l_2 norm to measure the distance between vectors and the operator $[\cdot]_+ = \max(0, \cdot)$ denotes the hinge function. $m > 0$ is the relaxing parameter constraining the maximum margin distance.

We use an example from two-dimensional space to explain the intuition shown in Fig. 2. The triplet loss penalizes the shorter edge e_{hn}, so that difference between $\boldsymbol{h}_{u,t}$ and $\boldsymbol{n}_{u,t}$ are significantly large. While it will reward the shorter edge e_{hc} to make $\boldsymbol{h}_{u,t}$ more similar to $\boldsymbol{c}_{u,t}$. By introducing margin m, we control the maximum difference between e_{hn} and e_{hc} by enforcing $e_{hc} + m \leq e_{hn}$. It keeps the *intermediate feedback* property of unclicked items between clicked and random recommended items. The introduction of hinge function is to avoid the further correction of those "qualified" triplets.

However, we find that current optimization may lead to undesirable situations, as shown in the Fig. 2. The movement of $\boldsymbol{n}_{u,t}$ to $\boldsymbol{n}_{u,t}'$ meets the optimization in

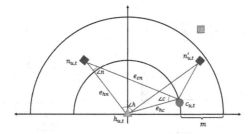

Fig. 2. Triplet structure. Red hollow squares represent embedding that have met the constraints and no longer need to be optimized. (Color figure online)

Eq. 4, but $n'_{u,t}$ is closer to the $c_{u,t}$, which leads to weak distinction between clicked and unclicked item representations. In order to eliminate the effect, we derive a symmetrical triplet constraint by increasing e_{hn} and e_{cn} at the same time, *i.e.*, adding constraint term $e_{hc} + m' \leq e_{cn}$. Symmetrical constraints are incorporated into Eq. 4 and the new optimization objective is defined as:

$$
\begin{aligned}
\mathcal{L}_{tri} &= \sum_{u \in \mathcal{U}} \left[\|h_{u,t} - c_{u,t}\|_2^2 - \|h_{u,t} - n_{u,t}\|_2^2 + m \right]_+ \\
&+ \sum_{u \in \mathcal{U}} \left[\|h_{u,t} - c_{u,t}\|_2^2 - \|c_{u,t} - n_{u,t}\|_2^2 + m' \right]_+ \\
&= \sum_{u \in \mathcal{U}} \left[2\|h_{u,t} - c_{u,t}\|_2^2 - \|h_{u,t} - n_{u,t}\|_2^2 - \|c_{u,t} - n_{u,t}\|_2^2 + m^* \right]_+
\end{aligned}
\tag{5}
$$

Here we use m^* to represent the addition of two margins in symmetrical losses.

Fusion Network. To make better use of unclicked sequences, we attempt to explicitly combine $n_{u,t}$ with base DSR. We first come up a simple method which directly adopts the difference between $n_{u,t}$ and $h_{u,t}$. The final representation $z_{u,t}$ could be formulated as:

$$
z_{u,t} = h_{u,t} - n_{u,t}
\tag{6}
$$

Further we elaborately design a confidence neural network as an activation unit in the fusion process:

$$
\begin{aligned}
G_{u,t} &= \sigma(W \operatorname{concat}([h_{u,t}, n_{u,t}]) + b) \\
\hat{z}_{u,t} &= h_{u,t} - G_{u,t} \odot n_{u,t}
\end{aligned}
\tag{7}
$$

where $G_{u,t} \in \mathbb{R}^{L_e}$ is used to determine the weight, which indicates how to dynamically combine $h_{u,t}$ and $n_{u,t}$. $\odot$ is element-wise multiplication. W is the weight matrix and σ is sigmoid function.

Overall Structure. Fig. 3 illustrates the model structure. Different colors represent different data resources, *i.e.*, clicked item sequences, unclicked item sequences and label data. The representation of $\mathcal{S}_{u,t}^+$ and $\mathcal{S}_{u,t}^-$ are concatenated

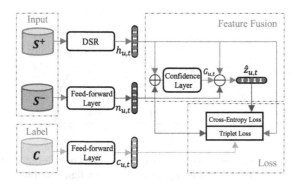

Fig. 3. Network structure.

($\oplus$) as the input of the confidence network. Then the confidence network outputs the activation unit $\boldsymbol{G}_{u,t}$ for feature fusion. The recommendations are made based on the final user representation $\hat{\boldsymbol{z}}_{u,t}$. The optimization of triplet metric learning for unclicked sequences is added to the final loss function.

Loss Function. Besides the triplet loss $\mathcal{L}_{tri}$, we use the sampled-softmax [5] method to calculate the cross-entropy loss $\mathcal{L}_{ce}$ over the large amount of items in our real-world dataset for the sake of efficiency. Importance sampling (*e.g.*, log-uniform sampler *w.r.t.* items frequencies) are conducted to obtain j random negative samples $\mathcal{C}_{u,t}^{-}$ from unobserved item set $\mathcal{I}/\mathcal{C}_{u,t}$ as most DSR models do [9]. The model performs joint optimization according to the overall loss defined as follows:

$$\mathcal{L}_{\text{XDM}} = \mathcal{L}_{ce} + \lambda\mathcal{L}_{tri} =$$
$$\sum_{u \in \mathcal{U}} \text{CrossEntropy}\big(\mathcal{C}_{u,t}, \text{SampledSoftmax}(\hat{\boldsymbol{z}}_{u,t}, \mathcal{C}_{u,t}, \mathcal{C}_{u,t}^{-})\big)$$
$$+ \lambda \sum_{u \in \mathcal{U}} \Big[2\|\boldsymbol{h}_{u,t} - \boldsymbol{c}_{u,t}\|_2^2 - \|\boldsymbol{h}_{u,t} - \boldsymbol{n}_{u,t}\|_2^2 - \|\boldsymbol{c}_{u,t} - \boldsymbol{n}_{u,t}\|_2^2 + m^*\Big]_+ \tag{8}$$

where $\mathcal{C}_{u,t}$ is the positive labels from real behaviors of user u after time t. The sampled-softmax takes final user representation, positive and negative samples as input, which outputs the prediction probability distribution over items in $\mathcal{C}_{u,t}$. λ is the trade-off coefficient of two loss terms.

3 Experimental Setup

3.1 Datasets

As we have discussed, incorporating unclicked sequences into sequential recommendation is a novel exploration, where few of benchmark datasets exist. Hence we construct two large-scale datasets collected from the logs of running recommender

Table 1. Statistics of experimental datasets

Type	Taobao dataset	Tmall dataset
Num. of Users	358,978	446,464
Num. of Items	1,078,723	908,214
Num. of Train Data	1,041,094	1,229,271
Num. of Test Data	17,048	20,134
Avg Short-term Clicked Seq Len	13.49	12.30
Avg Long-term Clicked Seq Len	20.87	18.74
Avg Unclicked Seq Len	71.32	64.26

systems from Mobile Taobao and Tmall platforms[1], within the time period from 2019/12/27 to 2020/01/03. The collected data contains user portrait features, user complete behavior sequences including clicked and unclicked items. Note that dataset in [9] is also sampled from Taobao, but they do not contain unclicked items and the data is not public available. For the training/validation/test dataset split and evaluation pipeline, we directly followed the well-defined procedure in [9] (Table 1).

3.2 Compared Methods

We used the following state-of-the-art sequential recommenders to compare with XDM: **DNN** [1], **GRU4REC** [4], **NARM** [7], **SHAN** [14], **BINN** [8], and **SDM** [9]. We conducted ablation experiments by gradually adding our proposed modules and compared with the baseline models above. We employ *SDM* (the best baseline) as the base DSR for modeling clicked sequences. We name several XDM variants with abbreviated terms.

- **XDM.** Proposed algorithm of this paper includes both symmetric triplet metric learning (Eq. 5) and confidence fusion network (Eq. 7).
- **XDM (w/o sym).** The only difference with XDM is using asymmetric triplet metric learning algorithm (Eq. 4).
- **XDM (w/o fusion+sym).** XDM only employs asymmetric triplet metric learning algorithm (Eq. 4) without any explicit feature fusion.
- **XDM (w/o metric).** XDM only combines unclicked sequences via the fusion network (Eq. 7) to improve feature fusion without metric learning.
- **XDM (w/o conf+metric).** XDM only combines unclicked sequences via simple feature fusion (Eq. 6) without metric learning.

Although models in [13,15] are applied in other tasks, they also use unclicked sequences. But their methods are similar to *XDM (w/o conf+metric)*, which simply regard unclicked sequences as features of neural networks. Hence we do not involve them as the baselines for fair comparisons.

---

[1] Popular E-commerce websites with ten millions of active items (www.taobao.com and www.tmall.com).

Evaluation Metrics. To evaluate the effectiveness of different methods, we use HR (Hit Ratio), MRR (Mean Reciprocal Rank), R (Recall), and F_1 metrics for the Top-k recommendation results, which are also widely used in the previous works [4, 9, 12]. We chose $k = \{50, 80\}$ to report the Top-k performance as [3]. The reason for setting larger k is the huge number of item set $\mathcal{I}$ in our datasets and results over smaller k have larger variance thus uncomparable for the matching stage. We calculated averaged metrics for the test sets.

3.3   Implementation Details

We used the distributed Tensorflow[2] to implement all the methods. Results of the baselines and our models on test datasets are reported according to optimal hyper-parameters tuned on validation data. We used 2 parameter severs (PSs) and 5 GPU (Tesla P100-pcie-16 GB) workers with average 30 global steps per second to conduct training and inference. The embedding size $L_e = 128$. For training, the learning rate was set to 0.1 and the sequences with similar length were selected in a mini-batch whose size is set to 256. Adagrad was used as the optimizer and the gradient clipping technique was also adopted. The next $k = 5$ clicked items after a sequence were taken as the label items in $\mathcal{C}_{u,t}$ in our experiments. The sampled-softmax used $j = 20,000$ random negative samples. All input feature representation and model parameters were initialized randomly. For parameters of XDM, we set the margin parameter m^* in the triplet loss to 5, the trade-off parameter λ between cross-entropy loss and triplet loss to 10. These two parameters were the best results obtained by parameter selection experiment.

4   Experiment Analysis

4.1   Overall Performances

The experimental results are reported in Table 2 as well as the relative improvement based on the best baseline model. DNN performs worst since the average pooling operation ignores the inheritance correlation between items. The performance of GRU4REC and NARM are far beyond the original DNN by modeling the evolution of short-term behavior. Compared to GRU4REC, SHAN and BINN encode more personalized user information, which are significantly better than GRU4REC and beat NARM. *SDM* performs well due to jointly modeling long-term and short-term behavior. Also it simulates multiple interests in users' short-term session and combine the long-term preference using a gating network. *XDM* takes *SDM* as the base model. Two modules *i.e.*, confidence fusion network and symmetric triplet metric learning, are added to the base model. Results of all metrics are substantially improved. *XDM* outperforms it by **6.21%** in MRR@50 and **5.63%** in F_1@50 on the Taobao dataset. Similar trends are also observed on the Tmall dataset. This confirms the effectiveness of overall proposed method.

---

[2] https://www.tensorflow.org/guide/distributed_training.

Table 2. Top-k recommendation comparison of different methods. The relative improvements compared to the best baseline (SDM) are appended on the right starting with "+/−". (k is set to 50, 80). * indicates significant improvement of XDM over the baselines in Sect. 3.2. ($p < 0.05$ in two-tailed paired t-test).

Taobao Dataset

Methods	HR@50		MRR@50		R@50		F_1@50		HR@80		MRR@80		R@80		F_1@80	
DNN	29.95%	-17.42%	6.65%	-24.94%	1.44%	-20.44%	1.14%	-19.72%	36.74%	-15.93%	6.99%	-24.43%	2.01%	-18.29%	1.19%	-17.93%
GRU4REC	32.39%	-10.70%	7.85%	-11.40%	1.62%	-10.50%	1.26%	-11.27%	39.25%	-10.18%	8.20%	-11.35%	2.22%	-9.76%	1.30%	-10.34%
NARM	32.68%	-9.90%	8.20%	-7.45%	1.66%	-8.29%	1.30%	-8.45%	40.27%	-7.85%	8.52%	-7.89%	2.29%	-6.91%	1.34%	-7.59%
SHAN	34.00%	-6.26%	8.84%	-0.23%	1.81%	-0.00%	1.40%	-1.41%	40.93%	-6.34%	9.20%	-0.54%	2.45%	-0.41%	1.41%	-2.76%
BINN	36.24%	-0.08%	8.70%	-1.81%	1.73%	-4.42%	1.38%	-2.82%	43.30%	-0.92%	8.64%	-6.59%	2.34%	-4.88%	1.39%	-4.14%
SDM	36.27%	-	8.86%	-	1.81%	-	1.42%	-	43.70%	-	9.25%	-	2.46%	-	1.45%	-
XDM	37.97%*	+4.69%	9.41%*	+6.21%	1.92%*	+6.08%	1.50%*	+5.63%	45.44%	+3.98%	9.75%*	+5.41%	2.61%*	+6.10%	1.53%*	+5.52%
- w/o conf+metric	36.77%	+1.38%	9.12%	+2.93%	1.82%	+0.55%	1.43%	+0.70%	44.43%	+1.67%	9.45%	+2.16%	2.48%	+0.81%	1.46%	+0.69%
- w/o metric	37.07%	+2.21%	9.12%	+2.93%	1.84%	+1.66%	1.45%	+2.11%	44.80%	+2.52%	9.53%	+3.03%	2.54%	+3.25%	1.49%	+2.76%
- w/o fusion+sym	37.13%	+2.37%	9.09%	+2.60%	1.89%	+4.42%	1.47%	+3.52%	44.87%	+2.68%	9.51%	+2.81%	2.58%	+4.88%	1.52%	+4.83%
- w/o sym	37.37%	+3.03%	9.29%	+4.85%	1.87%	+3.31%	1.47%	+3.52%	45.28%	+3.62%	9.71%	+4.97%	2.58%	+4.88%	1.52%	+4.83%

Tmall Dataset

Methods	HR@50		MRR@50		R@50		F_1@50		HR@80		MRR@80		R@80		F_1@80	
DNN	30.45%	-19.47%	7.01%	-26.37%	1.68%	-21.86%	1.27%	-21.12%	37.64%	-16.71%	7.38%	-25.23%	2.36%	-18.34%	1.33%	-17.39%
GRU4REC	34.38%	-9.07%	8.94%	-6.09%	1.98%	-7.91%	1.49%	-7.45%	41.17%	-8.90%	9.14%	-7.40%	2.65%	-8.30%	1.48%	-8.07%
NARM	34.75%	-8.09%	8.96%	-5.88%	2.03%	-5.58%	1.52%	-5.59%	41.89%	-7.30%	9.30%	-5.78%	2.74%	-5.19%	1.52%	-5.59%
SHAN	35.12%	-7.11%	9.48%	-0.42%	2.15%	-0.00%	1.59%	-1.24%	42.29%	-6.42%	9.87%	-0.00%	2.83%	-2.08%	1.55%	-3.73%
BINN	37.20%	-1.61%	9.17%	-3.68%	2.04%	-5.12%	1.54%	-4.35%	45.10%	-0.20%	9.65%	-2.23%	2.83%	-2.08%	1.59%	-1.24%
SDM	37.81%	-	9.52%	-	2.15%	-	1.61%	-	45.19%	-	9.87%	-	2.89%	-	1.61%	-
XDM	38.91%*	+2.91%	9.89%*	+3.89%	2.21%*	+2.79%	1.66%*	+3.11%	46.57%*	+3.05%	10.21%	+3.44%	3.04%*	+5.19%	1.69%*	+4.97%
- w/o conf+metric	38.22%	+1.08%	9.75%	+2.42%	2.19%	+1.86%	1.64%	+1.86%	45.80%	+1.35%	10.18%	+3.14%	2.97%	+2.77%	1.66%	+3.11%
- w/o metric	38.56%	+1.98%	9.67%	+1.58%	2.20%	+2.33%	1.64%	+1.86%	46.39%	+2.66%	10.11%	+2.43%	3.00%	+3.81%	1.67%	+3.73%
- w/o fusion+sym	38.69%	+2.33%	9.81%	+3.05%	2.21%	+2.79%	1.65%	+2.48%	46.37%	+2.61%	10.25%	+3.85%	3.00%	+3.81%	1.67%	+3.73%
- w/o sym	38.80%	+2.62%	9.81%	+3.05%	2.21%	+2.79%	1.66%	+3.11%	46.81%	+3.58%	10.25%	+3.85%	3.02%	+4.50%	1.68%	+4.35%

4.2 Ablation Analysis

To disentangle the capability of each module, we further conducted ablation study and results are also shown in Table 2. *XDM (w/o conf+metric)* attempts to eliminate noises contained in clicked sequences by using unclicked representation directly, as shown in the Eq. 6. The results show that all indicators of this method are slightly improved compared with *SDM*. *XDM (w/o metric)* introduces a confidence network and applies it to weight the unclicked representation in feature fusion process. Results show that almost indicators increase by about 2%–3% on average compared with the base model *SDM* on two datasets. These two experiments demonstrate that the unclicked items does reflect negative interest of users, and it plays an important role of denoising in user preference modeling, though equipped with DSR for clicked sequences.

XDM (w/o fusion+sym) only adds an asymmetric triplet loss shown in Eq. 4 without explicit feature fusion operation. The result is positive. It reveals the effect of metric learning. We considered the combination of two modules (confidence fusion and asymmetric metric learning) stated above denoted as *XDM (w/o sym)*. The average results increase about 3%~4% in almost indicators of two dataset, which indicates that the combination of metric learning and feature fusion can make better use of unclicked data. Metric learning provides higher-quality representations of unclicked sequences for the feature fusion network. Along this way, *XDM* further added symmetry constraints, as shown in the Eq. 5. The average results show it achieves the highest improvement over *SDM* and *XDM (w/o sym)* in almost evaluations. This result shows that symmetric constraints are very important for model learning. From comparison results of all variants, we can conclude that significant improvement is produced by the

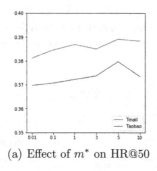

(a) Effect of m^* on HR@50

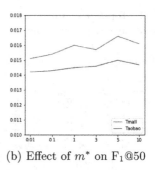

(b) Effect of m^* on F_1@50

Fig. 4. The effect of margin parameter m^*.

introduction of the confidence network and the triplet metric learning with symmetric constraint. Further, the analysis of metric learning as one of the important modules is included in the Appendix.

4.3   The Effect of Margin

The threshold m^* is a parameter for distance control between clicked and unclicked sequence representation in a certain range, so that the unclicked representation has differentiation with random items. It also fits the hinge function to avoid correcting "already correct" triplets within the threshold. A comparison experiment is performed on changes caused by m^*. Figure 4 shows the change of the parameters m^*. Result is the best when the parameter m^* is 5, and it performs worse if m^* is too large or too small. Similar observations could be drawn for MRR@50 and R@50, thus omitted due to space limitation.

4.4   Online A/B Test

We further conducted experiments on a much larger online dataset collected by Mobile Taobao App (recommendation logs within one week), which contains about *4 billion* user behavior sequences, *30 million* high-quality items, and *150 million* users. Distributed training was executed over 20 parameter servers and 100 workers (P100 GPU with 16 G B memory) considering the scalability of models. We kept the other parameters the same as offline experiments in the Sect. 3.3 and the training steps took more than 30 h. We conducted the online A/B test for several weeks between our proposed XDM and SDM [9], which was the previous deployed EBR model at Taobao. We used a fast nearest neighbor embedding retrieval method from Eq. 2, to retrieve Top-k items from the large-scale item pool. The detailed deployment architecture followed SDM and we compared the same evaluation metric pCTR (the Click-Through-Rate per page view where each page can recommend 20 items for a user). The results show that XDM improves **3%–4%** averagely compared to SDM, which demonstrates the advantages of incorporating unclicked behavior sequences and our proposed method. Moreover, XDM has been successfully deployed on EBR system of several recommendation scenarios at Taobao since April, 2020.

5   Conclusion

In this paper, we study users' unclicked sequence modeling in sequential recommender in order to enrich user representations. The importance of unclicked items is emphasized and then incorporated into our new recommendation model. For modeling sequential behaviors with unclicked data, we design a novel model XDM, which adopts the symmetric metric learning with a triplet structure as well as confidence fusion network. The experiment results demonstrate the effectiveness of the proposed XDM and verify the importance of unclicked sequences in the sequential recommendation. XDM has been fully deployed on EBR system of recommendation at Taobao. For sake of the space, the appendix is provided in the external link: https://github.com/alicogintel/XDM.

References

1. Covington, P., Adams, J., Sargin, E.: Deep neural networks for youtube recommendations. In: RecSys, pp. 191–198 (2016)
2. Ding, J., Quan, Y., He, X., Li, Y., Jin, D.: Reinforced negative sampling for recommendation with exposure data. In: IJCAI, pp. 2230–2236 (2019)
3. Gao, C., et al.: Learning to recommend with multiple cascading behaviors. TKDE (2019)
4. Hidasi, B., Karatzoglou, A., Baltrunas, L., Tikk, D.: Session-based recommendations with recurrent neural networks. arXiv preprint arXiv:1511.06939 (2015)
5. Jean, S., Cho, K., Memisevic, R., Bengio, Y.: On using very large target vocabulary for neural machine translation. arXiv preprint arXiv:1412.2007 (2014)
6. Li, C., et al.: Multi-interest network with dynamic routing for recommendation at Tmall. In: CIKM, pp. 2615–2623 (2019)
7. Li, J., Ren, P., Chen, Z., Ren, Z., Lian, T., Ma, J.: Neural attentive session-based recommendation. In: CIKM, pp. 1419–1428 (2017)
8. Li, Z., Zhao, H., Liu, Q., Huang, Z., Mei, T., Chen, E.: Learning from history and present: next-item recommendation via discriminatively exploiting user behaviors. In: KDD, pp. 1734–1743 (2018)
9. Lv, F., et al.: SDM: sequential deep matching model for online large-scale recommender system. In: CIKM, pp. 2635–2643 (2019)
10. Pazzani, M.J., Billsus, D.: Content-based recommendation systems. In: Brusilovsky, P., Kobsa, A., Nejdl, W. (eds.) The Adaptive Web. LNCS, vol. 4321, pp. 325–341. Springer, Heidelberg (2007). https://doi.org/10.1007/978-3-540-72079-9_10
11. Sarwar, B., Karypis, G., Konstan, J., Riedl, J.: Item-based collaborative filtering recommendation algorithms. In: WWW, pp. 285–295 (2001)
12. Tang, J., Wang, K.: Personalized top-n sequential recommendation via convolutional sequence embedding. In: WSDM, pp. 565–573 (2018)

13. Xie, R., Ling, C., Wang, Y., Wang, R., Xia, F., Lin, L.: Deep feedback network for recommendation. In: IJCAI, pp. 2519–2525 (2020)
14. Ying, H., et al.: Sequential recommender system based on hierarchical attention networks. In: IJCAI (2018)
15. Zhao, X., Zhang, L., Ding, Z., Xia, L., Tang, J., Yin, D.: Recommendations with negative feedback via pairwise deep reinforcement learning. In: KDD, pp. 1040–1048 (2018)

Mitigating Popularity Bias in Recommendation via Counterfactual Inference

Ming He[✉], Changshu Li, Xinlei Hu, Xin Chen, and Jiwen Wang

Faculty of Information Technology, Beijing University of Technology, Beijing, China
heming@bjut.edu.cn,
{lichangshu,huxl,chenxin,wangjiwen}@emails.bjut.edu.cn

Abstract. Popularity bias is a common problem in recommender systems. Existing research mainly tracks this problem by re-weighting training samples or leveraging a small fraction of unbiased data. However, the effect of popularity bias in user behavior data may lead to sacrifices in recommendation. In this paper, we exploit data bias from click behavior to derive popularity bias representation, and investigate how to mitigate its negative impact from a causal perspective. Motivated by causal effects, we propose a novel counterfactual inference framework named *Mitigating Popularity Bias in Recommendation via Counterfactual Inference* (MPCI), which enables us to capture the popularity bias as the direct causal effect of the prediction score, and we eliminate popularity bias by subtracting the direct popularity bias effect from the total causal effect. In this way, MPCI reduces popularity bias by decreasing the influence of popular items on model training. Extensive experiments on two real-world datasets demonstrate the superiority of our methods over some strong baselines and prove the effectiveness of mitigating popularity bias in recommender systems.

Keywords: Recommender system · Popularity bias · Counterfactual inference

1 Introduction

Recommender systems (RS) have been widely used to provide personalized suggestions to individual users. Nowadays, a large number of recommendation models have been developed, most of which are designed to fit user behavior data (i.e., click data). However, interaction data often exhibits severe popularity bias since it is observational rather than experimental, and the distribution over items is quite imbalanced or even long-tailed [2]. Consequently, recommendation models trained on such biased data may not only hurt user experience, but also may make popular items even more popular [2].

A line of existing research mainly mitigates popularity bias from the aspect of data. For example, causal embedding-based methods [1] utilize small unbiased

© The Author(s), under exclusive license to Springer Nature Switzerland AG 2022
A. Bhattacharya et al. (Eds.): DASFAA 2022, LNCS 13247, pp. 377–388, 2022.
https://doi.org/10.1007/978-3-031-00129-1_32

data by intervention to eliminate popularity bias. Another line of research focuses on removing the influence of popularity bias from the aspect of the model. For these, two types of methods are mostly used: IPS-based methods [8] aiming to re-weight each instance as the inverse of the corresponding item popularity score, and re-rank methods [4] that utilize regularization in the training to mitigate popularity bias in RS.

Despite their effectiveness in some scenarios, we argue that they suffer from two limitations: (1) For the causal embedding-based methods, the extra unbiased data are required for alleviating bias problems. However, these data are often too expensive to obtain in real-world recommender systems; (2) For IPS-based methods and re-rank methods, both types of methods are heuristically designed to intentionally increase the scores of less popular items. However, they do not consider the inherent bias in user behavior data (i.e., click data), which could harm the recommendation performance.

In this work, to address these problems, we propose a novel framework *Mitigating Popularity Bias in Recommendation via Counterfactual Inference* (MPCI) to mitigate popularity bias from the perspective of both the data and model. From the perspective of data, we utilize only click data to learn popularity bias representations, which can be applied to reduce the negative effect of popularity bias during inference. From the perspective of model, we formulate recommendation as a causal problem and construct a causal graph to reflect the causal relations of popularity bias in the recommendation model. Then we exploit the bias in click data to measure the causal effect of popularity bias on the prediction score, and mitigate its negative influence by counterfactual inference [10]. Lastly, we instantiate our framework on a representative multi-modal recommendation model called MMGCN [12] to utilize multi-modal item features. Extensive experiments over two benchmarks demonstrate that our framework not only mitigates popularity bias effectively but also improves recommendation accuracy over backbone models.

To summarize, the main contributions of our work are as follows:

- We study popularity bias in recommender systems from a causal view, and construct a new causal graph to analyze the causal effect of popularity bias for recommendation.
- We propose an MPCI framework to capture popularity bias representation by using click data only, and perform counterfactual inference to mitigate popularity bias in the inference stage of recommendation.
- We implement the proposed framework on MMGCN and conduct extensive experiments on two datasets, which show the effectiveness of our proposal.

2   Related Work

Popularity Bias. Some methods have been proposed to mitigate popularity bias from the perspective of the model itself. Among them, Inverse Propensity Scoring (IPS)-based methods [8] have mostly been adopted and have achieved state-of-the-art performance. IPS re-weights each instance as the inverse of the corresponding item popularity value in the loss of a model. Further, Ranking adjustment is another type of method [4] that directly re-ranks the recommendation list to improve the recommendation opportunity of unpopular items by modifying the predictions of the model. Some other methods have been explored to mitigate popularity bias from the perspective of the data. For example, CausE [1] performed unbiased learning on a large, biased dataset and a small, unbiased dataset respectively. Further, DICE [15] designed a framework with cause-specific data to disentangle user interest and conformity into two sets of embedding to handle popularity bias. *In contrast to previous works, we undertake a new attempt of counterfactual inference to solve the popularity bias issue.*

Causal Recommendation. Most studies focused on debiasing [2], such as language bias [5], conformity bias [15], and popularity bias [11]. The most popular methods can be divided into three types: the first comprises the aforementioned IPS-based methods. The second includes causal intervention methods [9,14] that eliminate the influence of biases from prediction score. However, because the sample space is too large, their approximation of scores over the intervention terms is subject to large variance and lacks stability. The last methods [5,7] are counterfactual methods that adjust rating prediction by reducing the effect of bias. *Different from prior works, we adopt biased data to learn popularity bias representation and analyze the causal effect of it in causal graph.*

3 Preliminaries

Causal Graph. The causal graph is a directed acyclic graph $G = \{N, E\}$ where a node (i.e., N) denotes a variable and an edge (i.e., E) denotes a causal relation between two nodes. As shown in Figs. 1(a) and (b), capital letters (e.g., X, M) and lowercase letters (e.g., x, m) represent the random variables and specific realizations of capital letters, respectively. $X \to Y$ means that the variable X has a direct effect on variable Y, and $X \to M \to Y$ means that M acts as a mediator between an indirect effect of X and Y [6].

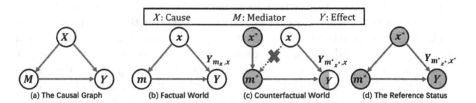

Fig. 1. (a) An example of a causal graph where Y is affected by X and M. (b) A causal graph with specific realizations of random variables. (c) A causal intervention *do (X = x*)*, where x^* denotes the no-treatment value of X in counterfactual situation (gray nodes). (d) Gray nodes mean the variables are at reference status (e.g., $x = x^*$).

Counterfactual Inference. Counterfactual inference [10] is a useful tool for comparing outcomes of interventions on complex systems, which essentially gives humans the inference ability to predict certain features of the world if the world had been different. The counterfactual world is shown in Fig. 1(c), and X changes to a reference value x^* for M. Under this situation, M is set to the value $m^*_{x^*}$ (i.e., m^*). Meanwhile, the color of nodes x^* and m^* change to gray, and Y will be $Y_{m^*_{x^*},x}$. i.e., $Y_{m^*_{x^*},x} = Y(M = M(X = x^*), X = x)$. The color of node Y is set half a gray due to the existence of both m^* and x, which is called a counterfactual situation because it does not really happen in the real world. It is only imagined in order to investigate what the final outcome would be if X was simultaneously set to different values for x and x^*.

Causal Effect. Formally, the causal effect can be defined as the difference between the outcome in counterfactual world and the outcome coming from the real-world observation value. Supposing that Fig. 1(b) stands for under treatment condition that $X = x$, and further Fig. 1(c) denotes under no-treatment condition that $X = x^*$. Note that the no-treatment condition is defined as blocking the signal from X, i.e., x is not given. The Total Effect (TE) of the treatment $X = x$ on Y is denoted as: $TE = Y_{m_x,x} - Y_{m^*_{x^*},x^*}$. Furthermore, total effect can be decomposed into natural direct effect NDE and total indirect effect (TIE) [6]. NDE denotes the effect of X on Y with the mediator M blocked. NDE expresses the increase in the outcome Y with X changing from x^* to x while M is set to the value m^* at $X = x^*$, which is denoted as: $NDE = Y_{m^*_{x^*},x} - Y_{m^*_{x^*},x^*}$. TIE can be obtained by subtracting NDE from TE, denoted as: $TIE = TE - NDE = Y_{m_x,x} - Y_{m^*_{x^*},x}$, which represents the effect of x on Y through the indirect path $x^* \rightarrow m^* \rightarrow Y$. Figure 1(d) denotes the reference status of all nodes when x is set to x^*.

4   Methodology

4.1   Causal View in Recommendation

As illustrated in Fig. 2(a), the direct paths $C \rightarrow (U, I)$ denote that the click data affects the representations of users and items since most models are trained on historical interaction records. Furthermore, the indirect paths i.e., $B \rightarrow (U, I) \rightarrow Y$ mean that U, I always act as mediators between B and Y. In fact, such effects might be beneficial. For example, a popular movie always has a good plot, vivid acting, and profound meaning, etc. Such a movie deserves a higher position in the recommendation list. In contrast, the direct path $B \rightarrow Y$ means that popularity bias will directly affect the prediction score, Such effect causes bias amplification, which should be avoided since an authentic recommendation model should estimate user preferences reliably. Therefore, the direct path $B \rightarrow Y$ should be eliminated in formulating the recommendation model.

4.2   Mitigating Popularity Bias

In this section, we propose the extraction of biased data to learn popularity bias representations for estimating the causal effect of popularity bias on prediction

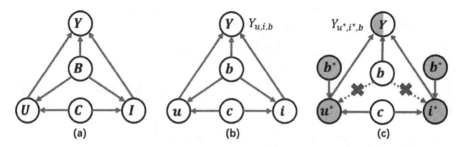

Fig. 2. (a) The proposed causal graph. B: popularity bias representation. C: click data. U: user representation. I: item representation. Y: prediction score. (b) Factual World. (c) Counterfactual World.

score. A user's click behavior on an item primarily indicates two factors: (1) the user's interest in the item's features, and (2) the user's conformity towards the item's popularity. Indeed, a click can come from one or both of the two factors [15]. In practice, if a user clicks a popular item, with no other feedback (e.g., 'like'), then we consider that the click behavior data is severely affected by popularity bias (i.e., **biased**) and can not truly reflect the user's preferences. By utilizing biased data, MPCI measures the causal effect of popularity bias on the prediction score and then mitigates its influence (i.e., the direct effect). Figure 3 illustrates the workflow of the MPCI framework. Taking popularity bias into account, we aggregate the three modules into a bias-aware prediction score:

$$Y_{u,i,b} = f_y(U = u, I = i, B = b) = f_y(f_u(\boldsymbol{u}, \boldsymbol{b}), f_i(\boldsymbol{i}, \boldsymbol{b})), \tag{1}$$

where $\boldsymbol{u}$, $\boldsymbol{i}$ and $\boldsymbol{b}$ denote the representation of user, item and popularity bias respectively, and f_u, f_i are functions used to aggregate the representation of user preferences and item features with popularity bias, respectively.

Compared to the factual world in Fig. 2(b), the indirect path is blocked by setting b as the reference value of b^* in Fig. 2(c). For the formulation of causal, we replace $f(.)$ with $Y(.)$, and the NDE is given as:

$$\begin{aligned} NDE &= Y_{u^*,i^*,b} - Y_{u^*,i^*,b^*} \\ &= Y(U = u^*, I = i^*, B = b) - Y(U = u^*, I = i^*, B = b^*), \end{aligned} \tag{2}$$

where u^* and i^* denote the reference values of U and I, i.e., the representation of user and item, respectively. $Y(U = u^*, I = i^*, B = b^*)$ denotes the outcome of counterfactual world (see Fig. 2(c)) where the treatment variable B is changed from b^* to b on the direct path (i.e., $B \rightarrow Y$) while remains its reference value on the indirect path (i.e., $B \rightarrow (U,I) \rightarrow Y$). This requires a counterfactual thinking: *what would the prediction score be if the item had only the item popularity affecting the recommendation*, where the reference value b^* is treated as the status that b is not given. Since we have obtained the NDE, we measure the causal effect of TIE which can be calculated by subtracting the NDE from TE:

$$\begin{aligned} TE &= Y(U = u, I = i, B = b) - Y(U = u^*, I = i^*, B = b^*) \\ &= Y_{u,i,b} - Y_{u^*,i^*,b^*}, \end{aligned} \tag{3}$$

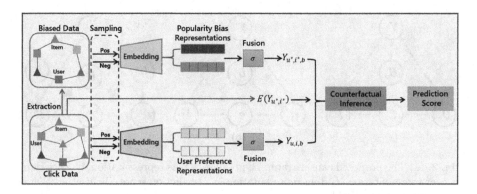

Fig. 3. The framework of MPCI. Specifically, the MPCI captures user preference representations from click data, and then extracts biased data from click data to learn popularity bias representations. Each representation has a set of positive and negative samples, which are embedded to calculate the bias-aware prediction score $Y_{u^*,i^*,b}$ and preference prediction score $Y_{u,i,b}$ by a fusion function σ, respectively. Next, we aggregate the two scores and expectation constant $E(Y_{u^*,i^*})$ as the input of counterfactual inference. Finally, MPCI obtains the prediction score of the user for each item.

$$
\begin{aligned}
TIE &= TE - NDE \\
&= Y(U = u, I = i, B = b) - Y(U = u^*, I = i^*, B = b) \\
&= Y_{u,i,b} - Y_{u^*,i^*,b}.
\end{aligned}
\tag{4}
$$

4.3 Debiased Recommendation Model

Model Design. According to Eq. (1), prediction score $Y_{u,i,b}$ and $Y_{u^*,i^*,b}$ are calculated by the score function $f_y(U = u, I = i, B = b)$ and the aggregation function $f_u(u, b_u)$ and $f_i(i, b_i)$. With the aim of improving the generality and leveraging the advantages of existing models, the score function is implemented by using a classical fusion strategy, formulated as:

$$
Y_{u,i,b} = Y(U = u, I = i, B = b) = Y_{u,i} * (\sigma(Y_{u,b}) + \sigma(Y_{i,b})),
\tag{5}
$$

$$
Y_{u^*,i^*,b} = Y(U = u^*, I = i^*, B = b) = Y_{u^*,i^*} * (\sigma(Y_{u^*,b}) + \sigma(Y_{i^*,b})),
\tag{6}
$$

where σ represents the non-linear sigmoid function, and $Y_{u,i}$, $Y_{u,b}$ and $Y_{i,b}$ denote the result of the conventional recommendation models with different input pairs (i.e., user/item representation, user/popularity bias representation, item/popularity bias representation, respectively). Thus, we can simply change the input of any existing RS model.

Model Training. Based on the theory of causality discussed in the prior section and the framework workflow in Fig. 3, our model requires two predictions: $Y_{u,i,b}$ and $Y_{u^*,i^*,b}$. Therefore, we adopt a multi-task learning schema that applies additional supervision over task $\mathcal{L}_2$. Formally, the training loss is given as:

$$
L = \mathcal{L}_1(Y_{u,i,b}, Y_{u^*,b}) + \alpha * \mathcal{L}_2(Y_{u,i,b}, Y_{i^*,b}).
\tag{7}
$$

where α is a hyper-parameter to tune the relative weight of $\mathcal{L}_1$ and $\mathcal{L}_2$. $Y_{u^*,b}$ and $Y_{i^*,b}$ can be obtained by forcing the model not to take u and i as the input.

Counterfactual Inference. To eliminate the direct effect of causal path $B \to Y$ from the prediction score, we calculate the predictions $Y_{u,i,b}$ and Y_{u,i,b^*}, and then perform the following reduction to obtain the counterfactual-inference result:

$$Y_{CR} = Y_{u,i,b} - Y_{u^*,i^*,b} = Y_{u,i,b} - c * (\sigma(Y_{u^*,b}) + \sigma(Y_{i^*,b}))). \qquad (8)$$

where c denotes the expectation constant of Y_{u^*,i^*}, formulated as $c = E(Y_{u^*,i^*})$, which indicates that all users and items share the same score c. In this way, we can eliminate the effect of popularity bias during the stage of inference.

5 Experiments

In this section, we describe the conducted experiments to answer the following research questions:

- **RQ1:** Does MPCI outperform state-of-the-art debiasing methods?
- **RQ2:** How does the hyper-parameter and percentage of biased data affect the performance of MPCI?
- **RQ3:** How does MPCI mitigate popularity bias?

5.1 Experimental Settings

Dataset. We evaluate the proposed MPCI model with two multi-modal real-world datasets: Tiktok[1] and Adressa[2]. **Tiktok** [10] is a large-scale micro-video dataset released by Tiktok, and **Adressa** [3] is a news dataset published by Norwegian University of Science and Technology and Adressavisen. Following prior studies [10], multi-modal item features have already been extracted, and actions of favorite or finish in Tiktok are used as the positive post-click feedback (i.e., like). Moreover, dwell time $> 30\,\mathrm{s}$ in Adressa reflects the likes of users. The statistics are shown in Table 1.

Table 1. Statistics of two datasets.

Dataset	#Users	#Items	#Clicks	#Likes
Tiktok	18,855	34,756	1,493,532	589,008
Adressa	31,123	4,895	1,437,540	998,612

For each dataset, we randomly split the historical interaction records into the training, validation, and testing subsets with a ratio of 80%, 10%, and 10%,

---

[1] http://ai-lab-challenge.bytedance.com/tce/vc/.
[2] http://reclab.idi.ntnu.no/dataset/.

respectively. We utilized the validation set to tune hyper-parameters and choose the best model for the testing phase. To investigate the causal effect of popularity bias, we extracted part of the click data as biased data from the original interaction data and we further analyzed how the percentage of biased data relates to performance. Moreover, for each click, we randomly choose an item the user has never interacted with as the negative sample (see Fig. 3) to create the triples for training, and we treat the post-click feedback data (i.e., like) as the test set in this work. We admit that the sparsity of post-click feedback might restrict the scale of the evaluation [10]. However, we still cover a large group of users for evaluation.

Evaluation Metrics. We adopted three widely used evaluation metrics: Precision@K (P@k), Recall@K (R@k), and Normalized Discounted Cumulative Gain (NDCG@K). We set K = {10, 20} and report the result achieved for each user over all the items. The higher values indicate better performance.

Baselines. To utilize the multi-modal features of items, we implemented our MPCI on the multi-modal recommendation model MMGCN, and compared our method with the following state-of-the-art methods that might mitigate popularity bias. For a fair comparison, most methods are instantiated on MMGCN.

- **MMGCN** [12] is a graph-based algorithm to learn representations of user preferences on different modalities.
- **CT** [10] is based on the clean training (CT) setting [10], in which only the clicks that end with likes are viewed as positive samples to train MMGCN.
- **NR** [13] aims to leverage post-click feedback, which treats "click-skip" and "not-click" items as negative samples. We adopt Negative Feedback Reweighting (NR) to reweight the negative samples during the training stage.
- **RR** [4] is the normal training of MMGCN, we propose a strategy to re-rank (RR) the top 20 items during inference for the whole dataset based on the like/click ratio of items.
- **IPS** [8] is a classical method of causal recommendation, which utilizes the inverse propensity score to reweight samples to alleviate the bias problem.
- **PD** [14] is a SOTA method that performs deconfounded training while intervenes the popularity bias during inference.

Hyperparameter Settings. We strictly followed the original settings of MMGCN [12] including loss function, learning size and embedding size, etc. For our MPCI framework, the additional weight α in the multi-task loss function was tuned in {0, 0.5, 1, 1.5, 2, 2.5, 3}.

5.2 Performance Comparison (RQ1)

Table 2 summarizes the best results of all the models on two benchmark datasets. From the table, we have the following observations:

RR can achieve a better result than MMGCN or other conventional methods, which validates the effectiveness of leveraging the like/click ratio of items

Table 2. Top-K recommendation performance of compared methods on Tiktok and Adressa. Bold scores are the best in each column, while underlined scores are the best results of the baselines. %Improve. denotes the relative performance.

Method	Tiktok						Adressa					
	P@10	R@10	N@10	P@20	R@20	N@20	P@10	R@10	N@10	P@20	R@20	N@20
MMGCN	0.0256	0.0357	0.0333	<u>0.0231</u>	<u>0.0635</u>	<u>0.0430</u>	0.0501	0.0975	0.0817	<u>0.0415</u>	<u>0.1612</u>	<u>0.1059</u>
CT	0.0217	0.0295	0.0294	0.0194	0.0520	0.0372	0.0493	0.0951	0.0799	0.0418	0.1611	0.1051
NR	0.0239	0.0346	0.0329	0.0216	0.0605	0.0424	0.0499	0.0970	0.0814	0.0415	0.1610	0.1058
RR	<u>0.0264</u>	<u>0.0383</u>	<u>0.0367</u>	0.0231	0.0635	0.0430	<u>0.0521</u>	<u>0.1007</u>	<u>0.0831</u>	0.0415	0.1612	0.1059
PD	0.0201	0.0271	0.0270	0.0182	0.0482	0.0343	0.0337	0.0604	0.0501	0.0307	0.1103	0.0692
IPS	0.0230	0.0334	0.0314	0.0210	0.0582	0.0406	0.0419	0.0804	0.0663	0.0361	0.1378	0.0883
MPCI	**0.0274**	**0.0386**	**0.0372**	**0.0251**	**0.0686**	**0.0478**	**0.0540**	**0.1054**	**0.0888**	**0.0444**	**0.1725**	**0.1143**
%Improve	3.79%	0.78%	1.36%	8.66%	8.03%	11.16%	3.65%	4.67%	6.86%	6.99%	7.01%	7.93%

to alleviate the effect of popularity bias. Although post-click feedback data are introduced into both NR and CT, they perform worse than MMGCN on both Tiktok and Adressa, e.g., the R@10 of CT decrease by 17.4% and 2.46%, which can be explained by the sparsity of post-click feedback. Moreover, the performance of IPS is inferior on two datasets, showing that the popularity bias may not be mitigated by simply inversing the weights of items with more popularity. In addition, the result indicates the importance of accurate propensity estimation to mitigate the bias.

In all cases, MPCI outperforms all baselines. First, MPCI achieves comparable performance to MMGCN, which indicates that mitigating popularity bias effectively improves the recommendation performance. Second, MPCI performs better than CT and NR. This result validates the superiority of biased data in capturing popularity bias and mitigating it compared to post-click feedback data. Third, the relative improvements of MPCI over RR and IPS $w.r.t$ R@20 are 11.16% and 17.73% on Tiktok dataset, respectively. This is attributed to the causal analysis of popularity bias in recommendation model through causal graph. Finally, MPCI outperforms PD, which verifies the effectiveness of counterfactual inference in mitigating popularity bias and capturing users' preferences.

5.3 Case Study (RQ2)

Effect of Hyper-parameter. As formulated in the loss function Eq. (7), α is the hyper-parameter balance of the two tasks of real-world and counterfactual-world recommendation. We tuned α in $\{0, 0.5, 1, 1.5, 2, 2.5, 3\}$ to explore the influence of this additional weight. As shown in Fig. 4, our model performs worst as $\alpha = 0$ and obtains the best performance when α is around 1.5, which reveals the effectiveness of multi-task training. Additionally, the performance of MPCI is enhanced with an increase of α from 0 to 1.5, which demonstrates the importance of estimating the effect of popularity bias through the biased data. However, when α surpasses a threshold, performance deteriorates with the increase of the parameter. We ascribe such inferior performance to the overthinking of popularity bias affecting the capturing of user preferences.

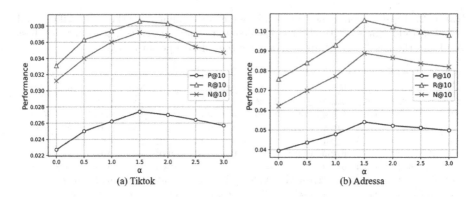

Fig. 4. Effect of hyper-parameter α

Impact of Biased Data. To study the effectiveness of the percentage of biased data, we ranked the items in descending order by the count of clicks (i.e., popularity) and extracted the top-ranked items without 'like' feedback at a certain proportion where a smaller proportion leads to a dataset with higher popularity. Note that we obtained biased data by using click data only. Due to space limitations, we show the results on Table 3 with the percentage changing from 0 to 10%, and results on other percentages show the decreasing trend. We found the following: 1) The model achieves its best performance when the percentage of biased data is 3%, which further validates the effectiveness of biased data; 2) Performance is close to the result of MMGCN when the percentage is 0%, and increases dramatically under a larger percentage then decreases when the percentage is over 3%. The result indicates that ignoring or overthinking popularity bias leads to recommendation performance degradation.

Table 3. Results of varying percentages of biased data.

Percentage	Adressa			Tiktok		
	P@20	R@20	N@20	P@20	R@20	N@20
0	0.0415	0.1612	0.1059	0.0231	0.0635	0.0430
0.5%	0.0423	0.1632	0.1076	0.0233	0.0653	0.0450
1%	0.0434	0.1671	0.1100	0.0231	0.0673	0.0452
3%	**0.0444**	**0.1725**	**0.1143**	**0.0251**	**0.0686**	**0.0478**
5%	0.0432	0.1679	0.1107	0.0240	0.0650	0.0460
10%	0.0406	0.1571	1.1034	0.0222	0.0623	0.0431

5.4   Case Study (RQ3)

We then investigated whether our model mitigates popularity bias. We compared MMGCN with MPCI on the Adressa dataset. First, we ranked all items on the final recommendation list (comprising the top-100 items) in ascending order by item popularity. Next, we divided the number of items into five groups, and each group had a similar number of items. Finally, we calculated the numbers of recommendation items and the relative improvement of the metric in each group. In Fig. 5(a), the background histograms indicate the number of recommended items in each group involved in the dataset. The horizontal axis indicates the item groups with a certain number of interactions. The left vertical axis is the value of the background histograms.

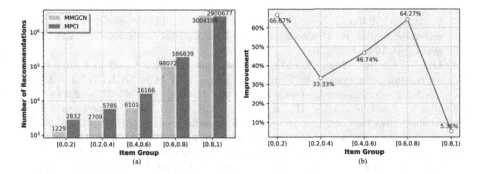

Fig. 5. We group items by popularity. (a) Numbers of different item groups recommended by MMGCN and MPCI. (b) Relative improvement achieved by MPCI over MMGCN for different item groups on Adressa. Note that we define comparison metric as the count of hit items divided by the total recommended items.

In Fig. 5(b), the left vertical axis is the value of the polyline, which corresponds to the improvement achieved by MPCI over MMGCN. As we can see, the metric of MPCI increase by only around 5% in the most popular item group. We attribute this to MMGCN recommends more popular items to users without considering item popularity (illustrated in Fig. 5(a)). However, the unpopular item groups shows relatively significant improvement. This improvement is mainly due to the fact that we recommend more unpopular items to users. This means our MPCI reduces the direct effect of popularity bias and recommends unpopular items which can satisfy users. This confirms our model's capability of capturing users' preferences and mitigating popularity bias in recommender systems.

6 Conclusion

In this present work, we present a novel cause-effect view for mitigating the popularity bias issue in RS. We propose a framework, MPCI, which utilizes the biased data extracted from interaction records and performs multi-task training according to the causal graph to assess the contribution of popularity bias to the prediction score. Counterfactual inference is performed to eliminate the direct effect of popularity bias on the prediction score. Extensive experiments on two real-world recommendation datasets demonstrated the effectiveness of MPCI.

Acknowledgement. This work is supported by the Beijing Natural Science Foundation under grants 4192008.

References

1. Bonner, S., Vasile, F.: Causal embeddings for recommendation. In: Proceedings of the 12th ACM Conference on Recommender Systems, pp. 104–112 (2018)
2. Chen, J., Dong, H., Wang, X., Feng, F., Wang, M., He, X.: Bias and debias in recommender system: a survey and future directions. arXiv preprint arXiv:2010.03240 (2020)
3. Gulla, J.A., Zhang, L., Liu, P., Özgöbek, Ö., Su, X.: The adressa dataset for news recommendation. In: Proceedings of the International Conference on Web Intelligence, pp. 1042–1048 (2017)
4. Liu, Y., Ge, K., Zhang, X., Lin, L.: Real-time attention based look-alike model for recommender system. In: Proceedings of the 25th ACM SIGKDD International Conference on Knowledge Discovery & Data Mining, pp. 2765–2773 (2019)
5. Niu, Y., Tang, K., Zhang, H., Lu, Z., Hua, X.S., Wen, J.R.: Counterfactual VQA: a cause-effect look at language bias. In: Proceedings of the IEEE/CVF Conference on Computer Vision and Pattern Recognition, pp. 12700–12710 (2021)
6. Pearl, J.: Direct and indirect effects. arXiv preprint arXiv:1301.2300 (2013)
7. Qian, C., Feng, F., Wen, L., Ma, C., Xie, P.: Counterfactual inference for text classification debiasing. In: Proceedings of the 59th Annual Meeting of the Association for Computational Linguistics and the 11th International Joint Conference on Natural Language Processing (Volume 1: Long Papers), pp. 5434–5445 (2021)
8. Rosenbaum, P.R., Rubin, D.B.: The central role of the propensity score in observational studies for causal effects. Biometrika **70**(1), 41–55 (1983)
9. Wang, W., Feng, F., He, X., Wang, X., Chua, T.S.: Deconfounded recommendation for alleviating bias amplification. arXiv preprint arXiv:2105.10648 (2021)
10. Wang, W., Feng, F., He, X., Zhang, H., Chua, T.S.: Clicks can be cheating: counterfactual recommendation for mitigating clickbait issue. In: Proceedings of the 44th International ACM SIGIR Conference on Research and Development in Information Retrieval, pp. 1288–1297 (2021)
11. Wei, T., Feng, F., Chen, J., Wu, Z., Yi, J., He, X.: Model-agnostic counterfactual reasoning for eliminating popularity bias in recommender system. In: Proceedings of the 27th ACM SIGKDD Conference on Knowledge Discovery & Data Mining, pp. 1791–1800 (2021)
12. Wei, Y., Wang, X., Nie, L., He, X., Hong, R., Chua, T.S.: MMGCN: multi-modal graph convolution network for personalized recommendation of micro-video. In: Proceedings of the 27th ACM International Conference on Multimedia, pp. 1437–1445 (2019)
13. Wen, H., Yang, L., Estrin, D.: Leveraging post-click feedback for content recommendations. In: Proceedings of the 13th ACM Conference on Recommender Systems, pp. 278–286 (2019)
14. Zhang, Y., Feng, F., He, X., Wei, T., Song, C., Ling, G., Zhang, Y.: Causal intervention for leveraging popularity bias in recommendation. arXiv preprint arXiv:2105.06067 (2021)
15. Zheng, Y., Gao, C., Li, X., He, X., Li, Y., Jin, D.: Disentangling user interest and conformity for recommendation with causal embedding. In: Proceedings of the Web Conference 2021, pp. 2980–2991 (2021)

Efficient Dual-Process Cognitive Recommender Balancing Accuracy and Diversity

Yixu Gao[1], Kun Shao[3(✉)], Zhijian Duan[4], Zhongyu Wei[1,2(✉)], Dong Li[3], Bin Wang[3], Mengchen Zhao[3], and Jianye Hao[3]

[1] Fudan University, Shanghai, China
[2] Research Institute of Intelligent and Complex Systems, Fudan University, Shanghai, China
{yxgao19,zywei}@fudan.edu.cn
[3] Huawei Noah's Ark Lab, Beijing, China
{shaokun2,lidong106,wangbin158,zhaomengchen,haojianye}@huawei.com
[4] Peking University, Beijing, China
zjduan@pku.edu.cn

Abstract. In this paper, we propose a dual-process cognitive recommendation system for sequential recommendations. The framework includes an intuitive representation module (System 1) and an inference module (System 2). System 1 is designed to understand the user's historical interaction sequences with external knowledge graph. System 2 is built to make recommendations by reinforcement learning to consider long-term returns and diversity. We demonstrate the performance of our method on a wide range of recommendation datasets. Experiments show significant improvement over the state-of-the-art models regarding both relevance and diversity.

Keywords: Recommendation system · Knowledge graph · Reinforcement learning · Diversity in recommendation · Dual-process theory

1 Introduction

Traditional recommendation systems mainly follow the fashion of supervised learning. Methods including convolutional neural network [29] and graph neural network [26] are adopted to represent the item sequences and generate a probability distribution to support the recommendation. Researchers further propose attention mechanism [24], memory network [5], and hybrid model [21] for better representation learning of item sequence and obtain positive results. Recently, SANS [30] adds a neural similarity module and a setwise attention module to address the few-shot recommendation problem.

In order to deal of the scenario of dynamic sequential recommendations, which aim to recommend next-item or next-session according to historical interaction records, researchers formulate the task as sequential decision problem

© The Author(s), under exclusive license to Springer Nature Switzerland AG 2022
A. Bhattacharya et al. (Eds.): DASFAA 2022, LNCS 13247, pp. 389–400, 2022.
https://doi.org/10.1007/978-3-031-00129-1_33

and reinforcement learning (RL) is used to characterize each user's personalized preferences considering the long-term returns [23]. DRN [32] presents a complete framework, and DEERS [31] balances both the positive and negative feedback of users.

Various model-free techniques are proposed to improve the recommendation performance [3]. Model-based RL is also widely used to learn the user model [15]. Despite the great improvements the above works achieved, the training efficiency of RL is still an unavoidable problem [28].

In this paper, we also focus on dynamic sequential recommendation task, and we explore to use supervised learning technique to accelerate the training efficiency of RL. Inspired by the dual-process theory [7], we propose a dual-process cognitive recommendation system, $CogRec_{Div}$, for sequential recommendations. We first design an intuitive representation module (System 1) to simulate the cognitive process. Moreover, for the reason that the users with similar interactive behaviors may click on completely different items in the next action according to the behavior polymorphism, our proposed model takes the relevance-to-diversity balance issue into account, that is the inference process (System 2). Specifically, System 2 balances the issue of diversity while considering long-term returns with the support of RL and determinant point process (DPP) methods. Among them, RL is adopted to balance the long-term and short-term relevance of interactive items, while the DPP serves as the description of diversity since the determinant can express the degree of similarity between items. Experiments are designed from three aspects: accuracy, diversity, and training efficiency. The results show that our proposed model achieves an improvement of up to 4.1% on four datasets compared with current state-of-the-art models.

2   The Proposed Method

2.1   Framework

In order to balance the trade-off between the accuracy and the diversity of recommendation systems, we propose $CogRec_{Div}$, whose framework is shown in Fig. 1. Given an interactive sequence and the knowledge graph, the representation of the sequence is extracted by System 1. An intuition model is used to make sure that the necessary information of the recommendations is gained and gives the intuitive recommendations quickly. Then, the system recommend the information extracted by System 1 through System 2. In the recommendation process, the recommendation accuracy is improved through RL, and the recommendation diversity is enriched through the DPP.

2.2   Preliminary

The item sequence of a user u is denoted as $i_{m:n}^u \doteq \left(i_m^u, i_{m+1}^u, \ldots, i_{n-1}^u, i_n^u \right)$, where i_t^u is the tth item in item set $\mathcal{I}$. m and n $(1 \leq m \leq n)$ are the start and end item respectively that the user u interacted with. Given a user u's interactive sequence

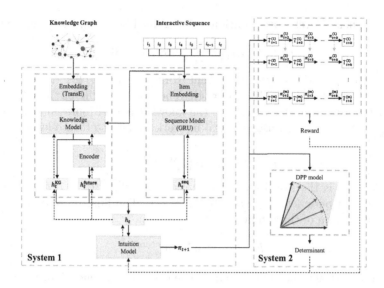

Fig. 1. The whole framework of our proposed method. The solid arrow represents the forward operation, and the dashed arrow represents the gradient back propaganda. System 1 efficiently extracts the intuitive representation from interactive sequence and the knowledge graph. Then, System 2 improves the accuracy by RL and enriches the diversity by DPP.

$i_{1:t}^u$, the goal is to predict the next k items $i_{t+1:t+k}^u$ that the user u would interact with. Each item i has its item embedding $e^{\text{item}}(i) \in \mathbb{R}^{d^{\text{item}}}$ and its knowledge graph embedding (KG embedding) $e^{\text{KG}}(i) \in \mathbb{R}^{d^{\text{KG}}}$, where $d^{\text{item}} \in \mathbb{Z}_+$ is the dimension of item embedding and $d^{\text{KG}} \in \mathbb{Z}_+$ is the dimension of KG embedding. For the sequence $i_{1:t}^u$, the item embedding sequence is $e^{u,\text{item}}(i_{1:t})$ and the KG embedding sequence is $e^{\text{KG}}(i_{1:t}^u)$.

At each step t, $\mathbf{y}_t \doteq \left[y_t^{(1)}, y_t^{(2)}, \cdots, y_t^{(|\mathcal{I}|)} \right]$ is the user's ground-truth interactive vector, where $y_t^{(i)}(i = 1, 2, \cdots, |\mathcal{I}|)$ is the indicator function of the actions:

$$y_t^{(i)} = \begin{cases} 1, & \text{if the user interact with the item } i \text{ at time } t \\ 0, & \text{otherwise.} \end{cases} \tag{1}$$

2.3   System 1: Intuitive Representation Module

The extraction capacity of System 1 model is fundamental for recommendation, thus a powerful model is needed. Inspired by KERL [23], the representation h_t for an interactive sequence $i_{1:t}$ is consisted of three parts: sequence-level representation h_t^{seq}, knowledge-level representation h_t^{KG}, and future-level representation h_t^{future}. Specifically, h_t^{seq} is used to extract the sequence information of items; h_t^{KG} combines external knowledge; and h_t^{future} is for further inference.

The sequence-level representation h_t^{seq} is gained by a standard recurrent neural network:

$$h_t^{\text{seq}} = \text{GRU}\left(h_{t-1}^{\text{seq}}, e^{\text{item}}(i_t); \theta^{\text{GRU}}\right), \tag{2}$$

where $\text{GRU}(\cdot)$ is the Gated Recurrent Unit [6], and θ^{GRU} denotes all the related parameters of the GRU network. The average value of the KG embedding of the item sequence is used as the knowledge-level representation $h_t^{\text{KG}} = \frac{1}{t}\sum_{j=1}^{t} e^{\text{KG}}(i_j)$. And we use a neural network to get the future-level representation h_t^{future}

$$h_t^{\text{future}} = \text{MLP}\left(h_t^{\text{KG}}; \theta^{\text{MLP}_1}\right), \tag{3}$$

where $\text{MLP}(\cdot)$ is the Multi-Layer Perception, and $\theta^{\text{MLP}_i}(i = 1, 2, \cdots)$ denotes all the related parameters of the MLP network. The final representation h_t is the concatenation of three representation vectors:

$$h_t \doteq \left[h_t^{\text{seq}}, h_t^{\text{KG}}, h_t^{\text{future}}\right]. \tag{4}$$

In order to extract the information needed for recommendation, we first use supervised learning to learn sequence representation

$$\pi_{\theta^{\text{MLP}_2}}(\cdot|h_t) = \text{MLP}\left(h_t; \theta^{\text{MLP}_2}\right), \tag{5}$$

where $\pi_{\theta^{\text{MLP}_2}}(\cdot|h_t)$ is the probability distribution over the complete item set of the next item that the user may interact with at time t.

Cross entropy is adopted as the loss of the supervised learning:

$$\mathcal{L}^{\text{sys}_1} = \text{CE}\left(\pi_{\theta^{\text{MLP}_2}}(\cdot|h_t), \mathbf{y}_t\right), \tag{6}$$

where $\text{CE}(\cdot)$ denotes the cross entropy loss function.

2.4   System 2: Inference Module

System 2 is used to inference based on the information extracted by System 1 and make the recommendations. Two components are used in system 2. RL is used to improve the accuracy and consider the long-term benefits of recommendation and DPP is used to enrich the diversity of items.

A Markov Decision Process Formulation for Our Task. The recommender is an *agent* that interacts with the *environments*, which are the users. At each time step, the *state* is the hidden state h of the previous interacted items, and the *actions* are the recommended items. The *transition* function is the representation networks. The *reward* can be obtained from the user and the knowledge. The *policy* π is the probability distribution over all the possible actions.

Reinforcement Learning. Strategies learned by RL perform better in the long-term cumulative reward. The reward R is consist of two parts: the relative part R^{rel}, and the KG part R^{KG}.

$$R = R^{\text{rel}} + R^{\text{KG}}. \tag{7}$$

The relative reward is bulit by discounted cumulative gain (DCG), which is a metric widely used in recommendations $R^{\text{rel}} = \text{DCG}\left(\pi_t, \mathbf{y}_t\right)$, where the relative score is the probability of each item. To make the recommended items and the previous items be as close as possible on the knowledge graph, the reward from KG is the similarity between the embeddings of the recommended items and the previous items $R^{\text{KG}} = \text{Distance}\left(e^{\text{KG}}(m_t), e^{\text{KG}}(i_t)\right)$, where $m = \arg\max_{j \in \mathcal{I}} \pi_t^j$ is the item with the highest probability at step t, and here we use cosine similarity as the distance metric.

We use the truncated policy gradient as the RL algorithm, and the RL loss at time t is

$$\mathcal{L}^{\text{RL}} = -\sum_{j=t}^{t+k} \gamma^{j-t} R_j \cdot \log p_i, \tag{8}$$

where γ is the discount factor, p_i is the probablity of the item that the user interacts with at the next step.

Determinantal Point Process. A DPP $\mathcal{P}$ on the whole item set $\mathcal{I}$ is a probabilistic model on $2^{\mathcal{I}}$, which is the set of all subsets of $\mathcal{I}$ [2]. When the empty set has nonzero probability, there exists a matrix $\mathbf{L} \in \mathbb{R}^{|\mathcal{I}| \times |\mathcal{I}|}$ such that for every subset $C \subseteq \mathcal{I}$, the probability of C is $\mathcal{P}(C) \propto \det(\mathbf{L}_C)$, where $\mathbf{L}$ is a real, positive semidefinite (PSD) kernel matrix indexed by the elements of $\mathcal{I}$, and $\mathbf{L}_C$ is the sub-matrix of $\mathbf{L}$ indexed by C in both rows and columns. $\det(\mathbf{L})$ is the determinant of $\mathbf{L}$, and $\det(\mathbf{L}_\varnothing) = 1$ by convention.

Kernel Matrix Construction. The kernel matrix can be written as a Gram matrix, $\mathbf{L} = \mathbf{B}^{\top}\mathbf{B}$, where the columns of $\mathbf{B}$ are items representations. Each column vector $\mathbf{B}_i$ can be constructed by the product of the item score $r_i \geq 0$ and a normalized vector $\mathbf{f}_i \in \mathbb{R}^{d^{\text{kernel}}}$ with $\|\mathbf{f}_i\|_2 = 1$, where $d^{\text{kernel}} \in \mathbb{Z}_+$ is the dimension of item representation. Here, we use the normalized KG embedding as $\mathbf{f}_i = \text{Norm}\left(e^{\text{KG}}(i)\right)$, where $\text{Norm}(\cdot)$ is the batch norm operation [12]. The item relevant score $r_{i,t}$ is constructed by the cosine similarity between the embeddings of the item i and the average of previous item embeddings before step t

$$r_{i,t} = \cos\left(e^{\text{KG}}(i), h_t^{\text{KG}}\right), \tag{9}$$

where $r_{i,t}$ is the relevant score of the item i at step t. The entries of kernel $\mathbf{L}_t$ is

$$(\mathbf{L}_t)_{i,j} = \langle \mathbf{B}_i, \mathbf{B}_j \rangle = \langle r_i\mathbf{f}_i, r_j\mathbf{f}_j \rangle = r_i r_j \langle \mathbf{f}_i, \mathbf{f}_j \rangle, \tag{10}$$

where $\langle \cdot, \cdot \rangle$ is the inner product of two vectors. The kernel matrix at step t is

$$\mathbf{L}_t = \text{Diag}(\mathbf{r}_t) \cdot \mathbf{S} \cdot \text{Diag}(\mathbf{r}_t), \tag{11}$$

where $\mathbf{S}$ is the similarity matrix of items that measures the distance between two items, and $\mathbf{S}_{ij} = \langle \mathbf{f}_i, \mathbf{f}_j \rangle$. Diag($\cdot$) is the diagonal matrix of the given vector.

Diversity with DPP. We hope that the final recommendation set can have as much diversity as possible. Therefore, we use the method of maximizing the determinant of the top k recommended items to achieve this goal

$$\mathcal{L}^{\mathrm{DPP}} = -\det\left(\mathbf{L}_{t,D}\right), \tag{12}$$

where D is the item set with top k probability, which includes the top k items that the user will interact with next predicted by the recommendation system.

2.5 CogRec$_{\mathbf{Div}}$

We combine the two systems, System 1 and System 2 including RL and DPP parts together with different weights $\lambda_1, \lambda_2, \lambda_3$ to construct the final loss $\mathcal{L}$,

$$\mathcal{L} = \lambda_1 \cdot \mathcal{L}^{\mathrm{sys}_1} + \lambda_2 \cdot \mathcal{L}^{\mathrm{RL}} + \lambda_3 \cdot \mathcal{L}^{\mathrm{DPP}}. \tag{13}$$

3 Experiments

We evaluate the accuracy, diversity, and efficiency of our proposed approach. We first describe the experimental settings.

3.1 Sequential Settings

Datasets. We conducted experiments on four datasets, including three e-commerce datasets, and a music recommendation dataset. We adopt three e-commerce categories from **Amazon** [17]: Books, Beauty, and CDs with different diverse sizes and sparsity levels. We take the subset of **Last.FM** [19] where the timestep is from Jan, 2015 to June, 2015. For all datasets, we remove users and items with less than three interaction records.

Evaluation Metrics. Following [23], we use Hit-Ratio@k and Normalized Discounted Cumulative Gain (NDCG@k) to evaluate the proposed method. Following [2], we use intra-list average distance (ILAD), and intra-list minimal distance (ILMD) to measure the diversity. Higher metrics are desirable. All the metrics were calculated based on the top-k recommendations to each user for each test case. k is set to 10. For each test case, we randomly sample 100 negative items and rank them with the ground-truth item.

Parameter Settings. We optimize all models with the Adam optimizer by setting the batch size as 2048, the coefficients used for computing running averages of gradient and its square as 0.001, the betas as (0.9, 0.999) without weight decay. The hidden layer sizes of the model used for Amazon dataset are set to 50, and 100 for Last.FM. All the hyper-parameters are obtained using grid search. The weights $\lambda_1, \lambda_2, \lambda_3$ are set to 1, 0.1 and 1 based on the analysis of experimental results. The path number and length are both set to 3 when sampling.

Table 1. Performance comparison between the baselines and our model. The best performance in each row is in bold font, and the starred numbers represent best baseline performance. The last column shows the absolute improvement of our results against the best baseline, which is significant at p-value $\leq$ 0.05.

Dataset	Metric ↑	KGAT	Ripple	GRU4Rec	KSR	KERL	CogRec$_{\text{Div}}$	Improvement
Beauty	Hit-Ratio@10	44.0	42.2	39.4	51.0	54.1*	**55.9**	3.3%
	NDCG@10	27.6	21.4	29.5	32.2	36.5*	**37.9**	3.8%
CDs	Hit-Ratio@10	63.4	58.4	50.5	68.3	73.7*	**75.1**	1.9%
	NDCG@10	41.7	37.6	32.9	45.0	50.8*	**52.6**	3.5%
Books	Hit-Ratio@10	70.2	63.8	56.2	75.1	80.0*	**80.7**	0.9%
	NDCG@10	45.8	42.8	38.5	52.4	57.1*	**58.7**	2.8%
Last.FM	Hit-Ratio@10	55.8	52.5	52.8	62.7	64.2*	**64.8**	0.9%
	NDCG@10	42.1	38.2	40.7	48.1	50.1*	**52.8**	5.4%

Compared Methods. Our model is compared with these competitive baselines:

- **KGAT** [25] explores the high-order connectivity with semantic relations in collaborative knowledge graph for knowledge-aware recommendation.
- **Ripple** [22] models users' potential interests along links in knowledge graph for recommendation through embedding.
- **GRU4Rec** [10] utilizes GRU to capture users' long-term sequential behaviors in session-based recommendation.
- **KSR** [11] integrates a memory network to a GRU-based sequential recommender with Knowledge-Enhanced Memory Networks.
- **KERL** [23] explicitly discusses and utilizes knowledge graph information for the exploration process in sequential recommenders with RL.

For KERL, we use the released code. For other baselines, we followed the same settings and used the results in the KERL paper.

3.2 Main Results

Table 1 shows our evaluation results of recommendation models. On the four datasets, our model outperforms all baselines both with respect to the two metrics. On *Books* and *Last.FM*, although our model does not improve the Hit-Ratio@10 much, it improves NDCG@10 a lot. That is to say, although the top-10 items recommended by our model and the baselines all hit the next item in the user interaction, the target item of the user ranks higher in the list recommended by our model.

Figure 2(a) shows the training process of KERL and our model CogRec$_{\text{Div}}$. Our model is more efficient than KERL during training, and the training process is more stable. Figure 2(b) shows the diversity metrics through training process. The diversity of the items recommended by KERL without constrains of the determinant is very variate, while the diversity of the recommendations produced by CogRec$_{\text{Div}}$ is limited in a small range. Notes that the lowest points of the diversity curves appears earlier than the highest points of the accuracy curves, which means that the diversity is helpful to the improvement of accuracy.

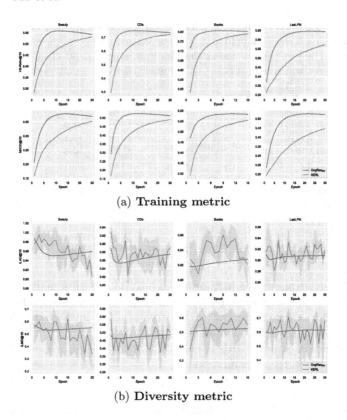

(a) **Training metric**

(b) **Diversity metric**

Fig. 2. Accuracy and diversity metric curves in training process. An early stop signal has been triggered during the training process of the CogRec$_{Div}$ model shown by the blue line. (Color figure online)

3.3 Ablation Study

Recall that we have three losses in Eq. 13, namely $\mathcal{L}^{sys_1}$ (Eq. 6), $\mathcal{L}^{RL}$(Eq. 8), $\mathcal{L}^{DPP}$(Eq. 12). Therefore, we consider three variants for comparison by examining the effect of each part for sequential recommendation, including: (1) Rec using only the representation got from system 1; (2) CogRec using two systems, but only uses RL in system 2; (3) Rec$_{Div}$ using two systems, but only uses DPP in system 2.

Table 2 shows the results of CogRec$_{Div}$ model and its three variants on four datasets. In general, System 1 can extract the information needed for recommendation better than other baseline models. System 2 considering long-term returns and the diversity can effectively improve the accuracy of recommendations. And according to the substantial improvement of NDCG, it can be seen that apart from Hit-Ratio, our model can rank the items that users will click next in the recommendation list higher.

Table 2. Performance comparison of Rec, CogRec, Rec$_{Div}$ and CogRec$_{Div}$.

Dataset	Metric	Rec	CogRec	Rec$_{Div}$	CogRec$_{Div}$
Beauty	Hit-Ratio@10	55.7	55.9	55.8	**55.9**
	NDCG@10	37.6	37.7	37.9	**37.9**
CDs	Hit-Ratio@10	74.6	75.1	74.5	**75.1**
	NDCG@10	51.9	52.5	51.8	**52.6**
Books	Hit-Ratio@10	80.3	80.5	80.7	**80.7**
	NDCG@10	58.3	58.5	58.4	**58.7**
Last.FM	Hit-Ratio@10	61.6	64.5	61.4	**64.8**
	NDCG@10	50.9	52.5	50.9	**52.8**

For *CDs* and *Last.FM*, System 2 with RL (CogRec) performs better than System 2 with DPP (Rec$_{Div}$) in both Hit-Ratio@10 and NDCG@10. However, for *Books*, the diversity part contributes a lot in Hit-Ratio@10, which leads to the conclusion that the both parts in System 2 are effective in improving recommendation accuracy.

3.4 Case Study

Figure 3 presents an example of the items recommended by the models based on a user's interactive sequence consisting of 11 items. In the interactive sequence, nail polish accounted for nearly half of the user's clicks. The target item for user's next interaction is a dark purple nail polish, which is expected to be included in the recommendation lists and ranked as high as possible. Among the recommendation lists of the four models, only the Rec model fails to hit the target, which means that it is infeasible to predict this user's interactive preference based on simple intuition system. Fortunately, the recommended lists of the remaining three models all include the target item. Among them, CogRec$_{Div}$, which includes two complete systems, not only hits the target, but also ranks it first. This shows that considering the diversity of long-term returns, in this example, the recommendation accuracy is significantly improved.

Although Rec's recommendation failed to hit the target, the variety of recommendation was not low, which was consistent with the trade-off relationship of accuracy and diversity. In the recommendation list of CogRec with the long-term benefit (System 2), there are more nail polishes recommended than the other three ones. Correspondingly, the variety of items is relatively lower than others, which concludes that the model with RL is more inclined to recommend items that is similar to the previous ones. The outcome of Rec$_{Div}$ with System 2 has the highest recommendation diversity. It not only recommends items in previous categories, but also adds a lot of items in categories that have not appeared before, such as hair care and styling tools. CogRec$_{Div}$'s recommendations balance the effects of accuracy and diversity at the same time.

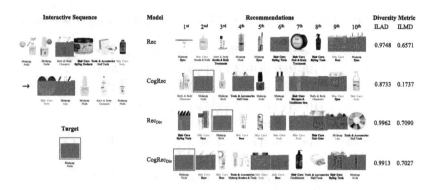

Fig. 3. A case where different models recommend items for the same sequence. The left side is the item sequence of the user's historical interaction and the target the user will click next. The middle one is the top 10 recommended items given by the four models. The category and subcategory of each item are marked below with bold and regular fonts. The diversity scores of each model's recommendation list are given on the right.

4   Related Works

The question of how the recommended satisfaction is affected by diversity and relevance has been deeply explored, and experiments show that diversity may have a negative net effect on user's satisfaction [13]. In terms of algorithms, the frequency of occurrence is considered, and at the same time, all projects are given fair exposure opportunities to strengthen the robustness of the system [18].

As for sequential recommendations, HISR [9] brings sequential dynamics into heterogeneous recommendations with textual and social information. SRec-GAN [16] combines the integration of adversarial training with BPR. Some scholars have recently used the DPP [1] to improve the diversity of the recommendation system. The adoption of DPP overcomes the NP-hard calculation problem and greatly accelerates the greedy MAP inference for DPP, which has a better performance in the relevance-to-diversity trade-off [2,8].

Currently, there are some approaches that combines knowledge graph and RL together. PGPR [27] introduces a sampling multi-path policy gradient method, and Ekar [20] expresses the problem as a sequential decision-making process. KGTN [4] designs the original path to join methods, and CPR [14] adds an attention mechanism to the actor network and proposes an interactive path reasoning algorithm on a heterogeneous graph.

The closest work to ours is KERL [23], which is a sequential recommendation system that combines RL and knowledge graphs. Unlike KERL, our model considers the **diversity** of recommendations and avoids the lack of user interest capture caused by frequent recommendations of similar items. On the other hand, we adopt the method of RL to assist **supervised** learning label prediction. And through multi-task learning, we can achieve better results more efficiently.

5 Conclusions

In this paper, we construct a recommendation system $CogRec_{Div}$ based on the dual-process theory in cognitive science. The dual-process cognitive recommender divides the recommendation into two subsystems. System 1 is used to quickly give intuitive representation based on the user's historical interactive items and the KG. System 2 analyzes the information and make further inferences, considering the long-term cumulative returns with RL and enriching the diversity of recommendation with DPP. Experiments show that our model has achieved state-of-the-art performance on all four data sets, while ensuring the diversity, the long-term returns, and the learning efficiency.

References

1. Calandriello, D., Derezinski, M., Valko, M.: Sampling from a k-DPP without looking at all items. Adv. Neural. Inf. Process. Syst. **33**, 6889–6899 (2020)
2. Chen, L., Zhang, G., Zhou, E.: Fast greedy MAP inference for determinantal point process to improve recommendation diversity. In: Advances in Neural Information Processing Systems, vol. 31, pp. 5622–5633. Curran Associates, Inc. (2018)
3. Chen, S.Y., Yu, Y., Da, Q., Tan, J., Huang, H.K., Tang, H.H.: Stabilizing reinforcement learning in dynamic environment with application to online recommendation. In: SIGKDD, pp. 1187–1196 (2018)
4. Chen, X., Huang, C., Yao, L., Wang, X., Liu, W., Zhang, W.: Knowledge-guided deep reinforcement learning for interactive recommendation. arXiv preprint arXiv:2004.08068 (2020)
5. Chen, X., et al.: Sequential recommendation with user memory networks. In: WSDM, pp. 108–116 (2018)
6. Cho, K., et al.: Learning phrase representations using RNN encoder-decoder for statistical machine translation. In: EMNLP, pp. 1724–1734 (2014)
7. Evans, J.S.B.: Dual-processing accounts of reasoning, judgment, and social cognition. Annu. Rev. Psychol. **59**, 255–278 (2008)
8. Gan, L., Nurbakova, D., Laporte, L., Calabretto, S.: Enhancing recommendation diversity using determinantal point processes on knowledge graphs. In: SIGIR, pp. 2001–2004 (2020)
9. Geng, H., Yu, S., Gao, X.: Gated sequential recommendation system with social and textual information under dynamic contexts. In: Jensen, C.S., et al. (eds.) DASFAA 2021. LNCS, vol. 12683, pp. 3–19. Springer, Cham (2021). https://doi.org/10.1007/978-3-030-73200-4_1
10. Hidasi, B., Karatzoglou, A., Baltrunas, L., Tikk, D.: Session-based recommendations with recurrent neural networks. CoRR abs/1511.06939 (2016)
11. Huang, J., Zhao, W.X., Dou, H., Wen, J.R., Chang, E.Y.: Improving sequential recommendation with knowledge-enhanced memory networks. In: SIGIR, pp. 505–514 (2018)
12. Ioffe, S., Szegedy, C.: Batch normalization: accelerating deep network training by reducing internal covariate shift. In: ICML, pp. 448–456 (2015)
13. Kunaver, M., Požrl, T.: Diversity in recommender systems-a survey. Knowl.-Based Syst. **123**, 154–162 (2017)
14. Lei, W., et al.: Interactive path reasoning on graph for conversational recommendation. In: SIGKDD, pp. 2073–2083 (2020)

15. Liu, Q., et al.: Building personalized simulator for interactive search. In: IJCAI, pp. 5109–5115 (2019)
16. Lu, G., Zhao, Z., Gao, X., Chen, G.: SRecGAN: pairwise adversarial training for sequential recommendation. In: Jensen, C.S., et al. (eds.) DASFAA 2021. LNCS, vol. 12683, pp. 20–35. Springer, Cham (2021). https://doi.org/10.1007/978-3-030-73200-4_2
17. McAuley, J., Targett, C., Shi, Q., Van Den Hengel, A.: Image-based recommendations on styles and substitutes. In: SIGIR, pp. 43–52 (2015)
18. Miyamoto, S., Zamami, T., Yamana, H.: Improving recommendation diversity across users by reducing frequently recommended items. In: Big Data, pp. 5392–5394. IEEE (2018)
19. Schedl, M.: The LFM-1b dataset for music retrieval and recommendation. In: ICMR, pp. 103–110 (2016)
20. Song, W., Duan, Z., Yang, Z., Zhu, H., Zhang, M., Tang, J.: Ekar: an explainable method for knowledge aware recommendation. arXiv preprint arXiv:1906.09506 (2019)
21. Tang, J., et al.: Towards neural mixture recommender for long range dependent user sequences. In: The World Wide Web Conference, pp. 1782–1793 (2019)
22. Wang, H., et al.: RippleNet: propagating user preferences on the knowledge graph for recommender systems. In: CIKM, pp. 417–426 (2018)
23. Wang, P., Fan, Y., Xia, L., Zhao, W.X., Niu, S., Huang, J.: KERL: a knowledge-guided reinforcement learning model for sequential recommendation. In: SIGIR, pp. 209–218 (2020)
24. Wang, S., Hu, L., Cao, L., Huang, X., Lian, D., Liu, W.: Attention-based transactional context embedding for next-item recommendation. In: AAAI, pp. 2532–2539 (2018)
25. Wang, X., He, X., Cao, Y., Liu, M., Chua, T.S.: KGAT: knowledge graph attention network for recommendation. In: SIGKDD, pp. 950–958 (2019)
26. Wu, S., Tang, Y., Zhu, Y., Wang, L., Xie, X., Tan, T.: Session-based recommendation with graph neural networks. In: AAAI, vol. 33, pp. 346–353 (2019)
27. Xian, Y., Fu, Z., Muthukrishnan, S., De Melo, G., Zhang, Y.: Reinforcement knowledge graph reasoning for explainable recommendation. In: SIGIR, pp. 285–294 (2019)
28. Xin, X., Karatzoglou, A., Arapakis, I., Jose, J.M.: Self-supervised reinforcement learning for recommender systems. In: SIGIR, pp. 931–940 (2020)
29. Yuan, F., Karatzoglou, A., Arapakis, I., Jose, J.M., He, X.: A simple convolutional generative network for next item recommendation. In: WSDM, pp. 582–590 (2019)
30. Zhang, Z., Lu, T., Li, D., Zhang, P., Gu, H., Gu, N.: SANS: setwise attentional neural similarity method for few-shot recommendation. In: Jensen, C.S., et al. (eds.) DASFAA 2021. LNCS, vol. 12683, pp. 69–84. Springer, Cham (2021). https://doi.org/10.1007/978-3-030-73200-4_5
31. Zhao, X., Zhang, L., Ding, Z., Xia, L., Tang, J., Yin, D.: Recommendations with negative feedback via pairwise deep reinforcement learning. In: SIGKDD, pp. 1040–1048 (2018)
32. Zheng, G., et al.: DRN: a deep reinforcement learning framework for news recommendation. In: Proceedings of the 2018 World Wide Web Conference, pp. 167–176 (2018)

Learning and Fusing Multiple User Interest Representations for Sequential Recommendation

Ming He[✉], Tianshuo Han, and Tianyu Ding

Beijing University of Technology, Beijing, China
heming@bjut.edu.cn, {hants,dingtianyu}@emails.bjut.edu.cn

Abstract. Deep learning is known to be effective at automating the generation of representations, which has achieved great success by learning efficient representations from data, especially for user and item representations in sequential recommendation. However, most existing methods usually represent a user's interest by one independent representation vector, which is inadequate to capture user's diverse interests. We aim to fully characterize the diversity of user interests in this study to improve the recommendation performance. We propose a *Multiple User Interest Representations* (MUIR) model that learns and fuses user's interests from different aspects. To learn different levels of user interests, we specifically leverage two self-attention-based modules that better capture user's local and global interests respectively. By considering the information of the recent interacted items, we further design a gating module to balance the interests, which is capable of modeling how user interests evolve and interact in recent behavior sequence. Extensive experiments conducted on the real-world datasets demonstrate that our model, MUIR, outperforms existing state of-the-art methods significantly.

Keywords: Recommender systems · Sequential recommendation · Attention mechanism

1 Introduction

With the advancement of Internet, the huge volume of data generated a problem of information overload, and it becomes difficult to identify what interests them. To alleviate the impact of the abovementioned problem, recommender systems play a crucial role in helping users target their interests.

Recent studies formalize [13] the recommender systems as a sequential recommendation problem. Sequential recommendation regards a user's behavior history as a sequence of items, aiming to predict the next-interacted item. *Recent studies usually provide an overall embedding (with large dimensionality) to represent user's diverse interests [7]; however, an overall embedding is hard to represent multiple interests.* For example, history records show a user's interests in books, games, and electronic products, as shown in Fig. 1, so we argue that

A. Bhattacharya et al. (Eds.): DASFAA 2022, LNCS 13247, pp. 401–412, 2022.
https://doi.org/10.1007/978-3-031-00129-1_34

Fig. 1. A simple example of the items which are interacted by user at different time. The behavior list shows that the user has a diverse interest in games, electronics and books.

existing sequential recommendation models may be sub-optimal for next-item prediction due to the bottleneck of learning a single embedding to represent the user's multiple interests.

The existing recommendation methods represent and model user's interestsin various ways. MIND [7] introduces capsule networks into recommendation, but because of the fixed internal structure, its representation performance is still limited. SASRec [6] models the entire user sequence without any recurrent or convolutional operations. *It must be noted that these methods balance diverse interests without considering how user's interests evolve in recent behaviors, and neglect the overall representation of the whole sequence.* In this study, we propose a model MUIR to represent a user with multiple representations vectors and address user diverse interests. Each of the vectors could be used to retrieve items independently. The key contributions of this study are summarized as follows:

- We propose a novel model named MUIR that captures user's diverse interests, and combine these interests to generate the user embedding that represents different levels of user interests.
- To effectively learn user's local-level and global-level interest representations, we apply two self-attention-based modules that combine side information (e.g., ratings) and session contextual information in the learning process, respectively, which are suitable for long sequence of user-item interactions.
- We design an MLP-based gating module, which balances the local-level and global-level representations by considering the recently interacted items.
- We have conducted extensive experiments to evaluate the performance of our model, MUIR, on the basis of three real-world datasets. These results reveal that our method significantly outperforms baseline methods.

2   Related Work

2.1   Sequential Recommendation

Different from previous methods, BERT4Rec [12] introduced a Bidirectional Encoder Representations model and try to predict the masked items in the sequence. However, the structures used by previous SASRec-based [2,6,8] works may not be applied directly and even cause some problems in accuracy in our task for their complex internal structure. *To the best of our knowledge, this paper is the first work that proposes leveraging self-attention-based multiple user interests extraction method for sequential recommendation. Compared with the current multi-interest models, our model is also suitable for dataset with long sequence of user-item interactions.*

2.2   User Representation

Recent works usually obtain user representation by neural networks [7]. The DIN-based methods [14] modeled evolving user interests and extracts latent temporal interests from user behavior sequence. S3-Rec [15] devised four self-supervised learning objectives to learn the correlations within the raw data. *Unlike most existing methods like [6,14], we generate multiple representation vectors to model user's diverse interests, and different from the previous self-attentive model, we utilizing the overall session contextual information and side information in the attention scores calculation.*

2.3   Representations Fusion

Some of the existing methods [5] employ LSTM structure to decide contribution percentages of different representations. Previous literature [11] fused global and local representations under the auto-encoder framework. We find that most of the existing fusion methods represent user interests by a weight summation and neglect how user interests evolve. *In this study, we consider combining both the local-level and global-level user preferences embeddings with a fusion layer, and utilize the recently interacted items when generate the final representation.*

3   Proposed Method

3.1   Problem Formulation

Let U and I represent the users and items set. In general, for each user $u \in U$, we denote the user's interaction sequence in a short period as $S^u = (s_1^u, s_2^u, ..., s_L^u)$ where s_t^u represents the t-th item interacted by u and L is the length of the sequence S^u. We also introduce side information [9] (such as ratings and positions) into our model, which can be defined as information sequence $R^u = (r_1^u, r_2^u, ..., r_L^u)$. The core task of our model, MUIR, is to generate interests representation $V_u \in \mathbb{R}^{K \times d}$ for user u, which represents the K interests of u with dimension d, and it can be formulated as:

$$V_u = \left(v_u^1, ..., v_u^K \right). \tag{1}$$

3.2   Embedding

As shown in Fig. 2, the input of our model, MUIR, is the user behavior sequence $S^u = (s_1^u, s_2^u, ..., s_L^u)$. We define a learnable item embedding matrix $M \in \mathbb{R}^{|I| \times d}$ where d is the latent dimension and $|I|$ is the number of items. We generate an embedding matrix $E^u = \left[m_{s_1}^u, m_{s_2}^u, ..., m_{s_L}^u\right]$ from M after the embedding lookup operation where $E^u \in \mathbb{R}^{L \times d}$. With the R^u, we could also obtain the side information embedding matrix $F^u = \left[f_{s_1}^u, f_{s_2}^u, ..., f_{s_L}^u\right] \in \mathbb{R}^{L \times d}$. Since the self-attention model is not aware of the position information, we also specify a learnable positional embedding matrix $P = [p_1, p_2, ..., p_L] \in \mathbb{R}^{L \times d}$ to capture position information. The method of initializing P is similar to that of initializing E. Then we could add P to E^u to obtain the **item embedding matrix** $X^u = \left[x_{s_1}^u, x_{s_2}^u, ..., x_{s_L}^u\right] \in \mathbb{R}^{L \times d}$:

$$x_{s_i}^u = m_{s_i}^u + p_i, i \in \{1, 2, ..., L\}. \tag{2}$$

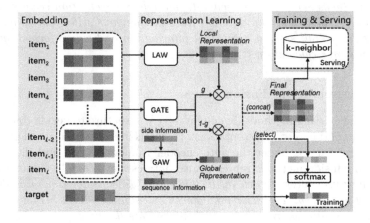

Fig. 2. An overview of our proposed model, MUIR. The input of our model, is a list of item IDs of one user. We feed them into the embedding layer (see Cap. (3.2)). Local-level and global-level interest representations are generated through two learning modules, LAW and GAW (see Cap. (3.3, 3.4)). To calculate the balanced weights and generate the final representation, we further design the GATE module (see Eqs. (11, 12)), and discuss training and serving details

3.3 Local-Level Representation Learning

The following example explains how to obtain one single interest for user u. Given the input matrix X^u, we first use the self-attention based module $(LAW(\cdot))$ to obtain a vector of local-level attention weights $a \in \mathbb{R}^{1 \times L}$ and then compute one of the user's interest $v_u^l = aX^u$ where $v_u^l \in \mathbb{R}^{1 \times d}$. This vector v_u^l focuses on and reflects specific interest of u. We need generate multiple v_u^l to capture user's diverse interests from interaction sequence to represent user u. As shown in Fig. 2, we need to obtain the **attention matrix** $A \in \mathbb{R}^{(K-1) \times L}$ and compute the user's diverse **local-level interest** representations $V_u^l \in \mathbb{R}^{(K-1) \times d}$ by:

$$V_u^l = AX^u. \tag{3}$$

Note that the main difference between our model and the literature [6] is that we do not adopt layer normalizations, residual connections, stack blocks or dropout regularization techniques, and we only utilize one feedforward layer when generating the attention matrix in Eq. (6), and it is helpful in our model. To improve the extraction performance of the user's diverse interests, we also employ a modified attention method. Our $LAW(\cdot)$ can be viewed as a self-attention layer $SAL(\cdot)$ followed by an interest-extraction layer $IEL(\cdot)$ as follows:

$$A = LAW\left(X^u\right) = IEL\left(SAL\left(X^u\right)\right). \tag{4}$$

We will introduce the two layers in detail. Different from previous work [6], we employ a modified no-softmax self-attention method to improve the learning ability, which will be discussed in Equations RQ3. We use a causality mask to make sure that the model considers only the first $(t-1)$ items when calculating the weight of item t following [6]. The self-attention layer $SAL(\cdot)$ and its output X_{sal}^u are defined as follows:

$$X_{sal}^u = SAL(X^u) = FFL((\frac{QK^\top}{\sqrt{d}}) \triangle V), \tag{5}$$

$$FFL\left(X\right) = tanh\left(XW_1 + b_1\right)W_2 + b_2, \tag{6}$$

where $X_{sal}^u \in \mathbb{R}^{L \times d}$ and will be used in $IEL\left(\cdot\right)$, $Q = X^u W_Q, K = X^u W_k$ and $V = X^u W_V$ with $W_Q, W_K, W_V \in \mathbb{R}^{d \times d}$ are the projected query, key, and value matrices, $W_1, W_2 \in \mathbb{R}^{d \times d}$ and $b_1, b_2 \in \mathbb{R}^d$ are the parameter matrices and vectors, $\triangle$ is the unit lower triangular matrix with size $L \times L$ and $\top$ denotes the transpose of the vector or the matrix.

The $IEL\left(\cdot\right)$ is introduced to calculate the user's attention weights of items for interest-extraction. Recent work SINE [13] also employs similar network structures. With the input $X_{sal}^u \in \mathbb{R}^{L \times d}$ calculated by Eq. (5) and parameter matrices $W_3 \in \mathbb{R}^{(K-1) \times w}$, $W_4 \in \mathbb{R}^{d \times w}$, we can employ $IEL\left(\cdot\right)$ to generate the **attention matrix** A used in Eq. (3):

$$A = IEL\left(X_{sal}^u\right) = softmax(W_3(tanh(X_{sal}^u W_4)^\top)). \tag{7}$$

3.4   Global-Level Representation Learning

In a real recommendation scenario, we need to discover the overall preferences of users and filter the noise in the behavior sequence, and integrate the side information into our model. So, to learn a vector of global-level attention weights $a \in \mathbb{R}^L$, we use the sequence information in the self-attention-based module $(GAW\left(\cdot\right))$ and get the Global Representation in Fig. 2, and this is different from the attention mechanism in the previous chapter. Different from the SASRec-based methods [6], we model the relationships between each item and take the overall contextual information into consideration. The input $X_{gal}^u \in \mathbb{R}^{L \times d}$ of $GAW\left(\cdot\right)$ is defined as follows:

$$S_u = \frac{1}{L}\sum_{i=1}^{L} S_i^u,$$

$$X_{gal}^u = FFL'((\frac{Q'K'^\top}{\sqrt{d}})V'), \tag{8}$$

where $Q' = X^u W'_Q, K' = S'_u W'_k$ and $V' = X^u W'_V$ with $W'_Q, W'_K, W'_V \in \mathbb{R}^{d \times d}$, $S'_u \in \mathbb{R}^{L \times d}$ is the broadcast result of $S_u \in \mathbb{R}^d$, and $FFL'(\cdot)$ is similar to $FFL(\cdot)$ defined in Eq. (6). We also try to use the attention matrix A defined in Eq. (7) to generate a weighted summation of S_u, but it does not work, which will be discussed in Eq. (16).

When we model the whole behavior sequence, the side information (such as ratings, browse, purchase, etc.) can also reflect user's preferences. For example, the products purchased by users can better reflect user's interests than the browsed products. The side information only participants in the attention matrix calculation [9]. With the side information F^u (ratings or behavior type, padding with *zeros*) defined in Cap. (3.2), we could obtain **the attention matrix** $a \in \mathbb{R}^{1 \times L}$ and compute the user's **global-level interest** representations $V_u^g \in \mathbb{R}^{1 \times d}$ with $GAW(\cdot)$ as follows:

$$X = [X_{gal}^u W_x, S'_u W_s] W_{xs},$$

$$a = GAW(X, F^u) \tag{9}$$

$$= softmax(W_5(tanh((X + F^u) W_6)^\top)),$$

$$V_u^g = a X^u, \tag{10}$$

where $W_x, W_s, \in \mathbb{R}^{d \times d}$, $W_{xs} \in \mathbb{R}^{2d \times 2d}$, $W_5 \in \mathbb{R}^{1 \times w}$, and $W_6 \in \mathbb{R}^{2d \times w}$, $[\cdot, \cdot]$ denotes the concatenation operation, we set w to 1 in the global-level module.

3.5   Representation Gating

To balance the local representation and the global representation, existing methods usually use summation or concatenation. Note that both of these methods balance the diverse interests by the extracted interest representations, which neglects how user interests evolve and interact in the recent behaviors.

We design an MLP-based gating module to calculate the weights of the local-level and global-level representations and balance them. Inspired by the literature [4], which balances the two representations by considering the consistency of the interaction sequence, we propose to model the consistency between the recently interacted items and the target item to balance the representations. In the serving stage, we replace the target item with the last interacted item. In particular, we obtain the representation of the target item $i \in I$ and the last $(t + 1)$ items in S^u from the item embedding matrix M, i.e., $m_i^u, m_{L-t}^u, ..., m_L^u$, and the gating module is defined as follows:

$$g = \sigma \left(\left[m_i^u, m_{L-t}^u, ..., m_L^u \right] W_g + b_g \right), \tag{11}$$

where $g \in \mathbb{R}$, $\sigma(\cdot)$ is the sigmoid activation function, $[\cdot, \cdot]$ denotes the concatenation operation, $W_g \in \mathbb{R}^{(t+2)d \times 1}$ and $b_g \in \mathbb{R}$. We set t to 3. Then we could obtain the interests of user u (V_u) used in Eq. (1) by the concatenation of the results obtained from Eq. (10) and Equation (3):

$$V_u = [g V_u^l, (1 - g) V_u^g]. \tag{12}$$

In training stage, we only take one representation into consideration. We update the interest extraction weights of this representation, so that each independent interest representation will obtain a specific semantics. We choose this interest representation v^u from V^u by an *argmax* operation:

$$arg \max_{v^u \in V^u} (V^u t_i^u), \tag{13}$$

where t_i^u denotes the embedding of the target item, and minimize the following negative log-likelihood:

$$loss = \sum_{u \in U} \sum_{i \in I_u} -\log \frac{exp(v^u t_i^u)}{\sum_{k \in I} exp(v^u m_k)}, \tag{14}$$

where I_u is the set of interacted items of user u, and m_k is the item embedding defined in Cap. (3.2). In serving stage, the representation vectors in $V_u \in R^{K \times d}$ are used to retrieve top N/k items from I.

4 Experiments

4.1 Experimental Setup

Datasets. We conduct experiments on three public datasets. We split all the users and their behavior sequences into groups by the proportion of 8:1:1 for train, valid and test. Each user only appears in one dataset. The statistics of the processed datasets are shown in Table 1.

- **MovieLens-1M**[1] is a widely used benchmark dataset containing 1M ratings with 6040 users and 3900 movies (items).
- **Amazon-Games**[2] contains games and other electronic products.
- **Amazon-Books**[3] consists of product reviews and metadata from Amazon website. Note that this 5-core version is more challenging than the version used in [10], due to its large volume and sparsity.

Table 1. Dataset statistics (after preprocessing)

Dataset	Users	Items	Interactions	Density
MovieLens-1M	6,040	3900	1,000,209	4.2%
Video Games	90,901	46,681	560,458	0.013%
Books	459,133	313,966	8,898,041	0.0061%

Baselines. Due to the space limitation, we ignore the classic models [6]. To make a fair comparison, we adjust the hyperparameters in the first two datasets, and report the original results of Books in [1] with six methods:

---

[1] https://grouplens.org/datasets/movielens/.
[2] http://jmcauley.ucsd.edu/data/amazon/.
[3] http://jmcauley.ucsd.edu/data/amazon/.

- **SASRec** [6] is widely used SOTA self-attention based sequential model for recommendation. We choose it because it is widely used as baseline model by previous works [8,13,15], and our model **obtains a more significant performance improvement than these methods.**
- **YouTube DNN** [3] is one of the most successful deep learning methods used in recommendation system.
- **GRU4Rec** [5] is the first seminal method that uses RNN to model user action sequences for sequential recommendation.
- **MIND** [7] proposes a new structure of capsule routing mechanism.
- **ComiRec-DR** [1] follows the original dynamic routing method to generate multiple user interests.
- **ComiRec-SA** [1] is an attention version of ComiRec-DR, and performs better than ComiRec-DR in some cases.

Evaluation Metrics and Implementation Details. We report the **Recall@N** ($R@N$), **Hit@N** ($H@N$), as well as one ranking-specific metrics **NDCG@N** ($N@N$) [1]. Please note that the evaluation metrics used in SAS-Rec [6] are different from ours, so the experimental results are also different. Our evaluation metrics method is more interpretable [1]. Some recent works like SINE [13], also reported similar results. For a fair comparison with the baseline models, the number of dimension d is set to 64 for every models. We set the learning rate $lr = 0.002$ for Amazon-Books and 0.004 for others. The number of interests k is set 6 for all the multiple interests models following [1]. We set the length of training sequence L to 20 for all the models. We set dropout rate to 0.05, and batch size to 128 based on our experiment results.

4.2   Experimental Results

Recommendation Performance. As shown in Table 2, our model achieves the superior performance in all the three experiments compared with all the baselines, where the improvements range from 4.9% to 55.4% in Recall@20 and from 3.4% to 50.0% in NDCG@20, and from 5.9% to 37.6% in Recall@50 and from 5.9% to 32.4% in NDCG@50, which demonstrates that our model significantly improves the recommendation performance. Therefore, exploring multiple user-embedding vectors may be an more effective way of modeling users' diverse interests.

Parameter Sensitivity. We study the effect of two hyperparameters (embedding dimension d, interests number k) in $R@20(50)$ and $N@20(50)$ ON Book and Game, and report the results. Our model, MUIR, achieves better results as the embedding dimension d gets bigger, as shown in Table 3a. We can find that when the dimension is low (64), we could also get similar results compared with the higher dimension (128), and We find that the improvement of interests number k could also make our model, MUIR, perform better in most cases, as shown in Table 3b.

Table 2. Recommendation performance. Due to space constraints, we omit the '@' symbol and use CR to represent the baseline ComiRec. The best result in each column is marked in bold, and the second-best result is marked using an underline. 'Impro. (%)' denotes the percentage improvement of the proposed model with respect to the best performing value in the baselines. All the numbers are in percentage

	MovieLens						Video games						Books					
	R20	N20	H20	R50	N50	H50	R20	N20	H20	R50	N50	H50	R20	N20	H20	R50	N50	H50
DNN	20.2	38.9	54.7	35.8	56.3	73.5	8.2	10.3	11.7	13.3	16.4	18.3	4.6	7.7	10.3	7.3	12.1	15.9
GRU	12.5	27.2	31.4	21.6	38.5	43.6	7.1	9.0	12.0	10.2	14.5	16.5	4.1	6.8	8.9	6.5	10.4	13.7
SASRec	—	31.9	44.3	—	49.3	66.4	—	10.9	13.2	—	18.4	19.4	—	8.1	10.9	—	11.4	15.9
MIND	17.2	34.7	50.6	31.8	52.9	70.4	10.2	12.5	13.9	15.7	19.1	20.1	4.9	7.9	10.6	7.6	12.2	16.2
CR-SA	14.8	30.7	43.7	30.3	51.0	68.1	9.8	11.9	13.2	16.1	19.3	21.3	5.5	9.0	11.4	8.5	13.6	17.2
CR-DR	17.5	35.3	41.7	31.7	52.2	69.9	10.0	11.8	15.3	15.9	19.1	20.8	5.3	9.2	12.0	8.1	13.5	17.6
MUIR	**21.2**	**40.2**	**56.5**	**37.9**	**59.6**	**76.5**	**13.0**	**15.9**	**17.7**	**20.4**	**24.4**	**27.0**	**8.5**	**13.8**	**17.0**	**11.7**	**18.0**	**22.5**
Impro.(%)	4.9	3.4	3.3	5.9	5.9	4.1	27.5	27.2	27.3	26.7	26.4	26.7	55.4	50.0	41.7	37.6	32.4	27.8

Table 3. (a) Recommendation performance with different embedding dimension d. All the numbers are in percentage (b) Recommendation performance with different k. All the numbers are in percentage

d	R20@B	N20@B	R20@G	N20@G
32	7.9	12.3	12.7	15.5
64	8.5	13.8	13.0	15.9
96	8.9	14.4	13.8	16.3
128	9.2	15.0	14.3	16.8

k	R50@B	N50@B	R50@G	N50@G
4	10.3	17.0	19.7	23.7
6	11.7	18.0	20.4	24.4
8	12.1	18.6	21.5	25.1
10	12.7	19.1	21.8	25.9

Ablation Study. We also discuss some options of structures in Amazon Books in the left of Fig. 3, and the variant models of are designed as follows:

- **MUIR(+s)** adds softmax function when compute the self-attention scores in Eq. (5, 8), it makes the model hard to find the weights of items in the interactions sequence.
- **MUIR(+f)** adopts the stack structure similar to Transformers to learn the matrix A used in Eq. (4).
- **MUIR(-)** removes the unit global module $GAW(\cdot)$ used in Eq. (8, 10), which reduces the model's ability to capture sequential information.

We can see that if we employ softmax or multi-layer feedforward structures like SASRec, which are denoted as '**MUIR(+s)**' and '**MUIR(+f)**', our model will perform worse, which indicating that the **original multi-layers transform-based structures can not be directly used in this experiment**. Model performance will also decrease slightly if we do not use the global module, which is denoted as '**MUIR(-)**', this result shows that the overall session contextual information and side information are more helpful in exploring the user's interests.

From the right of Fig. 3, we can find that only considering the last one interacted item or the interacted items with weights, which are denoted as '**gate-1**'

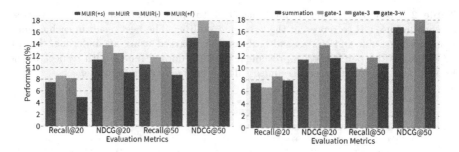

Fig. 3. Recommendation performance with different designs

and 'gate-3-w', our model will achieve worse results. '**summation**' (adding the global-level representation to each local-level representation) achieves slightly worse performance compared with the default model, which is denoted as '**gate-3**', possibly because the summation operation will cause some information loss. The variant models are designed as follows:

- **gate-1** only considering the last one interacted item in the gating module, which may not be able to reflect the user's recent interests:

$$g = \sigma \left(m_L^u W_g + b_g \right) \tag{15}$$

- **gate-3-w** attempts to use the attention weight scores learned in global-level module when computing gating score g, which may also be unable to reflect the user's recent interests:

$$g = \sigma \left([a_{L-2}m_{L-2}^u, a_{L-1}m_{L-1}^u, a_L m_L^u] W_g + b_g \right). \tag{16}$$

- **summation**, instead of computing the gating scores, it adds the learning results of the two level directly as many other models do.

Efficiency. In Fig. 4, we show the training results of the first 100,000 iterations in terms of Recall@50 in the Amazon Books dataset. When the training progressed to 50,000 iterations, our model achieves the best result, at the same time, other models still have not achieved their best results. When the training progressed to 20,000 iterations, our model's performance has exceeded that of the other models.

Case Study. To visualize the learned conceptual prototypes of our model, we conducted case study experiments in the Video Games dataset, following previous work [1]. As shown in Fig. 5, we can find that U_1 has more diverse interests, whereas U_2 pays more attention to games. For U_1, we can find that his interests are about headsets, Xbox controllers, shooting games and role-playing games. Our recommendation list covers all their interests.

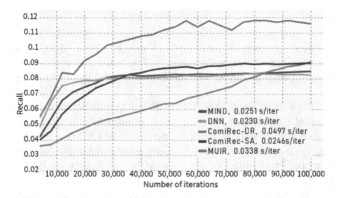

Fig. 4. In the line chart, we show the relationship between model performance and training iterations; our model achieved its best result in 50,000 iterations

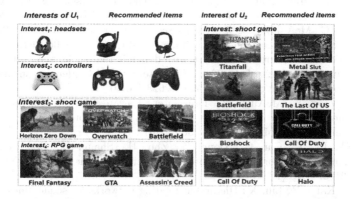

Fig. 5. Case study. We report items in interaction sequence to represent their interests on the left, and show the related recommended items on the right

5 Conclusion

We propose a novel model MUIR for sequential recommendation in this study. With the help of attention-based learning modules and fusion module, our model, MUIR, could capture multiple user interests and generate recommendation lists that match their diverse interests. The results on three challenging real-world datasets confirm the effectiveness and efficiency of our model.

Acknowledgements. This work is supported by Beijing Natural Science Foundation under grant 4192008.

References

1. Cen, Y., Zhang, J., Zou, X., Zhou, C., Yang, H., Tang, J.: Controllable multi-interest framework for recommendation. In: Proceedings of the 26th ACM SIGKDD International Conference on Knowledge Discovery and Data Mining, pp. 2942–2951 (2020)

2. Cho, S.M., Park, E., Yoo, S.: MEANTIME: mixture of attention mechanisms with multi-temporal embeddings for sequential recommendation. In: Fourteenth ACM Conference on Recommender Systems, pp. 515–520 (2020)

3. Covington, P., Adams, J., Sargin, E.: Deep neural networks for YouTube recommendations. In: Proceedings of the 10th ACM Conference on Recommender Systems, pp. 191–198 (2016)

4. He, Y., Zhang, Y., Liu, W., Caverlee, J.: Consistency-aware recommendation for user-generated item list continuation. In: Proceedings of the 13th International Conference on Web Search and Data Mining, pp. 250–258 (2020)

5. Hidasi, B., Karatzoglou, A., Baltrunas, L., Tikk, D.: Session-based recommendations with recurrent neural networks. arXiv preprint arXiv:1511.06939 (2015)

6. Kang, W.C., McAuley, J.: Self-attentive sequential recommendation. In: 2018 IEEE International Conference on Data Mining (ICDM), pp. 197–206. IEEE (2018)

7. Li, C., et al.: Multi-interest network with dynamic routing for recommendation at Tmall. In: Proceedings of the 28th ACM International Conference on Information and Knowledge Management, pp. 2615–2623 (2019)

8. Li, J., Wang, Y., McAuley, J.: Time interval aware self-attention for sequential recommendation. In: Proceedings of the 13th International Conference on Web Search and Data Mining, pp. 322–330 (2020)

9. Liu, C., Li, X., Cai, G., Dong, Z., Zhu, H., Shang, L.: Non-invasive self-attention for side information fusion in sequential recommendation. arXiv preprint arXiv:2103.03578 (2021)

10. Ma, C., Kang, P., Liu, X.: Hierarchical gating networks for sequential recommendation. In: Proceedings of the 25th ACM SIGKDD International Conference on Knowledge Discovery and Data Mining, pp. 825–833 (2019)

11. Ma, C., Kang, P., Wu, B., Wang, Q., Liu, X.: Gated attentive-autoencoder for content-aware recommendation. In: Proceedings of the Twelfth ACM International Conference on Web Search and Data Mining, pp. 519–527 (2019)

12. Sun, F., Liu, J., Wu, J., Pei, C., Lin, X., Ou, W., Jiang, P.: BERT4Rec: sequential recommendation with bidirectional encoder representations from transformer. In: Proceedings of the 28th ACM International Conference on Information and Knowledge Management, pp. 1441–1450 (2019)

13. Tan, Q., et al.: Sparse-interest network for sequential recommendation. In: Proceedings of the 14th ACM International Conference on Web Search and Data Mining, pp. 598–606 (2021)

14. Zhou, G., et al.: Deep interest network for click-through rate prediction. In: Proceedings of the 24th ACM SIGKDD International Conference on Knowledge Discovery and Data Mining, pp. 1059–1068 (2018)

15. Zhou, K., et al.: S3-rec: self-supervised learning for sequential recommendation with mutual information maximization. In: Proceedings of the 29th ACM International Conference on Information and Knowledge Management, pp. 1893–1902 (2020)

Query-Document Topic Mismatch Detection

Sahil Chelaramani, Ankush Chatterjee, Sonam Damani, Kedhar Nath Narahari, Meghana Joshi, Manish Gupta[(⊠)], and Puneet Agrawal

Microsoft, Hyderabad, India
{sachelar,anchatte,sodamani,kedharn,mejoshi,gmanish,
punagr}@microsoft.com

Abstract. Query-document topic match is one of the central signals for ranked information retrieval. Returning documents which are too broad or insufficient or off-topic compared to the query leads to a poor user experience. Thus, given a query and a document, it is critical to estimate degree of topical mismatch between the two. Previous work has either focused on very broad topics (like health, education, science, etc.) or predicted extremely-fine-level topic mismatch using query reformulations which suffer from sparsity. Predicting query-document topic mismatch is difficult because it needs semantic understanding of both the query and the document. In this paper, we model the problem as a five-class classification problem using a novel Transformer-based architecture. Our technique takes the query as input along with a detailed document representation including title, URL, snippet, key-phrases and topic distribution, and outputs one of these five grades of topic match: Very Unsatisfactory, Unsatisfactory, Neutral, Satisfactory, and Very Satisfactory. On a large dataset of ~2.43M query-document pairs, we show that our proposed method can provide an AUC of 0.75.

1 Introduction

Multiple signals play a very important role in web ranking: query-document relevance match, document authority, document freshness, document credibility, query-document topic match, location match for location sensitive queries, etc. Owing to the diversity of documents that modern search engines index, different degrees of topic mismatch may happen while retrieving results. For example while searching for "iphone X headphones", the results may return documents related to "iphone 6s headphones", which are semantically related but have distinctly different intents. When presented to users, such cases where documents are topically similar are preferable to cases where there is a complete topic mismatch between query and document. Table 1 shows examples of query-document pairs with different degrees of topic mismatch along with reasons. Typically, topic mismatch occurs because of dropping, substitution, shuffling or addition of entities, intents or other qualifiers in a document compared to a query.

In this paper, we model the problem of automatically predicting the degree of topic mismatch between the query and the document as a 5-class classification problem. Specifically, given a query-document pair, we want to classify it into one of these five classes: Very Unsatisfactory, Unsatisfactory, Neutral, Satisfactory, Very Satisfactory. Very Satisfactory implies that the document is almost completely about/related

A. Bhattacharya et al. (Eds.): DASFAA 2022, LNCS 13247, pp. 413–424, 2022.
https://doi.org/10.1007/978-3-031-00129-1_35

Table 1. Examples of Query-Document topic mismatch, along with the reasons for their mismatch. (VU = Very Unsatisfactory, U = Unsatisfactory, N = Neutral, S = Satisfactory, VS = Very Satisfactory)

Query	Document topic	Label	Reason of topic mismatch
Mail settings in iphone	Settings in iPhone	U	Doc is not about *mail* settings
iphone 6 earphones	iPhone earphones	N	Doc is not about iphone *6*
Harvard graduates at apple	Apple harvesting season	VU	Doc is about a completely different meaning of *apple*
Windows vista display settings	Windows XP display settings	N	Doc is about *XP* and not about *vista*
Convert pdf to ppt	Convert ppt to pdf	N	Doc is about *ppt to pdf* and not about *pdf to ppt*
Windows types and prices	Types of windows	S	Doc is not about *prices* of windows
Anxiety in teens	Symptoms of anxiety in teens	S	Doc is about symptoms only and does not discuss other aspects of anxiety
Juvenile arrests	Juvenile crime trends	U	Doc is about *crime trends* and not about *arrests*
Vinyl flooring	Flooring	S	Doc is too broad about all kinds of flooring and not specifically about *vinyl flooring*
Desserts requiring refrigeration	Desserts	U	Doc is too broad about all kinds of desserts and not specifically about those that *require refrigeration*
Good effects of running	Bad effects of running	U	Doc is about *bad effects* and not about *good effects*
ww2 italian battles	ww2 italian battles	VS	Doc is exactly on same topic as query

and almost exclusively about an interpretation of the query, and there is no topic mismatch. Very Unsatisfactory implies that the document is off-topic and not about/related to any possible interpretation of the query, resulting into a complete topic mismatch. We refer to the problem as the topic mismatch detection (TMD) problem. Although this is such a critical problem for effective ranking, there is hardly any published work in this area, mainly due to lack of labeled data.

One possible way of solving the TMD problem is based on query-URL click-through graph. Click history accumulated for query-URL pairs can be leveraged to serve as a proxy for user-defined topic match between queries and documents. Unfortunately, click signals are very sparse, and hence our experiments with clicks and impressions features show poor results (Sect. 3). Another approach is to perform topic modeling for both the query and the document, and build models that check for topic mismatch based on discrepancy in topic distributions. In this work, we investigate the effectiveness of deep learning based semantic methods for topic mismatch prediction. Document-side input to the deep learning models could comprise of document metadata (like title, URL or snippet[1]) or signals from entire content (like topics or key-phrases). Our experiments show that both kinds of signals (document metadata as well as content) lead to improved results.

First, we use BERT (Bidirectional Encoder Representations from Transformers) [8] to infer topic mismatch by finetuning the pretrained BERT on a ranking corpus, and then further finetuning for topic mismatch prediction. Second, we propose a novel modification (task-specific input smoothing) to BERT model for better generalization to

---

[1] A snippet is a short summary of webpage that appears in the Bing search results. Snippets are generated based on the search term and are presented as part of a search result list.

semantically related variants of input words. This modification helps us improve TMD accuracy while remaining within model size and latency budgets. Third, besides representing the document using URL, title and snippet tokens, we also experiment with using key-phrases [21] and topic distribution to obtain a richer document representation. Our experiments show that these novel modification to the vanilla-BERT input lead to significantly better results.

Overall, we make the following contributions. (1) We propose and concretely define the query-document topic mismatch detection (TMD) problem. We model it as a 5-class classification problem, where the classes indicate the degree of topic mismatch. (2) We propose a novel BERT-based model for this task using a comprehensive document representation (based on title, URL, snippet, key-phrases, topic distribution) and an overall smoothed input where smoothing weights are learned in a task-specific manner. (3) Our experiments with $\sim$2.43M query-document pairs provide an AUC of 0.75, an accuracy of $\sim$48% and a weighted F1 of $\sim$44% for the 5-class prediction task.

2   Topic Mismatch Detection (TMD) Approach

In this section, we first present our concrete TMD problem definition. Next, we present our basic model architectures like BiLSTMs (Bidirectional Long Short-Term Memory Networks) [12] and BERT [8]. Further, we explore three input representations: (1) Baseline document shallow representation (using URL, title, snippet) along with the query as input, (2) Comprehensive document representation (using URL, title, snippet, key-phrases, topic distribution) along with the query (and its topic distribution) as input, (3) Novel approach for learning task-specific smoothing of input embeddings for BERT.

2.1   TMD Problem Definition

Let x represent an input sequence of length N corresponding to a (query, document) pair, and $y \in L$ denote its class label, where $L = \{$Very Unsatisfactory, Unsatisfactory, Neutral, Satisfactory, Very Satisfactory$\}$. Our goal is to design a function f such that $y \approx f(x)$. "Very Satisfactory" implies that the document is almost completely about/related to an interpretation of the query. Thus, there is a strong connection between the document and an interpretation of the query; the document content is almost exclusively about the query and there is little content that is not related to the query. "Satisfactory" implies that the document is mostly about/related to an interpretation of the query. It is not completely about the query as it has some content about other topics/subject beyond the query, or it lacks some content about the query. "Neutral" implies that the document is somewhat about/related to an interpretation of the query. It still has some content about the query but much of its content is about other topics or it lacks sufficient content about the query. "Unsatisfactory" implies that the document is only superficially about/related to an interpretation of the query. It may contain some related keywords or thin content about the query, but it is hard to see a useful connection, however it is not completely unrelated. "Very Unsatisfactory" implies that the document is off-topic. It may contain some keywords, but the document is not about any possible interpretation of the query.

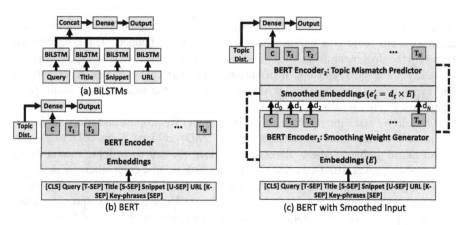

Fig. 1. Architecture of our proposed methods.

2.2 BiLSTM Baseline

As illustrated in Fig. 1(a), in this approach, the query, document URL, document title and document snippet are processed using four bidirectional LSTMs [12]. None of the BiLSTMs share weights. We use BiLSTMs with attention [1]. The final output from the last hidden layer of each of the BiLSTMs is concatenated and fed into a fully connected (dense) layer. Finally, the output softmax layer has five neurons (one for each of the classes) across all the architectures. We use the standard categorical cross entropy loss L_{TMD} which is computed as follows for an instance. $L_{\text{TMD}} = -\sum_{i=1}^{|L|} y_i(x) \log(p_i(x))$ where $p \in R^{|L|}$ and $y \in R^{|L|}$ are the prediction and ground-truth vectors respectively for the instance x.

2.3 BERT

BERT [8] is a Transformer [20] encoder with 12 layers. The BERT architecture for our task is illustrated in Fig. 1(b). We used the pre-trained model which has been trained on Books Corpus and Wikipedia using the masked language model (MLM) and the next sentence prediction (NSP) loss functions. The query and document (title, URL, snippet, key-phrases) tokens are concatenated into a sequence and are separated with special "SEP" tokens. The "SEP" tokens are different for each of the components ("T-SEP" for title separator, "S-SEP" for snippet separator, etc.). The sequence is prepended with a "CLS" token. The representation C for the "CLS" token from the last encoder layer is used for classification by connecting it to an output softmax layer with five neurons. We finetune the pre-trained model using labeled training data for the topic mismatch detection (TMD) task. We call this as the "BERT+TMD" method.

The accuracy of our model for the TMD task can possibly be improved by pretraining the model with labeled data for another closely related task: ranking. The ranking task aims at classifying every query-document pair as $M = \{$Perfect, Excellent, Good, Fair and Bad$\}$ [4,6]. To study correlation between the ranking and topic mismatch prediction tasks, on a separate labeled dataset of ∼631K (query, document) pairs, we obtain manual annotations using a popular crowdsourcing platform with 3-way redundancy for both ranking as well as TMD task. Table 2 shows the percentage overlap across various ranking class labels versus TMD class labels. Clearly, query-document ranking has a good positive correlation to topic-mismatch prediction (χ^2 test p-value is $< 10^{-5}$). Of course, this raises the question whether the ranking model can be directly used for TMD task? Table 6 shows that just the ranking model is not enough for the TMD task.

Table 2. Correlation between the ranking task and topic mismatch prediction task (in percentages).

Ranking↓ TMD→	Very unsatisfactory	Unsatisfactory	Neutral	Satisfactory	Very satisfactory
Bad	19.7	1.7	0.5	0.8	0.5
Fair	0.0	7.7	3.5	3.0	1.5
Good	0.0	0.0	12.8	13.4	5.5
Excellent	0.0	0.0	0.0	15.6	9.8
Perfect	0.0	0.0	0.0	0.1	3.6

Hence, we also experiment with two-stage BERT finetuning where we first finetune BERT for the ranking task and then for the TMD task. In the first stage, pre-trained BERT is finetuned using manually annotated ∼5.3M (query, document) pairs for the ranking task. The (query, document) pairs were annotated using 3-way redundancy on a popular crowdsourcing platform. Similar to the TMD task, the query and document are passed as input, and the representation C for the "CLS" token from the last encoder layer is used for classification by connecting it to an output softmax layer with five neurons. In the second stage, the model is further finetuned for the TMD task. For both the stages, categorical cross entropy loss is used to finetune the model. We refer to the model after the first stage as "BERT+Ranking" or "BERT+R" in short, and the model after the second stage finetuning as "BERT+R+TMD".

2.4   BERT with Smoothed Input (SI)

Each token in our (query, document) representation can be thought of as a point in the H dimensional space. For better generalization to semantically related variants of these words, one would want to rather use a smoothed representation for each point. This smoothing can be done by representing each point as a weighted summation of representations of its own as well as that of its semantically related variants. Consider a query "iphone5 price". "iphone6" can be considered as a variant for "iphone5". MLM in BERT is an established device to identify such variants conditioned on other words in the sequence.

Hence, we propose to use the first BERT encoder (BERT Encoder$_1$ in Fig. 1(c)) to learn similarity weights d_t for a word at position t in the sequence using MLM. The query and document representation is encoded in the same way as in the "BERT" method and fed as input to this BERT encoder$_1$. To enable MLM, each token in the input is *masked* with a probability p before being passed to the BERT Encoder$_1$.

Softmax is applied to the output at every position t from the BERT encoder$_1$ to get an output $d_t \in R^{|V|}$ where $|V|$ is the vocabulary size. These outputs d_t can be regarded as similarity weights for smoothing. Hence, we call BERT Encoder$_1$ as "Smoothing Weight Generator". Next, we compute a smoothed embedding at position t as $e'_t =$

$d_t \times E$ where $E \in R^{|V| \times H}$ is the embedding matrix, H is the hidden dimension for BERT model and $e'_t \in R^H$. These smoothed embeddings are then passed on as input to another BERT encoder (BERT Encoder$_2$ in Fig. 1(c)). The representation C for the "CLS" token from the last encoder layer is used for classification by connecting it to an output softmax layer with five neurons. Note that the two BERT encoders share weights. Also, the embedding matrix used for both the encoders is shared. This is indicated using dotted connections in Fig. 1(c). We refer to this method as "BERT+Smoothed Input". The overall loss for finetuning this model is computed as addition of the cross-entropy MLM loss at the output of Encoder$_1$ and TMD cross entropy loss at the output of Encoder$_2$. $L_{\text{MLM}} = -\sum_{t=1}^{N} \sum_{i=1}^{|V|} y_t^i(x) \log(d_t^i(x))$, where $d_t \in R^{|V|}$ and $y_t \in R^{|V|}$ are the prediction and ground-truth vectors respectively for the word at position t in the original unmasked input sequence x. The MLM loss at the output of Encoder$_1$ ensures that the smoothed embeddings generated by the Encoder$_1$ are not very different from its input embeddings.

2.5   BERT with Key-Phrases (KP) and Topic Distribution

The document URL, title, snippet are all shallow document signals. The quadratic complexity BERT model is not amenable to long documents. To harness the semantics of the document body, we resort to key-phrases and topic distribution obtained from the document body. Key-phrases are important phrases from the document body which were also contained in the queries leading to clicks to the document. For example, for the URL, https://www.thewindowsclub.com/surface-not-turning-battery-not-charging, extracted key-phrases include "Windows 10 Surface Pro, charging port, battery, power cord, USB charging port, LED light, power connector status". They were obtained using a key-phrase extraction model [21].

We remove key-phrases which are subsumed by other key-phrases. We also remove key-phrases which are redundant with respect to document URL, title or snippet. Individual key-phrases for the document are delimited by commas and used as part of input to our BERT-based models.

For topic modeling, we used LDA (Latent Dirichlet Allocation [2]) to compute topic distributions for the query and the document. We tried multiple configurations for LDA: (1) train on concatenated query + document text versus train separately for queries and for documents. (2) vary number of topics. Our experiments show that training separately for queries with 10 topics and for documents with 40 topics works best. Hence, we experiment with those as extra features in the last dense layer.

3   Experiments

3.1   Dataset

The dataset was created using 490K queries sampled from Bing's query log for Aug 2019 to Jan 2020. Figure 2 shows the size distribution of various input components in terms of sub-word tokens. Note that all the components show a power law beyond a length cutoff. As expected, snippets and key-phrases are much longer on average compared to query, title and URL. The top five URLs for each query were judged using three-way redundancy using the guidelines based on the definition mentioned in Sect. 1, leading to ~2.43M query-document pairs. Each query-document pair is labeled as one of these 5 classes: Very Unsatisfactory, Unsatisfactory, Neutral, Satisfactory, Very Satisfactory using 3-way redundancy on a popular crowdsourcing platform. Table 3 presents class distribution across train, validation and test splits of our dataset.

Table 3. Dataset statistics for our train, validation and test sets.

Category	Train	Validation	Test
VeryUnsat	414579	5936	48117
Unsat	239763	1028	13149
Neutral	402567	1691	26080
Sat	758534	4406	78131
VerySat	380127	2595	56504

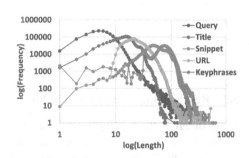

Fig. 2. Length distributions for Query, URL, Title, Snippet, Key-phrases tokens.

3.2 Traditional Baseline Methods

We experiment with two initial baseline methods: clicks and impressions based as well as topic model based method.

Clicks and Impressions (CI) Based Method. For every (query document) pair, we compute three sets of features as follows.

- Frequency features: QC = total clicks for query, QI = total impressions for query, DC = total clicks for document, DI = total impressions for document.
- Normalized features: QNC = p(document clicked|query), QNI = p(document shown|query), DNC = p(query|document clicked), DNI = p(query|document shown).
- CTR = click through rate.

Using these features, we train a 2-layer MLP (with ReLU non-linearity) for TMD. Table 4 shows results using these set of features for TMD. We observe that providing the raw frequency features seems to hurt performance most likely because of scaling issues which the MLP spends time normalizing for, hence only providing normalized features with CTR helps.

Topic Models Based Method. We trained LDA models on joint query + document text, as well as separately for queries and for documents with varying number of topics. In all cases, we use the concatenated query and document topic distributions as features, and train a 2-layer MLP (with ReLU non-linearity) for TMD. Table 5 shows results using these set of features for TMD. We observe that topic modeling based results are better compared to the clicks and impressions baseline. The best AUC method uses 50 topics to represent the document topic distribution, while the best accuracy and F1 method uses 40 topics. We use 10 topics for queries and 40 for documents in all further experiments.

Table 4. AUC, accuracy and F1 comparison using clicks and impressions based method.

Feature set	AUC	Acc.	F1
Frequency features + CTR	0.5011	15.74	11.12
Normalized features + CTR	**0.5031**	**17.74**	**12.84**
All features	0.5023	17.32	12.82

Table 5. AUC, accuracy and F1 comparison using topic modeling (LDA) based method.

LDA Configuration	AUC	Acc.	F1
Joint $T_{Q+D} = 40$	0.5133	25.11	17.46
$T_Q = 5, T_D = 20$	0.5094	26.31	18.64
$T_Q = 10, T_D = 20$	0.5276	27.10	19.88
$T_Q = 20, T_D = 20$	0.5136	26.17	17.80
$T_Q = 10, T_D = 30$	0.5369	28.02	19.99
$T_Q = 10, T_D = 40$	0.5409	**28.92**	**20.23**
$T_Q = 10, T_D = 50$	**0.5419**	28.62	20.11

3.3   Results Using Deep Learning Methods

Experiment Settings. Experiments were done on a machine with 4 T V100-SXM2-32 GB GPUs. We use the AUC, accuracy and weighted F1 for this five-class TMD task. AUC is computed by considering one-vs-rest paradigm in a macro-averaged manner. For Transformer based methods, the vocabulary size was fixed to 119581 and maximum sequence length was set to 128. For BiLSTMs in all baseline methods, the hidden layer size was 768 and we used the pretrained BERT embeddings for input representation. All models are trained using AdamW optimizer [18] with cross entropy loss with a batch size of 24 and the initial learning rate of $5e-6$. Key-phrases also have an importance score. We used an importance threshold of 0.1 which was tuned on the validation set.

Table 6. AUC, accuracy and F1 comparison across various methods. KP = Key-phrases; SI = Smoothed Input; CI = Clicks and impressions; R = Ranking; QTSU = Query, title, snippet and URL; LDATD = LDA topic distribution.

#	Model	Training data	Inputs	AUC	Acc.	F1
1	MLP	TMD	CI	0.5031	17.74	12.84
2	MLP	TMD	LDATD	0.5409	28.92	20.23
3	BiLSTMs	TMD	QTSU	0.6034	35.33	26.61
4	BERT	TMD	QTSU	0.6944	42.95	38.01
5	BERT	R	QTSU	0.6194	26.94	25.39
6	BERT	R+TMD	QTSU	0.7265	45.45	41.66
7	BERT	R+TMD	QTSU+KP	0.7268	45.46	41.70
8	BERT+SI	R+TMD	QTSU	0.7311	45.65	41.95
9	BERT	R+TMD	QTSU+LDATD	0.7346	46.12	41.24
10	BERT+SI	R+TMD	QTSU+KP	0.7332	45.84	42.07
11	BERT	R+TMD	QTSU+LDATD+KP	0.7346	46.10	42.41
12	BERT+SI	R+TMD	QTSU+LDATD	0.7346	46.10	42.44
13	BERT+SI	R+TMD	QTSU+LDATD+KP	0.7347	46.30	42.57
14	BERT+SI	R+TMD	QTSU+LDATD+KP+CI	0.7281	45.88	41.95
15	BERT-L	R	QTSU	0.7426	41.11	39.46
16	BERT-L	R+TMD	QTSU	0.7444	46.27	41.63
17	BERT-L	R+TMD	QTSU+KP	0.7448	47.12	42.89
18	BERT-L	R+TMD	QTSU+KP+LDATD	**0.7492**	**47.22**	**43.99**

Model Comparison. Table 6 shows accuracy comparison across various methods. We observe the following from these results. (1) Across most configurations, LDA topics lead to significant gains, KP typically lead to minor gains only which are not statistically significant, while CI do not lead to any gains. (2) The proposed BiLSTMs and BERT based method significantly outperform the traditional clicks, impressions and topic models based methods by huge margins. (3) As expected, on the TMD task, BERT models perform better than BiLSTMs since self-attention in BERT can capture complex relationships across query, URL, title and snippet (QTSU). Hence, we used BERT as our base model for further experiments. Also, note that experiments using a single BiLSTM with query+URL+title+snippet as input led to worse results compared to our 4 BiLSTM architecture. (4) BERT+Ranking performs worse compared to BERT+TMD which is finetuned for topic mismatch labels. (5) However, as shown in Sect. 2, TMD and ranking are highly correlated tasks which makes BERT finetuned for ranking and then for topic mismatch perform better than BERT+TMD. (6) Addition of LDA topics to the BERT+Ranking+TMD+QTSU model improves the results (rows 6 and 9). (7) Improving the input to BERT using task-specific learned smoothing results into a significant accuracy gains compared to the BERT+Ranking+TMD+QTSU model (rows 6 vs 8). (8) BERT+SI+ R+TMD with QTSU+LDATD+KP (row 13) leads to best results using a 12-layer model; adding CI features do not help (row 14). (9) BERT-Large models (rows 15–18) lead to better results compared to BERT-Base (rows 4–14) as expected. (10) Overall, best results are obtained using the BERT-Large (24 layer) model finetuned on both ranking and TMD tasks using QTSU+KP+LDA topics as features (row 18). However, BERT-large model needs almost double the RAM as well as inference latency compared to the BERT-base model in row 13.

Table 7. Confusion matrix for the best model.

		Predicted				
		VeryUnsat	Unsat	Neutral	Sat	VerySat
Actual	VeryUnsat	26544	1827	2965	15999	782
	Unsat	4697	1751	2115	4330	256
	Neutral	5865	800	4007	14792	616
	Sat	8453	1305	5104	59213	4056
	VerySat	3700	371	2081	37031	13321

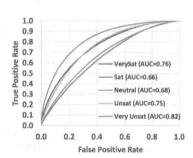

Fig. 3. Receiver Operating Characteristic (ROC) curve for the best model for each class.

Error Analysis. To understand the behavior of our best model better, we also present its confusion matrix in Table 7 and the ROC curve in Fig. 3. From the ROC, we observe that the system works best for the extreme VerySat and VeryUnsat classes, and worst for the Sat class. The model confuses a lot of other class instances to be Sat leading to a bad recall. Best precision is for the Sat class, while the best recall is for the VerySat class.

3.4 Case Studies

The top part of Table 8 shows a few examples where our model predictions were accurate, while the bottom part shows error cases. Many of the error cases happen because of the mismatch wrt an important entity between the query and document. E.g., "pc" versus "iphone" in the "how to see whatsapp backup on googledrive on pc" example. Similarly, in the "fleece sweater" example, although the URL contains the word "sweater", most of the products on the URL are hoodies.

Table 8. Examples of query-URL instances correctly (top)/incorrectly (bottom) predicted by our model.

Query	URL	Actual label	Pred. label
Online iphone 8plus screen	https://www.techjunkie.com/ when-your-screen-keeps-freezing-on-iphone-8-and-iphone-8-plus/	Very Unsat.	Very Unsat.
Fashion black pants blue jerseymens	https://www.amazon.com/Mens-Football-Clothing/b?node=2419333011	Unsat.	Unsat.
Purple blue paint colors	https://www.sherwin-williams.com/ homeowners/color/find-and-explore-colors/ paint-colors-by-family/family/purple	Neutral	Neutral
How to start a book club	https://www.wikihow.com/Start-a-Book-Group	Sat.	Sat.
Tourism tropical north queensland	https://tourism.tropicalnorthqueensland.org.au/	Very Sat.	Very Sat.
How to see whatsapp backup on google drive on pc	https://mobiletrans.wondershare.com/whatsapp/ backup-whatsapp-to-google-drive-on-iphone. html	Very Unsat.	Sat.
Google	https://podcasts.google.com/	Sat.	Unsat.
Japan rugby team height	https://www.rugbyworldcup.com/teams/japan? lang=en	Neutral	Very Unsat.
Fleece sweater	https://www.ebay.co.uk/b/Fleece-Fishing-Sweaters-Hoodies/179983/bn_9855717	Unsat.	Sat.
Rupees donated for coronavirus by sachin	https://www.india.com/sports/sachin-tendulkar-to-donate-rs-50-lakh-in-fight-against-coronavirus-pandemic-3982617/	Very Sat.	Unsat.

4 Related Work

There is no directly comparable work on the task of topic mismatch detection. Hence, in this section, we discuss related work broadly in the areas of topic analysis for queries and document representations.

4.1 Topic Analysis for Queries

Topic analysis for queries has been studied in sub-areas like topic shift detection in user sessions, query intent detection and query reformulation. Topic shift has been used as a strong signal for session segmentation [9, 14] beyond typical session timeouts. Recently, correlation between query reformulation patterns and topic shift queries has also been studied [17]. Brenes et al. [3] provide a good survey of traditional query intent detection methods. Recently, deep learning methods [11] have also been proposed for intent detection. To understand query-document mismatch, understanding query intent is essential but not sufficient. Considering the popularity of deep learning for capturing query intent, we also use BERT for learning semantic query representation jointly with the document representation.

4.2 Document Representation

Traditionally documents have been represented using syntactic representations like bag of words and Term Frequency-Inverse Document Frequency (TF-IDF). But these are very high dimensional. Topic models [13] like Latent Dirichlet Allocation (LDA) [2] help find low-dimensional description of high dimensional text. Documents could also be represented using average of word embeddings obtained using word2vec [19]. Further, distributed embedding models like Doc2Vec [15], Universal Sentence Encoder [5] and InferSent [7] have been proposed to encode sentences directly. Moreover, researchers also proposed several hierarchical neural network models [16,22] to represent documents. Transformer-based models like BERT [8] focus on the representation learning at the sentence level. Transformer models like BERT suffer from quadratic complexity of attention computation making it impossible to pass large document body as input. Although recent efficient Transformer methods [10] have shown promise, we plan to explore them as part of future work. In this paper, we represent a document using URL, title and snippet tokens. Further, we resort to a key-phrase extraction model [21] and LDA topic distributions to extract relevant information from documents.

5 Conclusion

We proposed the query-document topic mismatch prediction problem. We investigated the effectiveness of standard BERT model for this task using comprehensive query and document representations. For queries, we use query words and topic distribution using LDA. For documents, we leverage URL, title, snippet, key-phrases and topic distribution using LDA. We proposed a novel task-specific way to train smoothed embeddings. We observe that our best BERT model which is finetuned for both ranking and topic mismatch prediction task gives us an AUC of 0.75, an accuracy of $\sim$48% and a weighted F1 of $\sim$44% for the 5-class prediction task.

References

1. Bahdanau, D., Cho, K., Bengio, Y.: Neural machine translation by jointly learning to align and translate. arXiv preprint arXiv:1409.0473 (2014)

2. Blei, D.M., Ng, A.Y., Jordan, M.I.: Latent Dirichlet allocation. JMLR **3**(Jan), 993–1022 (2003)
3. Brenes, D.J., Gayo-Avello, D., Pérez-González, K.: Survey and evaluation of query intent detection methods. In: Workshop on Web Search Click Data, pp. 1–7 (2009)
4. Cao, Z., Qin, T., Liu, T.Y., Tsai, M.F., Li, H.: Learning to rank: from pairwise approach to listwise approach. In: Proceedings of the 24th International Conference on Machine Learning, pp. 129–136 (2007)
5. Cer, D., et al.: Universal sentence encoder. arXiv preprint arXiv:1803.11175 (2018)
6. Chapelle, O., Chang, Y.: Yahoo! learning to rank challenge overview. In: Proceedings of the Learning to Rank Challenge, pp. 1–24 (2011)
7. Conneau, A., Kiela, D., Schwenk, H., Barrault, L., Bordes, A.: Supervised learning of universal sentence representations from natural language inference data. In: Proceedings of the 2017 Conference on Empirical Methods in Natural Language Processing, pp. 670–680 (2017)
8. Devlin, J., Chang, M.W., Lee, K., Toutanova, K.: Bert: pre-training of deep bidirectional transformers for language understanding. arXiv preprint arXiv:1810.04805 (2018)
9. Gayo-Avello, D.: A survey on session detection methods in query logs and a proposal for future evaluation. Inf. Sci. **179**(12), 1822–1843 (2009)
10. Gupta, M., Agrawal, P.: Compression of deep learning models for text: a survey. arXiv preprint arXiv:2008.05221 (2020)
11. Hashemi, H.B., Asiaee, A., Kraft, R.: Query intent detection using convolutional neural networks. In: International Conference on Web Search and Data Mining, Workshop on Query Understanding (2016)
12. Hochreiter, S., Schmidhuber, J.: Long short-term memory. Neural Comput. **9**(8), 1735–1780 (1997)
13. Jelodar, H., et al.: Latent Dirichlet allocation (LDA) and topic modeling: models, applications, a survey. Multimedia Tools Appl. **78**(11), 15169–15211 (2019)
14. Jones, R., Klinkner, K.L.: Beyond the session timeout: automatic hierarchical segmentation of search topics in query logs. In: CIKM, pp. 699–708 (2008)
15. Le, Q., Mikolov, T.: Distributed representations of sentences and documents. In: International Conference on Machine Learning, pp. 1188–1196 (2014)
16. Li, J., Luong, M.T., Jurafsky, D.: A hierarchical neural autoencoder for paragraphs and documents. arXiv preprint arXiv:1506.01057 (2015)
17. Li, X., de Rijke, M.: Do topic shift and query reformulation patterns correlate in academic search? In: Jose, J.M., et al. (eds.) ECIR 2017. LNCS, vol. 10193, pp. 146–159. Springer, Cham (2017). https://doi.org/10.1007/978-3-319-56608-5_12
18. Loshchilov, I., Hutter, F.: Decoupled weight decay regularization. arXiv preprint arXiv:1711.05101 (2017)
19. Mikolov, T., Sutskever, I., Chen, K., Corrado, G.S., Dean, J.: Distributed representations of words and phrases and their compositionality. In: Advances in neural information processing systems, pp. 3111–3119 (2013)
20. Vaswani, A., et al.: Attention is all you need. In: NIPS, pp. 5998–6008 (2017)
21. Xiong, L., Hu, C., Xiong, C., Campos, D., Overwijk, A.: Open domain web keyphrase extraction beyond language modeling. In: EMNLP-IJCNLP, pp. 5178–5187 (2019)
22. Yang, Z., Yang, D., Dyer, C., He, X., Smola, A., Hovy, E.: Hierarchical attention networks for document classification. In: Proceedings of the 2016 Conference of the North American Chapter of the Association for Computational Linguistics: Human Language Technologies, pp. 1480–1489 (2016)

Beyond QA: 'Heuristic QA' Strategies in JIMI

Shuangyong Song(✉), Bo Zou, Jianghua Lin, Xiaoguang Yu, and Xiaodong He

JD AI Research, Beijing 100176, China
{songshuangyong,cdzoubo,cdlinjianghua,cdyuxiaoguang,hexiaodong}@jd.com

Abstract. JD Instant Messaging intelligence (JIMI) is an intelligent service robot designed for creating an innovative online shopping experience in E-commerce. We will introduce a framework that combines the 'intelligent prediction' and 'user click' to facilitate the user input and to format the user inputs as standard queries, which we call as 'heuristic QA'. It consists of three strategies: *intent prediction* before user querying, *query auto-completion* during user querying and *next query prediction* after user querying. Until now about 1/3 user queries are from *heuristic QA* in JIMI, which significantly improves the user experience.

Keywords: Intelligent service robot · Conversational recommendation

1 Introduction

Due to the increasing use of service robots in E-commerce platforms in recent years, *customer satisfaction degree* and *problem resolution rate* have been most important evaluating indicators, and they are mainly impacted by the *conversational convenience* of users and *reply accuracy* of service robots [1]. In this paper, a heuristic QA framework is proposed to improve the service level of an intelligent service robot, JIMI. This framework aims to facilitate the user input and to format the user inputs as standard queries, with 'intelligent prediction' and 'user click' of user queries before, during and after the user querying process.

In this paper, we call each query in a manually written query-answering knowledge base as a 'kbQue' for short. Our heuristic QA framework is mainly based on conversational recommendation models, which rank candidate kbQues to find best matching ones on specific conversational states. The three strategies of heuristic QA are briefly described: 1) before user querying: Intent Prediction. This strategy is for predicting proper kbQues before users entering the conversational interface (still on the order page) or before users inputting the first query on the conversational interface. 2) during user querying: Query Auto-Completion. This strategy is for predicting proper kbQues during users inputting queries in the input box. 3) after user querying: Next Query Prediction. With detecting relations between different user intents, we can present a menu list of related kbQues of the current user intent.

Several conversational recommendation models and multiple feature combinations are tested in our experiments, and the online application effects are given. Besides, an automatic model retraining mechanism is assembled in JIMI for updating heuristic QA models when data drift occurs.

2 Heuristic QA

All the experiments in our work are mainly based on the dataset published in [2], supplemented by user information from JD[1] with data masking. Besides, all the experimental results are based on Bert embeddings or Bert distillation.

Table 1. Model comparison results on intent prediction.

Models	TopN	XGBoost				DeepFM	W&D	CN	DN	**DCN**
Features	–	$f_{(p+d+o)}$	$f_{(p+d+o+b)}$	$f_{(p+d+o+b+c)}$	$f_{(p+d+o+b)}$					
Acc	23.00	34.30	39.50	37.50		36.39	40.24	39.11	40.47	**40.73**

2.1 Before User Querying: Intent Prediction

This strategy is for predicting proper kbQues before users describing problems. Two examples are respectively shown in Fig. 1(a) and Fig. 1(b), about the intent prediction results on the order page and the conversational interface.

The models we tried in experiments are: top N highest frequency results (TopN), XGBoost, DeepFM, wide and deep neural network (W&D), cross network (CN), deep network (DN), Deep & Cross Network (DCN). Meanwhile, the features we used are: user profiles (f_p); historical dialogues (f_d); order information (f_o); user browsing logs (f_b); user click data (f_c). XGBoost is used to select a best feature collocation, and other models are all tested on the best feature collocation. Model comparison results are shown in Table 1, and the DCN model gets the best performance, while the user click rate of intent prediction is 23.38%.

2.2 During User Querying: Query Auto-completion

Table 2. Model comparison results on query auto-completion.

Features		$f_{(t+c)}$	$f_{(t+c+i)}$	$f_{(t+c+s)}$	$f_{(t+c+i+s)}$
Acc	top_1	0.391	0.585	0.631	0.655
	top_3	0.775	0.862	0.891	0.907
	top_5	0.835	0.910	0.917	0.921

The function of query auto-completion is to help users formulate queries fast and precisely during the user inputting, and an example is shown in Fig. 1(c). In this part, a proper feature selection is more important than the model selection, so we just try with the classic LambdaMART model. The features we used are: 1) user click data (f_c); 2) intent prediction result with HAN [3] model (f_i); 3)

---

[1] https://www.jd.com/.

Fig. 1. User interfaces of heuristic QA: (a) intent prediction on the order page; (b) intent prediction on the conversational interface; (c) query auto-completion in the input box; (d) the menu list of the next query prediction.

features of text (f_t), such as text length, etc.; 4) similarity features (f_s), such as similarity between a candidate kbQue and the user input, etc.

Table 2 shows the model comparison results on query auto-completion, and a richer set of features lead to better performance. Finally, the user click rate of the best query auto-completion is 27.20%.

2.3   After User Querying: Next Query Prediction

Table 3. Model comparison results on next query prediction.

Models	Acc on manually data			Acc on user click data			Acc on combined data		
	top_1	top_3	top_5	top_1	top_3	top_5	top_1	top_3	top_5
Abcnn	0.454	0.910	0.925	0.296	0.824	0.868	0.360	0.859	0.890
ECIM	0.505	0.919	0.939	0.347	0.887	0.903	0.411	0.901	0.921
XGBoost	0.655	0.977	0.979	0.462	0.915	0.929	0.541	0.940	0.956
XGBoost+abcnn	0.658	0.979	0.983	0.483	0.927	0.933	0.553	0.948	0.958
XGBoost+ECIM	0.709	0.982	0.991	0.534	0.934	0.941	0.604	0.950	0.964

An example of next query prediction is shown in Fig. 1(d), where with detecting relations between different user intents, we can present a list of related kbQues of the current user intent. Different from intent prediction and query auto-completion which run in every possible situation, the next query prediction is restrictively activated, since a full-coverage activation will severely reduce user experience. We detect user intents which are frequently converted to other ones, and the next query menu lists are just shown after those intents.

The model comparison results are shown in Table 3, and the SOTA model ECIM [4] can help get the best performance, while the online activation rate of next query prediction is 10.01%, and the user click rate is 19.76%.

3   Application Effects and Automatic Model Retraining

With the heuristic QA framework, the average *time per input* of users (a factor reflecting the *conversational convenience*) is shortened from 14.39 s to 10.93 s, and the *reply accuracy* of JIMI is raised from 89.63% to 93.08%. Meanwhile, the *customer satisfaction degree* has been raised 1.85%, and the *problem resolution rate* has been raised 6.46%.

Besides, to address the issue of online data drifts, we come up with a solution to implement automatic model retraining (*AMR*) mechanism. The *Kullback-Liebler divergence* between intent distributions of user click and user input is calculated weekly, and if the value is above the preset threshold, the *AMR* mechanism is activated.

References

1. Yao, R., et al.: Session-level user satisfaction prediction for customer service chatbot in E-commerce. In: AAAI 2020, pp. 13973–13974 (2020)
2. Chen, M., et al.: The JDDC corpus: a large-scale multi-turn Chinese dialogue dataset for E-commerce customer service. In: LREC 2020, pp. 459–466 (2020)
3. Yang, Z., Yang, D., Dyer, C., He, X., Smola, A., Hovy, E.: Hierarchical attention networks for document classification. In: NAACL 2016, pp. 1480–1489 (2016)
4. Song, S., Wang, C., Pu, X., Wang, Z., Chen, H.: An enhanced convolutional inference model with distillation for retrieval-based QA. In: Jensen, C.S., et al. (eds.) DASFAA 2021. LNCS, vol. 12683, pp. 511–515. Springer, Cham (2021). https://doi.org/10.1007/978-3-030-73200-4_35

SQLG+: Efficient k-hop Query Processing on RDBMS

Li Zeng[✉], Jinhua Zhou, Shijun Qin, Haoran Cai, Rongqian Zhao,
and Xin Chen

Huawei Technologies Co., Ltd., Shenzhen, China
{zengli43,zhoujinhua1,qinshijun,caihaoran1,zhaorongqian,
chenxin}@huawei.com

Abstract. Graph algorithms (e.g., k-hop queries) are widely used to find the deep association of data in various real-world applications such as business recommendation and fraud detection. However, most of the data are still stored in relational database (i.e., RDBMS) and the performance is rather limited when processing graph queries on RDBMS due to the inherent hardness of complicated table join. In this paper, we propose a fast interactive engine $SQLG+$, which can be integrated to any RDBMS and enable them to process k-hop graph queries efficiently. Different from naive table-join implementations, SQLG+ caches important nodes with their adjacency lists in memory (i.e., *graph cache*) and generates a hybrid query plan which combines the ability of graph cache and RDBMS. Also, SQLG+ removes duplicates in the end of each hop (using *AdaptiveSet*) and expands the frontiers in different ways. Furthermore, dynamic BFS/DFS switch is adopted to achieve the balance between query performance and memory occupation. Extensive experiments show that SQLG+ outperforms the state-of-the-art RDBMS-based implementations by up to several orders of magnitude and is even comparable to the fastest graph databases.

Keywords: k-hop query · Graph cache · Gremlin · RDBMS

1 Introduction

Graphs have become increasingly important in modeling complicated structures and schema-less data such as financial transactions, social network and communication network. The growing popularity of graphs has generated many interesting data management problems which can help find the deep association of data. Among these, k-hop query is a fundamental problem: *how to efficiently enumerate all k-hop neighbors of a vertex u over a data graph*. It is the basis of many graph algorithms such as subgraph matching [22,23] and reachability [4,11]. Different from BFS (breadth-first search), k-hop query limits the depth of exploration, thus it is a kind of local search rather than complete graph traversal. k-hop query is the focus of this work, which has many applications, e.g., fraud detection [9], friend recommendation [17] and service influence

A. Bhattacharya et al. (Eds.): DASFAA 2022, LNCS 13247, pp. 430–442, 2022.
https://doi.org/10.1007/978-3-031-00129-1_37

analysis [18]. Figure 1(a) illustrates the results of 3-hop query starting from 'A' in a small portion of twitter network [12]. By finding out the k-hop friends of a person, targeted recommendation can be achieved in social network applications. Another example is fraud detection, where the k-hop subgraph of user u is extracted from the financial network and checked whether there is any pattern of fraud behavior.

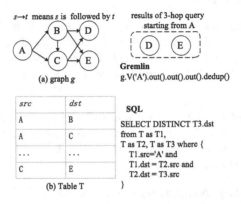

Fig. 1. A small portion of twitter network

Although existing graph databases (e.g., TigerGraph [7]) can process k-hop queries naturally by graph traversal, it does not always work. In many situations, most of the data are still stored in relational databases. On the one hand, the cost of transferring data from RDBMS to graph databases is rather high. On the other hand, graph databases are not as robust as RDBMS. Besides, we hope to support both relational data and graph data, and process both SQL and graph queries.

However, the performance is terrible when processing graph queries on RDBMS due to the inherent hardness of complicated table join in SQL [5]. In Fig. 1(b), a 3-hop Gremlin query can produce a heavy SQL with three table joins (T is the relational representation of the graph in Fig. 1(a)), which is prohibited for RDBMS. Therefore, we need to optimize the processing of graph queries based on RDBMS.

In this paper, we propose a novel system $SQLG+$, which can be integrated to any RDBMS and enable them to process graph queries efficiently. Currently, we focus on optimizing k-hop graph queries, which can be represented by Gremlin or other graph query languages. Different from naive table-join implementations, SQLG+ caches important nodes with their adjacency lists in memory (i.e., *graph cache*) and generates a hybrid query plan which combines the power of graph cache and RDBMS. Besides, SQLG+ leverages *AdaptiveSet* to remove duplicates in the end of each hop and classifies the frontiers in three categories which are expanded in different ways. Furthermore, dynamic BFS/DFS switch is adopted to achieve the balance between query performance and memory occupation, which can help eliminate OOM (out of memory) problems on large graphs. Note that our proposed solution is orthogonal to accelerative techniques in RDBMS.

2 Background

In this section, we first present the formal definition of our problem, then list the related work.

2.1 Problem Definition

Definition 1. *(Graph)* *A graph is denoted as $G = \{V, E, L\}$, where V is the set of vertices; $E \subseteq V \times V$ is the set of undirected edges; L is a labeling function that maps a vertex (of $V(G)$) to a label. The label function of G can also be specified as L_G. $V(G)$ and $E(G)$ are used to denote vertices and edges of graph G, respectively.*

Definition 2. *(Subgraph)* *Given a graph $G = \{V, E, L\}$, a subgraph of G is denoted as $G' = \{V', E', L'\}$, where vertex sets V' and edge sets E' in G' are subsets of V and E, respectively, denoted as $V' \subseteq V$ and $E' \subseteq E$. Furthermore, for vertex label functions, $L' \subseteq L$.*

Definition 3. *(k-hop Neighbors)* *Given a graph g, let u be the source node and k be the number of hops, k-hop neighbors of u is the vertex set KN such that $\forall u' \in KN$, u' is connected to u via a path of k edges.*

Definition 4. *(k-hop Subgraph)* *Given a graph g, let u be the source node and k be the number of hops, the k-hop subgraph of u is the subgraph of g induced by vertex set KS such that $\forall u' \in KS$, u' is connected to u via a path of no more than k edges.*

Definition 5. *(Problem Statement)* *Given a graph g, let u be the source node and k be the number of hops, the k-hop query problem is to find out all k-hop neighbors of u.*

A running example is given in Fig. 1(a). According to Definition 3, the 3-hop neighbors of 'A' are 'D' and 'E'. Obviously, if a breadth-first search (or depth-first search, DFS) is performed on g to extract the k-hop neighbors of u, the k-hop subgraph of u is acquired immediately. The only difference is that the vertices and edges visited needed to be recorded in the results.

This paper aims to support fast k-hop query processing (Definition 5) on RDBMS. Without loss of generality, we assume the graph g is connected and the result set of k-hop query is not empty. Though our solution can be easily extended to process directed graphs, vertex/edge labels or label sets, that is not our focus. Unless otherwise specified, we use u, $N(u)$, $deg(u)$, $num(L)$, and $|A|$ to denote a vertex, the neighbor set of u, degree of u, the number of currently valid elements in set L, and the size of set A, respectively.

2.2 Related Work

Existing work related to k-hop query can be mainly divided into three categories: relational table join, breadth-first search (BFS) and depth-first search (DFS).

Relational Table Join. Some earlier solutions (Jena [16], Virtuoso [8], and SQLG[1]) store the graph data as relational tables and answer k-hop queries by joining these tables. The structure of relational table and the algorithm of join are different in various systems.

Breadth-First Search. BFS starts from a source node and explores the graph layer by layer. It traverse the entire graph and each vertex is expanded only once. A large frontier queue needs to be maintained, thus BFS has high memory occupation. On the other hand, this enables it to support parallel processing naturally. A lot of accelerative techniques are proposed: direction optimization [3], GPU [15], FPGA [21], etc.

Depth-First Search. DFS follows backtracking paradigm, which recursively searchs each path. Though DFS has nearly linear memory occupation, it is hard to be parallelized [10]. Existing work [20] all targets at coarse-grained parallelism with rather limited speedup.

Modern graph databases also adopts these algorithms. For example, gStore [24] uses BFS to perform graph search. In contrast, GalaxyBase [6] generates a hybrid query plan of BFS and table join.

Variants. The k-hop reachability problem [4] checks whether the source node s can be connected to the target node t via a path of no more than k edges. It can be solved by any k-hop query algorithm, or accelerated with precomputed indices. The shortest k-hop query problem [6] imposes stronger restrictions on the length of path. Let u and RS be the source node and the result set respectively, it requires that $\forall u' \in RS$, the length of the shortest path between u and u' is exactly k. This enables more powerful pruning techniques such that all previous visited nodes can not occur in the final result.

3 Approach

SQLG+ leverages graph cache to store the adjacency lists of some important nodes (favoring high degree and betweenness centrality), while the full data still resides in RDBMS. Note that the memory usage of graph cache is rather limited ($<$4G). The cache can be easily updated if there is any change in the database.

SQLG+ is based on the original architecture of SQLG. Figure 2 shows the differences of SQLG and SQLG+. SQLG+ leverages a new middle layer, which analyzes the input gremlin query and generates a hybrid query plan which combines the power of graph cache and RDBMS. The framework of k-query processing is shown in Fig. 3. To find the k-hop neighbors of u in graph g, SQLG+ first computes the evaluation cost of this query by the formula:

---

[1] https://github.com/pietermartin/sqlg.

$cost(u, k) = out\_deg(u) \times avg\_deg(g)^{k-1}$. If the cost is low enough, this query is still parsed into SQL and answered by RDBMS. Otherwise, a BFS process is launched from the source node u, i.e., hop-by-hop expansion. Note that in our framework duplicates are eliminated in each hop, while in SQLG duplicates can only be removed from the final result.

During BFS, a frontier queue is maintained in each hop. The results of next hop are expanded from current frontier queue. For example, in Fig. 1, the 2-hop neighbors are {'C','D','E','D','E'}. After deduplication, the frontier queue fq is {'C','D','E'}. Finally, the 3-hop neighbors are expanded from fq, which yields {'D','E'}.

Based on the framework, SQLG+ proposes three techniques to optimize both query performance and memory occupation. The sections below will detail them.

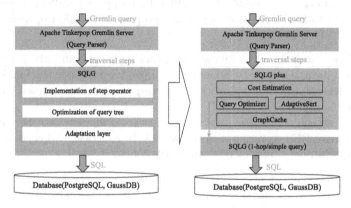

Fig. 2. The architecture of SQLG and SQLG+

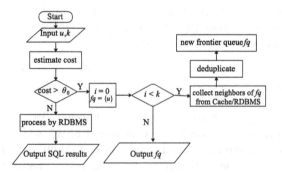

Fig. 3. The framework of query processing in SQLG+

3.1   Frontier Classification

The frontiers generated in each hop may also contain vertices that are already cached. Besides, the subprocesses of different frontiers may be in different scale as the degree of a super node may be $10^6 \times$ larger than an ordinary node. The performance is terrible if expanding all frontiers in the same scheme. On the one hand, the exploration of low-degree nodes is dragged down by super nodes. On the other hand, the size of frontier queue increases sharply, which brings high pressure on memory.

Consequently, SQLG+ classifies the frontier queue into three categories: small queue sq, medium queue mq, large queue lq. A simple example is shown in Fig. 4. These categories favor different expansion scheme. Vertices in lq has high degree, thus their neighbors are collected directly from graph cache. In contrast, vertices in sq and mq can only be probed in RDBMS.

Recall that we estimate the cost for source node u in the beginning. Also, we can estimate the cost of each frontier fu in each hop, which represents the search space of the subprocess of fu. Vertices with tiny cost belong to lq, while others are in mq. It is better to process each vertex of lq in one-time exploration via a single SQL. Thus, the vertices of lq are eliminated from current frontier queue. For vertices in mq, hop-by-hop expansion is still the best way.

We process lq with RDBMS directly because RDBMS is good enough on 1-hop query or simple queries. In power-law graphs [1], more than 90% nodes are classified as low-degree nodes. During each hop, more than 50% frontiers belong to lq. These nodes are completely eliminated from current frontier queue and do not expand to the new frontier queue. Thus, the peak memory occupation of frontier queues is reduced by 10%.

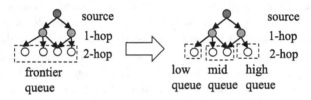

Fig. 4. An example of frontier classification

3.2   Deduplicate by AdaptiveSet

In the end of each hop, the adjacency lists of different frontiers need to be union merged and all duplicates should be removed. Though this process helps eliminate the redundant search of the same vertex, it may bring in high pressure on performance if using an inefficient implementation. The naive solution is the *HashSet* container in *Java*. Theoretically, the amortized complexity of each probe operation in *HashSet* is $O(1)$.

However, *HashSet* performs terrible when processing large graphs. On twitter, the frontier queue may have tens of millions of vertices, thus there are many operations of set union and deduplication. The inefficiency of *HashSet* is in three folds: (1) Let n be the set size, the memory occupation is larger than $36n$ (sometimes $100n$), thus on large graphs the Java GC (Garbage Collection) encounters

STW (stop the world) frequently; (2) When the set size grows, the resize operation (requiring rehash mechanism) occurs frequently, which brings unneglected cost; (3) When the set size grows, the conflict list extends and the probe cost is not $O(1)$. Therefore, SQLG+ leverages *AdaptiveSet* to merge sets and remove duplicates efficiently. The intuition is that sets in different scale should be stored in different structures and the implementation of set operations should adapt to the structures.

AdaptiveSet adopts three formats to store sets: *HashSet* (implementation of hash table), *SortedArr* (sorted array), *BitSet* (each bit represents a vertex). The format of storage, along with the method of set operation, is dynamically transformed according to current set size. DFA (Deterministic Finite Automation) is used to represent this transformation (see Fig. 5). Small sets are still processed via *HashSet*, while medium sets and large sets are processed by *SortedArr* and *BitSet* respectively. Let *HashSet* be the current storage format, the transformation to *SortedArr* is performed if the next set's size exceeds threshold θ_1. Similarly, if the next set's size exceeds threshold θ_2, the transformation to *BitSet* works.

The set operations on HashSet hs is performed by probing each vertex u of the new set S: if u is not in hs, u is inserted into hs; otherwise, u is discarded and no duplicate exist in hs. As for SortedArr sa, union operations are done by merging two sorted arrays. Meanwhile, the duplicates can be eliminated when merging two identical vertices. The amortized time complexity is $O(1+\frac{|S|}{|sa|})$, and the worst-case space complexity is $4 \times (|S| + |sa|)$. When it comes to BitSet bs, each vertex u of S is processed by setting $bs[u] = 1$. This simple bitwise operation is efficient and it finishes two operation directly (union and deduplication). The amortized time complexity is $O(1)$, and the space complexity is always $\frac{|V(G)|}{8}$.

The benefit of AdaptiveSet is two fold. On the one hand, the computation of set union and deduplication is much more efficient. On the other hand, the memory occupation of intermediate results is much lower. Even when the frontier queue contains 40 m nodes, the memory cost is lower than 5 MB.

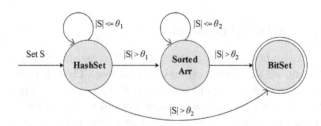

Fig. 5. The routine of AdaptiveSet

3.3 Dynamic BFS/DFS Switch

Generally, graph traversal is performed by BFS as it is more efficient and has the potential of massive parallelism. However, the computation is memory bounded in many situations. Thus, BFS does not work when the memory is rather limited or the intermediate result is too large. A naive method is leveraging compress strategies to reduce the memory cost, but both graph structure and frontier queue are frequently accessed, which marks the compress/uncompress process inefficient.

To solve this problem, dynamic BFS/DFS switch is adopted to achieve the balance between query performance and memory occupation, which can help eliminate OOM problems on large graphs. BFS is used in the beginning and it is transformed to DFS when the estimated size of frontier queue exceeds current available *JVM* (Java Virtual Machine) memory. Furthermore, naive DFS is not favored as it leads to low cache utilization and too many SQLs. In contrast, batched DFS is proposed. Frontiers are divided into small batches and each batch is processed by backtracking paradigm (see Fig. 6). By adjusting the value of batch size (denoted by β), a balance can be achieved between query performance and memory occupation. For example, setting $\beta = 1$ yields naive DFS, while $\beta = \infty$ is the case of naive BFS.

This technique eliminates the OOM problem when processing large graphs. It enables the engine to run on machines with small memory capacity. Besides, *limit k* queries can be responded immediately with batched DFS. The extension of pipeline processing can also be easily implemented in this case.

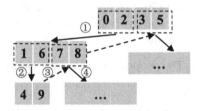

Fig. 6. The routine of batched DFS

4 Experiment

In this section, we evaluate our method (SQLG+) against state-of-the-art engines, such as RDBMS-based solutions (Jena [16], Virtuoso [8] and SQLG), and graph databases (gStore [24], GalaxyBase [6], GAIA [19] and TigerGraph [7]). All experiments are carried out on a workstation running CentOS 7 and equipped with Intel Xeon E5-2620 2.40 GHz CPU, 64 GB host memory and 256 GB disk.

4.1 Datasets and Queries

The experiments are conducted on both real and synthetic datasets. The statistics are listed in Table 1. The patent citation network (*patent*) and the user-follow network (*twitter*) are downloaded from SNAP [14]. Large RDF graphs, such as DBpedia [13] and WatDiv (a synthetic RDF benchmark [2]), are also used.

For each dataset, we select 100 source nodes and report the average running time. The result size of these source nodes is required to be in the same scale. The query set is classified into four categories: 1-hop queries, 2-hop queries, 3-hop queries, 6-hop queries. Note that the performance of 3-hop and 6-hop queries is the most significant metric in the following experiments. The time limit is set to two hours, i.e., a query fails if it is not responded within 2 h. Due to space limit, the performance on DBpedia and WatDiv is only reported in Sect. 4.5.

Table 1. Statistics of datasets

Name	$|V|$	$|E|$	MD[1]	Type[2]
Patent	3,774,768	16,518,948	793	rs
Twitter	41,652,230	1,468,365,182	3M	rs
DBpedia	22,623,324	170,481,066	2.2M	rs
WatDiv	10,899,920	109,959,180	671K	s

[1] Maximum degree of the graph.
[2] Graph type: r:real-world, s:scale-free, and m:mesh-like.

4.2 Evaluation of Frontier Classification

We report the performance of frontier classification on patent and twitter. The result is shown in Table 2. The improvement is remarkable: >1.2× speedup on patent and 7× speedup on twitter. On complex queries, the speedup is always larger than 2×. Besides, the scale of frontier queue decreases by 10% due to the special processing of small nodes.

Table 2. The performance of frontier classification

System	Patent				Twitter			
	1-hop	2-hop	3-hop	6-hop	1-hop	2-hop	3-hop	6-hop
SQLG	35	127	327	9 K	68 K	710 K	OOM	OOM
SQLG+FC[1]	30	97	242	4.7 K	612	99 K	481 K	3 M
Speedup	1.2×	1.3×	1.4×	2×	111×	7×	∞	∞

* the unit of time is ms; 1,2,3,6 denote the number of hops.
[1] SQLG+FC represents the implementation of SQLG+ framework with the optimization of frontier classification.

4.3 Evaluation of AdaptiveSet

The performance of AdaptiveSet is compared in Table 3. The minimum speedup is 1.3× and 4× on patent and twitter, respectively. This technique performs the best when the frontier queue is very large. For example, on 6-hop queries, the speedup is larger than 4.7×. On simple cases (e.g., 1-hop queries), the performance is also good enough. If we use BitSet instead of AdaptiveSet, the performance is terrible on lightweight queries. This proves the necessity of maintaining hybrid data structures in AdaptiveSet. Meanwhile, the flexible structures of AdaptiveSet reduce the size of intermediate results to $\frac{1}{10}$.

Table 3. The performance of AdaptiveSet

System	Patent				Twitter			
	1-hop	2-hop	3-hop	6-hop	1-hop	2-hop	3-hop	6-hop
SQLG+FC	30	97	242	4.7 K	612	99 K	481 K	3 M
SQLG+AS[1]	5	46	188	1 K	10	9 K	120 K	590 K
Speedup	6×	2×	1.3×	4.7×	61×	11×	4×	5×

* the unit of time is ms; 1,2,3,6 denote the number of hops.
[1] SQLG+AS represents the implementation of SQLG+FC with the optimization of AdaptiveSet.

4.4 Evaluation of Dynamic BFS/DFS Switch

We evaluate the technique of dynamic BFS/DFS switch in different sceneries to quantify how much memory is saved. In Table 4, the result size of k-hop query is varied to compare the naive BFS implementation and the novel BFS/DFS implementation. On 10M and 40M, the percentage of memory reduction is 0% because the memory usage is still under the JVM capacity. On 70M and 100M, the percentage is higher than 99% and the OOM problem is absolutely eliminated. In extreme sceneries, the memory occupation can be lowered to several MB. Note that this technique does not bring performance gain directly.

Table 4. Comparison of memory occupation (MB)

Result size	10 M	40 M	70 M	100 M
BFS	1200	7135	10123	OOM
BFS+DFS	1200	7135	6	8
Reduce	0%	0%	99%	100%

4.5 Overall Performance

In this section, we compare SQLG+ with Jena, Virtuoso and SQLG and list the result in Table 5. Obviously, SQLG+ is the single winner. The rank of other systems is hard to be distinguished, but they all fails to process complex queries on twitter. Among the other three systems, SQLG claims the best on small graphs, while Virtuoso is good at processing lightweight queries. SQLG+ shows absolute superiority due to its adaptive data structures and query plans. Compared with the original SQLG, SQLG+ achieves >1.7× speedup in all cases and >9× speedup on complex queries. Furthermore, the peak memory occupation of SQLG+ is several orders of magnitude lower than counterparts.

On DBpedia and WatDiv, SQLG+ is still the best. On WatDiv, It shows >5× speedup when compared with SQLG. On DBpedia, the improvement is more prominent: >6.5× on lightweight queries and >10× on complex queries. The speedup on twitter is the highest (>79×) because it has the largest graph size and maximum degree. There are more than 100 super nodes in twitter, thus every search process encounters super nodes within 3 hops. All these properties make it hard for other systems to process k-hop queries efficiently.

Table 5. Comparison of k-hop queries

System	Patent				Twitter			
	1-hop	2-hop	3-hop	6-hop	1-hop	2-hop	3-hop	6-hop
Jena	61	279	17 K	50 K	1 K	990 K	OOM	OOM
Virtuoso	12	25	3 K	14 K	327	233 K	OOM	OOM
SQLG	35	127	327	9 K	68 K	710 K	OOM	OOM
SQLG+	5	46	188	1 K	10	9 K	120 K	590 K

* the unit of time is ms; 1,2,3,6 denote the number of hops.

4.6   Comparison with Graph Databases

The performance comparison of SQLG+ and modern graph databases is shown in Table 6. Generally, TigerGraph is the most powerful one among four selected databases. Though graph databases has native graph storage and direct graph traversal algorithms, SQLG+ is still comparable to them on all queries. In some cases, SQLG+ is even better. For example, SQLG+ is superior to TigerGraph on lightweight queries. On twitter, SQLG+ outperforms all by 1–2 orders of magnitude except for TigerGraph. Besides, the memory occupation of SQLG+ is three times lower than these databases.

Table 6. Comparison with graph databases

System	Patent				Twitter			
	1-hop	2-hop	3-hop	6-hop	1-hop	2-hop	3-hop	6-hop
gStore	12	17	3.4 K	14 K	1.1 K	233 K	OOM	OOM
GalaxyBase	44	83	131	241	3.2 K	6.5 M	6.8 M	7.2 M
GAIA	7	69	113	1.2 K	2.6 K	280 K	TLE	TLE
TigerGraph	30	66	165	862	2 K	3 K	26 K	103 K
SQLG+	5	46	188	1 K	10	9 K	120 K	590 K

* the unit of time is ms; 1,2,3,6 denote the number of hops; TLE means "Time Limit Exceeded".

5   Conclusion

We introduce an efficient interactive query engine (SQLG+), which supports fast k-hop query processing on RDBMS. SQLG+ designs an advanced framework of query processing, which can combine the power of RDBMS and graph cache, and generate better query plans to accelerate k-hop query processing. Based on the framework, optimization techniques include frontier classification, deduplication by AdaptiveSet, and dynamic BFS/DFS switch. Experiments on real-world graphs show that SQLG+ outperforms the state-of-the-art RDBMS-based implementations by up to several orders of magnitude and is even comparable to modern graph databases. In practical, SQLG+ has been used in communication network, supporting the fault management and service influence analysis.

References

1. Aiello, W., Graham, F.C., Lu, L.: A random graph model for power law graphs. Exp. Math. **10**(1), 53–66 (2001)
2. Aluç, G., Hartig, O., Özsu, M.T., Daudjee, K.: Diversified stress testing of RDF data management systems. In: Mika, P., et al. (eds.) ISWC 2014. LNCS, vol. 8796, pp. 197–212. Springer, Cham (2014). https://doi.org/10.1007/978-3-319-11964-9_13
3. Beamer, S., Asanovic, K., Patterson, D.A.: Direction-optimizing breadth-first search. In: SC (2012)
4. Cheng, J., Shang, Z., Cheng, H., Wang, H., Yu, J.X.: K-reach: who is in your small world. Proc. VLDB Endow. **5**(11), 1292–1303 (2012)
5. Cheng, Y., Ding, P., et al.: Which category is better: benchmarking relational and graph database management systems. Data Sci. Eng. **4**(4), 309–322 (2019)
6. Create Link co., L.: Galaxybase: a high-performance graph database (2016). https://www.galaxybase.com/
7. Deutsch, A., Yu, X., Wu, M., Lee, V.: Tigergraph: a native MPP graph database. arXiv (2019)
8. Erling, O.: Virtuoso, a hybrid RDBMS/Graph column store. IEEE Data Eng. Bull. **35**(1), 3–8 (2012)
9. Fan, W., He, T., et al.: GraphScope: a unified engine for big graph processing. Proc. VLDB Endow. **14**(12), 2879–2892 (2021)
10. Jenkins, J., Arkatkar, I., Owens, J.D., Choudhary, A., Samatova, N.F.: Lessons learned from exploring the backtracking paradigm on the GPU. In: Jeannot, E., Namyst, R., Roman, J. (eds.) Euro-Par 2011. LNCS, vol. 6853, pp. 425–437. Springer, Heidelberg (2011). https://doi.org/10.1007/978-3-642-23397-5_42
11. Jin, R., Xiang, Y., Ruan, N., Fuhry, D.: 3-HOP: a high-compression indexing scheme for reachability query. In: Çetintemel, U., Zdonik, S.B., Kossmann, D., Tatbul, N. (eds.) SIGMOD, pp. 813–826. ACM (2009)
12. Kwak, H., Lee, C., Park, H., Moon, S.B.: What is Twitter, a social network or a news media? In: WWW, pp. 591–600. ACM (2010)
13. Lehmann, J., et al.: Dbpedia—a large-scale, multilingual knowledge base extracted from wikipedia. Semantic Web (2015)
14. Leskovec, J., Krevl, A.: SNAP datasets: stanford large network dataset collection (2014). http://snap.stanford.edu/data

15. Liu, H., Huang, H.H.: Enterprise: breadth-first graph traversal on GPUs. In: SC (2015)
16. McBride, B.: Jena: a semantic web toolkit. IEEE Internet Comput. **6**(6), 55–59 (2002)
17. Meng, T., Cai, L., He, T., Chen, L., Deng, Z.: K-hop community search based on local distance dynamics. KSII Trans. Internet Inf. Syst. **12**(7), 3041–3063 (2018). https://doi.org/10.1007/978-3-319-70139-4_3
18. Mi, Z., Yang, Y.: Connectivity restorability of mobile ad hoc sensor network based on k-hop neighbor information. In: ICC, pp. 1–5. IEEE (2011)
19. Qian, Z., Min, C., et al.: GAIA: a system for interactive analysis on distributed graphs using a high-level language. In: NSDI, pp. 321–335. USENIX Association (2021)
20. Rao, V.N., Kumar, V.: On the efficiency of parallel backtracking. IEEE Trans. Parallel Distrib. Syst. **4**(4), 427–437 (1993)
21. Umuroglu, Y., Morrison, D., Jahre, M.: Hybrid breadth-first search on a single-chip FPGA-CPU heterogeneous platform. In: FPL, pp. 1–8. IEEE (2015)
22. Zeng, L., Jiang, Y., Lu, W., Zou, L.: Deep analysis on subgraph isomorphism. arXiv (2020)
23. Zeng, L., Zou, L.: Redesign of the gStore system. Front. Comput. Sci. **12**(4), 623–641 (2018). https://doi.org/10.1007/s11704-018-7212-z
24. Zou, L., Mo, J., Chen, L., Özsu, M.T., Zhao, D.: gStore: answering SPARQL queries via subgraph matching. VLDB (2011)

Modeling Long-Range Travelling Times with Big Railway Data

Wenya Sun[1(✉)], Tobias Grubenmann[3], Reynold Cheng[1], Ben Kao[1], and Waiki Ching[2]

[1] Department of Computer Science, The University of Hong Kong, Hong Kong, People's Republic of China
{wysun,ckcheng,kao}@cs.hku.hk
[2] Department of Mathematics, The University of Hong Kong, Hong Kong, People's Republic of China
wching@hku.hk
[3] University of Bonn, Bonn, Germany
grubenmann@cs.uni-bonn.de

Abstract. Big Railway Data, such as train movement logs and timetables, have become increasingly available. By analyzing these data, insights about train movement and delay can be extracted, allowing train operators to make smarter train management decisions. In this paper, we study the problem of performing *long-range analysis* on Big Railway Data, such as estimating the remaining journey time, i.e., the amount of time for a given train to reach the terminal station. We study how existing statistical and machine learning methods, designed for short-range analysis (e.g., estimating the traveling time between two adjacent stations), can be extended to perform long-range analysis. We further design a method, called *a-LSTM*, based on LSTM (long short-term memory) neural network and attention models. Extensive evaluation on a large amount of train movement data provided by a train service provider in Hong Kong shows that *a-LSTM* is more effective than other solutions in predicting traveling times.

Keywords: Big Railway Data · Machine learning · Journey time estimation

1 Introduction

Railway and metro systems are indispensable public transportation tools in many parts of the world. A lot of "Big Railway Data" can be found in these systems, such as (1) signal data logs, describing the train arrival and departure times at each station [14]; (2) train timetables, which depict the scheduled arrival and departure times of each train at every station; and (3) the topology of railway station networks. Performing analysis on this data enables understanding of train behaviour (e.g., mining journey traveling times), thereby allowing train operators to make sensible decisions and provide better traveling experience.

© The Author(s), under exclusive license to Springer Nature Switzerland AG 2022
A. Bhattacharya et al. (Eds.): DASFAA 2022, LNCS 13247, pp. 443–454, 2022.
https://doi.org/10.1007/978-3-031-00129-1_38

Along with the rapid development of research on roadway transportation [3–5], the topic of analysis on Big Railway Data has also attracted a lot of interest in recent years [9–11]. A few machine learning models have been studied. For example, Kecman et al. [11] conducted feature engineering and used statistical machine learning models (e.g., linear regression, decision trees and random forests) to predict train dwell times (i.e., the amount of time a train stayed at a station) and train running times (i.e., the amount of time a train travelled between two adjacent stations). In [9], Huang et al. proposed a hybrid deep learning framework that integrates different neural network models, and used this model to predict the time required for a train to go from one station to the next one.

The above research work focus on *short-range analysis*, i.e., the forecasting of traveling times from one station to the next one. In this paper, we address the problem of *long-range analysis*, i.e., the derivation of traveling times for journeys that span one or more stations. We encountered this problem during our collaboration with the Mass Transit Railway Corporation (MTRC) [1], the only company in Hong Kong that provides subway services for more than 4 million passengers (or 40% of daily travel population). The company asked us to use about 4 billion signal data records to predict the *remaining journey time* of a train, i.e., given a time instant t_0, the amount of time needed for a train to reach the last station of the railway line (i.e., terminal). As MTRC pointed out, this is important to them because (1) various incidents (e.g., signalling faults, crowds at station platforms during peak hours, intentional blocking of train doors) can cause delay to a train; (2) a train delay of more than 5 min can lead to a hefty monetary charge from the Hong Kong government. By knowing the remaining journey time ahead, the operators of MTRC can perform appropriate actions to shorten this time needed, for instance, by ordering the train to skip a few stations.

Performing long-range analysis on railway data is challenging, MTRC demands a highly accurate solution, or they can face penalty charges. Unfortunately, current Big Railway Data solutions, tailored for short-range analysis, often only consider the past events of the target train t [11], and the information of the station s that t has just passed (e.g., arrival and departure delay of t and trains before t at station s [9]). They typically do not consider other factors, such as the movements of other trains at stations in front of t, incidents, and the extra delay incurred at interchange stations. However, these additional factors can also be crucial for long-range analysis. As our experiments show, incorporating these new factors into existing solutions leads to a mediocre performance, which cannot be accepted by MTRC that requires a very accurate solution.

In this paper, we develop a solution called *a-LSTM*, based on Long Short Term Memory (LSTM) and attention mechanisms. The main benefit of a-LSTM is that it carefully weighs various factors that can affect long-range analysis. We have performed a detailed experimental evaluation of a-LSTM and existing Big Railway Data solutions, ranging from simple ones (e.g., using timetables and historical averages) to more advanced ones (e.g., based on statistical learning and deep learning models). We found that a-LSTM achieves the highest accuracy in terms of remaining journey time prediction on the MTRC data.

The rest of our paper is organized as follows. We discuss related work in Sect. 2, and the problem definition in Sect. 3. We discuss the detail of the MTRC data and an analysis of this data in Sect. 4. Section 5 explains how existing Big Railway Data solutions can be adapted to solve the long-range analysis problem. We then present a-LSTM in Sect. 6. We report our experimental results in Sect. 7.

2   Related Work

With the availability of large amounts of data in railway systems, there is a lot of interest in analysing Big Railway Data for discovering insights from them. Knowing train movement situation and predicting delay information is of great importance for both passengers and train operators. Passengers can have a better planning of their journeys, and train operators can provide a better and more punctual service by managing the trains timely.

Big Railway Data. A real-time data collection system and at-crossing arrival time prediction model was developed in [2]. The studies of train traveling and dwell time prediction were discussed in [6]. Kecman et al. [11] studied the use of linear regression, regression tree, and robust forest for making predictions for running time between two continuous stops and dwell time (or staying time at a station platform). They trained models for each tunnel (for running time prediction) and stop (for dwell time prediction) by using features such as arrival or departure delay. Additional features indicating the uniqueness of the tunnel or stop (i.e., stop type, distance to terminal, and distance from the origin) were also considered to make prediction. We also use these features to predict remaining journey time.

Historical train visiting records at stops contain rich information about traffic situation, such as arrival and departure delays at stops, historical running and dwell time of different trains. Neural networks, especially recurrent neural networks, have shown its superior performance in handling those temporal data sequences [12,13]. Huang et al. [9] used a combination of neural network composing fully connected neural network (FNN), long short-term memory (LSTM) neural network and convolutional neural network (CNN) to make short-range prediction, i.e., arrival delay time at next stop. In this framework, historical train records at the past m stations before the predicted station was encoded and fed as the input for the three neural units. In our paper, we adapt this framework into the remaining journey time prediction task, by adding supplementary features indicating a journey and its current location.

As mentioned above, there is a research gap between existing work and a model which can make prediction of remaining journey travel time (including running time on track and dwell time at stop). In this paper, we fill in this gap by adapting short-range prediction method to long-range remaining journey time prediction.

3   Problem Definition

We now define the terms related to railway operations and used in our paper. Then we give a formal definition of the remaining journey time prediction problem.

Definition 1 (Journey). *A journey is a number $j \in \mathbb{N}$ indicating a specific sequence of stations which are visited by a train at a specific time of the day and day of the week. We denote with $S_j = <s_1, \ldots, s_{n_j}>$ the sequence of stations of that journey and with $\hat{t}^{arr}_{j,s} \in \mathbb{R}, s \in (s_2, \cdots, s_{n_j})$ and $\hat{t}^{dep}_{j,s} \in \mathbb{R}, s \in (s_1, \cdots, s_{n_{j-1}})$ the scheduled arrival time and departure time in timetable.*

Note that in the above definition, for the first (final) station, s_1 (s_{n_j}), the arrival (departure) time, $\hat{t}^{arr}_{j,s_1}$ ($\hat{t}^{dep}_{j,s_{n_j}}$) is undefined. A journey typically starts from a terminal station of a train line and ends at another terminal station of the line. We obtain the station and timing information of a journey from the "train movement logs" provided by MTRC, which contains the movement information of a train on each day.

Definition 2 (Visit Event). *Let $j \in \mathbb{N}$ be a journey. Let $t^{arr}_{j,s}$ and $t^{dep}_{j,s} \in \mathbb{R}$ denote the actual arrival time and departure time according to the train log. The four-tuple $e = (t^{arr}_{j,s}, t^{dep}_{j,s}, d^{arr}_{j,s}, d^{dep}_{j,s})$ is called a visit event, where $d^{arr}_{j,s} := t^{arr}_{j,s} - \hat{t}^{arr}_{j,s}$ and $d^{dep}_{j,s} := t^{dep}_{j,s} - \hat{t}^{dep}_{j,s}, s \in (s_2, \cdots, s_{n_{j-1}})$ are the arrival delay and departure delay at station s, respectively.*

Definition 3 (Remaining Journey Time). *Let $j \in \mathbb{N}$ be a journey. Let $t^{dep}_{j,s_i} \in \mathbb{R}$ denote the departure time at station s_i and $t^{arr}_{j,s_{n_j}} \in \mathbb{R}$ the arrival time at the destination station s_{n_j} of the train on journey j. The remaining journey time from station s_i to destination s_{n_j} for j is defined as $t^{jour}_j = t^{arr}_{j,s_{n_j}} - t^{dep}_{j,s_i}$. Similarly, the scheduled remaining journey time from s_i to s_{n_j} is defined as $\hat{t}^{jour}_j = \hat{t}^{arr}_{j,s_{n_j}} - \hat{t}^{dep}_{j,s_i}$.*

Given the definitions above, we can now define the problem of remaining journey time prediction:

Definition 4 (Remaining Journey Time Prediction). *Let $j \in \mathbb{N}$ be a journey and s_i be a station of the journey before destination s_{n_j}. The goal of remaining journey time prediction is to predict $t^{jour}_j = t^{arr}_{j,s_{n_j}} - t^{dep}_{j,s_i}$.*

4   Features for Remaining Journey Time Prediction

We now discuss the important features for the remaining journey time estimation task.

The train travelling time of a journey in a railway system is affected by many factors, such as scheduled journey time, the day of the week, the hour of

Fig. 1. KTL diagram

the day, delays caused by incidents, among others. In this section, we will discuss the MTRC datasets, and the feature selection procedure for remaining journey time prediction.

We focus on a single operation line in the Hong Kong MTR: the Kwun Tung Line (KTL). Figure 1 shows all the stations of KTL. We indicate different stations with their official three-letter abbreviation (e.g., WHA for the *Whampoa* station).

Because trains follow various timetables on different days of the week, the day of the week is an important factor. In Hong Kong, the subway serves as a main transportation mode for citizens for daily commute. Because of the significantly higher amount of people going to and from work in the morning and evening peak hours, the hour of the day is another critical feature. Figure 2b shows the average journey time from Whampoa (WHA) to Tiu Keng Leng (TIK) station. The figure shows that journey traveling times during the morning peak (9am to 12am) are often higher than those during the non-peak hours. In the evening peak (6pm to 8pm), a significant increase in the journey time is observed, especially at 7pm, where trains have the longest journey time (around 1,955 s). In the weekend, no notable peak hours are observed; this is reasonable, because Hong Kong follows a 5-working-day policy.

We need to extract journey data from train movement log records for journey time prediction. To obtain these journey data, we group log records by trains and order them in a temporal order. Suppose that there are R trains servicing KTL, we then have R temporally ordered log record sequences representing the trajectories of each train. We denote the trajectory of a train r as $H^r = [e_1^r, \cdots, e_n^r]$, where e_i refers to the ith visit event of train r. We then split the trajectory of each train into journeys, which are usually from one terminal station to another terminal station in the train line.

5 Baselines

We now discuss how basic methods (Timetable and Historical Grouped Average), machine learning methods (Linear Regression, Regression Tree, Random Forest and CLF-net [8]), can be used to predict the remaining journey time of a train. The prediction outcome is affected by several features, such as: (1) origin-destination pair, (2) scheduled journey time in the timetable, (3) the day of the week, the hour of the day and (4) train arrival and departure delays at a station. For example, the train timetable dictates the movement of a train along the network in normal situations. However, unusual high passenger demands in peak hours and accidents can cause a longer travelling time.

The machine learning methods above were originally proposed to make short-term travelling time prediction. We adapted them from making short-term prediction to long-term prediction by adding more features. In the following, we will discuss the adaptions, and discuss the drawbacks of existing methods.

5.1 Basic Methods

Timetable. We first consider to use the scheduled remaining journey time $\hat{t}_j^{\text{jour}}$ in the timetable as our predictor for the remaining journey time prediction. If everything runs perfectly and smoothly in the network, this method might achieve a very good performance in the prediction task. However, in real cases, there are many unexpected variations during the daily operation in the urban railway system, which makes it hard for trains to comply with the scheduled time.

Historical Grouped Average (HGA). As discussed before, categorical features such as origin-destination pair for a trip, the day of the week, and the hour of the day can impact the actual journey time of a train. Here, we predict the actual journey by organizing trips into groups. The prediction is made by the averaged journey time of the corresponding group. The features we used for grouping are as follows: (1) the origin-destination pair, (2) day of the week, and (3) hour of the day.

5.2 Machine Learning Methods

Robust Linear Regression. The robust linear regression model was originally developed for making short-range predictions (e.g., running time estimation based on departure delay, and dwell time estimation based on arrival delay) in the railway system in [11]. To adapt the above solutions in the remaining journey time prediction task, we supplement a few features as indicated by the following vector G:

$$G = [OD, h_{\text{now}}, d_{\text{now}}, \hat{t}_j^{\text{jour}}, t_j^{\text{pass}}, l_j^{\text{remain}}, \sin(2\pi h_{\text{now}}/24),$$
$$\cos(2\pi h_{\text{now}}/24), \sin(2\pi d_{\text{now}}/7), \cos(2\pi d_{\text{now}}/7)] \tag{1}$$

where $OD, h_{\text{now}}, d_{\text{now}}, \hat{t}_j^{\text{jour}}$ are the origin-destination pair, current hour of the day, current day of the week, and scheduled remaining journey time, respectively. Besides, the passed journey time t_j^{pass} and remaining track length l_j^{remain} of the journey are also taken as the inputs to give more information about the journey. Due to the cyclical features of the hour of the day and the day of week, the sine and cosine functions of these two features are also used. The target for the prediction now becomes the remaining journey time t_j^{jour}.

Regression Trees and Random Forest. Similar to the robust linear regression method mentioned above, the decision tree and random forest [11] methods were also built for short-range predictions. To relax the assumption of linear relationship between various factors and the output, these two methods segment the prediction space into different regions.

CLF-Net. Huang et al. [9] developed a deep learning framework which takes the traffic situation at previous stations as inputs, to predict the delay at next station. To adapt this model for the remaining journey time prediction task, the global vector G (Eq. 1 above) is added to the last dense layer. This provides additional information about the journey and the train's current location. In our experiments, CLF-net suffers from a high prediction error, especially for the journey's first three or four stations, which only have few stations preceding them. However, the effectiveness of CLF-net can be affected, when the number of stations preceding the station being predicted is small.

6   Attention Based LSTM (a-LSTM)

The Long Short-term Memory network (LSTM) [7], developed to address the issues of gradient vanishing or exploding in traditional Recurrent Neural Networks (RNN), has been found to achieve superb performance in sequence modelling. In this section, we study the adaptation of LSTM for the long-range journey time prediction task, which attempts to learn the dependence among the visit events of different journeys.

We now present the *attention-based long short-term memory* (a-LSTM), which is designed for remaining journey time prediction (Definition 4). We first denote V^s as a list of visit events for all journeys at station s, sorted in reverse order of the arrival times of the visit events. The ith visit event in a V^s is called V_i^s. We also denote V^{train} to be a list of visit events, sorted in reverse chronological order of the arrival times, for all journeys of a given train. We denote the i_{th} event of V^{train} as V_i^{train}.

The lists V^s and V^{train} for visit events that happens before current time t_{now} are fed to station-LSTM units and a train-LSTM unit respectively. In particular, the visit events that occurred at each station, i.e., V^s, are inputted to the station-LSTM unit corresponding to station s, in order to learn the travel pattern at s. The list of visit events of the train predicted, i.e., V^{train}, is fed to a train-LSTM, to learn the traffic pattern of *train*. To give different degrees of emphasis on the information contributed from each station, we apply an attention mechanism to the outputs of station-LSTM units that represent traffic situation of each station. The framework of a-LSTM is shown in Fig. 2a.

We denote the MTR line that we studied here by $\mathcal{L} = [s_0, s_1, \cdots, s_k, s_{k+1}]$, where s_0 (s_{k+1}) is the first (last) station in MTR line $\mathcal{L}$. An LSTM layer takes the historical visit events over all the middle stations at MTR line $[s_1, \cdots, s_k]$ as input:

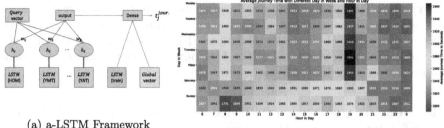

(a) a-LSTM Framework

(b) Travelling time at Different hour of a Day

Fig. 2. a-LSTM framework and travelling time at different hour of a day

$$E^s = [V^s_{N_{hist}}, \cdots, V^s_2, V^s_1] \tag{2}$$

$$[E^{s_1}, E^{s_2}, \cdots, E^{s_k}] \to [LSTM^{s_1}, LSTM^{s_2}, \cdots, LSTM^{s_k}] \tag{3}$$

where N_{hist} is the number of visit events used to train the model, E^s is the latest N_{hist} historical visit events before t_{now} ordered temporally. Note that $LSTM^{s_i}$ is the station-LSTM unit that models the temporal relationship of event sequence at station s_i.

Besides the historical visit event recorded at each station, we also consider the trajectory of the target train:

$$E^{train} = [V^{train}_{N_{hist}}, \cdots, V^{train}_2, V^{train}_1] \tag{4}$$

$$E^{train} \to LSTM^{train} \tag{5}$$

The output from the above k station-LSTM units are concatenated and fed into an attention layer. Then, the attention output is fed into a dense layer to merge with the $LSTM^{train}$ and the global vector G, as shown in Fig. 2a.

6.1   Attention Mechanism

After getting information from k station-LSTM units, a multi-head attention mechanism is utilized to associate weights with these station-LSTM outputs, according to a query vector Q. The query vector contains the latest visit event of the journey where we want to predict the remaining journey time. The key K and value V vector are the outputs of the k station-LSTM units for the k middle stations:

$$K = V = [LSTM^{s_1}, LSTM^{s_2}, \cdots, LSTM^{s_k}] \tag{6}$$

We first encode the query vector Q into the same embedding dimension d_k as the key/value vector. Then, each attention head does the following:

$$\text{Attention}(Q, K, V) = \text{softmax}\left(\frac{QK^T}{\sqrt{d_k}}V\right) \tag{7}$$

A dense layer is used to merge the outputs from $LSTM_{train}$, global vector G, and the attention output from k-station LSTM units for prediction of remaining journey time t_j^{jour}, which is Dense($LSTM^{train}$, Attention, G). The predicted remaining journey time t_j^{jour} will then used to evaluate the model performance in Sect. 7.

7   Evaluation

We now present the experimental results for different solutions that predict remaining journey time on the MTRC data. Sect. 7.1 describes the experimental setup. We discuss the results in Sect. 7.2.

7.1   Experimental Settings

We randomly select 47 days between November 2019 and December 2019 to train our models and test the model performances on the remaining 13 days. For each day, we extract journey data from the train movement log, and predict the remaining journey time at each station s, where $s \in (s_1, \cdots, s_{n_{j-1}})$ before the journey ends. The experiments were run on a machine with two Intel® Xeon® Silver 4215R CPUs @3.20 GHz, and one GV100GL (Tesla V100 PCLe 32 GB) 33 MHz GPU card.

We use mean absolute error (MAE), root mean squared error(RMSE), and mean absolute percentage error(MAPE) to evaluate the remaining journey time prediction task:

$$\text{MAE} = \frac{\sum_{i=1}^{n} |A_i - P_i|}{n}; \text{RMSE} = \sqrt{\frac{1}{n} \sum_{i=1}^{n} (A_i - P_i)^2}; \text{MAPE} = \frac{1}{n} \sum_{i=1}^{n} \left| \frac{A_i - P_i}{A_i} \right| \tag{8}$$

where n is the number of samples in the test set, A_i is the actual value, and P_i is the predicted value for the remaining journey time in seconds. We remark that MAE and MAPE are popular evaluation metrics for regression tasks: MAE indicates the averaged absolute difference between actual and estimated travel time, while MAPE measures the relative prediction error based on the percentage distance between real and predicted time. The RMSE is a regression metric, which is more sensitive to outliers and can amplify large prediction errors.

We executed the timetable baseline solutions on the test days, because there is no need for training for this method. For other methods, we have trained the model on the training set, and report the performance on the test set. Because the performance of the machine learning models can be sensitive to their parameters, we conduct grid search and a 5-fold cross validation on the training set to tune the parameter values.

Table 1. Experimental results

Method	MAE	RMSE	MAPE(%)	< 1 min (%)	< 2 min (%)	< 3 min (%)
Timetable	76.4	103.43	15.7	48	80	98
HGA	39.02	74.74	10.7	83	96	98
Robust linear regression	37.57	75.87	16.32	86	**98**	99
Regression tree	32.09	71.95	4.0	89	97	99
Random forest	30.81	68.68	**3.7**	89	**98**	99
CLF-net	31	50.13	4.3	87	**98**	99
A-LSTM	**24.86**	**40.21**	3.8	**91**	**98**	**100**

7.2   Results

Table 1 shows the MAE, RMSE, and MAPE scores of different solutions studied in this paper. We also display the proportion of samples predicted with MAE less than 1, 2, and 3 min. The MTRC told us that a prediction error in MAE below 1 min is acceptable. We observe that the timetable method performs the worst for the overall performance in MAE, RMSE, and second-worst in MAPE. Also, the percentage of samples below 1, 2, and 3 min error is the least. The poor performance of timetable method reveals that trains are not following the timetable schedule closely. We can also see that a-LSTM performs the best in MAE and RMSE, and have the largest portion of samples with MAE less than 1, 2, and 3 min.

To have a comprehensive understanding of the prediction capability of different methods, we study the prediction results of those models at varying quantiles in Fig. 3a. The MAE error at the 0.0 and 1.0 quantile show the minimum and maximum errors, respectively. We observe that the top MAE of the a-LSTM method is the lowest, showing its robustness. The black line indicates the 1-min threshold. From Fig. 3a, we observe that a-LSTM performs the best in predicting samples at less than 1-min error, as its point of intersection between the quantile line and the 1-min threshold line is the rightmost, at the 0.91 quantile point. It is worth notice that for any other error bound which is not 1 min, it also has the best performance, as shown by the rightmost position of its quantile line. We also see that timetable method performs the worst.

In Fig. 3b, the prediction power of models at different hours of the day is presented. It shows that the predictive power of Linear Regression, Regression Tree, Random Forest and HGA is not very stable during peak hours, especially in the morning peak (8 am–10 am). The performance of the timetable method is also not very robust over different hours of the day, but its performance variation does not only come from peak hours. In contrast, CLF and a-LSTM achieved a better performance at different hours of the day. Moreover, a-LSTM often performs the best over the day.

The prediction error at the different predicted stations of a journey is reported in Figs. 3c and 3d. The completion level of a journey increases when the train leaves from a terminal to another one. We notice that the closer a train is to the

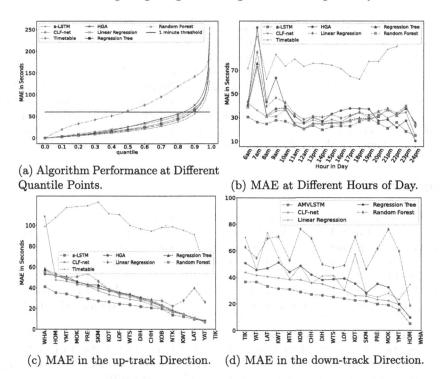

(a) Algorithm Performance at Different Quantile Points.

(b) MAE at Different Hours of Day.

(c) MAE in the up-track Direction.

(d) MAE in the down-track Direction.

Fig. 3. Performance of remaining journey time prediction.

end terminal station, the more accurate we can predict the remaining journey time. An interesting finding is that the performance improvement of a-LSTM in early stations over other baselines is more significant than in the later stations. The reason can be rooted in constructing data inputs; a-LSTM takes the visit events at stations both behind and ahead of the train so that it can still capture the traffic situation even at the early part of the journey. More specifically, in the up-track direction, we observed a large MAE for CLF-net when predicting WHA station. As discussed in Sect. 6, such bad performance is due to lack of information for the first few stations of a journey.

8   Conclusions

Big Railway Data enables the extraction of insights from a huge number of train-related records. In this paper, we examine the problem of predicting long-range remaining journey time of a train based on train movement records provided by the Hong Kong MTR Corporation. Our experiments show that the proposed a-LSTM method performs better than state-of-the-art solutions, which were designed for short-range predictions. Our future work is to assist MTRC to make better decisions for train scheduling and incident handling operations.

Acknowledgement. Wenya Sun and Reynold Cheng were supported by MTR (project 200009153) and the University of Hong Kong (Projects 104005858, 10400599, 207300392). Tobias Grubenmann was supported by the Federal Ministry of Education and Research (BMBF), Germany, under Simple-ML (01IS18054), and the European Commission under PLATOON (872592) and Cleopatra (812997). We would like to thank the MTR Corporation, especially Mr. Leo Cheng, for providing their data and advice. We thank Huawei corporation (and Ms. Kathy Ng) for providing a high-performance server for our study. We also thank Prof. W. K. Li, Prof. Philip Yu, and Mr. W. K. Kwan for their valuable suggestions in the early phase of this work.

References

1. Mtr corporation. http://www.mtr.com.hk/en/customer/main/index.html
2. Chen, Y., Rilett, L.R.: Train data collection and arrival time prediction system for highway-rail grade crossings. Transp. Res. Rec. **2608**(1), 36–45 (2017)
3. Han, X.: Traffic incident detection: a deep learning framework. In: MDM, pp. 379–380. IEEE (2019)
4. Han, X., Cheng, R., Grubenmann, T., Maniu, S., Ma, C., Li, X.: Leveraging contextual graphs for stochastic weight completion in sparse road networks. In: SDM (2022)
5. Han, X., Grubenmann, T., Cheng, R., Wong, S.C., Li, X., Sun, W.: Traffic incident detection: a trajectory-based approach. In: ICDE, pp. 1866–1869. IEEE (2020)
6. Hansen, I.A., Goverde, R.M.P., van der Meer, D.J.: Online train delay recognition and running time prediction. In: ITSC, pp. 1783–1788 (2010)
7. Hochreiter, S., Schmidhuber, J.: Long short-term memory. Neural Comput. **9**(8), 1735–1780 (1997)
8. Huang, P., Wen, C., Fu, L., Peng, Q., Li, Z.: A hybrid model to improve the train running time prediction ability during high-speed railway disruptions. Saf. Sci. **122**, 104510 (2019). https://doi.org/10.1016/j.ssci.2019.104510
9. Huang, P., Wen, C., Fu, L., Peng, Q., Tang, Y.: A deep learning approach for multi-attribute data: a study of train delay prediction in railway systems. Inf. Sci. **516**, 234–253 (2020)
10. Kecman, P., Goverde, R.M.P.: Online data-driven adaptive prediction of train event times. IEEE Trans. Intell. Transp. Syst. **16**(1), 465–474 (2015)
11. Kecman, P., Goverde, R.M.P.: Predictive modelling of running and dwell times in railway traffic. Public Transp. **7**(3), 295–319 (2015). https://doi.org/10.1007/s12469-015-0106-7
12. Lipton, Z.C., Berkowitz, J., Elkan, C.: A critical review of recurrent neural networks for sequence learning (2015)
13. Sutskever, I., Vinyals, O., Le, Q.V.: Sequence to sequence learning with neural networks (2014)
14. Wang, P., Zhang, Q.P.: Train delay analysis and prediction based on big data fusion. Transp. Saf. Environ. **1**(1), 79–88 (2019)

Multi-scale Time Based Stock Appreciation Ranking Prediction via Price Co-movement Discrimination

Ruyao Xu[1,4], Dawei Cheng[2], Cen Chen[1(✉)], Siqiang Luo[3], Yifeng Luo[1], and Weining Qian[1]

[1] East China Normal University, Shanghai, China
ryxu@stu.ecnu.edu.cn, {cenchen,yfluo,wnqian}@dase.ecnu.edu.cn
[2] Tongji University, Shanghai, China
dcheng@tongji.edu.cn
[3] Nanyang Technological University, Singapore, Singapore
siqiang.luo@ntu.edu.sg
[4] Seek Data Inc., Shanghai, China

Abstract. The prediction of the stock market trends is an important problem and has attracted tremendous research interest. However, previous methods often consider modeling each stock separately and rarely leverage the information between different stocks to jointly train a model. In this paper, we address the problem of predicting the stock market trends and bring two key insights. First, we show that a better prediction model can be trained by simultaneously considering the features of correlated stocks. Unlike previous methods, our model does not rely on any prior manual input knowledge. Second, we observe that stock trend information on a single time scale is confined and not sufficient because the holding period can be different among investors. We thus design an encoder with multiple time scales to capture features for different time granularity. On top of these, we present a novel stock trend prediction framework called MPS. Extensive experiments are conducted on both the China A-Shares and NASDAQ markets, and results show that MPS outperforms baselines on different holding periods.

Keywords: Time series · Stock embedding · Multi-task learning

1 Introduction

Stock trading is important in financing and investment. Traders and investors aim to purchase shares at low prices and sell them at higher prices for profits. However, markets are dynamic, nonlinear, non-parametric, and chaotic in nature [3], making stock prediction extremely challenging.

Experienced investors treat technical analysis as a key method to make precise decisions in stock trading [4]. Technical analysis does not consider the company data, assuming that information capable of affecting the market is reflected

© The Author(s), under exclusive license to Springer Nature Switzerland AG 2022
A. Bhattacharya et al. (Eds.): DASFAA 2022, LNCS 13247, pp. 455–467, 2022.
https://doi.org/10.1007/978-3-031-00129-1_39

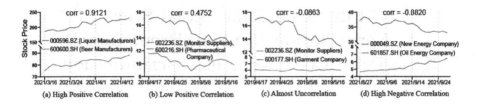

Fig. 1. Price trends and the fluctuations correlation among stocks.

in the price [18] and that historical behavior has a verifiable influence on future price movements. Researches use historical price series as input data to predict price with statistical methods and optimization algorithms like step-wise regression analysis [2] and genetic programming [13]. Recently, a new trend is to enhance the prediction by deep learning algorithms [12], entailing the technical indicators as input features, which promotes the wide use of technical analysis.

Most existing models focus on leveraging single stock's historical information to forecast movement trend [8]. Previous studies either applied gradient boosting to extract features for stock prices forecasting [20] or leveraged convolutional and recurrent networks to process time series stock data [19,23]. However, they ignore the natural correlation between stocks caused by industry chain and company attributes. Similar or converse movement trends often happen between two stocks, and the prediction can be more accurate if such useful information can be extracted effectively. To our knowledge, there are only a few studies that tried to employ the information among different stocks. However, they are too simplistic or require heavy manual efforts. For example, [22] simplified the relationship among stocks and ignored the complex information between stocks with different industry types. [10] employed explicit domain knowledge such as stock industry types or the cooperation among companies. Nevertheless, they require manual efforts to establish stock relationships that are complicated.

Moreover, we also discover that the information about stock trends of a single time scale can be deceptive. Because the performance of stocks varies over time windows of different sizes. Therefore, the information of different time scales can more accurately reflect the trend of stock fluctuations and help investors to have a holistic understanding. Besides, investors may have different preferences for *stock holding period* (i.e., the duration that an asset is held by an investor), thus it is meaningful to judge the revenue and risk of specific stock with investment cycles of different granularity. However, state-of-the-art models have not considered stock trends information of multiple time granularity [9,21,26].

As pointed out in [10], the relationships between stocks are reflected in the price fluctuations of stocks. We discover that the degree of correlation between stocks can be effectively measured by the correlation coefficient. As is shown in Fig. 1, stocks with higher fluctuation similarities have higher correlation scores, such as 000596.SZ and 600600.SH which have extremely similar trends due to the industry types. Negative values represent negative correlation contrarily, whose causes include competition like 000049.SZ and 601857.SH. Inspired by this, we

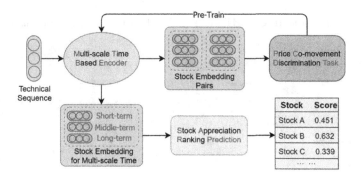

Fig. 2. Overview of our proposed MPS Framework.

design a *self-supervised learning* based stock technical series model which learns the relationships between different stocks by mining the correlation coefficient of future price movements. It can automatically identify correlated stocks through the correlation of stock trends and determine whether the relationship is positive or negative. Besides, our model concerns stock features in multiple time scales of future information and is trained through multi-task joint learning, which discriminates stocks' co-movement trends in different spans.

In this paper, we propose MPS framework, which consists of the above-mentioned three components: Multi-scale time based encoder, Price co-movement discrimination, and Stock appreciation ranking prediction. It learns multi-period correlation price features by multi-task soft parameter sharing [24] and forecast the stock appreciation ranking, i.e., the rank of stock's price appreciation proportion among all the stocks. Our contributions are summarized as follows.

- We propose an MPS[1] framework to learn intrinsic relations among different stocks in a self-supervised approach, by correlation coefficient instead of artificial rules such as industry and market value.
- We present a multi-scale time encoding method that integrates stock embeddings for multiple time spans and use a multi-tasking approach to train them simultaneously. The result shows that the model benefits from different time scale information and performs well in various stock holding periods.
- We assess the effectiveness of MPS on two real-world datasets in the NASDAQ and Chinese market. We further deploy our methods at Emoney, a leading financial service provider in China. The result of financial evaluation metrics demonstrates that our framework outperforms baseline methods.

2   The Proposed Framework

In this section, we present the details of MPS. As illustrated in Fig. 2, the multi-scale time based encoder (abbrev. as MTB Encoder) extracts embeddings from

---

[1] We will release the source codes online: https://github.com/ECNU-CILAB/MPS.

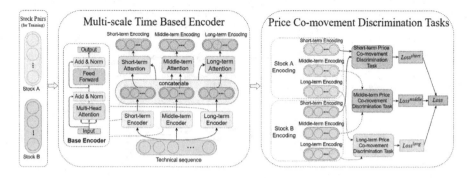

Fig. 3. The structure of MTB Encoder with PCD tasks.

technical sequence data, while price co-movement discrimination (PCD) task pre-trains MTB Encoder to learn correlation of stock pairs over different time scales. Finally, the stock appreciation ranking prediction module employs embeddings generated by MTB Encoder to predict appreciation scores for stocks.

Our model uses daily technical sequence data as input, including open price p^o, close price p^c, the highest price p^h, the lowest price p^l, turnover rate r^t and trading volume v^t of a stock. We define $x_t^s = (p^o, p^c, p^h, p^l, r^t, v^t)$ as technical data for stock s at t-th day. We further define $X_t^s = \{x_{t-q+1}^s, ..., x_{t-i}^s, .., x_t^s\}$, indicating a sequence data for stock s of continuous q day.

2.1 Multi-scale Time Based Encoder

Base Encoder. Financial embedding acquired with a deep learning framework have strong generalization ability to predict stock market trend and other financial problems [8], also skilled at abstracting temporal features in time series forecasting [25]. We adopt Transformer [26] based encoder as *Base Encoder* to extract stock representation. Given stock technical sequence X_1 and X_2, consider $X_1 W_i^Q$, $X_2 W_i^K$, $X_2 W_i^V$ as Query, Key, Value. Scaled Dot-Product Attention mechanism computes attention between X_1 and X_2 by:

$$\text{Attention}(X_1, X_2) = \text{softmax}(\frac{X_1 W_i^Q (X_2 W_i^K)^T}{\sqrt{d_K}}) X_2 W_i^V, \tag{1}$$

where $W_i^Q \in \mathbb{R}^{d \times d_Q}$, $W_i^K \in \mathbb{R}^{d \times d_K}$, $W_i^V \in \mathbb{R}^{d \times d_V}$ are parameter matrices for Query, Key and Value. $\sqrt{d_K}$ is the scaling factor, prompting gradient variation in a more stable way. The attention of Query at Key is scored by the dot product of them. After scaling and activation of softmax, the output score are multiplied by Value to stress the significant value and weaken the less important ones.

The Base encoder encodes X_1 by the attention score between X_1 and X_2, and leverages multi-head to extend the ability of the model to focus on different locations and representation subspaces, named as Multi-Head Attention:

$$\text{MultiHead}(X_1, X_2) = \text{Concat}(head_1, ..., head_h)W^M$$
$$where\ head_i = \text{Attention}(X_1, X_2).$$

(2)

h is the number of heads. $W^M \in \mathbb{R}^{hd_V \times d}$ is weight matrices of heads and each of heads has different parameter matrices for Query, Key and Value.

Sequentially, the network is connected to residual layers with normalization and to a feed forward layer with ReLU activation. Residual and fully connected layers stop the output from degeneration and prevent a rank collapse [7]. Besides, in our framework, the attention mechanism employs self-attention when encoding a stock, where X_1 is equal to X_2. We use stock technical sequences X_t^s as input and present the base encoder as BaseEncoder($\boldsymbol{X_t^s}$).

Multi-scale Time Based Encoder. The influence of information on the stock market diminishes by time, thus weekly and monthly prediction is harder than few days prediction [6] and overwhelming majority of researches center on a short-term prediction horizon like a day or several minutes. Considering the challenge mentioned before, we combine multi-scale time based features by soft parameters sharing [24] and we name it as Multi-scale Time Based Encoder.

As shown in Fig. 3, we use Base Encoder to gain initial encoding for multi-scale time separately, i.e., short-term, middle-term and long-term encoders:

$$e_{s,t}^\alpha = \text{BaseEncoder}(X_t^s),$$

(3)

where $\alpha = \{short, middle, long\}$ and $e_{s,t}^{short}$ presents short-term encoding, $e_{s,t}^{middle}$ for middle-term and $e_{s,t}^{long}$ for long-term. Particularly, base encoders are of the same architecture and encodes the same input sequence across different terms, which are trained to capture information for different future period lengths.

Next, every encoding for a single term interacts on multi-term encoding to help enrich the information of encoders in different terms.:

$$e_{s,t} = \text{concatenate}(e_{s,t}^{short}, e_{s,t}^{middle}, e_{s,t}^{long}),$$

(4)

$$\epsilon_{s,t}^\alpha = \text{Attention}(e_{s,t}^\alpha, e_{s,t}),$$

(5)

2.2 Price Co-movement Discrimination Task

Inspired by self-supervised representation learning [15], we propose a contrast-based self-supervised method with Price Co-movement Discrimination Task to identify similarities between stock sequences pairs and train MTB Encoder.

Price Co-movement Metrics Definition. We formulate stock price comovement prediction as a ternary classification problem: positive, negative and low correlation. The close price sequence for stock s from t to the following k days is defined as $P_t^s = \{p_{s,t}^c, ...p_{s,t+j}^c, ...p_{s,t+k}^c\}$. We randomly choose stock s_1 and stock s_2 and measure the co-movement trend between them:

$$\rho(P_t^{s_1}, P_t^{s_2}) = \frac{Cov(P_t^{s_1}, P_t^{s_2})}{\sqrt{\sigma(P_t^{s_1})\sigma(P_t^{s_2})}},$$

(6)

where $Cov(\cdot)$ is covariance and $\sigma(\cdot)$ is variance of two price sequences. $\rho(\cdot)$ is in the range $[-1, 1]$ and generates ground truth δ_c for price co-movement prediction where $\delta_c = 1$ if $\rho(\cdot) > r_1$; $\delta_c = -1$ if $\rho(\cdot) < r_2$; else $\delta_c = 0$, which signify positive, no significant and negative correlation between stock price sequence $P_t^{s_1}$ and $P_t^{s_2}$. k is different when $\alpha = \{short, middle, long\}$ and $k_{(short)} < k_{(middle)} < k_{(long)}$.

Price Co-movement Discrimination Task. We design self-supervised tasks to train stock encoders for multi-terms by discriminating the co-movement trend of two stocks. MTB Encoder generates stock embedding pairs $\epsilon_{s_1,t}^{\alpha}, \epsilon_{s_2,t}^{\alpha}$, which are passed into PCD tasks to predict the correlation trends between stocks.

$$H^{\alpha} = \text{ReLU}([\epsilon_{s_1,t}^{\alpha}, \epsilon_{s_2,t}^{\alpha}]W^H + b^H), \tag{7}$$

$$O^{\alpha} = \text{Sigmoid}(H^{\alpha}W^O + b^O), \tag{8}$$

where O^{α} is the prediction of co-movement and W^O, W^H, b^O, b^H are parameters to be trained. Next, we define the objective function by cross-entropy to minimize the error between δ_c and co-movement prediction:

$$L^{\alpha} = \sum_{i=1}^{n} \delta_c(P_t^{s_1}, P_t^{s_2})^{(i)} \log\left(\frac{\exp(O^{\alpha(i)})}{\sum_{i=1}^{n} \exp(O^{\alpha(i)})}\right), \tag{9}$$

where $n = 3$ for ternary classification problem. Task predicts co-movement for a longer time span when $\alpha = long$, and it is shorter when $\alpha = middle$, and it is shortest when $\alpha = short$. The final loss $\mathcal{L}$ is defined as losses of multi-terms:

$$\mathcal{L} = \lambda L^{short} + \gamma L^{middle} + \mu L^{long}, \tag{10}$$

where λ, γ and μ are hyper-parameters to tune the importance of sub-tasks.

2.3 Stock Appreciation Ranking Prediction

In this component, we leverage stock embedding for multi-scale time obtained by MTB Encoder to predict the appreciation ranking score of stocks. We define stock prediction as a regression problem to predict the percentage of stocks appreciation ranking. The rank score is formulated as:

$$\zeta_r(p_{t,t+d}^s) = \frac{\text{rank}(p_{s,t+d}^c/p_{s,t}^c)}{\text{count}(p_{s,t+d}^c/p_{s,t}^c)}, \tag{11}$$

where the numerator is the ranking of stock s's appreciation ratio from t-th to $t + d$-th day and the denominator is the counting of all stocks.

For a stock, MTB Encoder encodes technical sequence X_t^s and generates embeddings for short, middle and long term, containing features of status in different forecast targets. Next, the importance of three embeddings is adjusted by a Fully Connected Layer. Then we compute the final prediction by a Gated Recurrent Unit (GRU) network, followed by an output layer.

Table 1. The detailed explanation of metrics in our experiments.

Metrics and detailed explanation	
ARR	**Annualized Rate of Return** is the equivalent annual return over a given period. It is defined as $ARR = (\frac{CV-IV}{IV})^{\frac{1}{N_y}} - 1$, where CV and IV is current value and initial value of portfolio, N_y is the number of holding years
AV	**Annualized Volatility** describes the average variation of portfolio's value over a year, defined as $AV = \sqrt{\frac{\sum_{t=1}^{T}(R_t - \overline{R_t})^2}{T}}\sqrt{N_d}$, where R_t is the daily return rate in t-th day and N_d is the number of trading days in a year
ASR	**Annualized Sharpe Ratio** is the average return earned per volatility, defined as $ASR = \frac{ARR}{AV}$ and to evaluate return of an investment compared to its risk
MDD	**Maximum Drawdown** is the maximum loss from peak to trough of a portfolio: $MDD = \frac{TV-PV}{PV}$, where TV and PV is the trough and peak value
CR	**Calmer Ratio** describes the relationship between return and maximum drawdown, defined as $CR = \frac{ARR}{\lceil MDD \rceil}$
IR	**Information Ratio** measures a portfolio returns beyond a benchmark: $IR = \frac{(PR-BR)}{TE}$, where PR is portfolio return, BR is benchmark return. TE is tracking error, meaning the standard deviation between PR and BR

3 Experiments

We use two real market datasets from Chinese A-Shares[2] and NASDAQ market[3] to conduct experiments. For two datasets, we take sequence from Jan 01,2019 to Dec 31, 2019 as training sets and from Jan 01,2020 to Dec 31,2020 as test sets. In the experiment, the window size of technical sequence q is 20, and the window size of co-movement discrimination $k_{(short)}, k_{(middle)}, k_{(long)}$ are separately set as 1, 5, 20. It means that the PCD Task trains the co-movement trend of stocks for the next 1, 5 and 20 days. The correlation boundary $r_1 = 0.5$ and $r_2 = -0.5$ for middle and long term, $r_1 = \frac{2}{3}$ and $r_2 = -\frac{1}{3}$ for short term. Additionally, the number of heads h is set to 1, and the loss weights of λ, γ, μ are set as $\frac{1}{3}$.

Evaluation Metrics. We use five standard financial indicators for evaluation shown in Table 1. ARR, ASR, CR and IR are metrics to estimate stock return, while the latter three take risk into account. They reveal a better revenue if higher. However, AV and MDD measure the volatility and risk of investment. The smaller the AV or larger the MDD suggests a lower risk of the portfolio.

---

[2] http://www.csindex.com.cn/en/indices/index-detail/000906.
[3] https://www.spglobal.com/spdji/en/indices/equity/sp-500/.

Table 2. Model performance comparison with baselines in Chinese A-Share and NAS-DAQ data. Bold and underlined denote the best and second best results respectively.

Methods	ARR	AV	ASR	MDD	CR	IR	ARR	AV	ASR	MDD	CR	IR
	China A-Shares data						NASDAQ Data					
Market	0.225	0.230	0.976	−0.155	1.447	1.000	0.153	0.345	0.444	−0.339	0.451	0.609
LR	0.294	**0.234**	1.254	−0.179	1.640	1.223	0.166	0.356	0.466	−0.370	0.449	0.629
CNN	0.336	0.276	1.218	−0.191	1.761	1.195	0.180	0.383	0.470	−0.362	0.497	0.634
LSTM	0.481	0.280	1.715	−0.200	2.406	1.549	0.178	0.367	0.485	−0.345	0.516	0.646
ALSTM	0.384	0.328	1.169	−0.205	1.875	1.160	0.290	0.374	<u>0.774</u>	−0.363	<u>0.799</u>	<u>0.902</u>
DARNN	0.472	0.302	1.561	−0.206	2.292	1.437	<u>0.318</u>	0.438	0.727	−0.418	0.761	0.863
SFM	0.477	0.279	1.710	−0.193	2.466	1.547	0.282	0.458	0.614	−0.388	0.726	0.787
B-TF	<u>0.497</u>	0.280	<u>1.776</u>	−<u>0.174</u>	<u>2.856</u>	<u>1.591</u>	0.249	**0.343**	0.726	−<u>0.333</u>	0.748	0.819
MPS	**0.508**	<u>0.273</u>	**1.858**	−0.164	**3.095**	**1.648**	**0.332**	<u>0.352</u>	**0.943**	−0.329	**1.009**	**0.925**

Baselines. We compare our method with various baselines for stock prediction. Market is the market performance of all stocks in index S&P500 for US market and CSI 800 for Chinese market. LR [11], CNN [23] and LSTM [19] predict stock price trends according to historical numerical records, based on logistic regression, convolution neural network and long short-term memory respectively. DARNN [21] uses a novel dual-stage attention-based RNN to predict the stock scores, which consists of encoder and decoder with attention mechanism. An adversarial learning-based method ALSTM [9] predicts stock trends based on a variant of LSTM model with adversarial training to add perturbations to simulate the stochasticity of price variables. Recently proposed SFM [29] uses a state frequency memory regression network learning trading patterns with multiple frequencies to predict the trend of stock prices. A state-of-art method B-TF [26] proposes a basic transformer structure with attention mechanism.

3.1   Performance Comparison with Baselines

We verify our model effectiveness on both Chinese A-Shares and the NASDAQ market datasets. As is shown in Table 2, We observe that the NASDAQ market is more unstable than the Chinese A-Shares market, because AV of the former (0.345) is higher than the latter (0.230) and MDD is on the contrary. LR model achieves the lowest value of ARR, ASR, and CR in both markets, which shows its limitation to encode time series data. LSTM, SFM, and B-TF performed differently in these two datasets, which perform noticeably well in China A-Shares Data but poorly in NASDAQ Data. However, ALSTM has the opposite performance and is more adapted to the NASDAQ Data. It confirms that the ALSTM model is designed to actively add noise to extend the generalization capability of the model, which gives it an advantage in high volatility data.

In China A-Shares dataset, our proposed MPS surpasses baselines in almost all metrics and leads to a wide gap with a value of 0.508 for ARR, signifying our method brings more net benefits. Our model also achieves a high level in ASR, CR and IR which means earning more profits under the same level of risk. In the metric of MDD, our model achieves similar performance with Market, also

Table 3. Performance of MPS and baselines in holding periods of 1, 5 and 20 days.

Metrics	ARR			ASR			CR			IR		
Holding days	1	5	20	1	5	20	1	5	20	1	5	20
LR	0.294	0.284	0.194	1.254	1.091	0.803	1.640	1.268	1.311	1.223	0.945	0.859
CNN	0.336	0.398	0.379	1.218	1.409	1.333	1.761	2.010	2.026	1.195	1.321	1.647
LSTM	0.481	0.413	0.444	1.715	1.541	1.683	2.406	2.618	2.794	1.549	1.432	1.533
ALSTM	0.384	0.380	0.261	1.169	1.152	0.968	1.875	1.937	1.443	1.160	1.147	1.000
DARNN	0.474	0.401	0.440	1.647	1.605	1.872	2.384	2.264	3.129	1.500	1.481	1.678
SFM	0.477	0.435	0.396	1.710	1.655	1.535	2.466	2.570	2.269	1.547	1.514	1.430
B-TF	0.482	0.391	0.643	1.829	1.363	2.340	2.887	1.801	3.602	1.634	1.301	1.956
MPS	**0.508**	**0.555**	**0.669**	**1.858**	**2.020**	**2.389**	**3.095**	**3.179**	**3.827**	**1.648**	**1.754**	**1.980**

exceeding the baselines. Though LR gets the best score when it comes to AV, its scores on other metrics are undesirable. By comparison, MPS keeps step with LR and gets second place on AV without sacrifice on the performance of other metrics. It is reasonable to bring a higher risk with a higher return for the positive correlation between sectional return and idiosyncratic risk of stock [17] and this value is within the acceptable range. What's more, the experiment conducted in NASDAQ also shows similar results that our model performs quite well in all metrics, demonstrating the effectiveness and robustness of our framework.

3.2 Performance in Different Stock Holding Period

In this set of experiments, we varied the number of stock holding cycles in 1 day, 5 days, and 20 days to study our model's performance in different holding periods. We predict the stock score of the next $1, 5, 20$ days by MPS and baseline methods to conduct stock investment simulations, and the results are in Table 3. It is evident that our model outperforms baselines in all holding periods in all metrics. The performance of B-TF is relatively outstanding in all baseline models, but the advantage of our model becomes even more significant when the holding period is 5. It clearly reveals that most of the baseline models show their weaknesses in the longer forecast interval and perform not as well as the 1-day forecast, which validates that weekly and monthly prediction is harder than a few days prediction [6]. Our model is stable in different holding periods and also has better performance in the longer forecast periods. This proves the effectiveness of our stock embeddings for multiple time scales.

3.3 Ablation Analysis

We performed an ablation study on Chinese market to evaluate the contributions of sub-tasks from three time scales, where 'noShort', 'noMiddle' and 'noLong' respectively denote that short-term, middle-term and long-term price co-movement discrimination tasks are detached from the model to predict the stock scores. As is shown in Fig. 4, we performed detailed ablation experiments on three different holding periods of 1, 5 and 20 days. It gives us confidence that

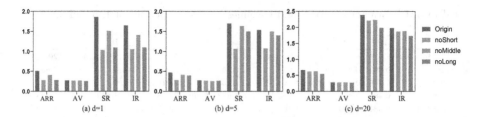

Fig. 4. Ablation study to verify the effectiveness of different terms, where d is the day of holding periods. 'noShort', 'noMiddle' and 'noLong' respectively denotes that long-term, middle-term and short-term PCD tasks are detached from origin model.

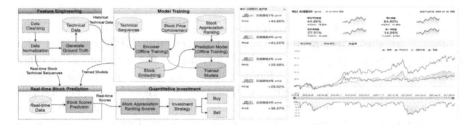

Fig. 5. The workflow of MPS model for quantitative investment.

Fig. 6. The interface of MPS in a web-based portfolio management product.

all tasks are effective in stock prediction of various holding periods. Surprisingly, the contribution of sub-task is complex and not intuitive. For example in Fig. 4(a), all three sub-tasks play a role in one-day forecasting, while short-term and long-term task contributes the most to it. It's also shown in Fig. 4(b) that short-term encoding has a strong influence on 5 days' prediction. And in terms of $d = 20$ in Fig. 4(c), the long-term task has the greatest impact on the model. In conclusion, ablation experiments validate our motivation that multi-scale time based encodings provide valuable information for the stock forecast.

4   Real-World Deployment

The workflow of the MPS model for quantitative investment is illustrated in Fig. 5, consisting of feature engineering, model training, real-time stock prediction, and quantitative investment modules. *Feature engineering* module preprocesses stock technical sequence and generates ground truth like stock co-movement trend and stock ranking score mentioned in Sects. 2.2 and 2.3. *Model training* module trains the encoder and prediction model sequentially. It produces trained models for real-world stock prediction and is retrained by the result of online investment. *Real-time stock prediction* module uses the trained models to predict stock ranking daily in nearly real-time, serving as a basis for judging stock performance. *Quantitative Investment* dynamically constructs the investment strategy and sends out buy/sell signals for every stock.

The interface of our method is presented in Fig. 6. As the figure shows, the profit of our method surpassed all deployed strategies in the past year. Compared with the annualized rate of return and accumulated income of CSI 800, which is 19.25% and 31.90%, our method leads by a wide margin achieving the results of 44.66% and 84.80% respectively. The line chart in Fig. 6 reveals a phenomenon that when there is a descending trend in the CSI 800 (e.g., Aug. 2020 - Jan. 2021), which can reflect the trend of the whole market to some extent, our method keeps an ascending trend, though has slight fluctuation. And our method leads to a higher growth rate contrary to the performance of CSI 800 from Mar. 2021 to Jul. 2021, which shows the stability and effectiveness of our model.

5   Related Work

Using historical stock sequences to predict trends is widely used by investors and researchers. These predictions are initially conducted by statistical methods, like ARIMA [28], GARCH [1] and optimization algorithms such as stepwise regression analysis [2], genetic programming [13]. The technical data can be further used as input features for machine learning and deep learning. Additionally, with the development of Natural Language Processing (NLP), and text including the financial news or twitters is used as input for prediction [27], sentiment analysis [5] has been applied with relative success in the financial market. In recent years, it is common to predict the stock trends by pre-training the stock representation, which is also called stock embeddings and stock encoding. [25] proposed the financial embedded vector model to represent the discretized financial time series into daily and weekly financial vectors. Similarly, [8] encoded news articles and price history through stock embedding. Considering that stock price movements have complex dynamic associations with other correlated stocks, [14] presents a multi-task recurrent neural network with high-order Markov random fields to predict stock price movement and [16] processes the time series of multiple related stocks simultaneously in a jointly forecasting approach.

6   Conclusions

In this paper, we present an MPS framework, which pre-trains the stock embeddings by self-supervised multi-task learning to address the stock trend prediction task. Since the stock market shows different fluctuation trends in different periods and investors may invest for different holding periods, it's rational to leverage the sequence data to predict the trends in a multi-view to supply more information. Hence, we fuse the multi-scale time information into the embeddings through parameter sharing mechanism, and results show the effectiveness and stability of our model for stock prediction in different holding periods.

Acknowledgement. This work was supported by the Joint Research Program of SeekData Inc. and East China Normal University.

References

1. Hung, J.C.: A fuzzy asymmetric GARCH model applied to stock markets. Inf. Sci. **179**(22), 3930–3943 (2009)
2. Hadavandi, E., Shavandi, H., Ghanbari, A.: Integration of genetic fuzzy systems and artificial neural networks for stock price forecasting. Knowl.-Based Syst. **23**(8), 800–808 (2010)
3. Abu-Mostafa, Y.S., Atiya, A.F.: Introduction to financial forecasting. Appl. Intell. **6**(3), 205–213 (1996)
4. Arevalo, R., Garcia, J., Guijarro, F., Penis, A.: A dynamic trading rule based on filtered flag pattern recognition for stock market price forecasting. ESWA **81**, 177–192 (2017)
5. Cambria, E., White, B.: Jumping NLP curves: a review of natural language processing research. IEEE Comput. Intell. Mag. **9**(2), 48–57 (2014)
6. Ding, X., Zhang, Y., Liu, T., Duan, J.: Deep learning for event-driven stock prediction. In: IJCAI (2015)
7. Dong, Y., Cordonnier, J.B., Loukas, A.: Attention is not all you need: pure attention loses rank doubly exponentially with depth. arXiv:2103.03404 (2021)
8. Du, X., Tanaka-Ishii, K.: Stock embeddings acquired from news articles and price history, and an application to portfolio optimization. In: ACL (2020)
9. Feng, F., Chen, H., He, X., Ding, J., Sun, M., Chua, T.S.: Enhancing stock movement prediction with adversarial training. In: IJCAI, pp. 5843–5849 (2018)
10. Feng, F., He, X., Wang, X., Luo, C., Liu, Y., Chua, T.S.: Temporal relational ranking for stock prediction. TOIS **37**(2), 1–30 (2019)
11. Gong, J., Sun, S.: A new approach of stock price prediction based on logistic regression model. In: NISS, pp. 1366–1371 (2009)
12. Jiang, W.: Applications of deep learning in stock market prediction: Recent progress. Expert Syst. Appl. **184**, 115537 (2021)
13. Kaboudan, M.A.: Genetic programming prediction of stock prices. Comput. Econ. **16**(3), 207–236 (2000)
14. Li, C., Song, D., Tao, D.: Multi-task recurrent neural networks and higher-order Markov random fields for stock price movement prediction. In: SIGKDD (2019)
15. Liu, X., et al.: Self-supervised learning: generative or contrastive. CoRR abs/2006.08218 (2020)
16. Ma, T., Tan, Y.: Multiple stock time series jointly forecasting with multi-task learning. In: IJCNN, pp. 1–8. IEEE (2020)
17. Merton, R.C.: A simple model of capital market equilibrium with incomplete information. J. Financ. **42**(3), 483–510 (1987)
18. Murphy, J.J.: Technical analysis of the financial markets: a comprehensive guide to trading methods and applications. New York Institute of Finance (1999)
19. Nelson, D.M., Pereira, A.C., De Oliveira, R.A.: Stock market's price movement prediction with LSTM neural networks. In: IJCNN, vol. 2017-May, pp. 1419–1426 (2017)
20. Vuong, P.H., Dat, T.T., Mai, T.K., Uyen, P.H., Bao, P.T.: Stock-price forecasting based on XGBoost and LSTM. Comput. Syst. Sci. Eng. **40**, 237–246 (2022)
21. Qin, Y., Song, D., Cheng, H., Cheng, W., Jiang, G., Cottrell, G.W.: A dual-stage attention-based recurrent neural network for time series prediction. In: IJCAI (2017)
22. Sawhney, R., Agarwal, S., Wadhwa, A., Shah, R.: Spatiotemporal hypergraph convolution network for stock movement forecasting. In: ICDM (2020)

23. Sayavong, L., Wu, Z., Chalita, S.: Research on stock price prediction method based on convolutional neural network. In: ICVRIS, pp. 173–176 (2019)
24. Sun, T., et al.: Learning sparse sharing architectures for multiple tasks. In: AAAI, vol. 34, pp. 8936–8943 (2020)
25. Sun, Y., Zhang, M., Chen, S., Shi, X.: A financial embedded vector model and its applications to time series forecasting. Int. J. Comput. Commun. Control **13**(5), 881–894 (2018)
26. Vaswani, A., et al.: Attention is all you need. In: NIPS (2017)
27. Xing, F.Z., Cambria, E., Welsch, R.E.: Natural language based financial forecasting: a survey. Artif. Intell. Rev. **50**(1), 49–73 (2017). https://doi.org/10.1007/s10462-017-9588-9
28. Zhang, G.: Time series forecasting using a hybrid ARIMA and neural network model. Neurocomputing **50**, 159–175 (2003)
29. Zhang, L., Aggarwal, C., Qi, G.J.: Stock price prediction via discovering multi-frequency trading patterns. In: SIGKDD, vol. Part F1296, pp. 2141–2149 (2017)

RShield: A Refined Shield for Complex Multi-step Attack Detection Based on Temporal Graph Network

Weiyong Yang[1,2], Peng Gao[2(✉)], Hao Huang[1], Xingshen Wei[2], Wei Liu[2],
Shishun Zhu[2], and Wang Luo[2]

[1] Nanjing University, Nanjing 210008, China
yangweiyong@sgepri.sgcc.com.cn, hhuang@nju.edu.cn
[2] NARI Group Corporation/State Grid Electric Power Research Institute, Nanjing NARI
Information and Communication Technology Co., Ltd., Nanjing 21003, China
gao.itslab@gmail.com

Abstract. Complex multi-step attacks (i.e., CMA) have caused severe damage to core information infrastructures of many organizations. The graph-based methods are well known as the ability for learning complex interaction patterns of systems and users with discrete graph snapshots. However, such methods are challenged by the computer networking model characterized by a natural continuous-time dynamic graph. In this paper, we propose RShield, a temporal graph network-based CMA detection and defense method. It first constructs the continuous-time dynamic graph based on interactions among users and entities from various log records. Then it trains the detection model offline and performs streaming detection for live online network events. A prototype of RShield has been implemented. The experimental evaluation shows that RShield can achieve superior detection performance than the state-of-the-art methods in both transductive and inductive settings.

Keywords: Complex multi-step cyber-attack · Insider attack · Advanced persistent threat (APT) · Continuous-time dynamic graph (CTDG) · Anomaly detection

1 Introduction

Over the last few years, the core information infrastructure of some organizations has been suffering critical cyber-attacks. As illustrated in Fig. 1, different from conventional attacks, the low-profile and long-time span, i.e., "low & slow", nature of such attacks make detection extremely difficult. These kinds of attacks are called complex multi-step attacks (CMA).

Extensive works have been conducted to deal with CMA detection. Most of them mainly focused on the techniques of Intrusion Detection System (IDS), which can be generally divided into signature-based and anomaly-based detections. The former approach is restricted to malicious activity signatures and thus cannot detect unknown attacks.

A. Bhattacharya et al. (Eds.): DASFAA 2022, LNCS 13247, pp. 468–480, 2022.
https://doi.org/10.1007/978-3-031-00129-1_40

While the latter anomaly detection approaches generally pay more attention on analyzing operation traces of network events, and can only model user and system behaviors in a chronological order, due to insufficiency of the associated dimension [1]. Moreover, it is not applicable for long time span sequences [2]. When applied to detect CMA, these approaches often have low success rates of detection, due to CMA usually would lurk for a long time, and may involve multi-dimension activities.

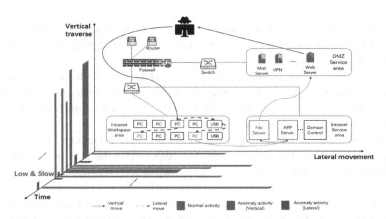

Fig. 1. The illustration of CMA attack on spatial and continuous temporal dimension. Supposing that the attack target is the file server. For instance, a vertical attack activity on a PC using Spear Phishing emails was successful. Then it was used as the "bridge" to attack other PCs within the Intranet workspace area to obtain more rights of the same level. At last, it stole sensitive data from the target and uploaded them through another tunnel exploited on the Web server.

Recently, graph-based deep learning methods for CMA detection have received much attention since it can directly learn complex interaction patterns of systems and users from the graph. The directed dynamic graph is used to describe operations (e.g., logon) between subjects (e.g., user) and objects (e.g., PC). Accordingly, multi-dimension relations (e.g., causality) among network entities are naturally represented. Current methods often represent a dynamic graph by a sequence of snapshots of the graph [3], which, however, cannot fully characterize the properties of computer networks, because in which the dynamic graphs are often continuous-time based (i.e., edges can appear at any time) and evolving (i.e., new nodes join the graph continuously) [4]. Thus, the performance of graph-based deep learning methods is still limited for CMA detection. The challenges mainly consist of two aspects. Firstly, due to the sparse distribution of CMA-events in time and space, the discrete graph (snapshot sequence) representation may lead to losing some important "bridge" events. Secondly, discrete graph -based method enforces detection on the whole graph of entire network topology, which not only requires large memory space for real-time streaming analysis but also will lead to a coarse-grained result with less contextual information. Above all, malicious attacks are hard to be predicted or traced due to the lack of contextual information between events.

In this paper, we propose a new CMA detection method: RShield (i.e., a refined CMA defense and detection shield). RShield first constructs a continuous-time dynamic graph (CTDG) from interactions between entities in the network, and then obtains the node embedding of the graph based on the temporal graph network (TGN) [4] model. Finally, based on the feature difference of node embedding, the loss function of CMA detection is constructed to predict the abnormal edges of the CTDG. To the best of our knowledge, RShield is the first to use continuous temporal graph construction to detect anomaly event in cyber security field. Specifically, the contributions of our approach are summarized as follows.

- It proposes a new anomaly detection method based on the CTDG, which correlates user and entity interactions on computer network where the graph is represented as a nonlinear sequence of events.
- It does not need to input the entire network graph and can locate the context of anomaly operation and traceback. From another perspective, it has a finer granularity than the state-of-the-art methods, making it able to predict the connection change at a certain future moment of a specific node.
- It improves the detection accuracy compared with the rival methods. The experimental evaluation shows the proposed method outperforms the previous techniques in multiple tasks and datasets under both transductive and inductive settings.

The rest of this paper is organized as follows. Section 2 discusses related work. A design and implementation of RShield is given in Sect. 3. Experiments and evaluations are presented in Sect. 4. Section 5 concludes the paper.

2   Related Work

Signature-based detection approaches analyze network events in which whose behaviors that match pre-defined rules will be detected as attacks [5, 6]. In industrial applications, a large number of rules have been accumulated based on expert experience. They are efficient and low false positives, however, cannot detect unseen attacks.

Anomaly-based intrusion detection approaches detect abnormal activities that deviates from pre-learned baseline of normal behaviors. Some researchers focus on extracting user behavior features and then feed them to machine learning models to discover malicious events [7, 8]. In [7, 9], multidimensional interaction patterns between users and entities, such as email communications, web browsing, server login and operations, were used for feature engineering. Researchers also try to convert the user's behavior into sequences or graphs and then analyze them [10–12]. Such methods construct normal patterns of user behavior, and then compare them with new behaviors to identify anomalies. The provenance tracking system are proposed to monitor and analyze the activities of the system [2, 13, 14]. The main advantage of this approach is the ability to detect unknown attacks whose signature is not available. The main disadvantage is that it is more likely to cause false positives. Because the process of CMA is very complex, "normal" and "abnormal" behavior are very hard to locate. Many legitimate programs perform the same abnormal activities as malicious ones and vice versa. The main differences between RShield and them are as follows. First, most of them are aimed at attack forensics rather than cyber threat event detection. Second, these systems use causality graphs to track process operations and interactions at the system call level. RShield analyzes logs that record user behavior in the information system (such as login remote host and browsing websites). It mainly captures and expresses various relationships between logs that reflect typical behaviors of users.

Deep learning on graph embedding has been widely applied in cyber security field. Backes et al. use this technique to conduct social relations reasoning attacks [15]. The log2vec [1] uses self-defined heuristic relationships to transform log records into heterogeneous graphs and then converts them into graph embedding. By applying clustering algorithms on the node embeddings, log2vec is able to separate malicious and benign activities into different clusters and identify malicious ones. Log2vec is a log record-level graph construction method that based on heuristic rules, while the graph construction of RShield is based on the interaction relations within the log. Therefore, RShield has a finer granularity in attack tracing and prediction. At the same time, a large number of feature engineering processes that depend on datasets are reduced.

3   RShield Design and Implementation

3.1   Overview

RSheild is a network-based intrusion detection system that can simultaneously detect intrusions on a collection of networked hosts. Figure 2 illustrates RShield's general pipeline, where the dotted line divides the two stages of offline training and online detection.

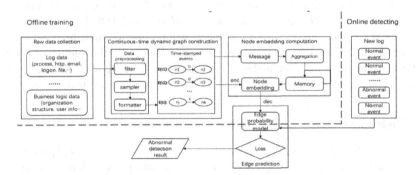

Fig. 2. The schematic diagram of RShield

As presented in the schematic diagram, a CTDG is first constructed based on extracted (interactive) behaviors of users and entities from historical log records. Second, the offline training of the CMA detection model is trained through the temporal graph network algorithm based on the node embedding. Finally, RShield performs streaming detection of online live network events.

3.2 Continuous-Time Dynamic Graph Construction

Various logs in computer and network systems are a valuable resource for monitoring anomaly events. In RShield, a *data preprocessing* module processes raw logs' data with the following procedure. The *filter* acquires data within a time window in historical logs and filters out invalid ones. Then, the *sampler* randomly selects a collection of users/entities belong to the time window. At last, the *formatter* constructs the continuous-time dynamic graph by the ordered list of time events.

Compared to the Log2vec [1] that is based on artificial heuristic rules, we use a natural *"subject - operation@time - object"* graph construction approach which can capture the raw interactive information hidden in the log record better and trace abnormal events with finer granularity (cf. Fig. 3).

Fig. 3. Graph construction rule

Different from the discrete-time dynamic graph (DTDG) which is a sequence of static graph snapshots taken at intervals of time, the continuous-time dynamic graph (CTDG) can be represented as a list of time-stamped events. We define the CTDG as $G = [E(t1), E(t2), ...]$ where an event $E(t)$ denotes one of three basic operations (*insert, delete* and *update*) is enforced on the node or edge at time t. For instance, log samples are shown in Fig. 4a, where each event on a separate line represents an authentication event at the given time in the form of *"time, source, destination, authentication type, logon type, authentication orientation, success/failure"*. Its CTDG is constructed as the Fig. 4b. Let *src* and *dst* be the token of source and destination node (e.g., user, entity) respectively, t the timestamp of the event, l the anomaly label of the interaction, *ns_feats* and *e_feats* the features of the nodes of both sides and the edge respectively. Then, the time-stamped event $E(t)$ can be defined as a tuple: (*src, dst, t, l, ns_feats, e_feats*).

3.3 Node Embedding Based on TGN

With the constructed temporal dynamic graph, the neural model TGN [4] is employed for computing node embeddings. The TGN model can be regarded as an encoder-decoder with memory, where the encoder is a function that maps from a CTDG to node embeddings and the memory is to represent the node's history in a compressed format.

The time-stamped events of CTDG are input to update the memory to keep long term dependencies for each node in the graph. The memory of a node is updated after an event (e.g., interaction with another node or node-wise change) as the following three modules together. The first one is the message function. For an interaction event $e_{ij}(t)$ between source node i and target node j at time t, two messages can be computed:

$$m_{i/j}(t) = msg_{s/d}s_{i/j}(t^-), s_{j/i}(t^-), \Delta t, e_{ij}(t)) \tag{1}$$

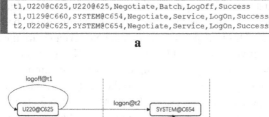

a

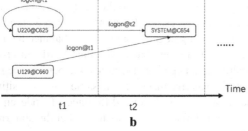

b

Fig. 4. a) Authentication event sample. b) Temporal graph of authentication events

Here, $s_{i/j}(t^-)$ is the memory of node i (or j) just before time t (i.e., from the time of the previous interaction involving i (or j). The $msg_{s/d}$ are simply the concatenation of the inputs. The Δt is the interval of time. The second one is the message aggregator which is to aggregate the existing information of each node. The aggregated message is notated as $\overline{m}_i(t)$ which keeps only most recent message for a given node. Finally, TGN uses the former messages to update the memory by

$$s_i(t) = mem(\overline{m}_i(t), s_i(t^-))\qquad(2)$$

For interaction events involving two nodes i and j, the memories of both nodes are updated after the event has happened. Here, *mem* is a learnable memory update function gated recurrent unit (GRU). The temporal embedding $z_i(t)$ of node i at any time t was defined as following:

$$z_i(t) = emb(i, t) = \sum_{j \in n_i^k([0,t])} h(s_i(t), s_j(t), e_{ij}, v_i(t), v_j(t))\qquad(3)$$

Here, let h be a learnable function, $v_i(t)$ a node-wise event, where i denotes the index of the node and v is the vector attribute associated with the event., $n_i = \{j: (i, j) \in \varepsilon\}$ represents the neighborhood of node i. A series of L graph attention layers compute i's embedding by aggregating information from its L-hop temporal neighborhood.

So far, we have got the temporal embedding of the nodes in the CTDG. Then they are used to perform edge prediction task by models got in the training phase.

3.4 Anomaly Edge Detection

The node embeddings are used to compute the loss and anomaly edge prediction. The goal is to predict "whether the edge between two nodes i and j at time t is anomaly"?

Regarding the real-world network data, anomalies rarely occur and the majority is normal data (i.e., anomalies only typically occur 0.0001–1% of the time), and learning from imbalanced data is still an open challenge for classification algorithm. In order to mitigate this problem, we use the negative sampling technique to balance positive and negative samples in the training process, so that the model can learn the characteristics of anomalies more effectively. Different from original TGN that only predicts the existence of edges (non-existent node pairs are sampled as negative sample edges), we develop a new negative sampling approach to adapt to the goal of our task. It takes the normal logs in datasets as positive examples, while taking the scarce anomaly log records as negative examples and performing negative sampling to balance the number of positive and negative examples in each training batch. In other words, in order to mitigate the imbalance between positive and negative categories, we perform negative sampling: at each training batch, sample negative instances the same size as positive instances.

On the other hand, based on the feature difference of node embedding, the loss function of CMA detection is defined as follows to predict the abnormal edges.

$$L\left(\widetilde{z}, z\right) = \begin{cases} 0, & \|\widetilde{z}_i(t) - z_i(t)\| < T \\ 1, & \|\widetilde{z}_i(t) - z_i(t)\| \geq T \end{cases} \tag{4}$$

$\widetilde{z}_i(t)$ represents the predicted embedding near to the temporal embedding $z_i(t)$, T is the threshold set in experience. The loss $L\left(\widetilde{z}, z\right)$ equals to zero while the distance between the two embedding (i.e., $\widetilde{z}_i(t)$ and $z_i(t)$) is less than T, whereas equals to 1 while the distance is not less than T. The gradient descent algorithm [16] is implemented to optimize the surrogate loss $L_s\left(\widetilde{z}, z\right)$ of the $L\left(\widetilde{z}, z\right)$.

4   Experiment and Evaluation

4.1   Experiment Setup

In this section, we set up experiments in different attack scenarios to verify the following **research questions**.

RQ1   Compared with state-of-the-art baselines, does RShield improve the detection capability of CMA attacks?

RQ2   As a supervised detection method, whether it can achieve better results through fewer training samples?

RQ3   What is the influence of the hyperparameters of neighbor nodes in the model on the detection results?

In order to measure the detection result mentioned in RQ1 and RQ2, we adopt the Area Under the Curve (AUC) as the main performance metric. AUC is insensitive to the imbalance of dataset, which reaches its best value at 1 and worst at 0. If a method has a higher AUC on the dataset, its predictions is regarded as more correct.

Datasets. Two cyber-security datasets are used in the experiments: a real-world dataset, Los Alamos National Lab's (LANL's) comprehensive cyber-security events dataset [17] and a synthetic dataset, CERT Insider Threat Test Dataset [18].

The LANL dataset represents 58 consecutive days of event data collected from five sources (*authentication, process, network flow, DNS* and *redteam*). The *authentication* events include 1,648,275,307 records collected over 58 days for 12,425 users and 17,684 computers. We only use the *authentication* data to form the CTDG. As mentioned in Sect. 3.1, in the preprocessing stage we randomly selected a subset of LANL which contains 9,918,928 edges generated from 10,895 nodes (user host pairs), and all 691 malicious interactions generated by 104 users.

The CERT dataset consists of event logs with labeled insider threat activities from a simulated organization's computer network, generated with sophisticated user models. It consists of five log files that record the computer-based activities for all employees in a simulated organization, including *logon/logoff* activity, *http/email* traffic, *file* operations, and external storage *device* usage. Over the course of 516 days, 4,000 users generate 135,117,169 events. Among these, we randomly sampled 41 users' 1,306,644 events including 470 ground truth, representing five insider threat scenarios taking place. Additionally, user attribute metadata is included; namely, the six categorical attributes: *role, project, functional unit, department, team, supervisor*.

Experiment Process. The experiment process is shown in Fig. 5. To answer RQ1 and RQ2, we compare RShield with the state-of-the-art baseline method on the LANL and CERT datasets. Baselines are Tiresias [12] and Log2vec [1]. The Tiresias is an advanced supervised approach for security event prediction in various events with noise. The log2vec is an unsupervised approach to separate malicious and benign activities into different clusters and identify malicious ones.

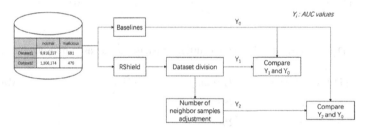

Fig. 5. Experiment process

The dataset is then divided into training set, validation set and test set to show the effectiveness of anomaly edge prediction. We perform both transductive and inductive settings. In the transductive setting, we predict future log entries whether they are malicious or not for all nodes (user and computer pairs) in test dataset. While in the inductive settings, we only predict future log entries for users and computers that are never combined as pairs in the training phase (which we call new nodes). This is aimed to test the detection effect of our method on network nodes that have never appeared before.

In order to answer RQ3, our setup of edge prediction model closely follows the TGN [4] and focuses on the tasks of future edge prediction. Our encoder is combined with a simple MLP decoder mapping from the concatenation of two node embeddings to the probability of the edge. We adjust the hyperparameter of the number of sampled neighbors in the graph neural network and observe its influence on the detection results.

For other parameters, in the memory update phase, we set message function to 'identity', message aggregator to 'last' and memory updater to 'GRU'. In the training phase, we set batch size as 200, embedding module to Temporal Graph Attention (attn) with 1 layer and 10 neighbors. At last, we set dimensions of node embedding and time embedding to 100, memory dimensions to the number of node features.

4.2 Experiment Results

In Table 1, we summarize the AUC results of baseline methods and RShield in LANL and CERT dataset respectively. In Table 2 and Table 3, we present the detection result of RShield on different division of train, validation and test set on the dataset. The result includes specific percentage of division, number of malicious logs in training, validation, and testing phases, and their corresponding AUC value.

Table 1. Detection result of different methods (AUC value)

Approach	LANL	CERT
Tiresias	0.85	0.39
Log2vec	0.91	0.86
Log2vec++	/	0.93
RShield (transductive)	0.98	0.94
RShield (inductive)	0.97	0.99

Table 2. Detection result of RShield on different dataset divisions in transductive setting

	Dataset division	#Training	#Validation	#Test	AUC
LANL	0.5:0.3:0.2	483	28	45	0.9736
	0.27:0.03:0.7	264	5	376	0.9736
	0.22:0.04:0.74	48	176	466	0.9631
	0.258:0.002:0.74	161	43	466	0.9810
CERT	0.5:0.24:0.26	68	63	313	0.8345
	0.6:0.19:0.22	80	176	200	0.8165
	0.78:0.07:0.15	226	120	101	0.9438

Parameter Sensitivity. In Fig. 6, we show the detection result of RShield on different setting of the number of sampled neighbors per layer (red polyline). For all settings and datasets, we perform the same chronological split (0.258:0.002:0.74). The test set includes 7,340,002 interactions involving 10,689 different nodes, and 466 of them are malicious logons. The Log2vec and Tiresias are shown by the dotted lines.

Table 3. Detection result of RShield on different dataset divisions in inductive setting

	Dataset division	#Training	#Validation	#Test	AUC
LANL	0.5:0.3:0.2	483	15	17	0.9722
	0.27:0.03:0.7	264	5	145	0.9714
	0.22:0.04:0.74	48	143	272	0.9535
	0.258:0.002:0.74	161	14	224	0.9790
CERT	0.5:0.24:0.26	68	63	222	0.9021
	0.6:0.19:0.22	80	7	53	0.9553
	0.78:0.07:0.15	226	27	11	0.9995

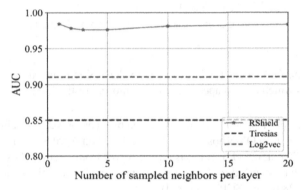

Fig. 6. Performance of RShield on different numbers of sampled neighbors per layer.

4.3 Discussion

Answer to the Research Questions. In Table 1, we see that in all datasets RSshield outperforms baseline methods which are the state-of-the-art log-entry level approaches to anomaly detection. In LANL dataset, with the transductive and inductive settings, RShield increased the AUC by 15.3% and 14.1% respectively compared to Tiresias, and 7.7% and 6.6% respectively compared to Log2vec. In CERT dataset, with the transductive and inductive settings, we increased the AUC by 141% and 154% respectively compared to Tiresias, and 1.1% and 6.5% respectively compared to Log2vec++.

Table 2 and Table 3 show that compared with Tiresias, which is also a supervised learning method, using our negative sampling method, under transductive and inductive settings, we do not need a large number of training samples to achieve high-level detection performance. Therefore, the answer to RQ2 is *Yes*.

For the RQ3, in Fig. 6, we compare detection result when increasing the number of sampled neighbors per layer and found that the fewer neighbor sampling can improve the efficiency of the model. The experimental result shows that as the parameter changes, the anomaly detection effect of the RShield model is still robust.

Application Case. In the power grid regulation business system, various programs running on the server carry the collection, transmission and display of the business function. Therefore, the security protection of the server is particularly important and various intelligent intrusion detection systems have been widely employed.

We apply the RShield to monitoring business servers (Linux host) in the I/II zone of a provincial company's dispatch operation management system (OMS). The detection target is the command line (CMD) logs on the server and the interface calls (URL requests) between the business terminal and the host. A large number of operation behavior sequences of OMS are generated in real-time. For instance, a CMD operation sequence and a URL request are as follows.

- CMD: [cd/2svnupdate.sh, ls, ll, chage, chage, ll, ifconfig, ll, chpasswd, chage, chage, more, exit, ll, chpasswd, exit, ll, more, chage, for, for]
- URL: ip:port/interface/login/login_frame.php?site   = default

We input the OMS data for half a year to the RShield and found the following anomaly events which had escaped existing monitoring of the IDS.

(1) RShield detect suspicious operations caused by an employee. The abnormal behavior occurred on an OMS host from 16:00 to 17:00 on April 8, 2021, which consists of 28 "SCP" command operations on the server from a remote host. In hindsight, no "SCP" has occurred in the past six months, except for this particular one-hour period (all 28 times "SCP" occurred).
(2) On May 20, 24, and 27, 2021, a server service was requested 13 times. The abnormal URL was "ip:port/_vti_bin/shtml.exe/_vti_rpc". It was verified that the call was used Microsoft FrontPage buffer overflow vulnerability which can cause the server crash and has the possibility of arbitrary code execution.

5 Conclusion

In this paper, we proposed RShield, a CTDG based CMA detection method. To the best of our knowledge, RShield is the first to use continuous temporal graph construction to detect anomaly event in network security field. The experimental evaluation demonstrates that our method outperforms other state-of-the-art methods. Furthermore, it not only has a finer granularity than the baseline methods, but also can predict the connection change at a certain future moment of a specific node. Therefore, it can detect complex cyber threats effectively in varied scenarios.

In future work, we think the following issue is important and interesting. Based on the TGN model, we have a very good effect on event correlation in the time dimension, however, there is no relevant processing in the space dimension. We believe that by introducing knowledge bases of malicious behavior in network, such as attack chain models, to map and weight network behaviors, certain network behaviors that were originally "inconspicuous" can be amplified and well correlated in the spatial dimension. Thereby it would further improve the detection performance.

Acknowledgments. This work is supported by the National Key Research and Development Program of China (No. 2021YFB3101700).

References

1. Liu, F., Wen, Y., Zhang, D., Jiang, X., Xing, X., Meng, D.: Log2vec: a heterogeneous graph embedding based approach for detecting cyber threats within enterprise. In: Proceedings of the 2019 ACM SIGSAC Conference on Computer and Communications Security, pp. 1777–1794. Association for Computing Machinery, New York (2019)
2. Han, X., Pasquier, T., Bates, A., Mickens, J., Seltzer, M.: UNICORN: runtime provenance-based detector for advanced persistent threats. In: Proceedings 2020 Network and Distributed System Security Symposium (2020)
3. Guo, J., Li, R., Zhang, Y., Wang, G.: Graph neural network based anomaly detection in dynamic networks. Ruan Jian Xue Bao/J. Softw. **31**(3), 748–762 (2020). (in Chinese)
4. Rossi, E., Chamberlain, B., Frasca, F., Eynard, D., Monti, F., Bronstein, M.: Temporal Graph Networks for Deep Learning on Dynamic Graphs. arXiv:2006.10637 [cs, stat]. (2020)
5. More, S., Matthews, M., Joshi, A., Finin, T.: A knowledge-based approach to intrusion detection modeling. In: 2012 IEEE Symposium on Security and Privacy Workshops, pp. 75–81 (2012)
6. Karim, I., Vien, Q.-T., Le, T.A., Mapp, G.: A comparative experimental design and performance analysis of snort-based intrusion detection system in practical computer networks. Computers **6**, 6 (2017)
7. Gavai, G., Sricharan, K., Gunning, D., Rolleston, R., Hanley, J., Singhal, M.: Detecting insider threat from enterprise social and online activity data. In: Proceedings of the 7th ACM CCS International Workshop on Managing Insider Security Threats, pp. 13–20. Association for Computing Machinery, New York (2015)
8. Legg, P.A., Buckley, O., Goldsmith, M., Creese, S.: Caught in the act of an insider attack: detection and assessment of insider threat. In: 2015 IEEE International Symposium on Technologies for Homeland Security (HST), pp. 1–6 (2015)
9. Senator, T.E., et al.: Detecting insider threats in a real corporate database of computer usage activity. In: Proceedings of the 19th ACM SIGKDD International Conference on Knowledge Discovery and Data Mining, pp. 1393–1401. Association for Computing Machinery, New York (2013)
10. Du, M., Li, F., Zheng, G., Srikumar, V.: DeepLog: anomaly detection and diagnosis from system logs through deep learning. In: Proceedings of the 2017 ACM SIGSAC Conference on Computer and Communications Security, pp. 1285–1298. Association for Computing Machinery, New York (2017)
11. Rashid, T., Agrafiotis, I., Nurse, J.R.C.: A new take on detecting insider threats: exploring the use of hidden Markov models. In: Proceedings of the 8th ACM CCS International Workshop on Managing Insider Security Threats, pp. 47–56. Association for Computing Machinery, New York (2016)
12. Shen, Y., Mariconti, E., Vervier, P.A., Stringhini, G.: Tiresias: predicting security events through deep learning. In: Proceedings of the 2018 ACM SIGSAC Conference on Computer and Communications Security, pp. 592–605. Association for Computing Machinery, New York (2018)

13. Hossain, M.N., et al.: {SLEUTH}: real-time attack scenario reconstruction from {COTS} audit data. Presented at the 26th {USENIX} Security Symposium ({USENIX} Security 17) (2017)
14. Milajerdi, S.M., Gjomemo, R., Eshete, B., Sekar, R., Venkatakrishnan, V.N.: HOLMES: real-time APT detection through correlation of suspicious information flows. In: 2019 IEEE Symposium on Security and Privacy (SP), pp. 1137–1152 (2019)
15. Backes, M., Humbert, M., Pang, J., Zhang, Y.: walk2friends: inferring social links from mobility profiles. In: Proceedings of the 2017 ACM SIGSAC Conference on Computer and Communications Security, pp. 1943–1957. Association for Computing Machinery, New York (2017)
16. Hinton, G., Srivastava, N., Swersky, K.: Neural networks for machine learning lecture 6a overview of mini-batch gradient descent. Cited on **14**, 2 (2012)
17. Kent, A.D.: Cyber security data sources for dynamic network research. In: Dynamic Networks and Cyber-Security, pp. 37–65. World Scientific (Europe) (2015)
18. The CERT Division: Insider Threat Tools. https://www.cert.org/insiderthreat/tools/. Accessed 17 Sept 2021

Inter-and-Intra Domain Attention Relational Inference for Rack Temperature Prediction in Data Center

Fang Shen[1,2], Zhan Li[1(✉)], Bing Pan[1], Ziwei Zhang[3], Jialong Wang[1], Wendy Zhao[1], Xin Wang[3(✉)], and Wenwu Zhu[3(✉)]

[1] Alibaba Group, Hangzhou, China
{ziru.sf,zhan.li,diyuan.pb,quming.wjl,wendy.zhao}@alibaba-inc.com
[2] Alibaba Group, Bellevue, WA 98004, USA
[3] Tsinghua University, Beijing, China
{zwzhang,xin_wang,wwzhu}@tsinghua.edu.cn

Abstract. In a data center, predicting the rack temperature then generating alarms when an exception is detected can prevent server failure caused by high rack temperature. Each measuring point records the temperature of the rack over time, and each pair of measuring points may be associated with services or locations. Therefore, the rack temperature prediction problem can be modeled as a graph-based prediction problem. In this case, the prediction of the rack temperature depends not only on its own historical temperature but also on the temperature of racks having the same service or located near each other. Furthermore, the temperature of the rack is actually determined by various factors such as IT workloads and cold aisle temperature. Existing graph-based prediction methods do not consider the influence of these domains during the prediction, but only consider the temperature domain itself. To overcome this challenge, we propose an Inter-and-Intra domain Attention Relational Inference (I2A-RI) model: an unsupervised model that learns the relations between time series variables from different domains and utilizes the inferred interaction structure to achieve accurate dynamical predictions. Two attention modules, the guidance domain attention (GDA) module and the intra-domain attention (IDA) module, are proposed in I2A-RI, which encodes the inter-and-intra domain information to guide the learning procedure. Experiments on the real-world rack temperature dataset show that I2A-RI outperforms other state-of-the-art models since it takes the advantage of the ability to infer the potential interactions across domains. The benefits of the two proposed attention modules are also verified in the experiments.

Keywords: Data center · Rack temperature prediction · Relational inference · Graph neural network

A. Bhattacharya et al. (Eds.): DASFAA 2022, LNCS 13247, pp. 481–492, 2022.
https://doi.org/10.1007/978-3-031-00129-1_41

1   Introduction

In the data center intelligent operation and maintenance (O&M) system, monitoring the temperature of racks is of significant to prevent server downtime due to high temperature. By predicting the rack temperature, the O&M center can sense anomalies and generate alarms in advance, enabling onsite personnel to intervene to prevent accidents promptly. There are many racks in a large data center and the temperatures of different racks may be related to each other due to service or location proximity. The temperature of each rack is recorded over time, so the prediction of rack temperature can be modeled as a graph-based multivariate time series prediction problem. Many efforts have been made over the decades to model the multivariate time series, including statistical learning methods [2], deep neural networks (DNNs) [11] and graph neural networks (GNNs) [13,15]. Though statistical learning and DNNs have shown values in the area, our work focuses on GNNs since they take the graphs as inputs, allowing the complex relations and interactions between variables [14] to be naturally expressed in the model.

In recent years, prediction algorithms based on GNNs have been widely studied. Yu et al. [14] proposed the spatio-temporal graph convolutional network to capture both temporal and spatial dependencies for mid-and-long term traffic prediction. Wu et al. [13] proposed the multivariate time Series forecasting with GNN (MTGNN) model which constructs the graph from time series by learning the uni-directed relations then the temporal and spatial dependencies are captured by the dilated inception layer and the mix-hop propagation layer. However, these existing graph-based prediction methods only focus on the historical correlation information of the prediction domain itself. In this way, in the scenario of rack temperature prediction, only the historical temperature of its own rack and that of associated racks are considered. Nevertheless, the temperature of a rack in a data center is not only related to its historical temperature but also affected by factors of other domains, such as IT workloads and cold aisle temperature. In terms of IT workloads, the temperature of different racks and the workloads are a two-domain dynamic system, in which the potential interactions include some servers are running services in a sequence and tend to reach their peak workloads in a fixed order, or a few racks are close to each other and influence the temperature of one another more noticeable. The existing graph-based methods do not take these other domains' important factors into account, thus affecting the accuracy of prediction. It is encouraging to consider both intra-domain and inter-domain relationships in the complex system interact and achieve dynamical predictions.

Thus, we propose a novel GNN model, inter-and-intra domain attention relational inference (I2A-RI), which addresses the problem of learning the latent relations among time series variables across domains. The model is in the form of a variational autoencoder (VAE) [6,8]: the encoder learns the implicit relations between variables and constructs multiple graphs in an unsupervised manner; while the decoder takes the constructed structure and the time series data for prediction. The graphs learned by the encoders are in the guidance domains or the predictive domain. Accordingly, we propose two attention [12] modules to make these graphs interact within and between domains: the guidance domain attention (GDA) module and the in-domain attention (IDA) module. The GDA module extracts information from other domains to guide the relational inference in the prediction domain. The IDA module is used to capture the asynchronous interactions in the prediction stage. For example, when the servers' workload jumps up, it takes a while for the heat to be fully spread into the server room and influence the temperature. In summary, our contributions are as follows:

- We propose a novel GNN framework (I2A-RI) to infer the variables' relations in multivariate time series modeling. The model automatically learns relations then combines them with the time series data to perform predictions. To the best of our knowledge, no prior work studies multivariate time series modeling problems from a relational inference perspective with multiple graphs representing relations of variables from different domains.
- We define two types of domains: prediction domain and guidance domain. The prediction domain contains the variables for forecasting, and the guidance domain contains the variables that influence those in the prediction domain. We also introduce two attention modules (GDA and IDA) in the model: the GDA module extracts information from the guidance domain into the prediction domain to guide the relational reasoning. The IDA module captures the asynchronous interactions in the prediction stage.
- We show that I2A-RI outperforms the state-of-the-art approaches in forecasting the temperature measurements of a server room on the real operation data.

2 Preliminaries

In this paper, our task is to model the multivariate time series. Given the multivariate time series with historical T time steps $X = [\mathbf{x}_1^i, \mathbf{x}_2^i, \cdots, \mathbf{x}_T^i], i = 1, 2, \cdots, N$, where N is the number of variables. Our goal is to predict the future value $Y = [\mathbf{x}_M^i]$, where M means M steps away or the future sequence $Y = [\mathbf{x}_{T+1}^i, \mathbf{x}_{T+2}^i, \cdots, \mathbf{x}_{T+M}^i]$, where M represents M time steps in the future. We aim to find a function f that maps from X to Y. From the graph's perspective, each variable in a multivariate time series can be regarded as a node in the graph. Connections between nodes are represented by an edge category matrix, which is expected to be learned and cannot be obtained in advance. Some definitions used in this paper are given as follows:

Definition 1 Graph. A graph is denoted as $G = (V, E)$, where V represents the set of nodes and E represents the set of edges. The number of all nodes is denoted as N.

Definition 2 Prediction Domain. The domain expected to obtain the predicted value.

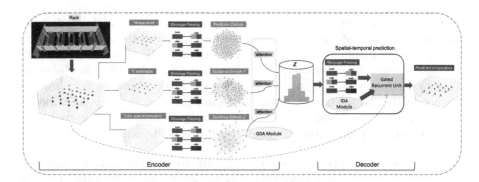

Fig. 1. I2A-RI architecture. The framework is composed of a multi-graph relational reasoning encoder and a spatial-temporal prediction decoder.

Definition 3 Guidance Domain. Factors affecting the prediction domain. For example, if we want to predict the future values of temperatures in data center operations, we need to consider the servers' workloads (guidance domain) and the cold aisle temperature (guidance domain) because the workloads are the indicator of the heat generated by the servers and the cold aisle temperature is the upstream measurement point of the rack temperature.

3 I2A-RI Model

The I2A-RI architecture is presented in Fig. 1. As illustrated in the figure, the whole framework of I2A-RI is based on VAE and mainly consists of two components: the multi-graph relational reasoning encoders and a spatial-temporal prediction decoder. The multi-graph relational reasoning encoder inputs data from multiple domains, and each domain forms a separate graph. The graphs of different domains are aggregated by a cross-domain attention layer. Note that although the encoder extracts features for multiple graphs, it only outputs one graph that integrates information of all domains. In summary, given historical time series $\mathbf{X}$, the encoder returns a factorized distribution $q_\psi(\mathbf{Z}|\mathbf{X})$ of the discrete relation type z_{ij} between nodes v_i and v_j. The decoder takes the learned graph from the encoder and the historical data to perform the spatial-temporal prediction:

$$p_\eta(\mathbf{X}|\mathbf{Z}) = \prod_{t=1}^{T} p_\eta(\mathbf{x}^{t+1}|\mathbf{x}^t,\dots,\mathbf{x}^1,\mathbf{Z}) \tag{1}$$

The details of these two parts, including cross-domain and intra-domain attention mechanisms, will be covered next. Note that our proposal is implemented based on neural relational inference (NRI), and we focus on the improved part here. For more details please refer to [7].

3.1   Multi-graph Relational Reasoning Encoder

The purpose of the encoder is to learn the relationship type z_{ij} between each pair of nodes from historical time series data. Since we don't have a graph to start with, we first perform GNN with a fully-connected graph without self-loops. In the application of rack temperature prediction, the encoder has a total of three inputs, one for the prediction domain, that is, the temperature to be predicted, and the other two for the guidance domain, that is, IT workloads and cold aisle temperature. Then three fully-connected graphs are used to extract the features of the three domains respectively.

Any domain in the encoder contains three message passing operations. For one single domain, given the time series of each node $\mathbf{x}_1, \mathbf{x}_2, \ldots, \mathbf{x}_N$, the message passing operations in the encoder are performed as follows:

$$h_j^1 = fc(\mathbf{x}_j) \tag{2}$$

$$n \longrightarrow e : h_{i,j}^1 = f_{n2e}^1(h_i^1 || h_j^1) \tag{3}$$

$$e \longrightarrow n : h_j^2 = f_{e2n}^1(\sum_{i \neq j} h_{i,j}^1) \tag{4}$$

$$n \longrightarrow e : h_{i,j}^2 = f_{n2e}^2(h_i^2 || h_j^2) \tag{5}$$

where $fc(\cdot)$ denotes the fully connected networks and $\cdot || \cdot$ denotes the concatenation of two feature vectors.

After the last layer of node-to-edge operations, we need to consolidate the information for all the domains. We propose the guidance domain attention (GDA) layer to achieve this goal.

Guidance Domain Attention. For any domain k, assume that the dimension of the edge feature output by the last layer of node-to-edge is F, the weight vector $W_k \in \mathbf{R}^F$ is applied to each edge feature $h_{i,j}$ then obtain the relation coefficient $Rel_{i,j}^k$.

$$Rel_{i,j}^k = W_k^T \cdot h_{i,j}^2 \tag{6}$$

$Rel_{i,j}^k$ represents the importance score between nodes j and i. For better comparison, the relation coefficients are normalized across all edges by employing the softmax function:

$$a_{i,j}^k = softmax(Rel_{i,j}^k) = \frac{exp(Rel_{i,j}^k)}{\sum_{s \in Neigh(i)} exp(Rel_{i,s}^k)} \tag{7}$$

To summary, the whole process can be described as,

$$a_{i,j}^k = \frac{exp(LeakyReLU(W_k^T \cdot h_{i,j}^2))}{\sum_{s \in Neigh(i)} exp(LeakyReLU(W_k^T \cdot h_{i,s}^2))} \tag{8}$$

where LeakyReLU($\cdot$) [12] is used as the nonlinearity function and it can produce either positive or negative relationship coefficients. $a^k_{i,j}$ is the final output attention coefficients for domain k. Finally, the guidance domain features are aggregated to aid the prediction domain:

$$\tilde{h}^2_{i,j} = \sigma(\frac{1}{K}\sum_{k=1}^{K} a^k_{i,j}\mathbf{W}^k_{agg}h^2_{i,j}) \tag{9}$$

where $\mathbf{W}^k_{agg}$ is the learned weight vector applied to $h^2_{i,j}$. Then there is a residual connection and an output layer after the GDA layer.

The encoder finally returns a distribution $q_\psi(\mathbf{z}_{ij}|\mathbf{x}) = softmax(f_{enc}(\mathbf{x})_{ij})$, where $f_{enc}(\mathbf{x})$ denotes all operations performed on the fully-connected graph in the encoder. Refer to [7] for details about VAE's reparametrization trick and the way to handle discrete variables.

3.2 Spatial-Temporal Prediction Decoder

The components of the decoder are the same with [7]. We only focus on the improvements in this section. Due to the influence of other domains, the prediction domain needs a certain amount of time to deal with these changes. Moreover, time series are frequently coherent to themselves and most of them can not change instantaneously. Therefore we design the IDA module to take the advantage of the information carried in its history. To predict the value of $\mathbf{x}^{t+1}$, not only the value of $\mathbf{x}^t$, but also the earlier observations such as $\mathbf{x}^{t-n}, \ldots \mathbf{x}^{t-2}, \mathbf{x}^{t-1}$ are considered. Thus, the input of the IDA module contains two parts: x^t and $[\mathbf{x}^{t-n}, \ldots \mathbf{x}^{t-2}, \mathbf{x}^{t-1}]$. In this paper, n is set to 3. Suppose that the shapes of the two input tensors are: $U_1 = [Batch, N, 1, channels]$ and $U_2 = [Batch, N, 3, channels]$, where N is the number of nodes in a graph. The two tensors are then input to two 1d convolution layers ($\alpha(\cdot)$ and $\beta(\cdot)$) in order to transform them into the same space. The dot-product is adopted to calculate the similarity between the two transformed tensors:

$$Similarity(U_1, U_2) = \alpha(U_1)^T \beta(U_2) \tag{10}$$

U_2 is also input into another function $\theta(\cdot)$. The final vector output by the IDA module is calculated as,

$$IDA = Similarity(U_1, U_2)\theta(U_2) \tag{11}$$

For simplicity, we consider $\theta(\cdot)$ in the form of a linear transformation: $\theta(U_2) = W_\theta U_2$, where W_θ is a learnable weight matrix. The details are described in Fig. 2, where the initial input channel is set to 256.

The decoder's inputs include the learned graph and the historical data of different domains. In general, GNN with the message passing operator is applied to capture the spatial feature, and GRU is used to capture the temporal feature. The cross-domain feature, the value at the current time step x^t_j, the output of the IDA module, and the hidden state of the previous time step x^{t-1} are fed into GRU:

$$h_j^{t+1} = GRU([\tilde{h}_{i,j}^2, \mathbf{x}_j^t], [h_j^{t-1}, IDA_j]) \tag{12}$$

Noted that we only learned the changes of x_j^t:

$$\mu_j^{t+1} = \mathbf{x}_j^t + fc(h_j^{t+1}) \tag{13}$$

And

$$p(\mathbf{x}^{t+1}|\mathbf{x}^t, \mathbf{Z}) = N(\mu^{t+1}, \sigma^2 I) \tag{14}$$

The loss function of the whole framework is defined as the ELBO [7]:

$$Loss = E[\log p_\eta(\mathbf{X}|\mathbf{Z})] - KL[q_\psi(\mathbf{Z}|\mathbf{X})||p_\eta(\mathbf{Z})] \tag{15}$$

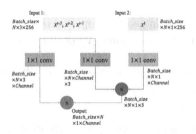

Fig. 2. The architecture of the IDA module.

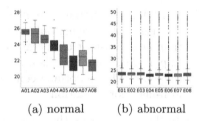

(a) normal   (b) abnormal

Fig. 3. The distributions of the two different data conditions.

4   Experiments

We create the RATEDC (Rack TEmperature data of the Data Center) dataset for the rack temperature prediction task.

4.1   RATEDC

The RATEDC dataset contains temperature, IT workloads, and cold aisle temperature data of 63 racks for one year. The data are collected every 2.5 min. A rack's temperature is a critical metric in data center daily operations: high temperature increases the equipment failure rate dramatically [4]. The workloads of a rack are the sum of the workloads of the servers in the rack. The workloads indicate how much heat will be generated. The cold aisle temperature is upstream of the rack temperature, thus they are closely related. The future rack temperatures are affected by various factors, such as the historical rack temperatures, other racks in the same server room, IT workloads of the rack itself, and the cold aisle temperature. Therefore, we set the temperature data as the prediction domain while the IT workload data and the cold aisle temperature data as the guidance domains. Since it takes time to exchange the heat in the air, the prediction targets are all racks' temperatures 10 min ahead.

4.2 Experiment Setup

The data preprocessing steps are as follows:

- Filter out abnormal data. The abnormal data are determined based on industry knowledge and data distribution. For example, the data is abnormal when the cold aisle temperature rises but the rack temperature does not, or the cold aisle temperature rises to 30 °C. According to the data distribution, the data points far from the distribution center are filtered out. Figure 3 shows the distribution of the normal and abnormal temperature. The horizontal coordinate represents the rack name. For example, A01 indicates the first rack in column A.
- Screening for fluctuating temperatures as training, validation, and test sets for the reason that we are only interested in predicting fluctuating temperatures than near-constant temperatures.
- Exponential smoothing is used to smooth the time series to further filter out the noise. The smoothing constant is set to 0.9.
- Min-max normalization.

In the experiments, the dimension of the weight vector in the attention layer is set to 256. Other network parameters are the same as those of the NRI. Adam [5] with the learning rate of 0.001, decayed by the factor of 0.5 every 100 epochs, is used as the optimizer. The maximum number of epochs is 500. The batch size is set to 16 and each batch has 48 time points for the RATEDC dataset. The reported results are averaged after 5 runs. Both of the datasets are split into training (80%), validation (10%), and testing sets (10%). The Mean Square Error (MSE) is used to evaluate the performance of the models.

4.3 Performance Comparison

To further study the effectiveness of the model, we compare I2A-RI with other advanced prediction algorithms as follows:

  * **VAR:** vector autoregression [9]
  * **ARIMA:** The auto-regressive integrated moving average [1]
  * **GRU:** Gated Recurrent Unit [3]
  * **TPA-LSTM:** A temporal pattern attention LSTM for multivariate time series forecasting [11]
  * **DARNN:** Dual-stage attention-based recurrent neural network [10]
  * **STGCN:** The spatio-temporal graph convolutional network [14]
  * **MTGNN:** The multivariate time Series forecasting with GNN [13]
  * **NRI:** Neural relational inference [7].

In Table 1, we present the performance of I2A-RI compared with VAR, ARIMA, GRU, TPA-LSTM, DARNN, STGCN, MTGNN, and NRI for the RAT-EDC datasets. The VAR model is a statistical method that represents a group of time-dependent variables as linear functions of their own past values and the past values of all other variables. The ARIMA model needs to transform the non-stationary time series into stationary time series first, then the predicted values depending on the past values, and the present and past values of the random error term. GRU, TPA-LSTM, and DARNN are deep learning models that can utilize the latent inter-dependencies among variables for prediction. STGCN, MTGNN, and NRI are graph-based prediction methods by modeling the relationships between variables as the graph to help make better predictions.

In the experiments, we divided the data into two categories: one is abnormal temperature caused by failure, and the other is normal data. In the data with the abnormal occurrence, we also evaluate the predicted data by two criteria: All conditions and *delta* > 1 °C. The former calculates MSE on all of the actual and predicted target values in the testing set, while the latter calculates MSE only when there was 1 °C or more temperature increase in 2.5 min. The "*delta* > 1 °C" criterion is added because we want to catch a more significant temperature change in data center operations.

As depicted in Table 1, I2A-RI achieves the best performance over other methods. I2A-RI reduces the MSE of the second-best model (STGCN) by 2.88% with the criterion "All" and 4.30% with the criterion "*delta* > 1 °C".

Table 1. Performance comparison (MSE) among different approaches.

	Abnormal		Normal
	All	*delta* > 1 °C	All
VAR	0.0334 ± 0.001	0.4290 ± 0.02	$2.58e-03 \pm 1.17e-04$
ARIMA	0.0354 ± 0.003	0.4016 ± 0.02	$8.90e-04 \pm 1.16e-05$
GRU	0.0291 ± 0.002	0.3591 ± 0.04	$8.50e-04 \pm 1.91e-05$
TPA-LSTM	0.0270 ± 0.006	0.3271 ± 0.05	$1.30e-03 \pm 1.29e-05$
DARNN	0.0254 ± 0.005	0.3267 ± 0.04	$7.92e-04 \pm 1.03e-06$
STGCN	0.0243 ± 0.001	0.2019 ± 0.00	$4.89e-04 \pm 1.39e-06$
MTGNN	0.0246 ± 0.001	0.2579 ± 0.02	$4.59e-04 \pm 1.14e-06$
NRI	0.0263 ± 0.006	0.3132 ± 0.04	$5.09e-04 \pm 2.11e-06$
I2A-RI	$\mathbf{0.0236 \pm 0.002}$	$\mathbf{0.1932 \pm 0.02}$	$\mathbf{4.54e-04 \pm 1.01e-06}$

4.4 Ablation Study

In this section, we aim to verify the effect of the two import modules: the GDA module and the IDA module. The methods without these two modules are denoted as follows:

- ⋆ + **GDA:** I2A-RI without IDA.
- ⋆ + **IDA:** I2A-RI without GDA.
- ⋆ **-GDA-IDA:** I2A-RI without either GDA or IDA

The performance of the three variants are shown in Table 2. The results show that both the GDA and IDA can improve the model's performance, which verifies the correctness of our conjecture about inter-and-intra domain attention.

Table 2. Effects of GDA and IDA modules (MSE).

	Abnormal		Normal
	All	$delta >1\,^{\circ}C$	All
-GDA-IDA	0.0263 ± 0.006	0.3132 ± 0.04	$5.09e-04 \pm 2.11e-06$
+GDA	0.0240 ± 0.001	0.2511 ± 0.01	$4.88e-04 \pm 1.95e-06$
+IDA	0.0226 ± 0.001	0.2700 ± 0.02	$4.90e-04 \pm 2.98e-06$

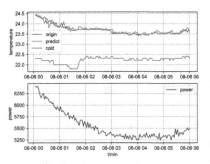 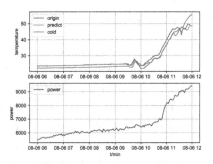

(a) in-distribution prediction   (b) out-of-distribution prediction

Fig. 4. Examples of the ground-truth and predicted time series.

4.5 Visualization

Additionally, we provide ground-truth and predicted time series trends of several selected variables from the dataset, as shown in Fig. 4. As can be seen from the

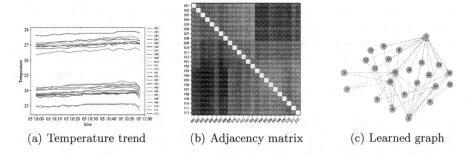

(a) Temperature trend   (b) Adjacency matrix   (c) Learned graph

Fig. 5. The adjacency matrix of the learned graph (same trend).

figures, when the predicted points are within the distribution of the training set, the model can predict them accurately; when the predicted points are beyond the distribution range, the model cannot capture this trend. We present the adjacency matrix and the learned graph of two columns of racks sharing the same cold aisle in Fig. 5 and 6.

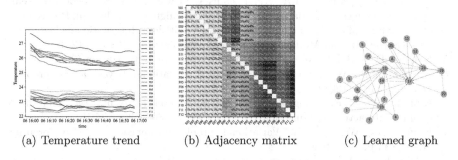

(a) Temperature trend   (b) Adjacency matrix   (c) Learned graph

Fig. 6. The adjacency matrix of the learned graph (different trend).

As can be seen from Fig. 5, when the temperature curves of different racks change in a similar way and the trend is consistent, the adjacency matrix presents similar relationship weights. As can be seen from Fig. 6, when the temperature curves of different racks fluctuate in different ways, the relationship between different racks is obviously different in the adjacency matrix. In the learned graph, each rack shows only the other two racks with which they are most closely associated. They may be responsible for the same business, thus these learned relationships may help troubleshoot when faults occur.

5 Conclusions

In the motivation of modeling interactions between time series from different domains, we propose the I2A-RI model to utilize the information of the guidance domain to learn a more accurate graph for prediction in this paper. Our results strongly prove that I2A-RI can learn underlying relations from data of different domains. With the learned structure, I2A-RI outperforms other state-of-the-art models for the RATEDC dataset. In the future, to better predict the values out of the data distribution, we will introduce more prior knowledge, including the pre-known dynamic and static information and the latest events.

Acknowledgements. This work is supported by Alibaba Post-doctoral Research Station and the National Key Research and Development Program of China No. 2020AAA0106300 and National Natural Science Foundation of China No. 62102222.

References

1. Adhikari, R., Agrawal, R.K.: An introductory study on time series modeling and forecasting. arXiv preprint arXiv:1302.6613 (2013)
2. Box, G.E., Jenkins, G.M., Reinsel, G.C., Ljung, G.M.: Time Series Analysis: Forecasting and Control. Wiley, Hoboken (2015)
3. Chung, J., Gulcehre, C., Cho, K., Bengio, Y.: Empirical evaluation of gated recurrent neural networks on sequence modeling. arXiv preprint arXiv:1412.3555 (2014)
4. El-Sayed, N., Stefanovici, I.A., Amvrosiadis, G., Hwang, A.A., Schroeder, B.: Temperature management in data centers: why some (might) like it hot. In: Proceedings of the 12th ACM SIGMETRICS/PERFORMANCE Joint International Conference on Measurement and Modeling of Computer Systems, pp. 163–174 (2012)
5. Kingma, D.P., Ba, J.: Adam: a method for stochastic optimization (2014)
6. Kingma, D.P., Welling, M.: Auto-encoding variational bayes. arXiv preprint arXiv:1312.6114 (2013)
7. Kipf, T., Fetaya, E., Wang, K.C., Welling, M., Zemel, R.: Neural relational inference for interacting systems. arXiv preprint arXiv:1802.04687 (2018)
8. Kipf, T.N., Welling, M.: Variational graph auto-encoders. arXiv preprint arXiv:1611.07308 (2016)
9. Lütkepohl, H.: Vector autoregressive models. In: Handbook of Research Methods and Applications in Empirical Macroeconomics. Edward Elgar Publishing (2013)
10. Qin, Y., Song, D., Chen, H., Cheng, W., Jiang, G., Cottrell, G.: A dual-stage attention-based recurrent neural network for time series prediction. arXiv preprint arXiv:1704.02971 (2017)
11. Shih, S.Y., Sun, F.K., Lee, H.Y.: Temporal pattern attention for multivariate time series forecasting. Mach. Learn. **108**(8–9), 1421–1441 (2019)
12. Veličković, P., Cucurull, G., Casanova, A., Romero, A., Lio, P., Bengio, Y.: Graph attention networks. arXiv preprint arXiv:1710.10903 (2017)
13. Wu, Z., Pan, S., Long, G., Jiang, J., Chang, X., Zhang, C.: Connecting the dots: multivariate time series forecasting with graph neural networks. In: Proceedings of the 26th ACM SIGKDD International Conference on Knowledge Discovery & Data Mining, pp. 753–763 (2020)
14. Yu, B., Yin, H., Zhu, Z.: Spatio-temporal graph convolutional networks: a deep learning framework for traffic forecasting. arXiv preprint arXiv:1709.04875 (2017)
15. Zhou, J., et al.: Graph neural networks: a review of methods and applications. arXiv preprint arXiv:1812.08434 (2018)

DEMO Papers

An Interactive Data Imputation System

Yangyang Wu[1], Xiaoye Miao[1(✉)], Yuchen Peng[2], Lu Chen[2], Yunjun Gao[2], and Jianwei Yin[1,2]

[1] Center for Data Science, Zhejiang University, Hangzhou, China
{zjuwuyy,miaoxy,zjuyjw}@zju.edu.cn
[2] College of Computer Science, Zhejiang University, Hangzhou, China
{zjupengyc,luchen,gaoyj}@zju.edu.cn

Abstract. In this demonstration, we propose an interactive data imputation system, called DITS, built upon ten state-of-the-art imputation algorithms and a novel variational generative adversarial imputation network. It consists of three modules, namely *source uploader*, *algorithm evaluation*, and *interactive imputation*. In the source uploader module, DITS allows users to register new imputation and prediction algorithms. Then, DITS is able to make users more aware of various integrated imputation algorithms via algorithm evaluation module, so as to support both lay and power users to operate a customized data imputation for their target dataset via the *interactive* imputation module. Using a public incomplete meteorologic dataset, we demonstrate that, DITS is capable of assisting users to effectively address real-life missing data issues.

Keywords: Missing data · Customized imputation · Interactive process

1 Introduction

Numerous data imputation algorithms have been developed to deal with the missing data problem [3]. These imputation algorithms can be categorized by mainly used models, including i) statistical ones [1] like the Mean imputation and k nearest neighbor imputation, ii) machine learning ones [2], e.g., XGBoost imputation, MissForest imputation, matrix factorization imputation, and soft-impute, and iii) deep learning ones [4], such as multiple imputation denoising autoencoder, variational autoencoder imputation, heterogeneous incomplete variational autoencoder model, and generative adversarial imputation network.

It is challenging to apply imputation methods to real-life scenarios. First, there is a lack of an imputation system to recommend proper algorithms in various scenarios. Second, users' domain knowledge is crucial for imputation.

There is therefore a strong need for an easy-to-use and interactive system. The main contributions of this imputation system are summarized as follows.

- We propose an interactive data imputation system, named DITS, that incorporates ten state-of-the-art imputation algorithms and a newly proposed variational generative adversarial imputation network VGAIN.

© The Author(s), under exclusive license to Springer Nature Switzerland AG 2022
A. Bhattacharya et al. (Eds.): DASFAA 2022, LNCS 13247, pp. 495–499, 2022.
https://doi.org/10.1007/978-3-031-00129-1_42

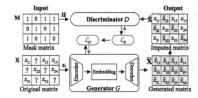

Fig. 1. The architecture of VGAIN   **Fig. 2.** The architecture of DITS

- The system DITS is composed of three functionality modules, i.e., source uploader, algorithm evaluation, and interactive imputation. The source uploader module not only allows users to upload their own data and integrate new algorithms, but also supports parameter tuning for different methods.
- The algorithm evaluation module subsumes an online data-specific *algorithm recommendation* and an offline example-based *algorithm exploration*, in order to guide users to learn more about the performance of imputation algorithms.
- The interactive imputation module allows users to conduct a customized imputation with an *interactive* process by using their domain knowledge.

2   VGAIN Imputation Algorithm

The generative adversarial networks (GANs) and autoencoders (AEs) are the two dominant approaches for deep learning imputation models. To this end, we develop a new variational generative adversarial imputation network (VGAIN for short), on top of GAN and AE, to prevent mode collapse in the generative model by ensuring that it is grounded in observed data.

Figure 1 illustrates the general procedure of VGAIN. It takes the data matrix $\mathbf{X}$ and the mask matrix $\mathbf{M}$ as inputs. When training the generator G, VGAIN not only ensures to minimize the probability (i.e., $\mathcal{L}_{pro}$) of the updated discriminator D identifying the missing values generated by G, but also has to minimize the sum of the reconstruction error $\mathcal{L}_{rec}$ on observed values and the prior regularization term $\mathcal{L}_{prt}$. Thus, G's loss function can be defined as $\mathcal{L}_G = \mathcal{L}_{pro} + \gamma \cdot (\mathcal{L}_{rec} + \mathcal{L}_{prt})$, where γ is a hyper-parameter.

On the other hand, the discriminator D is to maximize the probability of correctly predicting $\mathbf{M}$, i.e., D's loss function is defined as $\mathcal{L}_D = -\mathcal{L}_{pro} = -\mathbb{E}[\mathbf{M} \odot \log(D(\bar{\mathbf{X}}), \mathbf{H})) + (1 - \mathbf{M}) \odot \log(1 - D(G(\bar{\mathbf{X}}), \mathbf{H}))]$, where $\bar{\mathbf{X}}$ is the reconstructed matrix of G. The hint matrix $\mathbf{H}$, that contains a fraction of missing states in $\mathbf{X}$, is used by D to focus the attention on the imputation accuracy of other components, and thus it drives G to effectively learn the unique true data distribution. In the demonstration of the algorithm evaluation module, it is confirmed that, VGAIN outperforms state-of-the-art imputation methods.

3   System Overview

Figure 2 illustrates the workflow of the proposed system DITS. It consists of three modules, *source uploader*, *algorithm evaluation*, and *interactive imputation*.

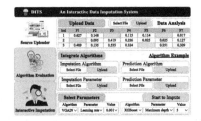

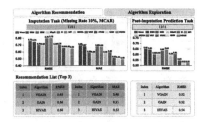

Fig. 3. Source uploader module   **Fig. 4.** Algorithm recommendation

Source Uploader Module. DITS allows users to upload their missing data. Moreover, users can integrate new imputation methods and the post-imputation prediction models into DITS. They can also tune the parameters for each model.

Algorithm Evaluation Module. To guide users to learn more about the algorithm performance, DITS offers *online* algorithm recommendation and *offline* algorithm exploration functionalities in this module. The imputation algorithm recommendation functionality uses the imputation and post-imputation prediction tasks to evaluate the algorithm performance over the uploaded data. If users are not willing to upload their dataset, the imputation algorithm exploration functionality allows users to explore the imputation ability of the integrated algorithms, based on real-world datasets in an offline and favorable way.

Interactive Imputation Module. To incorporate the domain expertise into the imputation, DITS provides a convenient way for users to operate a customized data imputation with an *interactive* imputation process. It considers two types of users: *lay* users who directly impute the uploaded data with the good-performance imputation algorithm; and *power* users who integrate their knowledge into imputation models by rejecting or rewriting the imputed data.

4   Demonstrations

In our demonstration, we utilize a public incomplete meteorologic dataset, to enable users to experience how to interact with DITS. DITS was implemented by Django 3.0.2 with Python 3.6. Bootstrap was utilized to develop interactive web pages. The source code and detailed references are available on GitHub[1].

First of all, users are able to visit the three functionality modules in DITS, i.e., *source uploader*, *algorithm evaluation*, and *interactive imputation* modules, as depicted on the left side of Fig. 3.

---

[1] https://github.com/zjuwuyy-DL/DASFAA22-DITS.

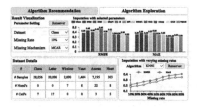

Fig. 5. Algorithm exploration   **Fig. 6.** Interactive imputation module

Source Uploader. Users can upload their dataset, which is visualized in a table, as plotted in Fig. 3. They can freely plug new imputation methods and prediction models into DITS. Users are also allowed to do parameter tuning for different methods, as exemplified in the bottom area of Fig. 3.

Algorithm Evaluation. We will invite users to experience the *algorithm recommendation* and *algorithm exploration* functionalities, as shown in Fig. 4 and Fig. 5, respectively. In the algorithm recommendation, DITS provides the performance evaluation for imputation methods over the uploaded data, in terms of the imputation and post-imputation prediction tasks. DITS runs all methods on the uploaded data. It visualizes the performances in histograms for the imputation task, as shown in Fig. 4. For the post-imputation prediction task, the prediction performances of the XGBoost model over the datasets imputed by different methods are presented, as exemplified in Fig. 4. In the algorithm exploration, users are also invited to study the offline performances of imputation methods on six real-world datasets. They can tune the dataset, missing rate, and missing mechanism, to evaluate the performance of imputation methods, as exemplified in the top left area of Fig. 5. After pressing the "Runserver" button, the corresponding performance under the parameter setting will be reported. Meanwhile, users can select one of the imputation methods and press the "Runserver" button to get the performance with varying missing rates.

Interactive Imputation. In this module, users are free to select an imputation method depending on the performance evaluated. They can press the "Impute" button to execute the chosen method on the uploaded dataset. If some power users are unsatisfied with the imputation result, they can select the first option in the context menu via the right click to reject the unsatisfied values, as shown in Fig. 6. Then, users can select one method to impute the rejected values, via pressing the "Impute" button again. Also, the second option can be selected to rewrite the rejected values during the interactive imputation process.

Acknowledgements. This work is partly supported by the NSFC under Grant No.61902343, the Zhejiang Provincial NSF under Grant No.LR21F020005, and the Fundamental Research Funds for the Central Universities. Xiaoye Miao is the corresponding author of the work.

References

1. Altman, N.S.: An introduction to kernel and nearest-neighbor nonparametric regression. Am. Stat. **46**(3), 175–185 (1992)
2. Mazumder, R., Hastie, T., Tibshirani, R.: Spectral regularization algorithms for learning large incomplete matrices. J. Mach. Learn. Res. **11**(11), 2287–2322 (2010)
3. Miao, X., Wu, Y., Wang, J., Gao, Y., Mao, X., Yin, J.: Generative semi-supervised learning for multivariate time series imputation. In: AAAI, pp. 8983–8991 (2021)
4. Yoon, J., Jordon, J., Schaar, M.: GAIN: missing data imputation using generative adversarial nets. In: ICML, pp. 5675–5684 (2018)

FoodChain: A Food Delivery Platform Based on Blockchain for Keeping Data Privacy

Rodrigo Folha[1]([✉])[iD], Valéria Times[1][iD], Arthur Carvalho[2][iD], André Araújo[3][iD], Henrique Couto[3][iD], and Flaviano Viana[1][iD]

[1] Center for Informatics, Federal University of Pernambuco, Recife, PE, Brazil
{rbf2,vct,fjlv}@cin.ufpe.br
[2] Farmer School of Business, Miami University, Oxford, OH 45056, USA
arthur.carvalho@miamioh.edu
[3] Computing Institute, Federal University of Alagoas, Maceió, AL, Brazil
{andre.araujo,henrique.couto}@ic.ufal.br

Abstract. Marketplace systems for delivery food retain valuable sensitive data besides charging high fees for each order. On the other hand, Blockchain has being adopted as a distributed database to reduce costs and remove intermediary actors. We have developed FoodChain that is a blockchain based system, decentralizes the food delivery solution and keeps data in public ledgers by ensuring the privacy of sensitive data. Thus, customers, restaurants and couriers can verify and track their sensitive data that are stored in the blockchain. The interactions that occur directly between the involved actors are tracked by smart contracts, which in turn, may contain rules and conditions about the order delivery agreement. Moreover, all sensitive data is encrypted before being sent to the blockchain and can be decrypted using asymmetric encryption. The proposed demo presents the following FoodChain features: smart contracts running on Fantom testnet blockchain, an API to connect with a public database and three web-based decentralized applications (DApp).

Keywords: Privacy · Decentralization · Food system delivery

1 Introduction

Since 2012, several marketplace platforms (Grubhub, UberEats, Rappi, iFood) have emerged to solve the food delivery problem, offering multiple menus from different restaurants in one place [1]. In addition to enabling the customer to compare and choose the order in a single place, all these orders can be delivered to the customer through autonomous couriers who aim at carrying out a delivery in areas close to him/her. These new platforms changed the traditional way of food delivery made by restaurants, which used to be made directly between customer and restaurant, via telephone or internet, and the restaurant was in charge of delivering the orders, or customers came to the restaurant's address to pick his/her order up. Customers connect with several restaurants simultaneously by using these platforms, which are responsible for providing restaurants with a greater visibility and giving a wide variety of options to customers.

The visibility caused by the adoption of marketplace systems enables restaurants to go beyond physical boundaries, reaching potential customers who have never visited their addresses. Also, customers can get to know the restaurants, see the restaurants menus and compare prices of restaurants that, occasionally, they have never visited before [4]. Despite intermediating each transaction made by users, these food delivery platforms may charge commissions from 5% to 25% of the order value for intermediating commercial transactions [2]. Another factor of great concern is that these platforms have been keeping all the information related to the orders, including customers' preferences and addresses, giving the companies that own these platforms the chance of earning even more money from customers and restaurants sensitive data.

To eliminate the intermediary actor by maintaining the privacy of those involved and eliminating costs and decentralizing data, this article presents a marketplace platform based on smart contracts hosted on a blockchain. The proposed platform, called FoodChain, enables customers to place orders directly to restaurants, by creating smart contracts that will be responsible for managing the agreement made between the parties. Once the order is placed, it becomes available for couriers to show their interest in the delivery.

A related work of ours is Bistroo.io [3] that is a marketplace system based on blockchain. Bistroo.io charges a commission fee of 5% for each order price; the order data is stored by Bistroo.io on a centralized database and Bistro.io rewards users that share their consumption and order data. The main differential of FoodChain is to build a decentralized system that guarantees the data privacy for the users while storing sensitive data in a public ledger. This demo article is designed to show how FoodChain connects the actors involved in the food delivery chain, using a decentralized and autonomous solution by maintaining the privacy of sensitive data through the use of smart contracts.

2   The FoodChain Scenario and Features

The FoodChain's scenario involves three actors: establishments, couriers and customers. The customer accesses the system, chooses one of the available establishments, and selects the wanted products. Then, the restaurant must accept the order and begin the dish preparation, while the courier must accept the delivery. Once the restaurant finishes the order, it must delivery it to the courier, who will receive and take it to the customer's address.

Sensitive data are classified as any data containing information about orders received by a restaurant, or about orders placed by a customer, such as: name and quantity of ordered items, order price, and customer's address and name. On the other hand, public data, stored in a relational database, are freely accessible to all users, may be updated by each restaurant and correspond to: the addresses and names of the restaurants, public key, the logo of each restaurant and the restaurants' menus containing all dishes and their corresponding prices. To store order data, which are mostly non-sensitive and stored as plain text (e.g. the status of the order, the time of the last update, the digital address of the customer who did the order, the restaurant that received the order and the delivery person who accepted the order and the delivery price), a smart contract is deployed for each order created with FoodChain. Despite indicating the amount of requests made by certain digital addresses, each actor may change his/her digital address, unlinking the entire order history from the new digital address.

However, there are some data that are stored in the order that are sensitive, such as the customer's name and address, as well as the ordered items. To maintain the privacy of these sensitive data that are kept in open and public smart contracts, asymmetric encryption tools are adopted. To encrypt the sensitive information that are necessary for each stakeholder involved, the respective public keys are used. First, the customer encrypts the order items with the restaurant's public key and her/him own public key. Thus, when the order is stored in the blockchain, customers and restaurants can decode the order with their own private key, identify the order and accept, or not, the order. At this point, the restaurant does not need to know the customer's name or address. Following this, when the customer's order is accepted by a delivery person, the customer encrypts his/her personal data (name and address) as well as the list of items of the order using the delivery person's public key. Then, the delivery person may decrypt the data using his/her private key to have access to the customer's address and the requested items, and to check the order correctness.

The main feature of FoodChain is the capability to be a plug and play solution where restaurants may create their own DApp and connect to the blockchain, offering a personalized experience. Moreover, FoodChain is composed of three different layers, which may be distributed over different internet nodes. In the Data Layer, to store order data and maintain business rules, smart contracts written in Solidity version 0.85 are hosted on the Fantom's test network. Also, to store the public data of restaurant menus, the PostgreSQL database in version 9.5 is used. The second layer, called Service Layer, is composed of two interfaces: (i) a server running on Node.js that makes the communication between the DApp and the lower layer for querying or storing public data; and (ii) Metamask, acessible in https://metamask.io/, that is a digital wallet that shows the balance of the user's digital address, manages the encryption operations and connects the decentralized application (DApp) with the lower layer through the Web3.js library in version 1.3.5. Finally, the third layer, DApp Layer, are web applications written in Javascript using the React.js framework in version 16.13.1.

3   The FoodChain System

FoodChain is a web based system developed according to the model design of Sect. 2 that enables the use of smart contracts on public ledgers for managing orders and taking into consideration the privacy of sensitive data. Figure 1 displays an example of an order made by a customer, how data are stored in a smart contract and the restaurant's orders page.

(a) The customer's check-out page.

(b) An example of a smart contract's order data.

(c) The restaurant's orders page.

Fig. 1. Ordering and accepting the order.

Figure 1a indicates that a customer is ordering a soda, a diet coke, an orange juice and a crispy chicken Parmesan. Then, the customer uses his public key and the restaurant's public key to encrypt the data before confirming the transaction and creating a smart contract to represent the referred order. Figure 1b shows how the order's sensitive and non-sensitive data are stored in smart contracts. The order items are encrypted with two different public keys (customer and restaurant) and stored into two separated variables. Lastly, in Fig. 1c, the restaurant's manager can see all orders linked to the restaurant's digital address. Thus, the he/she can get more information using the restaurant's private key to decrypt the items details stored in the order's smart contract. During the demo, examples of use of FoodChain will be shown through the creation, acceptance and delivery of orders using three distinct actors. The FoodChain modules[1][2][3], source code[4] and video demonstration[5] are available through the following links.

As future work, we plan to evaluate the behavior of our system prototype with a massive quantity of users, and to conduct a post evaluation with some restaurant's owners, customers and couriers.

References

1. Almunawar, M.N., Anshari, M.: Digital enabler and value integration: revealing the expansion engine of digital marketplace. In: Technology Analysis & Strategic Management, pp. 1–11 (2021)

---

[1] https://arthurcarvalho.info/apps/foodchain/customer/.
[2] https://arthurcarvalho.info/apps/foodchain/manager/.
[3] https://arthurcarvalho.info/apps/foodchain/delivery/.
[4] https://github.com/rodrigofolha/foodchain.
[5] https://youtu.be/xDSsFeFHXBY.

2. Erickson, A., Losekoot, E.: If covid-19 doesn't kill you, uber eats will: Hospitality entrepreneurs' views on food aggregators. Danish University Colleges, p. 375 (2021)
3. Roos, B., Dohmen, B., Geelen, B.: The Peer-2-Peer Food Marketplace. https:// bistroo.io/whitepaper, Accessed 19 Mar 2022
4. Subramanian, H.: Decentralized blockchain-based electronic marketplaces. Commun. ACM **61**(1), 78–84 (2017)

A Scalable Lightweight RDF Knowledge Retrieval System

Yuming Lin, Chuangxin Fang, Youjia Jiang, and You Li[(⊠)]

Guangxi Key Laboratory of Trusted Software, Guilin University of Electronic
Technology, Guilin 541004, China
liyou@guet.edu.cn

Abstract. Currently, there are numerous knowledge retrieval systems
available to researchers, among which the RDF retrieval system is the
most common. However, in practice, these systems are often plagued
with problems, such as long index constructions and loading times and
require large amounts of disk storage space, which are faults that make
the system non-conducive to the dynamic incremental updates of data.
This paper proposes a scalable lightweight RDF retrieval system, which
has the following characteristics: 1) optimization of the index structure
to reduce disk occupation and speed up the construction; 2) query optimizations based on query strategy selection; and 3) an interactive visual
operation interface. The final evaluation results show that our system
uses less disk space and has faster index construction times, and the
query performance is competitive with the latest RDF retrieval systems.

Keywords: Knowledge graph · RDF data · Knowledge retrieval

1 Introduction

A knowledge retrieval system is a software infrastructure system that enables
the construction, storage, querying, and analyses of knowledge graph data. An
RDF data format is used to represent the relationship between entities in the
semantic network and is a format very commonly used in existing knowledge
retrieval systems. RDF (resource description framework) data are a collection
of $\langle subject, predicate, object \rangle$ triples, and the SPARQL query statement can be
used to query knowledge data based on an RDF representation. Existing RDF
storage and retrieval systems are divided into two main categories [1]: relational-
based query systems (e.g. RDF-3X [2]) and graph-based query systems (e.g.
gStores [3]). As representative RDF retrieval systems, RDF-3X and gStore have
good query performances, but generated data index files occupy a large space in
actual use, and are not easy to handle in scenarios where knowledge data needs to
be updated dynamically; gSMat [4] converts query processing into approximates
multiplications for sparse matrices, but the sparse matrix-based storage strategy
in not conducive to the real-time incremental data update, furthermore, the

© The Author(s), under exclusive license to Springer Nature Switzerland AG 2022
A. Bhattacharya et al. (Eds.): DASFAA 2022, LNCS 13247, pp. 505–508, 2022.
https://doi.org/10.1007/978-3-031-00129-1_44

redundancy of the intermediate results will slow down the query. Therefore, a scalable lightweight knowledge retrieval system is required.

In this paper, we propose a scalable lightweight knowledge retrieval system, which greatly reduces the size of the index data by optimizing the index structure and adopts query optimizations based on a query selection strategy according to the index structure, which has better scalability. In addition, we provide an interactive visual interface.

2   System Architecture and Key Techniques

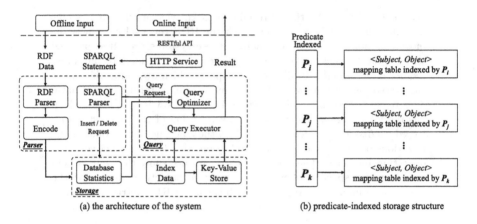

(a) the architecture of the system   (b) predicate-indexed storage structure

Fig. 1. The architecture of the system and an example of a knowledge graph

The architecture of the system is shown in Fig. 1(a). The core of the system is divided into three parts, the parser, storage and query. The parser parses the RDF data and SPARQL statements.

The storage module stores the parsed ID triples and the entry pairs mapped from the subject to the object based on the predicate indexed, and records the statistics of RDF data for generating the query plan during the query. The predicated-indexed storage structure as shown in Fig. 1(b), after loading the RDF data, the triplets will be transformed into the predicate-indexed entity pairs and map the strings to unique IDs, which makes the index structure not only more compact but also easy to update, also conforms to the structure of the directed graph.

The query module consists of a query optimizer and a query executor. The former generates query plans based on data statistics, and the latter adopts the binary join strategy. Since the entity list is the mapping data pair from subject to object by default, handling the join query by subject variables is simple (as shown in Fig. 2(I)); The join query by object variables only needs to reverse map the object to the subject when loading the index to obtain the lookup table to execute a hash join (as shown in Fig. 2(II)); A naive join query will

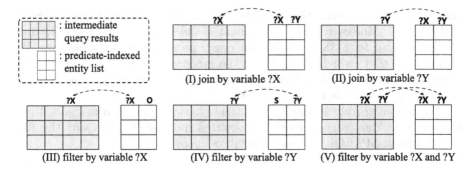

Fig. 2. Predicate-indexed data model and query strategy selection

make the intermediate results redundant, resulting in a slower query. Therefore, for the entity list of known objects, the intermediate results can be filtered by the subject variable (as shown in Fig. 2(III)), while for cases with a known subject, the intermediate results can be filtered by the object variable (as shown in Fig. 2(IV)). In the case of Fig. 2(V), the intermediate result contains the query variables of the entity list, so the lookup table can still be used to filter the data. In summary, in the query plan generation stage, different query strategies can be executed by marking the types of query triples.

3   Evaluation and Demonstration

We compare 3 indicators (i.e., the index size, index building time, and query time) using gStore and RDF-3X to evaluate the system's performance. The experiments are carried out on an inexpensive, commonly-available machine with an i5-9400F 2.90 GHz CPU, Linux Ubuntu 16.04 as the OS and 64 GB of main memory. The datasets are the standard RDF benchmarks, WatDiv10M and Wat-Div100M provided in [5], and contain approximately 10 million and 100 million triples respectively; four types of query statements are provided, complex (C), linear (L), snowflake (SF) and star (S).

Table 1. Comparison of index size (MB), build time (sec), and query time (msec) for WatDiv10M and WatDiv100M

	WatDiv10M						WatDiv100M					
	Index size	Build time	Query time				Index size	Build time	Query time			
			C	L	SF	S			C	L	SF	S
gStore	840	170	105.3	28.9	105.1	27.8	8294	1957	1036.7	40.2	219.0	38.2
RDF-3X	438	80	316.0	1.3	3.6	6.6	4608	1172	3150.5	13.5	27.3	21.7
ours	217	11	200.1	9.2	24.6	12.7	2457	163	2088.6	29.5	137.1	42.4

The comparison of the 3 indicators on the 2 WatDiv datasets is shown in Table 1. The experiment shows that our system completes the construction of the data index in a shorter time, and the performance is much better than gStore and RDF-3X. Moreover, when the amount of index data is much smaller than the gStore and RDF-3X amounts, our query performance is competitive. We also observe that an index size that is too large makes the actual query time-consuming and the index loading time too long. Therefore, our system can finish the query processing faster in actual use.

Finally, we provide an interactive visual interface, as shown in Fig. 3. Users can select the knowledge graph to be visualized and easily interact with the data and can input SPARQL statements for data operations.

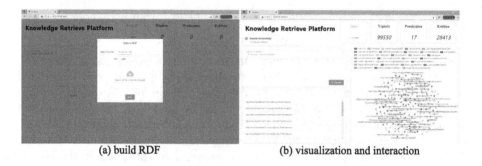

(a) build RDF   (b) visualization and interaction

Fig. 3. Interactive visual interface

Acknowledgment. This work was supported by National Natural Science Foundation of China (Nos. 62062027, U1811264), Guangxi Natural Science Foundations (No. 2018GXNSFDA281049), the Science and Technology Major Project of Guangxi Province (No. AA19046004), the Innovation Project of GUET Graduate Education (No. 2021YCXS052), the project of Guangxi Key Laboratory of Trusted Software.

References

1. Abdelaziz, I., Harbi, R., Khayyat, Z.: A Survey and experimental comparison of distributed SPARQL engines for very large RDF data. PVLDB **10**(13), 2049–2060 (2017)
2. Neumann, T., Weikum, G.: The RDF-3X engine for scalable management of RDF data. VLDB J. **19**(1), 91–113 (2010)
3. Zou, L., Özsu, M.T., Chen, L., Shen, X., Huang, R., Zhao, D.: gStore: a graph-based SPARQL query engine. VLDB J. **23**(4), 565–590 (2013). https://doi.org/10.1007/s00778-013-0337-7
4. Zhang, X., Zhang, M., Peng, P., et al.: A scalable sparse matrix-based join for SPARQL query processing. In: DASFAA, pp. 510–514 (2019)
5. Aluç, G., Hartig, O., Özsu, M.T., Daudjee, K.: Diversified stress testing of RDF data management systems. In: ISWC, pp. 197–212 (2014)

CO-AutoML: An Optimizable Automated Machine Learning System

Chunnan Wang[1], Hongzhi Wang[1(✉)], Bo Xu[1], Xintong Song[1], Xiangyu Shi[1], Yuhao Bao[1], and Bo Zheng[2]

[1] Harbin Institute of Technology, Harbin, China
{WangChunnan,wangzh,xyu.shi}@hit.edu.cn
{1190201620,20s103144,1180300605}@stu.hit.edu.cn
[2] CnosDB, Beijing, China
harbour.zheng@cnosdb.com

Abstract. In recent years, many automated machine learning (AutoML) techniques are proposed for the automatic selection or design machine learning models. They bring great convenience to the use of machine learning techniques, but are difficult for users without programming experiences to use, and lack of effective optimization scheme to respond to users' dissatisfaction with final results. To overcome these defects, we develop CO-AutoML, a user-friendly and optimizable AutoML system. CO-AutoML allows users to interact with the system in a customized mode. Besides, it can continuously optimize the search space of the AutoML technique based on reinforce policy and graph neural network (GNN), and thus provide users with more powerful machine learning schemes. Our system empowers ordinary users to easily and more effectively use AutoML techniques, which has a certain application value and practical significance. Our demonstration video: https://youtu.be/nGnmA7noeJA.

Keywords: Automated machine learning · Search space optimization

1 Introduction

The automated machine learning (AutoML) is currently one of the popular subfields within data science that aims at automating the time consuming and iterative tasks of machine learning model development. It allows data scientists and developers to build machine learning models with high efficiency while sustaining model quality, and has attracted great attention from researchers. Recently, many AutoML techniques, including hyperparameter optimization (HPO) algorithms [3], automatic algorithm selection (AAS) algorithms [4] and neural architecture search (NAS) algorithms [1], are proposed for the automatically set, select or design machine learning models respectively.

These AutoML techniques bring great convenience to the use of machine learning techniques, but have the following two shortcomings which limit their applications. On the one hand, they are implemented by various programming languages, which are difficult for users without programming experiences and ability to understand and use them.

On the other hand, they lack optimization strategy to respond to users' dissatisfaction with final results. Specifically, in the real applications, users may not be satisfied with final result of the AutoML algorithm, and hope to further improve the performance. The realization of this goal requires the AutoML algorithm to be able to optimize itself by analyzing the historical data. However, the existing AutoML techniques fail to do so. They lack effective optimization scheme, and can only increase the running time or the number of attempts to improve the final performance, which is inefficient and irrational.

To overcome the above shortcomings, we develop CO-AutoML, a user-friendly and optimizable AutoML system, to let more users to easily and more effectively use AutoML techniques. The contributions of our system are threefold as follows.

(1) Customized and User-friendly Mode. We design the user-friendly interface to guide ordinary users to customize dataset and running details according to their requirements, run algorithm and read performing results.
(2) Effective AutoML Optimization Strategy. We propose to utilize reinforce policy [5] and graph neural network (GNN) [2] technique to further optimize the search space of the AutoML algorithm. This strategy can assist AutoML algorithms to better serve users obtaining more powerful ML schemes.
(3) Running Effect Visualization. We also design the visualization module to enable users to monitor the running effect of the AutoML algorithm, and thus better understand its performance and decide its running times.

2   System Overview

Figure 1 shows the overall framework of CO-AutoML system, which has four modules: 1) Selection Module, 2) Setting Module, 3) Visualization Module and 4) Optimization Module.

Selection Module. We provide users with three types of well-known AutoML techniques, including HPO, AAS and NAS, so as to cope with different machine learning needs. For each type of AutoML technique, we implement one of the state-of-the-art algorithms in CO-AutoML system to handle the corresponding tasks. For example, we implement a knowledge-pruning-based HPO method ExperienceThinking [3], a knowledge-driven AAS method Auto-Model [4] and a reinforcement learning (RL) based NAS method GraphNAS [1].

Setting Module. After selecting a suitable AutoML technique, we allow users to customize the dataset, and running details, including desired running time and required search space, according to their requirements. We provide a user-friendly interface to guide users to complete the above settings.

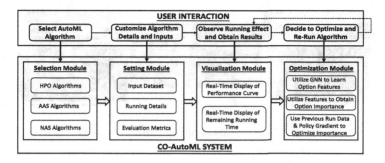

Fig. 1. The overall framework of CO-AutoML system.

Visualization Module. After inputting the customized information, CO-AutoML can start running the corresponding AutoML algorithm and jump to a visualization interface. In this interface, we display the performance curve of the searched solutions and the running time of the AutoML algorithm in real-time. With the real-time information, users can have a better understanding of the effect of the executed AutoML algorithm, and thus decide whether to optimize and re-run the AutoML algorithm after current executing process.

Optimization Module. In CO-AutoML system, we also design an optimization module for the AutoML algorithm. If users are not satisfied with the final solutions, they can click "Optimize" button. Then, CO-AutoML starts executing the optimization module to find a suitable way of improving AutoML algorithm's performance, and jumps to a new visualization interface to display the details and running results of the optimized algorithm.

We all know that the great amount of poor search areas in the search space brings great challenges to AutoML techniques. Therefore, the effective pruning strategy is necessary and significant for optimizing AutoML algorithm's performance. Our optimization module can remove useless and poor machine learning schemes from the original search space by analyzing the AutoML algorithm's historical data. This pruning strategy can make the AutoML algorithm pay more attention to more important search area in the next execution, and thus improve its search effect and search efficiency.

3   Optimization Strategy for AutoML Techniques

The optimization module is a bright spot of our CO-AutoML system. Considering the original search space is the Cartesian product of the related parameters, the solutions in the search space can be denoted as the complete paths of a directed graph network $\mathbb{R}$, where the options of two adjacent parameters are fully connected. Our optimization module aims to remove useless and less important directed edges from $\mathbb{R}$, so as greatly reduces the size of search space.

Fig. 2. The interfaces of CO-AutoML system.

The AutoML algorithm's historical run data provide us with valuable experience. We make full advantages of these data to learn importance of each directed edge, and thus prune the search space reasonably. We use an embedding layer to initialize the feature vector of each option node n_i. Then, we introduce GNN [2] to learn high-level node representations $X_{n_i} = GNN_\beta(Emb_\omega(n_i))$ and importance score θ_{n_i,n_j} of each directed edge e_{n_i,n_j}, i.e., $\theta_{n_i,n_j} = sum(X_{n_i} \times X_{n_j})$. Finally, we use reinforce policy [5] to optimize GNN and embeddings according to the historical run data: $\{< path^i_{1:T} = (n^i_1, ..., n^i_T), Reward(path^i_{1:T}) > \mid i = 1, ..., N\}$, where $path^i_{1:T}$ is a complete path in $\mathbb{R}$, and $Reward(path^i_{1:T})$ is its performance.

$$\nabla_{\omega,\beta}\mathbb{E}_{P(path_{1:T};\omega,\beta)}\Big[Reward(path_{1:T})\Big]$$
$$\approx \frac{1}{N}\sum_{i=1}^{N}\sum_{j=1}^{T}\nabla_{\omega,\beta}\ log\ \theta_{n^i_j,n^i_{j+1}}\ Reward(path_{1:T}) \tag{1}$$

In this reinforce policy, we use the importance score of edges in the complete path to approximate the complete path's selection probability, and learn effective importance scores by maximizing the expected reward. After getting the effective importance scores, we will remove less important directed edges in $\mathbb{R}$ according to importance, and thus achieve the rational pruning of the search space.

4   Demonstration Scenarios and Conclusion

Figure 2 demonstrates the four modules of CO-AutoML system. In this demonstration, the user aims to optimize the hyperparameters of IBk, a classification algorithm in Weka library. After selecting the classification dataset and setting the running details, the user observes the real-time performance of the HPO algorithm. The user sets the pruning ratio of the search space to 0.5 and re-runs the HPO algorithm, and the performance of the HPO algorithm increases 26%.

CO-AutoML fully considers users' usage requirements on AutoML technique. It empowers ordinary users to easily and effectively use AutoML techniques.

References

1. Gao, Y., Yang, H., Zhang, P., Zhou, C., Hu, Y.: Graph neural architecture search. In: IJCAI, pp. 1403–1409 (2020)
2. Klicpera, J., Bojchevski, A., Günnemann, S.: Predict then propagate: graph neural networks meet personalized pagerank. In: ICLR (2019)
3. Wang, C., Wang, H., Zhou, C., Chen, H.: Experience thinking: constrained hyperparameter optimization based on knowledge and pruning. Knowl.-Based Syst. (5), 106602 (2020)
4. Wang, C., Wang, H., Mu, T., Li, J., Gao, H.: Auto-model: utilizing research papers and HPO techniques to deal with the CASH problem. In: ICDE, pp. 1906–1909. IEEE (2020)
5. Williams, R.J.: Simple statistical gradient-following algorithms for connectionist reinforcement learning. Mach. Learn. 8, 229–256 (1992)

OIIKM: A System for Discovering Implied Knowledge from Spatial Datasets Using Ontology

Liang Chang[1], Long Wang[1], Xuguang Bao[1(✉)], and Tianlong Gu[2]

[1] Guangxi Key Laboratory of Trusted Software, Guilin University of Electronic Technology, Guilin 541004, China
changl@guet.edu.cn, bbaaooxx@163.com
[2] College of Cyber Security, Jinan University, Guangzhou 510632, China

Abstract. Spatial association analysis is an important task in spatial data mining. The co-location pattern, an important expression form of spatial association analysis, has guided users' decisions in many aspects, such as city service, business, etc. However, current methods aim at mining co-location patterns consisting of fine-grained features, they ignore the background knowledge including known relationships of things in a certain domain. Moreover, current methods generate numerous and independent patterns, which causes users to be confused when making the following decisions. Unlike existing works, in this demonstration, we present a system called *OIIKM* (**O**ntology-based **I**nteresting **I**mplied **K**nowledge **M**iner), which employs *ontology* to integrate user knowledge during the mining process, to discover more knowledge (represented it as co-location patterns consisting of *ontology concepts*) implied in the spatial datasets. Besides, to alleviate user confusion, *OIIKM* provides a visual *OntologyTree* for the user to select *ontology concepts* she/he is interested in, by which only a few patterns are displayed to provide more guidance for better decisions.

Keywords: Spatial data mining · Co-location patterns · Domain knowledge · Ontology · POI data

1 Introduction

The rapid increasing volume and the ever-deepening application requirements of spatial datasets promote the development of spatial association analysis. Spatial co-location pattern mining is an important branch in spatial data mining [1]. Co-location patterns can serve various fields, including ecosystem, city service, business, etc.

There are many methods focusing on discovering co-location patterns [2, 3]. However, most existing methods aim at discovering the co-location patterns consisting of fine-grained spatial features, and many relationships of things (background knowledge) are ignored.

To solve the above problems, *OIIKM* uses *ontology* to integrate the user's background knowledge. In the Semantic Web field, *ontology* is considered as the most appropriate

A. Bhattacharya et al. (Eds.): DASFAA 2022, LNCS 13247, pp. 514–517, 2022.
https://doi.org/10.1007/978-3-031-00129-1_46

representation to express the user knowledge [4]. Besides, the conceptual framework for spatial data mining driven by formal *ontology* has been verified and proposed in [5]. In *OIIKM*, a co-location pattern is a set of *ontology concepts* whose instances are frequently co-located in a certain space. Given a user-specified minimum distance threshold (*min_dis*), two instances are co-located if the distance between them is no more than *min_dis*. In co-location pattern mining, participation index (*PI*) is commonly used as the measure of the prevalence of a co-location pattern. Given a co-location pattern *c* and the minimum prevalence threshold (*min_prev*), if $PI(c) \geq min\_prev$ holds, *c* is considered as a prevalent co-location pattern.

OIIKM is proposed for users to discover co-location patterns consisting of *ontology concepts* derived from the domain *onology*. *OIIKM* discovers more patterns than traditional methods, therefore, to alleviate the user's confusion facing a huge number of patterns, *OIIKM* provides a visual *OntologyTree* for users to select *ontology concepts*. Co-location patterns that have hierarchical relationships [4] with the user's selections in *OntologyTree* are presented, which enables users to discover the natural relationships of *ontology concepts* and helps their next decision-making.

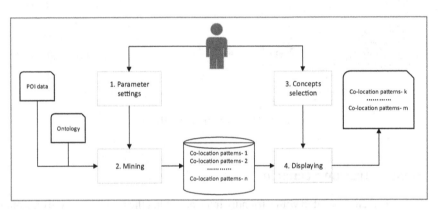

Fig. 1. Framework description of OIIKM

2   System Overview

Given the *ontology* and the POI (Points of Interest) dataset, Fig. 1 illustrates the mining procedure of *OIIKM*. There are 4 major steps described as follows.

1. **Parameter Settings**: For mining prevalent co-location patterns, the *min_dis* and *min_prev* are set by the user.
2. **Mining**: *OIIKM* first discovers the prevalent co-location patterns based on a clique-based method presented in [2]. Moreover, by the hierarchical relationships between ontology concepts [4], co-location patterns consisting of *ontology concepts* in different hierarchies are discovered in *OIIKM*.

3. **Selection**: Instead of presenting all generating co-location patterns to the user, *OIIKM* provides the user with a visual *OntologyTree* where the user can select his/her interested *ontology concepts*.

4. **Displaying**: *OIIKM* displays the patterns that have hierarchical relationships with user-selected *ontology concepts*, by which the user can quickly discover natural relationships of his/her interesting concepts.

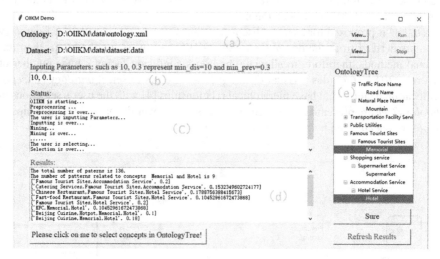

Fig. 2. Demonstration of OIIKM

3 Demonstration Scenarios

OIIKM is well encapsulated with a friendly interface, what the user faces is only a simple *UI*. We use part of the data from points of interests (POI) in Beijing [2–4] to show the demonstration of *OIIKM*.

Figure 2 shows the interface of *OIIKM*. In Fig. 2(a), an XML file with the description of *ontology* and a file with a spatial dataset are required. In Fig. 2(b), *min_dis* and *min_prev* are set by the user. The status bar displays the running status of *OIIKM* in Fig. 2(c). Co-location patterns are mined and displayed in Fig. 2(d). After the user selects *concepts* in *OntologyTree* shown in Fig. 2(e), Fig. 2(d) will display co-location patterns that have hierarchical relationships with selected concepts.

In Fig. 2(b), when the *min_dis* and *min_prev* were set as 10 and 0.1 respectively, *OIIKM* discovered 136 co-location patterns. After concepts *Memorial* and *Hotel* were selected by the user in *OntologyTree* in Fig. 2(e), only 9 patterns that have the hierarchical relationship with *Memorial* and *Hotel* were presented. *OIIKM* also presented the *PI* values of patterns to the user for facilitating his/her data analysis.

Richer knowledge can better reflect the natural relationships of things. *OIIKM* can mine richer knowledge than traditional methods focusing on discovering co-location patterns. As shown in Fig. 3, with the decrease of *min_prev*, the gap of the count of patterns

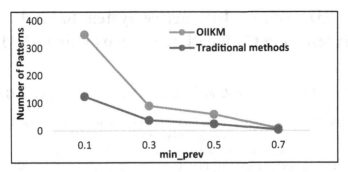

Fig. 3. Effects of OIIKM

generated by *OIIKM* and traditional methods increases rapidly. *OIIKM* will discover more co-location patterns but the user can find what they need through *OntologyTree*.

4 Conclusion

In this demonstration, a system *OIIKM* is proposed to discover more knowledge that can reflect the natural relationships of things than traditional methods in spatial datasets. To help users quickly discover their interesting knowledge, *OIIKM* presents patterns related to user-selected ontology concepts. By these ways, *OIIKM* can provide more meaningful guidance for users' decision-making in application fields of co-location patterns.

Acknowledgements. This work was supported in part by grants (No. U1811264, No. U1711263, No. 61966009, No. 62006057, 61762027) from the National Natural Science Foundation of China, in part by grants (No. 2018GXNSFDA281045, No. 2019GXNSFBA245059) from the Natural Science Foundation of Guangxi Province, and in parts by grants (No. AD19245011) from the Key Research and Development Program of Guangxi Province.

References

1. Huang, Y., Shekhar, S., Xiong, H.: Discovering colocation patterns from spatial data sets: a general approach. IEEE Trans. Knowl. Data Eng. **16**(12), 1472–1485 (2004)
2. Bao, X., Wang, L.: A clique-based approach for co-location pattern mining. Inf. Sci. **490**, 244–264 (2019)
3. Bao, X., Lu, J., Gu, T., Chang, L., Xu, Z., Wang, L.: Mining non-redundant co-location patterns. IEEE Trans. Neural Netw. Learn. Syst. (2021)
4. Bao, X., Gu, T., Chang, L., Xu, Z., Li, L.: Knowledge-based interactive postmining of user-preferred co-location patterns using ontologies. IEEE Trans. Cybern. (2021)
5. Hwang, S.: Using formal ontology for integrated spatial data mining. In: Laganá, A., Gavrilova, M.L., Kumar, V., Youngsong Mun, C.J., Tan, K., Gervasi, O. (eds.) Computational Science and Its Applications – ICCSA 2004, pp. 1026–1035. Springer Berlin Heidelberg, Berlin, Heidelberg (2004). https://doi.org/10.1007/978-3-540-24709-8_108

IDMBS: An Interactive System to Find Interesting Co-location Patterns Using SVM

Liang Chang[1], Yuxiang Zhang[1], Xuguang Bao[1(✉)], and Tianlong Gu[2]

[1] Guangxi Key Laboratory of Trusted Software, Guilin University of Electronic Technology, Guilin 541004, China
changl@guet.edu.cn, bbaaooxx@163.com
[2] College of Cyber Security, Jinan University, Guangzhou 510632, China

Abstract. Spatial co-location pattern mining is an important task in spatial data mining. However, traditional mining frameworks cannot help a particular user effectively discover interesting co-location patterns according to his specific interest because traditional mining algorithms decide the prevalence (frequency) of a co-location pattern only by a user-specified real number. Thus, in order to discover the user's real interesting co-location patterns, in this demonstration, we present IDMBS (Interactive data mining based on support vector machine), an interactive mining system, to discover user-preferred co-location patterns based on SVM. With IDMBS, users only need to go through a few rounds of interactions to efficiently discover the user-preferred patterns. IDMBML contains a filtering algorithm and an SVM model. The patterns selected by the filtering algorithm are annotated by the user, and the SVM model trains these patterns in order to discover more user-preferred co-location patterns. IDMBS can effectively and accurately discover the user-preferred patterns.

Keywords: Spatial data mining · Interesting co-location patterns · Machine learning · Support vector machines · Interactive system

1 Introduction

The extraction of spatial co-location patterns is a rising and promising field in spatial data mining. A spatial co-location pattern represents a subset of spatial features whose instances are frequently located in spatial neighborhoods. Spatial co-location pattern mining yields significant insights for various applications such as Earth science.

In spatial co-location pattern mining, a common problem existed in most mining algorithms is that only a small part of the co-location patterns mined are of interest to users. Typically, spatial co-location pattern mining methods use the frequencies of a set of spatial features participating in a co-location pattern to measure a pattern's prevalence (known as PI), and require a user-specified minimum prevalence threshold to filter prevalent co-location patterns [1]. However, user's interest is often subjective, and there is no consistent objective measure to represent user's interest. A co-location pattern could be interesting to one user but may be not to another. Thus, several methods were

A. Bhattacharya et al. (Eds.): DASFAA 2022, LNCS 13247, pp. 518–521, 2022.
https://doi.org/10.1007/978-3-031-00129-1_47

proposed in the literature to overcome this drawback such as Ontology-based interactive mining [2], Probability-based interactive mining [3]. But not all co-location patterns within the same semantics are of interest to users.

In the field of machine learning, support vector machine (SVM) is a supervised machine learning algorithm that can be used for both classification and regression challenges. SVM is one of the most standard classification tools and does not need to rely on all the data, i.e., only a part of the support vector is used to make hyperplane decisions. Thus, we can predict all patterns by learning the preferences of a few patterns. Xin, D. described a framework of discovering interesting patterns and used a ranking support vector formulation [4]. However, this system does not user the SVM model to make predictions. The system we designed used the SVM model to predict the user-preferred patterns.

IDMBS is a system to discover interesting patterns based on [4] and [5]. The system designs a filtering algorithm based on the similarities between different patterns and an SVM prediction model. Given a set of co-location patterns, our goal is to help users find interesting patterns in an interactive way based on their real interests. In the whole process, we do not require users to clearly express the pattern they are really interested in. IDMBS filters a small group of patterns as the training set per iteration, and the user decides whether each pattern in the training set is interesting or not, thus the pressure of user selection is reduced, and the user's decision-making process is facilitated. Users can use interactive mining methods to mine patterns of interest according to their own preferences, without relying on too many objective criteria.

2 System Overview

Figure 1 shows the description of IDMBS. IDMBS requires a set of prevalent co-locations as the input. IDMBS contains an interactive process to obtain the user's remarks on the filtered samples, and contains an SVM model to train the marked samples.

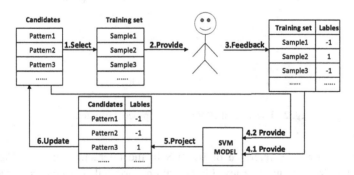

Fig. 1. Framework description.

Step 1: A small collection of co-location patterns is filtered from the input as the training set. The specific screening strategy is as follows: First, the filtering algorithm selects a pattern with the maximum PI value into the training set (when the training set is empty); second, each feature of the co-location patterns is united in the training set; third, the similarity between input co-location patterns and the union set generated from the previous step are calculated and it is denoted S (If the system is not interacting with the user for the first time, S = similarity + distance; distance: the pattern output by the SVM model to the hyperplane); finally, IDMBS chooses the co-location pattern corresponding to the minimum S and puts it into the training set.

Step 2: IDMBS visually presents the patterns in the training set to the user.

Step 3: The user judges whether he/she is interested in the samples, and makes a label on the corresponding patterns.

Step 4: IDMBS takes the training set and candidates as the input of the SVM model.

Step 5: The SVM model predicts candidates and outputs the distance from each pattern to the hyperplane.

Step 6: IDMBS will go back to step 1 to continue the interaction (step 1–step 5). In general, IDMBS interactions are limited to 10 times or terminated by the user at any time.

3   Demonstration Scenarios

IDMBS is well encapsulated with a friendly interface and users only need to interact through a simple interface. In this demonstration, we use spatial POI dataset of urban elements in Beijing to show the demonstration.

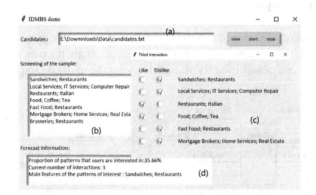

Fig. 2. Interface of IDMBS in the process of interaction.

Figure 2 shows the interface of IDMBS in the interaction process. Figure 2(a) shows that IDMBS requires a text file containing original prevalent co-location patterns. Figure 2(b) shows the training set that IDMBS filters from the text file. Given the co-location patterns, the interaction process can be performed once pressing the *start* button. If the user wants to interrupt the current process, the *stop* button can do it. Figure 2(c) shows the interaction interface which requires the user to judge whether he/she is interested in the pushed patterns. Figure 2(d) shows the analysis information including the proportion of the user's interest patterns, the current number of interactions, and the main features of interest to users.

We simulated the preference of a user and performed multiple interactions with IDMBS. The accuracy rate obtained is shown in Fig. 3. F-3 is the accuracy rate of interactive probabilistic post-mining method [3] under the same conditions. The accuracy rate of IDMBS is higher than that of F-3 under the same number of interactions. Thus, the prediction result shows that IDMBS is effective in predicting according to the preferences of different users.

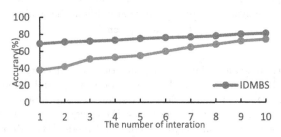

Fig. 3. Forecast result.

4   Conclusion

In many data mining approaches, the algorithm is pre-designed, and users passively accept the results. In this demonstration, we designed a system to find interesting patterns by interacting with users and solved the problem that the patterns under the same semantics was predicted to have the same result. The demonstration scenarios showed that the system can effectively discover the user-preferred patterns.

Acknowledgements. This work was supported in part by grants (No. U1811264, No. U1711263, No. 61966009, No. 62006057, 61762027) from the National Natural Science Foundation of China, in part by grants (No. 2018GXNSFDA281045, No. 2019GXNSFBA245059) from the Natural Science Foundation of Guangxi Province, and in parts by grants (No. AD19245011) from the Key Research and Development Program of Guangxi Province.

References

1. Huang, Y., Shekhar, S., Xiong, H.: Discovering co-location patterns from spatial data sets: a general approach. IEEE Trans. Knowl. Data Eng **16**(12), 1472–1485 (2004)
2. Bao, X., Gu, T., Chang, L., et al.: Knowledge-based interactive postmining of user-preferred co-location patterns using ontologies. IEEE Trans. Cybern. (2021)
3. Wang, L., Bao, X., Cao, L.: Interactive probabilistic post-mining of user-preferred spatial co-location patterns. In: 2018 IEEE 34th International Conference on Data Engineering (2018)
4. Xin, D., Shen, X., Mei, Q., et al.: Discovering interesting patterns through user's interactive feedback. In: Proceedings of the 12th ACM SIGKDD International Conference on Knowledge Discovery and Data Mining, pp. 773–778 (2006)
5. Yang, K., Gao, Y., Liang, L., et al.: Towards factorized SVM with gaussian kernels over normalized data. In: 2020 IEEE 36th International Conference on Data Engineering (2020)

SeTS³: A Secure Trajectory Similarity Search System

Yiping Teng[(✉)], Fanyou Zhao, Jiayv Liu, Mengfan Zhang, Jihang Duan, and Zhan Shi

School of Computer, Large-scale Distributed System Laboratory,
Shenyang Aerospace University, Shenyang, China
typ@sau.edu.cn

Abstract. To achieve the privacy protection on trajectory processing in outsourced environments, in this paper, we propose and demonstrate a Secure Trajectory Similarity Search System (SeTS³) to process the similarity search on encrypted trajectories. We design our system with four layers following the non-colluded dual-cloud model. Based on our previous work, three secure trajectory similarity search algorithms, which support different similarity metrics, are implemented. To assist the secure computations on encrypted trajectories, we develop a cryptographic toolkit in SeTS³ based on Paillier cryptosystem. To achieve a better demonstration, we apply map APIs to visualize the trajectories, and the search performance of the implemented algorithms can be displayed.

Keywords: Encrypted trajectories · Trajectory similarity · Secure search

1 Introduction

In recent years, techniques and applications on trajectory processing and analysis have been studied and developed in both academic and industrial communities following the development of the mobile devices and mobile Internet. To offload the massive costs of data management, data owners are motivated to outsource the trajectory search processing to the public cloud. However, direct outsourcing these search services may cause serious privacy issues. Moreover, trajectory data may also be considered as confidential data in business, of which the accessing is prohibited from any unauthorized parties. Therefore, to process the trajectory similarity search with privacy protection and data confidentiality, it calls for a secure similarity search system on encrypted trajectories.

In this paper, we propose and demonstrate a Secure Trajectory Similarity Search System (SeTS³) to process the similarity search on encrypted trajectories,

---

The video of this paper can be found in https://youtu.be/3f2Xjq8sGhc.
The code of this paper can be found in https://github.com/zfy1412/zzsource.

© The Author(s), under exclusive license to Springer Nature Switzerland AG 2022
A. Bhattacharya et al. (Eds.): DASFAA 2022, LNCS 13247, pp. 522–526, 2022.
https://doi.org/10.1007/978-3-031-00129-1_48

which mainly supports the secure search processing with three different trajectory similarity metrics under the assistance of our implemented cryptographic toolkit. In the system implementation, following the non-colluded dual-cloud model, we design four layers in SeTS^3, i.e., Data Layer, Service Layer, Communication Layer and User Layer. Data Layer is designed to take charge of the storage and access of the encrypted trajectory datasets. We develop two modules in Service Layer, of which Module I executes secure trajectory computation algorithms, and Module II consists of the cryptographic tools based on Paillier cryptosystem. Communication Layer is designed for transmitting parameters and results between Service Layer and User Layer, and User Layer is to process the user parameters and trajectory visualization by map APIs to achieve a better demonstration. The performance of the search algorithms can also be displayed.

2 System Architecture

2.1 System Model

SeTS^3 follows the cloud service framework widely applied in several important studies [1,2], which consists of three types of entities: authorized user (U), data owner (DO) and cloud servers (C_1 and C_2). DO generates and distributes the secret keys, and then encrypts and uploads the datasets to C_1. U encrypts and sends the search request to C_1 to issue a secure search, and receives the encrypted search results partially from C_1 and C_2. C_1 executes the secure search processing with C_2, and sends the encrypted results to the corresponding U. To achieve the security guarantees, C_1 and C_2 are assumed to be non-colluded. Due to space limitation, the details of the service model of SeTS^3 can be referred to [1–3].

2.2 System Design

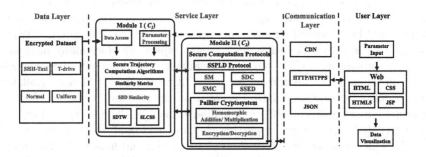

Fig. 1. The system design.

As shown in Fig. 1, based on the system model, we design 4 layers in SeTS^3 including Data Layer, Service Layer, Communication Layer and User Layer.

Data Layer. We design the Data Layer in charge of the storage of the different datasets of encrypted trajectories.

Service Layer. Service Layer consists two modules, of which Module I is designed for the secure trajectory computations of C_1 and Module II is designed for the cryptographic services of C_2. In this layer, Module I accesses the encrypted datasets from Data Layer, and processes the user parameters from Communication Layer. With them as the input, several secure trajectory computation algorithms supporting different similarity metrics are implemented in Module I. In Module II, we design a cryptographic toolkit for the secure computations of Module I including secure computation protocols based on Paillier cryptosystem.

Communication Layer. Communication Layer is designed to transmit data between Service Layer and Data Layer, which includes parameters and results.

User Layer. We design the web interfaces in User Layer, where the search parameters can be set and search results can be visualized through browsers.

3   System Implementation

We implement User Layer through HTML, CSS and JavaScript. To better demonstrate the search results of real trajectories on the map, we apply the *Bing Maps* APIs (https://www.microsoft.com/en-us/maps/). To achieve two isolated cloud servers, we implement Module I and Module II of Service Layer in two separated virtual machines deployed in a workstation, so that they can cooperate with each other through the inter-domain communication between the virtual machines. Communication Layer adopts the open-sourced *Gin* framework to transfer information with User Layer through JSON package.

Implementation of Secure Trajectory Computation Algorithms. To achieve the secure computation on trajectories, in Module I of Service Layer, we implement three secure trajectory computation algorithms, of which two (denoted as SDTW and SLCSS [2]) support Dynamic Time Warping (DTW) and Longest Common Sub Sequence (LCSS) [4], respectively. Originally introduced for signal processing, DTW allows time-shifting in comparison of trajectories. Based on the longest common subsequence, LCSS is robust to noise by allowing skipping of some sample points. Recently, as the previous work [3] of SeTS[3], we propose a signature-based secure similarity search scheme (denoted as SBD) over encrypted trajectories supporting the Bi-Directional (BD) similarity measure [5], which can capture more shape information and handle the trajectories sampled in different sampling rates. We implement SBD scheme in SeTS[3] and compare the performance to the other two algorithms.

Implementation of Secure Basic Protocols. In Module II of Service Layer, we implement the cryptographic toolkit based on Paillier cryptosystem. In the toolkit, we first develop the encryption and decryption of Paillier cryptosystem, and then implement several basic secure protocols [1,2] including Secure Multiplication (SM), Secure Division Computation (SDC), Secure Minimum Computation (SMC) and Secure Squared Euclidean Distance (SSED). On the strength of the basic protocols, we further develop Secure Squared Point to Line-segment Distance (SSPLD) [3], a essential secure component of SBD scheme.

Implementation Environment. We implement all the algorithms in Python 3.8 and perform experiments on a Tower Server with two 40-core Intel(R) Xeon(R) Bronze 3204 1.90 GHz CPUs and 256 GB RAM running Ubuntu 20.04.

4   Demonstration

We demonstrate the secure trajectory similarity search on different real and synthetic datasets in SeTS³. We provide four different datasets to perform the search processing. Two of them are real trajectory datasets (SHH-Taxi [6] and T-drive [7]) and the other two are randomly generated following the *Normal* distribution ($N(0, 0.002)$) and *Uniform* distribution ($U(-0.002, 0.002)$) with the random walk. After login the system, the input parameters can be selected on the menu bar including the similarity metrics (i.e., SDTW, SLCSS and SBD), the datasets and the input trajectory as the search request. Click *Run* button to issue the similarity search processing, and the search results, the similar trajectories w.r.t the input trajectory, will be visualized on the map as shown in Fig. 2. After the search processing, the performance comparison of the implemented algorithms can be illustrated in terms of the search processing time by varying different variables in SeTS³ as shown in Fig. 3.

Fig. 2. Trajectory visualization

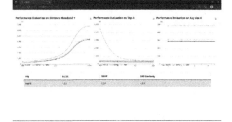

Fig. 3. Search results and performance

5   Conclusion

In this paper, we propose and demonstrate a secure trajectory similarity search system. Three secure trajectory similarity search algorithms are implemented supporting different similarity metrics. As the auxiliary of the secure computations, we develop a set of cryptographic tools based on Paillier cryptosystem. To achieve a better demonstration, we apply *Bing Maps* APIs to show the trajectories, and the performance comparison of the search algorithms can be shown.

Acknowledgement. The work is supported by National Natural Science Foundation of China (61902260), Scientific Research Project of Education Department of Liaoning Province (JYT2020026) and College Students' Innovative Entrepreneurial Training Project of Shenyang Aerospace University (Z202110143136).

References

1. Cui, N., Yang, X., Wang, B., Li, J., et al.: SVkNN: efficient secure and verifiable k-nearest neighbor query on the cloud platform. In: ICDE, pp. 253–264. IEEE (2020)
2. Liu, A., Zheng, K., Li, L., Liu, G., Zhao, L., Zhou, X.: Efficient secure similarity computation on encrypted trajectory data. In: ICDE, pp. 66–77. IEEE (2015)
3. Teng, Y., Shi, Z., Zhao, F., Ding, G., Xu, L., Fan, C.: Signature-based secure trajectory similarity search. In: IEEE TrustCom (2021, to appear)
4. Toohey, K., Duckham, M.: Trajectory similarity measures. ACM Sigspatial Spec. **7**(1), 43–50 (2015)
5. Ta, N., Li, G., Xie, Y., Li, C., Hao, S., Feng, J.: Signature-based trajectory similarity join. IEEE Trans. Knowl. Data Eng. **29**(4), 870–883 (2017)
6. Ni, L.M., Chen, L., et al.: SHH-Taxi data (2007). https://www.cse.ust.hk/scrg/
7. Zheng, Y.: T-drive trajectory data (2011). https://www.microsoft.com/en-us/research/publication/t-drive-trajectory-data-sample/

Data-Based Insights for the Masses: Scaling Natural Language Querying to Middleware Data

Kausik Lakkaraju[1](✉), Vinamra Palaiya[2], Sai Teja Paladi[1],
Chinmayi Appajigowda[2], Biplav Srivastava[1], and Lokesh Johri[2]

[1] University of South Carolina,Columbia, USA
kausik@emil.sc.edu
[2] Tantiv4, Milpitas, USA

Abstract. In this demonstration, we focus on middleware data obtained from devices like the network routers and power meters which may be of interest to a technician fixing a customer complaint or a user trying to self-diagnose their utility usage. The users in our case are often unaware of both the data details and database querying language which is in contrast to typical natural language to structured query (NL2SQL) situations where the business analyst knows their domain data but not the querying techniques. We adapt the rule-based NL2SQL approach to our problem and in particular, focus on queries about users, devices and spatio-temporal properties that are unique to this setting. We demonstrate an Alexa-based system, implemented using open-source Rasa, that can answer router usage queries in a home setting and easily extended for power usage or other utilities.

Keywords: Natural language query · Chatbots · Middleware

1 Introduction

Customer service is important for any business to retain its customers but is also costly to provide. As a result, businesses often look towards decision-support technology to improve customer satisfaction by improving the speed and accuracy of information they provide. The new technology can ideally be used by the customer to self-diagnose and fix the problem themselves, and if they cannot, a technician sent by the company has sufficient information to fix it onsite.

We are interested in information provided for middleware and utility services offered in residential or commercial spaces like internet connectivity, power and water. As an example, a smart home may have a router and the user is interested to query about network usage by different users and apps at that residence. The user may also have a smart electric meter and wants to know about power usage or water meter for corresponding information.

© The Author(s), under exclusive license to Springer Nature Switzerland AG 2022
A. Bhattacharya et al. (Eds.): DASFAA 2022, LNCS 13247, pp. 527–531, 2022.
https://doi.org/10.1007/978-3-031-00129-1_49

For improving access to the data, Natural Language Querying (NLQ) has emerged as a promising direction where a user's query is translated to a database language like structured query language(SQL). Common approaches for this problem, NL2SQL, are based on rules [3,4] and learning [2,5]. In the former, rules which are created are data independent (i.e., based just on SQL), domain dependent [3] or data dependent. In the latter, labeled examples of the translations are needed to train deep-learning architectures that learn the mapping and try to generalize across data sources [2].

In this paper, we focus on a rule-based approach. Specifically, we parse user's NLQ and translate them using (a) data-independent rules; (b) middleware specific rules which are aware of the concepts of users, devices, space (locations) and time; and (c) data source specific rules. We show the generality of our approach by implementing it for querying data generated by a commercial network router and by a commercial power meter. Our system, implemented as a chatbot using the open-source Rasa platform, can be invoked via embedded devices like Amazon's Echo Show. In the demonstration, a user will be able to query Echo Show's Alexa interface to pose query about a building's networking or power usage. The system will answer them using a common set of rules (data-independent or middleware) except when data-dependent fields are involved.

Our contributions are: (a) a robust NLQ tool that can be used by non-technical people to gain results from structured databases with middleware data like routers and power meters, (b) a general set of middleware rules for translating natural language queries involving users, devices, locations (space) and time to SQL, (c) demonstration using a commercial platform (Amazon's Alexa) and live data from an instrumented building.

2   Methodology

We now describe the data used, the implemented system and its query generation approach.

2.1   Data

The data collected and maintained in this study consists of two sources. IoT power sensors and networking (router) domains. **IoT Power Sensor data:** Here, an electronic sensor is placed in the path of the power source for each device connected to IoT, the sensor captures data such as phase based current, voltage, angle, power factor, apparent power, reactive power, active power, frequency, etc. Ten data points are clubbed together into a JSON packet and uploaded to database to minimize frequent database access. We have collected this data from about 100 servers which are being utilized at an University. **Router Data:** All of the Internet using devices in this study are connected to a router that stores and pushes network activity data to the database. Network activity data consists of Device ID, client ID, host name, downloaded and uploaded data size in bytes, SCID, etc. We have collected this data from a residence.

From the above meta data, various metrics such as user time per application, most used application, maximum download speed per user, maximum power consumption per device, time of maximum power consumption, etc. can be computed to inform the user of various power and network usage within this ecosystem.

2.2   System Overview

The proposed system architecture is shown in Fig. 1. Key components of our system are: 1. **Dialogue System (B1):** The dialogue system is the conversational agent with a dialogue pipeline which consists of different components for understanding human interaction and responding to it. The components that are used in the dialogue system pipeline are provided by the RASA framework [1]. The SQL Query Classifier present in the Intent Classifier is customized by us to recognize the user intent for which the system has to perform an SQL operation to get the relevant data from the database. 2. **Orchestrator (B2):** The orchestrator is responsible for giving the control to the right component depending on the user query. 3. **Action Generator (B3):** The action generator generates the appropriate action based on the user query. For example, if the user greets the conversational agent, it will greet back. This action does not require access to the database. If the intent is a query which requires the system to access the database, then the action generator will convert the user query to a MySQL query which would be used to get the relevant data from the database. 4. **Databases (B4):** We have used two databases, corresponding to the respective power and networking domains, to build our rule-based system as described in the previous section. The system can be easily expanded to different kinds of data. 5. **Executor (B5):** The executor is responsible for executing the final action based on the user query.

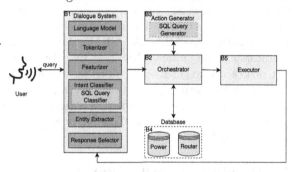

Fig. 1. System architecture

Table 1. Types of queries supported and illustrative examples.

Query type	Networking	Power
Data independent	What kind of data do you have?	
Single column	How much time did I spend on Youtube last week?	At what time power consumption was the highest last week?
Multiple columns	For how many applications was the download data more than 1000 bytes and upload data is less than 500 bytes?	For how many times the voltage value has crossed below 100 vrms and power factor has dropped below 6
Aggregation	What was my most used application last week?	At what time in the last week my power consumption was the least?
User aware	How many apps are connected?	How many devices are connected?
Temporal	How many days did my usage go above 100 mbps?	How many days did my power go above 100?

2.3 Rule Mapping

The user query is tokenized to produce the word tokens. The next step is to recognize entities and attributes from these tokens. In the entity extraction procedure, the numerical values, if present, will be extracted from the tokens along with different operators. The operators can be comparison operators like '¿' or '¡', aggregation operators like 'SUM', 'MIN', 'MAX', or 'COUNT', conditional operators can be either 'AND' or 'OR'. For attribute selection, we used token sort ratio score to get a score based on which we matched the tokens from user query with the actual attribute names from the tables present in the database. Once the entities and attributes are extracted, a MySQL query would be formulated combining all these. This query will be used to fetch the relevant data from the database.

3   System Demonstration

Table 1 shows the various types of queries supported by the system along with illustrations. It is easy to expand the query types and data sources seamlessly.

The demonstration will have a user start by asking about data-independent queries like knowing about the number of features in the data, and then asking about queries related to a domain and about some features. They can also request for aggregation of results and ask about multiple columns.

4   Conclusion

In this work, we presented a general, rule-based, system to convert natural language queries to SQL (NL2SQL) with a focus on middleware data collected by

IoT devices for networking and power usage. Such a capability allows people without database skills to benefit from their own data stored in the databases. Our approach not only helps any middleware vendor to easily offer NL2SQL benefits when they do not have training data in a domain but also the results so generated can serve as training data for advanced deep-learning based, but data-hungry, NL2SQL approaches for the future.

References

1. Bocklisch, T., Faulkner, J., Pawlowski, N., Nichol, A.: Rasa: open source language understanding and dialogue management (2017). https://arxiv.org/abs/1712.05181
2. Gan, Y., et al.: Natural SQL: making SQL easier to infer from natural language specifications. In: EMNLP Findings Preprint https://arxiv.org/abs/2109.05153 (2021)
3. Saha, D., Floratou, A., Sankaranarayanan, K., Minhas, U.F., Mittal, A.R., Özcan, F.: Athena: an ontology-driven system for natural language querying over relational data stores. In: Proceedings of the VLDB (2016)
4. Sheinin, V., Khorashani, E., Yeo, H., Xu, K., Vo, N.P.A., Popescu, O.: QUEST: a natural language interface to relational databases. In: Proceedings of the LREC, May 2018
5. Yu, T., et al.: Spider: a large-scale human-labeled dataset for complex and cross-domain semantic parsing and text-to-SQL task. In: Proceedings of the EMNLP (2018)

Identifying Relevant Sentences for Travel Blogs from Wikipedia Articles

Arnav Kapoor[1] and Manish Gupta[1,2(✉)]

[1] IIIT-Hyderabad, Hyderabad, India
`arnav.kapoor@research.iiit.ac.in`
[2] Microsoft India, Hyderabad, India
`gmanish@microsoft.com`

Abstract. Travel blogs are a rich source of information about tourist spots. Millions of travellers use them to decide where to go and what to do. However, they are often devoid of factual information, which leads to a lack of trust and credibility surrounding blogs. On the other hand, Wikipedia is rich with accurate factual descriptions but is often bland and unfit for blogs. We propose a system to identify blog-worthy sentences on Wikipedia to augment travel blogs with factual information to make them more reliable and trustworthy. We curate a dataset of over 1.83 M Wikipedia sentences from 234,305 travel-related articles and assign a blog-worthiness score to each sentence by comparing them against 20 M sentences from ∼600 K blogs. Our best BERT based model provides an F1 score of 0.74 to identify if a sentence is blog-worthy or not.

1 Introduction

Travel blogs are rich sources of information about tourist places. Millions of travellers use them to choose travel destinations and plan their activities. Travel blogs capture stories by real travellers and express the nuances and emotional aspects of a place, in an informal style. Often times, travel bloggers manually borrow relevant factual information from various sources like Wikipedia to supplement their story. Factual information from Wikipedia can help to make travel blogs more reliable and trustworthy while maintaining the informal and emotional component. Although Wikipedia is rich with factual descriptions as it is well moderated, it is often bland, and hence not all sentences are suitable for a blog. In this study, we propose a system to identify blog-worthy English sentences from Wikipedia travel articles automatically. Such a system would serve two main purposes. (1) It will allow travel bloggers to quickly identify relevant sentences from Wikipedia articles that can augment their blogs. (2) Wikipedia can provide a "watered-down" summary of blog-worthy sentences making it more approachable and valuable for travellers.

A. Bhattacharya et al. (Eds.): DASFAA 2022, LNCS 13247, pp. 532–536, 2022.
https://doi.org/10.1007/978-3-031-00129-1_50

A few popular travel bloggers already harness Wikipedia resources to enhance their blogs. We identify such Wikipedia sentences in travel blogs and create an auto-labeled corpus,WIKIBLOGS, which has a blog-worthiness score for each Wikipedia sentence. We use this dataset to train a binary classifier to identify whether a sentence is blog-worthy. Our contributions through this study are as follows: (1) We create a large corpus of Wikipedia sentences in the travel domain with their corresponding blog-worthiness scores. We make this dataset, WIKI-BLOGS, publicly available[1]. (2) We propose an interesting novel classification task to identify if the sentence is blog-worthy or not and demonstrate such a system using Wikipedia articles.

2 Related Work

The tourism industry is one of the largest in the world. Much work in this domain has focused on automatic itinerary generation [1,8]. Chang et al. [1] consider five important factors, user preference, popularity, cost, distance and time to plan an itinerary. Gurjar et al. [8] identify inclusion and exclusion phrases from reviews. However, an important aspect to consider is the trust travellers associate with such systems since they are built around user-generated content. Ayeh et al. [3] look at the perception of travellers towards user-generated content. Filieri et al. [6] look at how multiple aspects of source credibility, information quality, website quality, customer satisfaction, user experience impact credibility. Our proposed system is a way to enhance information quality and thus provide more credibility to user-generated content, specifically travel blogs.

3 Our WIKIBLOGS Dataset

We collect data from two main sources, Wikipedia and Travel blogs[2]. We use infobox type to identify travel-related articles on Wikipedia. We curate a list of 18 different infobox categories: NRHP, museum, historic site, religious building, park, ancient site, UNESCO world heritage site, UK place, building, theatre, Australian place, venue, river, a body of water, church, protected area, zoo, and amusement park. The number of Wikipedia articles in the top five categories is listed in Table 1.

---

[1] https://tinyurl.com/wiki2blogs.
[2] www.travelblog.org.

The categories are domain-specific (Theatre, Amusement Park) or location-specific (Australian Place, UK places). We crawled the travel blog website to retrieve all the links in multiple geographical locations, namely Africa, Antarctica, Asia, Central-America-Caribbean, Europe, MiddleEast, North-America, Oceania, Oceans and Seas and South-America. We collected a total of 599,981 blogs from these locations. The top 5 locations are listed in Table 2. Diverse set of infobox categories and locations were picked to ensure our system generalizes well.

To filter out relevant sentences, we only keep those containing named entities from both corpora (Wikipedia articles and travel blogs). We then set out to create correspondence from a Wikipedia article to a blog. Every Wikipedia article represents a unique location. However, multiple blogs could talk about the same location. We thus form a one to many correspondence from Wikipedia articles to blogs as follows. We use Okapi BM25 to get a set of k_i matching blogs with score greater than a threshold τ for each Wikipedia article i. We cap the maximum number of matching blogs for each Wikipedia article at 100. Thus, $0 \leq k_i \leq 100$. We fix τ to 10 for our experiments based on manual assessment of relevance on a small set of articles.

After a document level correspondence, we establish a sentence level alignment to identify the best set of blog sentences for each Wikipedia article sentence. We compute similarity between each Wikipedia article sentence s_{ij} from article i with each sentence from the matching k_i blogs. For each sentence, we use GloVe embeddings [10] to get word representations and combine them to get the sentence level embedding using SIF (Smooth Inverse Frequency) [2]. We use cosine similarity to compute the matching score. We then select one best scoring sentence from each of the matching k_i blogs, and the mean of these k_i cosine similarity scores gives us the final blog-worthiness score of a sentence s_{ij} in the Wikipedia article i. Overall, our WIKIBLOGS dataset has 1,825,897 Wikipedia article sentences and their corresponding blog-worthiness scores.

4   Blog Worthiness Score Prediction

We illustrate the distribution of blog-worthiness scores in Fig. 1 and observe an approximate normal distribution, $\mathcal{N}(\mu = 0.23, \sigma = 0.10)$. We plot the sorted blog-worthiness scores and note in Fig. 2 that over 1.5 m sentences have scored in a very small range between 0.1 and 0.4. To ensure a very objective task definition and a clean labeled data, we propose a classification task to differentiate between the extremes, i.e., high blog-worthy sentences and unworthy sentences. We filter the top 100K and bottom 100K sentences based on blog-worthiness scores for our classification task. Thus, both the classes are equally balanced, and a trivial classifier would have an F1 score of 50%.

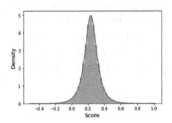

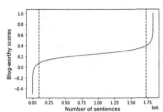

Fig. 1. Normal distribution of blog worthiness scores with μ = 0.23, σ = 0.10. A large proportion of sentences have scores between 0.1 and 0.4.

Fig. 2. Cumulative distribution of blog-worthiness scores. The two vertical dotted lines identify the top 100 K and bottom 100 K sentences based on blog-worthiness scores.

Table 1. Top categories of Wikipedia articles.

Category	# articles
NRHP	63086
UK Place	24602
Building	21291
Theatre	14791
Australian place	14657

Table 2. Top locations of travel blogs.

Location	# blogs
Asia	160660
Europe	139937
North America	95180
Oceania	77910
South America	56941

We used classical machine learning (ML) models like SVM and logistic regression (LR) with simple sentence embedding features created using doc2vec [9]. We also tried XGBoost [4] with doc2vec. The results across models were almost identical, and hence we next experimented with other embeddings like Sentence-BERT [11] and the vanilla BERT (Bidirectional Encoder Representations from Transformers) model [5]. We fine-tune the pre-trained BERT model, Adam optimizer with learning rate = 1e−5, and batch size = 8, with cross-entropy loss.

The results for all our experiments are captured in Table 3. SentenceBERT and doc2vec lead to F1 hovering ∼0.70. BERT outperforms the other deep learning models resulting into an F1 score of 0.74. We believe this is because BERT has been pre-trained on the travel-specific Wikipedia data, thus extracting better representation from the underlying WikiBlogs corpus. We also illustrate BERT F1 scores for some of the categories in Table 4.

Table 3. F1 scores

Model	F1
doc2Vec + LR	0.70
doc2Vec + SVM	0.70
doc2Vec + XGBoost	0.69
SentenceBERT + XGBoost	0.68
BERT	**0.74**

Table 4. Category wise F1 scores for BERT model.

	Categories	F1
Top scoring	Protected area	0.85
	UNESCO world heritage sites	0.78
	Museums	0.77
Bottom scoring	UK place	0.72
	Body of water	0.70
	Park	0.69

5 Conclusion

We proposed a novel task of identifying blog-worthy sentences from Wikipedia articles. We create a unique pipeline to build a large corpus, WikiBlogs, of

~1.8M Wikipedia sentences and their corresponding blog-worthiness scores. We propose the problem of classifying wikipedia sentences as blog-worthy or not. We test out various machine learning and deep learning methods to solve this challenging task and achieve a best F1 score of ~0.74. In the future, we plan to experiment with smaller models or compress it using popular distillation methods [7].

References

1. ATIPS: Automatic Travel Itinerary Planning System for Domestic Areas 2016, 1–13 (2016)
2. Arora, S., Liang, Y., Ma, T.: A simple but tough-to-beat baseline for sentence embeddings. In: ICLR (2017)
3. Ayeh, J.K., Au, N., Law, R.: "Do we believe in TripAdvisor?" examining credibility perceptions and online travelers' attitude toward using user-generated content. J. Travel Res. **52**(4), 437–452 (2013)
4. Chen, T., Guestrin, C.: Xgboost: a scalable tree boosting system. In: KDD, pp. 785–794 (2016)
5. Devlin, J., Chang, M., Lee, K., Toutanova, K.: BERT: pre-training of deep bidirectional transformers for language understanding. In: NAACL, pp. 4171–4186 (2019)
6. Filieri, R., Alguezaui, S., McLeay, F.: Why do travelers trust TripAdvisor? antecedents of trust towards consumer-generated media and its influence on recommendation adoption and word of mouth. Tour. Manage. **51**, 174–185 (2015)
7. Gupta, M., Agrawal, P.: Compression of deep learning models for text: a survey. ACM Trans. Knowl. Disc. Data (TKDD) **16**(4), 1–55 (2022)
8. Gurjar, O., Gupta, M.: Should i visit this place? inclusion and exclusion phrase mining from reviews. In: Hiemstra, D., Moens, M.-F., Mothe, J., Perego, R., Potthast, M., Sebastiani, F. (eds.) ECIR 2021. LNCS, vol. 12657, pp. 287–294. Springer, Cham (2021). https://doi.org/10.1007/978-3-030-72240-1_27
9. Le, Q., Mikolov, T.: Distributed representations of sentences and documents. In: ICML, vol. 32, pp. 1188–1196 (2014)
10. Pennington, J., Socher, R., Manning, C.D.: Glove: global vectors for word representation. In: EMNLP, pp. 1532–1543 (2014)
11. Reimers, N., Gurevych, I.: Sentence-BERT: sentence embeddings using Siamese BERT-networks. In: EMNLP-IJCNLP, pp. 3982–3992 (2019)

PhD Constorium

Neuro-Symbolic XAI: Application to Drug Repurposing for Rare Diseases

Martin Drancé[(⊠)]

Inserm 1219, Bordeaux Population Health Research Center, Team ERIAS,
University of Bordeaux, Bordeaux, France
martin.drance@u-bordeaux.fr

Abstract. An emerging challenge in high stakes fields such as health-care is to build a more transparent artificial intelligence. Among the fields which use AI techniques, is drug development, and more specifically drug repurposing. DR involves finding a new indication for an existing drug. The hypotheses generated by DR techniques must be validated. There-fore, the mechanism of generation must be understood. In this project, we focus on DR using link prediction algorithms. Link prediction consists of generating hypotheses about the relationships between a known molecule and a given target. More specifically, this PhD project aims at creating methods allowing to make transparent prediction in the context of drug repurposing but also to understand how the organization of data in a knowledge graph changes the quality of predictions.

Keywords: XAI · Drug repurposing · Knowledge graph

1 Research Problem

Financial constraints are the main limiting factors for the pharmaceutical indus-try, which is primarily concerned with cost-effectiveness. In particular, they do not invest in the development of new drugs for so-called rare diseases. Over the last decade, drug repurposing (DR), which consists of searching for a new indi-cation for an existing drug, has been attracting growing interest. Advances in Artificial Intelligence (AI) methods have made it possible to develop AI-based computational DR with very encouraging results.

However, the use of *black-box* models, which are completely opaque to the understanding of their users, is currently one of the main obstacles to the use of AI in the health field. In this context, a double challenge emerge around the use of innovative methods for drug repositioning. Firstly, the use of AI must be rethought in order to develop tools that are transparent, trustworthy and capable of providing an explanation of how they operate. Also, there is a need to think about these tools in the context of rare diseases, where data and its small quantity represent an additional constraint to take into account. This constraint must be analyzed from an AI point of view but also of data processing. The new

© The Author(s), under exclusive license to Springer Nature Switzerland AG 2022
A. Bhattacharya et al. (Eds.): DASFAA 2022, LNCS 13247, pp. 539–543, 2022.
https://doi.org/10.1007/978-3-031-00129-1_51

solutions considered in that field focus exclusively on the improvement of the underlying algorithms, starting from the erroneous assumption that real world data are unlimited and perfectly organized, which is not the case in reality.

2   Current Development and Related Work

Knowledge Graph. With the emergence of KGs, various AI methods for performing graph-based prediction and classification tasks have been proposed recently. These data structures are well suited for describing biomedical information, as they are usually derived from multiple databases and KGs are known to be suited to maintaining the semantic relationship between entities. The linking of data is done through triples, linking two entities (*head*) and (*tail*) by a relation, in the form (*head, relation, tail*). From this data structure came the task of *link prediction*, where an AI algorithm must be trained to answer a question of the form (*head, relation,* ?).

Drug Repurposing. Using a Bayesian network, Madhukar [8] proposed an approach to repurpose drugs based on a bipartite graph linking drugs and target. Computing similarity scores between pairs of drugs with shared targets was effective for predicting new targets for small orphan molecules. In 2017 Hetionet [5], a bio-medical knowledge graph, was developed from 29 different databases. From it, an automated learning approach was applied to predict new drug-protein relationships. To improve the interpretability of their predictions, a graph-oriented database using Neo4j[1] was implemented. This database enables to perform a post hoc visual verification of the veracity of the prediction by observing, for example, the nodes located in the surroundings of the repurposed drugs. This study allowed to propose two lists of drugs that can be repurposed in the context of epilepsy and nicotine addiction.

Explainable Artificial Intelligence (XAI). Link prediction problem was first solved using logical rules [6]. These purely symbolic methods are limited by their limited capacity of generalization. To overtake this problem, embedding-based methods [1] were proposed. As they showed better performances, they also lack interpretability. Aiming at using both strengths, link prediction methods based on multi-hop reasoning [3] are now widely used, using reinforcement learning coupled with logical rules. Today, one of the neuro-symbolic algorithm reaching state of the art results used on a bio-medical KG is PoLo (*policy-guided walks with logical rules*) [7]. Multi-hop reasoning methods have the advantage of proposing explicit reasoning paths during link prediction.

---

[1]   https://neo4j.com/.

3 Methodology

Neuro-Symbolic Algorithm for Link Prediction. Right now, the results given by these methods are very numerous and the explanations can be quite hard to synthesize. In order to build an algorithm adapted to the data on rare diseases, a more precise model should be designed, providing more targeted results around these diseases. Also, the interaction between the model and the clinicians should be facilitated, in particular by proposing to them to define a priori the known biological mechanisms on which the algorithm will be able to rely to make its predictions. Concerning the changes in the model, several tracks must be studied in order to select the best current statistical and symbolic tools that will constitute the final neuro-symbolic model. For example: (i) how to improve the reward mechanism in the case of spurious paths; (ii) how to improve the agent's search in the graph with a priori knowledge; (iii) how to improve prediction in the case of long-range dependency;

Implementation in the Context of Rare Diseases. As with any machine learning task, it is important to consider the format of the input data. Most of the research conducted today attempts to improve the results of links prediction by improving the model. However, in areas of AI that have grown in recent years, there has been an evolution of algorithms along with data processing methods. The evaluation will be carried out using the KG of the OREGANO framework [2]. The impact of these modifications in the data structure will be tested on our neuro-symbolic model, but also on the principal models currently used in the field. For example: (i) How node centrality concepts play a role when using multi-hop reasoning; (ii) How the quantity of data/training example affect the results. Thus, it will be possible to validate that these recommendations concerning the organization of the data have an impact on a larger scale than this project.

3.1 Validation and Exploitation of Results

In a first step, the developed model will be evaluated and compared to already existing models, thanks to the metrics already established in the field (MRR and Hits@K). This comparison will aim at evaluating the relevance of developing a very specialized model dedicated to the drug repositioning task. In the same way, the impact of data optimization will be measured thanks to the same prediction evaluation tools. A validation step based on a verification of the results from scientific publications will be tested to validate or invalidate results automatically. Indeed, an important amount of information can be found in the scientific abstracts available online, which can be analyzed thanks to natural language processing tools [9]. Finally, these results will be communicated and made available to rare disease specialists. This exchange will allow to validate or not the efficiency of the developed method, to propose new research tracks or to confirm existing ones.

4 Current Results and Future Work

Currently, first results have already been obtained from the use of PoLo on the KG of the OREGANO [4]. These results are 50% higher than the results previously obtained from these data following a more traditional approach (embedding-based). In these results, were found 682 drugs identified as candidates for repurposing and 447 diseases for which a drug was proposed for repurposing. The explanatory mechanism proposed by PoLo has been validated by the clinical genetics unit of the Bordeaux University Hospital. These results support the use of models using logical rules as an explanation mechanism. Concerning the problematic surrounding the organization of data within the knowledge graph, first results have shown that: (i) increasing the quantity of training data rather has a negative impact on the quality of the prediction; (ii) it is impossible to predict if the algorithm will give good results, even knowing exactly how data are structured in the KG.

The main hypothesis for this project is that neuro-symbolic AI approaches will make models more transparent and suitable for the healthcare field while showing better results. Thanks to a reinforced interaction with clinicians and the ability of these approaches to give good results from limited data, the two major issues surrounding drug repurposing for rare diseases will be addressed: clinicians must be able to trust the tools used and a limited amount of data on rare diseases is available.

References

1. Bordes, A., et al.: Translating embeddings for modeling multi-relational data. Adv. Neural Inf. Process. Syst. 26 (2013)
2. Boudin, M.: Computational approaches for drug repositioning: towards a holistic perspective based on knowledge graphs. In: Proceedings of the 29th ACM International Conference on Information and Knowledge Management, pp. 3225–3228 (2020). https://doi.org/10.1145/3340531.3418510
3. Das, R., et al.: Go for a walk and arrive at the answer: reasoning over paths in knowledge bases using reinforcement learning. In: International Conference on Learning Representations, April 2018
4. Drancé, M., Boudin, M., Mougin, F., Diallo, G.: Neuro-symbolic XAI for computational drug repurposing. In: Proceedings of the 13th International Joint Conference on Knowledge Discovery, Knowledge Engineering and Knowledge Management - Volume 2: KEOD, pp. 220–225. INSTICC, SciTePress (2021). https://doi.org/10.5220/0010714100003064
5. Himmelstein, D.S., et al.: Systematic integration of biomedical knowledge prioritizes drugs for repurposing. eLife **6**, e26726 (2017). https://doi.org/10.7554/eLife.26726
6. Landwehr, N., Kersting, K., De Raedt, L.: nfoil: Integrating naïve Bayes and Foil, vol. 8, pp. 795–800 (2005)
7. Liu, Y., Hildebrandt, M., Joblin, M., Ringsquandl, M., Raissouni, R., Tresp, V.: Neural multi-hop reasoning with logical rules on biomedical knowledge graphs. In: Verborgh, R., et al. (eds.) ESWC 2021. LNCS, vol. 12731, pp. 375–391. Springer, Cham (2021). https://doi.org/10.1007/978-3-030-77385-4_22

8. Madhukar, N.S., et al.: A Bayesian machine learning approach for drug target identification using diverse data types. Nature Commun. **10**(1), 5221 (2019). https://doi.org/10.1038/s41467-019-12928-6

9. Rindflesch, T.C., Fiszman, M.: The interaction of domain knowledge and linguistic structure in natural language processing: interpreting hypernymic propositions in biomedical text. J. Biomed. Inform. **36**(6), 462–477 (2003). https://doi.org/10.1016/j.jbi.2003.11.003

Leveraging Non-negative Matrix Factorization for Document Summarization

Alka Khurana$^{(\boxtimes)}$ (iD)

Department of Computer Science, University of Delhi, Delhi 110007, India
akhurana@cs.du.ac.in

Abstract. This paper outlines the doctoral research work carried out to develop unsupervised approaches for extractive single document summarization using Non-negative Matrix Factorization (NMF). NMF is a popular topic modeling technique, which divulges inter-relation between three prime semantic units of the document. We exploit the inter-relationship among the three semantic units, viz. *terms*, *sentences*, and *latent topics*, in a novel manner to extract informative sentences from the salient topics in the document. The three methods developed during this doctoral study are language-, domain- and collection-independent.

Keywords: Extractive summarization · Non-negative matrix factorization · Entropy · Semantic similarity · Language independence

1 Problem and Motivation

Automatic Document Summarization (ADS) aims to shorten the text without compromising its essence. Additionally, it reduces readers' time and cognitive effort to comprehend the information in the document. Recent advances in ADS confirm its popularity and appositeness among the research community. The current revolution in neural methods and resulting masked language models in NLP have remarkably enhanced the performance of document summarization methods. However, the advantage comes with the ancillary cost of preparation of annotated data and the time required to train the models. The trained models inherit the training data characteristics and unintentionally acquire domain- and collection-dependence. Further, these models lack transparency and interpretability. The overheads associated with neural models galvanized the research work presented here. The approaches presented in this work are unsupervised, language-, domain- and collection-independent. Moreover, the methods are explainable and fast enough to summarize documents in real-time.

Supervised by Vasudha Bhatnagar.

© The Author(s), under exclusive license to Springer Nature Switzerland AG 2022
A. Bhattacharya et al. (Eds.): DASFAA 2022, LNCS 13247, pp. 544–548, 2022.
https://doi.org/10.1007/978-3-031-00129-1_52

2   Related Work

The earlier work by Lee et al. [4], who applied NMF for extractive single document summarization, provided the original impetus to our study. Peyrard postulated the idea of employing Shannon's Entropy to capture *informativeness* required for document summarization, which is another stimulant for our research work [5]. We briefly describe NMF and Entropy in the present context.
(i) **Non-negative Matrix Factorization:** NMF is a matrix decomposition technique to obtain reduced rank approximation in lower dimensional space. NMF holds the promise to uncover the latent semantic space of the document. Consider a document D represented as a sequence of n sentences $(S_1, S_2, \ldots, S_n)$ and consisting of m terms $(t_1, t_2, \ldots, t_m)$. Let A be $m \times n$ term-sentence matrix for D, where an element a_{ij} in A denotes the occurrence of term t_i in sentence S_j. NMF decomposes A into W and H (i.e. $A \approx WH$), where W is $m \times r$ term-topic (feature) matrix, H is $r \times n$ topic-sentence (co-efficient) matrix and r is the number of latent topics $(\tau_1, \tau_2, \ldots, \tau_r)$ in the document. Starting with initial seed values for factors, W and H, NMF iteratively improves these matrices such that Frobenius norm, $\| A - WH \|_F^2$ is minimized. Element w_{ij} in W quantifies the strength of term t_i in latent topic τ_j and element h_{ij} in H denotes contribution of sentence S_j in latent topic τ_i.

Thus, Non-negative Matrix Factorization of *term-sentence* matrix A into *term-topic* feature matrix W and *topic-sentence* co-efficient matrix H, reveals the inter-relationship between the three semantics units, viz. terms, sentences, and latent topics (Fig. 1). Non-negativity constraints on NMF factor matrices enhance the interpretability of semantic units in the latent space. Random initialization of NMF factor matrices is the major caveat, which results in different W, H pairs (base models). We experimented with ensemble methods by combining the base models to create a consensus summary and found that ensembles often perform worse than the best base model [2].

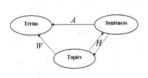

Fig. 1. Relationship among semantic units revealed by NMF decomposition. A:term-sentence, W:term-topic, H:topic-sentence matrix

(ii) **Entropy of Semantic Units:** *Informativeness* is the key attribute for capturing the essence of a document while generating the summary [5]. Inspired by the idea, we delve into the latent semantic space of the document revealed by NMF and mutate the semantic units into their corresponding probabilistic form. We provide different interpretations of topic and sentence entropy, and investigate their interplay.

3   Summarization Methods

This ongoing doctoral research proposes three novel approaches that excavate inter-relationship among the semantic units - *terms*, *sentences*, and *topics* in the document. The first approach *NMF-TR* is term-oriented, which leverages the information carried by the terms in the latent semantic space exposed in the NMF feature matrix. Next is *NMF-TP*, a topic-oriented approach that exploits the information inferred by the latent topics uncovered by NMF [1]. The third approach is *E-Summ*, which follows an information-theoretic approach for selecting the informative sentences [3]. The three approaches score the sentences differently by using distinct inter-relationship between semantic units. Top scoring sentences are included in the summary.

(i) **NMF-TR (Term-Oriented Sentence Scoring):** Short documents are characterized by the small amount of information, which is insufficient to generate demarcated latent topics. Accordingly, considering the terms as the prime carriers of information in the document, we propose an algorithm *NMF-TR*, which is term-oriented sentence scoring method. *NMF-TR* considers that a term with high relative contribution is more important and that the importance of a sentence is an additive combination of importance of terms in the sentence.

$$Score(S_q) = \sum_{i=1}^{m} a_{iq}\phi_i, \text{ where } \phi_i = \frac{\sum_{q=1}^{r} w_{iq}}{\sum_{p=1}^{m}\sum_{q=1}^{r} w_{pq}} \tag{1}$$

That is, the sentence consisting of terms with higher contribution in latent space is preferred for inclusion in the summary. *NMF-TR* scores a sentence in the document according to Eq. 1.

(ii) **NMF-TP (Topic-oriented Sentence Scoring):** Long documents contain adequate information and concepts to yield clearly separated latent topics, and that the summary should reflect these latent topics proportionate to their importance in the document. Grounded on this idea, we propose *NMF-TP*, which is a topic-oriented method that explicitly considers the importance of latent topics in the document to extract salient sentences for the summary.

$$Score(S_q) = \sum_{i=1}^{r} \omega_i h_{iq}, \text{ where } \omega_i = \frac{\sum_{q=1}^{m} w_{qi}}{\sum_{p=1}^{m}\sum_{q=1}^{r} w_{pq}} \tag{2}$$

Since sentences are carriers of the information reflected by the latent topics, therefore, *NMF-TP* scores a sentence in the document using the formula in Eq. 2.

(iii) **E-Summ:** *E-Summ* algorithm is grounded on the principle of selecting informative sentences from the document, which convey important topics. *E-Summ* follows an information-theoretic approach, exploits the information contained in topic and sentence entropies, and identifies candidates by scoring a sentence as follows.

$$Score(S_j) = \zeta^\tau(S_j) + \Psi^S(\tau_i), \text{ where}$$

$\zeta^{\tau}(S_j)$ is sentence entropy in topic space and $\Psi^S(\tau_i)$ is topic entropy in sentence space (Refer [3] for details). Subsequently, it uses Knapsack optimization algorithm to maximize the information conveyed by sentences selected for inclusion in the summary. Since *E-Summ* works on the premise of maximizing information contained in summary and does not explicitly consider the terms and topics as in the case of *NMF-TR* and *NMF-TP*, it is reasonable to employ *E-Summ* to summarize short and long documents.

4 Data-Sets and Evaluation

We use well known public data-sets - DUC2001, DUC2002, CNN and DailyMail for performance evaluation of the proposed algorithms. The proposed methods are language-, domain- and collection-independent. The methods are faster than state-of-the-art extractive summarization algorithms (Refer [1,3] for details) and are suitable for real time summarization of web documents.

We evaluate the performance of all the three algorithms viz. *NMF-TR*, *NMF-TP*, and *E-Summ* on DUC2001, DUC2002, and CNN/DailyMail data-sets. Additionally, we employ WikiHow data-set consisting of general Wikipedia articles and CL-SciSumm data-sets containing scientific articles, for justifying the domain independence of *E-Summ* algorithm. We demonstrate the language independence of *E-Summ* using two Indian and three European languages.

We use standard ROUGE measure for qualitative assessment of the algorithmic summaries. ROUGE is a measure of computing lexical overlap between algorithmic summaries and gold standard ground truth summaries. We additionally employ semantic similarity measure to evaluate the quality of summaries.

5 Conclusion and Future Directions

Growing demand of automatic summarization methods in diverse domains, genres and for resource poor languages necessitate development of unsupervised, language independent and fast methods. We use Non-negative Matrix Factorization to tease out the inter-relationship between terms, sentences and latent topics in the latent semantic space of the document and proposed three single document summarization algorithms. As expected, the performance of the proposed algorithms does not match with that of deep neural methods.

Existing research majorly focuses on generating generic summaries of documents. However, generic summaries ignore the subjectivity aspect as these summaries do not consider the background knowledge of the user. Our current focus is on automatically generating personalized summary of the document.

References

1. Khurana, A., Bhatnagar, V.: Extractive document summarization using non-negative matrix factorization. In: Hartmann, S., Küng, J., Chakravarthy, S., Anderst-Kotsis, G., Tjoa, A.M., Khalil, I. (eds.) DEXA 2019. LNCS, vol. 11707, pp. 76–90. Springer, Cham (2019). https://doi.org/10.1007/978-3-030-27618-8_6
2. Khurana, A., Bhatnagar, V.: NMF ensembles? not for text summarization! In: Proceedings of the First Workshop on Insights from Negative Results in NLP, pp. 88–93 (2020)
3. Khurana, A., Bhatnagar, V.: Investigating entropy for extractive document summarization. Expert Syst. Appl. **187**, 115820 (2021)
4. Lee, J.H., Park, S., Ahn, C.M., Kim, D.: Automatic generic document summarization based on non-negative matrix factorization. Inf. Process. Manage. **45**(1), 20–34 (2009)
5. Peyrard, M.: A simple theoretical model of importance for summarization. In: Proceedings of the 57th Annual Meeting of the ACL (2019)

Author Index

Agrawal, Puneet III-413
Ai, Zhengyang II-306
Amagata, Daichi I-224
Ao, Xiang I-353, I-387, II-166
Appajigowda, Chinmayi III-527
Araújo, André III-500
Au, Man Ho I-404

Bai, Chaoyu III-272
Bai, Luyi II-391
Bai, Ting II-102, II-423
Ban, Qimin II-85
Bao, Qiaoben III-238
Bao, Xuguang III-514, III-518
Bao, Yinan I-615
Bao, Yuhao III-509
Bi, Jingping I-722
Bi, Sheng I-162
Bi, Wenyuan I-96
Bian, Shuqing I-38
Blackley, Suzanne V. II-673

Cai, Desheng II-574
Cai, Haoran III-430
Cai, Xunliang II-298
Cao, Caleb Chen I-648
Cao, Shulin I-107
Cao, Yiming II-407, III-117
Cao, Zhi II-489
Carvalho, Arthur III-500
Chang, Liang II-248, II-281, III-281,
    III-514, III-518
Chao, Pingfu I-191
Chatterjee, Ankush III-413
Chelaramani, Sahil III-413
Chen, Cen III-306, III-455
Chen, Guihai I-552, II-3, II-615, II-706
Chen, Jiajun II-216
Chen, Jiangjie III-197
Chen, Lei I-648
Chen, Lu III-495
Chen, Qi III-331
Chen, Siyuan II-264
Chen, Tongbing II-590

Chen, Weitong II-289
Chen, Xiang II-556
Chen, Xin II-166, III-377, III-430
Chen, Xingshu III-133
Chen, Yan II-681
Chen, Yijiang II-375
Chen, Yueguo III-331
Chen, Yunwen III-36, III-197, III-238,
    III-340
Chen, Yuting I-341
Chen, Yuxing I-21
Chen, Zhigang I-137, III-149
Chen, Zihao I-309
Chen, Zongyi III-230
Cheng, Bing II-298
Cheng, Dawei III-306, III-455
Cheng, Reynold III-443
Cheng, Yunlong II-706
Chennupati, Saideep I-569
Chhabra, Vipul I-569
Chi, Jianfeng I-353, I-387
Ching, Waiki III-443
Chu, Xiaokai I-722
Chu, Yuqi II-574
Couto, Henrique III-500
Cui, Chuan II-150
Cui, Hang III-52
Cui, Lizhen II-315, II-407, III-117

Damani, Sonam III-413
Dao, Minh-Son I-569
Das, Souripriya I-21
Deng, Sinuo III-222
Deng, Sucheng II-556
Ding, Tianyu III-401
Ding, Yihua II-590
Dong, Linfeng I-387
Dong, Qiwen II-689
Dong, Xiangjun I-459
Dou, Wenzhou III-52
Draheim, Dirk I-596
Drancé, Martin III-539
Du, Liang II-681
Du, Ming III-289

Du, Wei   I-370
Du, Yingpeng   II-19
Duan, Jihang   III-522
Duan, Lei   II-681, III-165
Duan, Zhewen   II-656
Duan, Zhijian   III-389

Fan, Jiangke   II-298
Fan, Ju   I-587
Fan, Wei   I-604
Fan, Xinxin   I-722
Fan, Yu   II-472
Fan, Zhenfeng   II-332
Fan, Zhuoya   I-404
Fang, Chuangxin   III-505
Fang, Junhua   I-191, I-207
Fang, Ruiyu   II-199, III-351
Fang, Shineng   III-197
Feng, Jinghua   I-353, I-387
Feng, Luping   II-118
Feng, Shi   II-256, III-255
Folha, Rodrigo   III-500
Fu, Bin   II-697

Gao, Hanning   II-150
Gao, Peng   III-468
Gao, Shan   II-298
Gao, Xiaofeng   I-552, II-3, II-615, II-706
Gao, Yixu   III-389
Gao, Yuanning   II-615
Gao, Yunjun   III-495
Gao, Zihao   II-281
Goda, Kazuo   I-88
Gong, Zheng   III-213
Gong, Zhiguo   II-556
Grubenmann, Tobias   III-443
Gu, Hansu   II-359
Gu, Ning   I-333, II-359
Gu, Tianlong   III-514, III-518
Gu, Yu   I-325, I-731, I-739
Gudmundsson, Joachim   I-241
Gui, Min   III-297
Gui, Xiangyu   I-122
Guo, Deke   I-441
Guo, Jiayan   I-682
Guo, Shu   II-306
Guo, Tonglei   III-364
Guo, Zhiqiang   II-183
Gupta, Manish   III-413, III-532

Han, Baokun   I-309
Han, Ding   II-523
Han, Donghong   III-255
Han, Tianshuo   III-401
Hao, Fei   I-714
Hao, Jianye   III-389
Hao, Zhifeng   II-556
Hara, Takahiro   I-224
He, Liang   II-85, II-118
He, Ming   III-377, III-401
He, Qing   I-387, II-166
He, Xiaodong   III-425
He, Yi   II-323
He, Yihong   II-455
He, Ying   II-574
He, Zhenying   I-72, I-96, I-476
Ho, Shen-Shyang   I-509
Holub, Jan   I-509
Hong, Yu   III-340
Hou U, Leong   III-68
Hou, Lei   I-107
Hou, Yupeng   I-38
Hu, Huiqi   I-293
Hu, Jun   II-574
Hu, Maodi   II-199, III-351
Hu, Nan   I-162
Hu, Songlin   I-615, II-623
Hu, Wenjin   III-222
Hu, Wenxin   II-85
Hu, Xinlei   III-377
Hu, Xuegang   II-574
Hua, Yuncheng   I-162
Huang, Chen   III-238
Huang, Faliang   II-343
Huang, Hao   III-468
Huang, Junyang   III-238
Huang, Linpeng   I-341
Huang, Qiang   I-232, I-268
Huang, Wei   III-213
Huang, Weiming   II-407
Huang, Xiuqi   II-706
Huang, Yanlong   II-102
Huang, Yanyong   II-656

Ji, Yu   II-118
Jia, Siyu   II-306
Jia, Xueqi   I-714
Jian, Yifei   III-133
Jiang, Qi   I-413
Jiang, Rui   I-333

Jiang, Sihang   I-578
Jiang, Weipeng   III-165
Jiang, Xiaoqi   I-459
Jiang, Xueyao   I-180, I-578
Jiang, Youjia   III-505
Jiao, Pengfei   III-322
Jin, Beihong   I-268
Jin, Cheng   III-314
Jin, Hai   I-122, I-153, I-250
Jin, Peiquan   I-560
Jin, Taiwei   III-364
Jing, Yinan   I-72, I-96, I-476
Johri, Lokesh   III-527
Joshi, Meghana   III-413

Kankanhalli, Mohan   I-232
Kao, Ben   III-443
Kapoor, Arnav   III-532
Kaushik, Minakshi   I-596
Khurana, Alka   III-544
Kim, Hong-Gee   I-632
Kim, Junghoon   I-543
Kiran, R. Uday   I-569
Kitsuregawa, Masaru   I-88
Kou, Yue   III-52
Krčál, Luboš   I-509

Lakkaraju, Kausik   III-527
Lan, Michael   I-55
Lei, Yifan   I-232
Li, Ailisi   I-180, III-36
Li, Anchen   I-171, II-134
Li, Aoran   II-289
Li, Bohan   II-289
Li, Changshu   III-377
Li, Chuanwen   I-325, I-731
Li, Dong   III-389
Li, Dongsheng   II-359, III-3
Li, Fangfang   I-739
Li, Guohui   II-183
Li, Haihong   III-314
Li, Huichao   I-587
Li, Huilin   II-523
Li, Jianjun   II-183
Li, Jiaoyang   I-493
Li, Jingze   II-590
Li, Juanzi   I-107
Li, Kun   II-623
Li, Mengxue   III-364
Li, Peng   II-656

Li, Pengfei   I-191
Li, Renhao   III-165
Li, Ruixuan   II-623
Li, Shangyang   I-682
Li, Shuai   II-298
Li, Shuaimin   III-263
Li, Tao   II-199, III-351
Li, Wei   I-665
Li, Xiang   I-268
Li, Xiaohua   I-731
Li, Xiaoyang   II-689
Li, Xin   I-268
Li, Xinyu   I-404
Li, Xiongfei   III-247
Li, Yexin   II-656
Li, Yingying   I-413
Li, Yongkang   II-298
Li, You   III-505
Li, Yue   I-259
Li, Yunchun   I-665
Li, Zhan   III-481
Li, Zhao   I-459
Li, Zhen   III-117
Li, Zhi   II-183
Li, Zhisong   III-101
Li, Zhixin   II-248, II-281
Li, Zhixu   I-137, I-180, I-578, III-85, III-149, III-297
Li, Zonghang   II-455
Liang, Jiaqing   I-180, III-36, III-238
Liang, Yile   II-69
Liang, Yuqi   III-306
Lim, Sungsu   I-543
Lin, Hui   II-85
Lin, Jianghua   III-425
Lin, Junfa   II-264
Lin, Leyu   II-166
Lin, Longlong   I-250
Lin, Meng   II-623
Lin, Yuming   III-505
Lin, Ziyi   I-341
Liu, An   I-137, I-191, I-207, III-85, III-149
Liu, Baichuan   I-425
Liu, Bang   I-180, III-238
Liu, Chang   I-268
Liu, Chengfei   II-472, III-181
Liu, Dajiang   I-714
Liu, Gaocong   I-560
Liu, Haobing   II-53
Liu, Hongtao   III-322

Liu, Hongzhi   II-19, II-697
Liu, Huaijun   II-199
Liu, Jiayv   III-522
Liu, Kuan   II-53
Liu, Lixin   I-404
Liu, Meng   I-604
Liu, Ning   II-407, III-117
Liu, Qi   III-213
Liu, Qingmin   II-3
Liu, Rongke   II-391
Liu, Wei   III-468
Liu, Xiaokai   I-122
Liu, Ximeng   I-413
Liu, Xing   II-631
Liu, Xinyi   III-181
Liu, Yang   I-387
Liu, Yi   II-289
Liu, Yong   I-604, II-232, II-315
Liu, Yudong   III-230
Liu, Zhen Hua   I-21
Liu, Zhidan   II-375
Liwen, Zheng   III-230
Long, Lianjie   II-343
Lu, Aidong   I-370
Lu, Haozhen   I-552
Lu, Jiaheng   I-21
Lu, Jianyun   II-639
Lu, Tun   I-333, II-359
Lu, Xuantao   III-238
Lu, Yanxiong   I-632
Luo, Siqiang   III-455
Luo, Wang   III-468
Luo, Yifeng   III-306, III-455
Luo, Yikai   II-489
Luo, Yongping   I-560
Lv, Fuyu   III-364
Lv, Ge   I-648
Lv, Junwei   II-574
Lv, Xin   I-107

Ma, Denghao   III-331
Ma, Guojie   I-259
Ma, Huifang   II-248, II-281, III-281
Ma, Jianfeng   I-413
Ma, Ling   II-298
Ma, Rui   I-353
Ma, Weihua   II-289
Ma, Xinyu   II-631
Ma, Xuan   II-69
Ma, Yunpu   III-101

Meng, Liu   I-3
Meng, Xiaofeng   I-404
Miao, Chunyan   II-232
Miao, Hang   II-134
Miao, Xiaoye   III-495
Miao, Yinbin   I-413

Narahari, Kedhar Nath   III-413
Ng, Wilfred   III-364
Ni, Jiazhi   I-268
Nie, Tiezheng   III-20, III-52
Ning, Bo   III-181
Niyato, Dusit   II-455

Obradovic, Zoran   II-689
Ouyang, Kai   I-632

Paladi, Sai Teja   III-527
Palaiya, Vinamra   III-527
Pan, Bing   III-481
Pan, Xingyu   I-38
Pang, Jinhui   I-698
Pang, Yitong   II-150
Pavlovski, Martin   II-506, II-689
Pei, Hongbin   III-331
Peious, Sijo Arakkal   I-596
Peng, Qiyao   III-322
Peng, Yuchen   III-495
Pfeifer, John   I-241

Qi, Dekang   II-656
Qi, Guilin   I-162
Qi, Xuecheng   I-293
Qian, Shiyou   I-277
Qian, Tieyun   II-69
Qian, Weining   II-506, II-689, III-306,
    III-455
Qiao, Fan   II-664
Qiao, Zhi   I-665
Qin, Shijun   III-430
Qin, Zhili   II-639
Qu, Jianfeng   I-137, III-85, III-149

Reddy, P. Krishna   I-569
Ren, Ziyao   II-606

Seybold, Martin P.   I-241
Sha, Chaofeng   II-36, III-289
Shahin, Mahtab   I-596
Shang, Jiaxing   I-714
Shang, Lin   II-216
Shang, Mingsheng   II-323

Shao, Junming   II-639
Shao, Kun   III-389
Sharma, Rahul   I-596
Shen, Derong   III-20, III-52
Shen, Fang   III-481
Shen, Qi   II-150
Shen, Shirong   I-162
Shen, Xinyao   III-197
Shen, Zhiqi   II-315
Shi, Bing   II-489
Shi, Chuan   II-199, III-351
Shi, Dan   I-171
Shi, Ge   III-222
Shi, Jiaxin   I-107
Shi, Liye   II-118
Shi, Shengmin   II-590
Shi, Wanghua   I-277
Shi, Xiangyu   III-509
Shi, Yuchen   I-425
Shi, Zhan   III-522
Skoutas, Dimitrios   I-55
Song, Ailun   I-552
Song, Hui   III-289
Song, Kaisong   II-256
Song, Shuangyong   III-425
Song, Weiping   II-298
Song, Xintong   III-509
Song, Xiuting   II-391
Song, Yang   I-38, II-697
Song, Yiping   III-3
Song, Yumeng   I-731, I-739
Song, Zhen   I-739
Srivastava, Biplav   III-527
Su, Fenglong   I-698
Sun, Bo   II-323
Sun, Chenchen   III-20, III-52
Sun, Chuanhou   I-459
Sun, Fei   III-364
Sun, Ke   II-69
Sun, Tao   II-439
Sun, Weiwei   II-590
Sun, Wenya   III-443
Sun, Xigang   II-272
Sun, Yueheng   III-322

Takata, Mika   I-88
Tang, Chunlei   II-673
Tang, Daniel   II-523
Tang, Haihong   II-232
Tang, Jintao   III-3

Tang, Moming   III-306
Tang, Zhihao   III-230
Taniguchi, Ryosuke   I-224
Tao, Hanqing   III-213
Teng, Yiping   III-522
Theodoratos, Dimitri   I-55
Tian, Junfeng   III-297
Tian, Yu   II-590
Tian, Zhiliang   III-3
Times, Valéria   III-500
Tong, Shiwei   III-213
Tong, Yongxin   II-606
Tung, Anthony   I-232

Uotila, Valter   I-21

Van, Minh-Hao   I-370
Viana, Flaviano   III-500
Vinay, M. S.   I-395

Wang, Bin   I-3, III-389
Wang, Binjie   II-664
Wang, Can   II-216
Wang, Changyu   III-331
Wang, Chaoyang   II-183
Wang, Chunnan   III-509
Wang, Chunyang   II-53
Wang, Daling   II-256, III-255
Wang, Ding   I-615
Wang, Dong   II-199, III-351
Wang, Fangye   II-359
Wang, Fei   II-439
Wang, Guoxin   II-232
Wang, Haizhou   III-133
Wang, Hongya   III-289
Wang, Hongzhi   III-509
Wang, Jiaan   III-85, III-149
Wang, Jiahai   II-264
Wang, Jialong   III-481
Wang, Jie   II-272
Wang, Jingyu   I-341
Wang, Jiwen   III-377
Wang, Kai   II-639
Wang, Ke   II-53
Wang, Lei   II-681
Wang, Liping   I-259
Wang, Long   III-514
Wang, Meng   I-162, II-631
Wang, Peng   II-540, II-664
Wang, Pengsen   I-268

Wang, Qi   II-673
Wang, Qiang   III-314
Wang, Sen   II-631
Wang, Sheng   II-298
Wang, Shi   II-523
Wang, Shupeng   II-306
Wang, Wei   II-540, II-664, III-340
Wang, Weiping   I-493
Wang, Wenjun   III-322
Wang, Wentao   III-281
Wang, X Sean   I-72
Wang, X. Sean   I-96, I-476
Wang, Xiaofan   I-526
Wang, Xin   III-481
Wang, Xinpeng   III-101
Wang, Xuwu   I-578, III-297
Wang, Yang   I-325
Wang, Yansheng   II-606
Wang, Yifan   II-298
Wang, Yike   II-248
Wang, Yitong   I-747
Wang, Youchen   I-268
Wang, Yu   III-247
Wang, Yueyi   I-425
Wangyang, Qiming   II-639
Wei, Di   I-739
Wei, Lingwei   I-615
Wei, Xing   I-293
Wei, Xingshen   III-468
Wei, Yunhe   II-248
Wei, Zhihua   II-150
Wei, Zhongyu   III-389
Wen, Ji-Rong   I-38
Wen, Zhihua   III-3
Wu, Bin   II-102, II-423
Wu, Di   II-323
Wu, Han   III-213
Wu, Lifang   III-222
Wu, Lin   II-439
Wu, Longcan   II-256
Wu, Siyuan   III-68
Wu, Wei   I-180
Wu, Wen   II-85, II-118
Wu, Xiaoying   I-55
Wu, Xintao   I-370, I-395
Wu, Yangyang   III-495
Wu, Yiqing   II-166
Wu, Zhen   II-272
Wu, Zhenghao   III-3
Wu, Zhonghai   II-19, II-697

Xia, Tianyu   I-476
Xia, Xiufeng   I-3
Xiang, Ye   III-222
Xiao, Fu   II-648
Xiao, Ning   I-268
Xiao, Shan   III-165
Xiao, Yanghua   I-180, I-578, III-36, III-197,
    III-238, III-297, III-340
Xiao, Yiyong   II-590
Xie, Guicai   III-165
Xie, Rui   I-180
Xie, Ruobing   II-166
Xie, Yi   III-314
Xing, Lehao   III-222
Xing, Zhen   II-375
Xiong, Yun   III-314
Xu, Bo   III-289, III-509
Xu, Chen   I-309
Xu, Feifei   III-101
Xu, Haoran   II-656
Xu, Hongyan   III-322
Xu, Jiajie   I-207, II-472
Xu, Jungang   III-263
Xu, Ke   II-606
Xu, Minyang   II-375
Xu, Ruyao   III-455
Xu, Siyong   II-199
Xu, Tiantian   I-459
Xu, Xianghong   I-632
Xu, Yonghui   II-315, II-407, III-117
Xu, Yongjun   II-439
Xu, Yuan   I-207
Xu, Zenglin   II-455
Xu, Zheng   I-333

Yadav, Amrendra Singh   I-596
Yan, Cheng   I-153
Yan, Ming   III-297
Yang, Bin   I-747
Yang, Bo   I-171, II-134
Yang, Cheng   II-199
Yang, Deqing   I-425
Yang, Fanyi   III-281
Yang, Geping   II-556
Yang, Han   I-698
Yang, Hao   I-353, I-387
Yang, Qinli   II-639
Yang, Shiyu   I-259
Yang, Tianchi   II-199, III-351
Yang, Weiyong   III-468

Yang, Xiaochun   I-3
Yang, Xiaoyu   III-230
Yang, Yifan   II-631
Yang, Yiyang   II-556
Yang, Yonghua   II-315
Yao, Di   I-722
Ye, Jiabo   III-297
Yi, Xiuwen   II-656
Yin, Dawei   I-722
Yin, Hongzhi   II-216
Yin, Jianwei   III-495
Yin, Yunfei   II-343
Yu, Changlong   III-364
Yu, Fuqiang   II-407, III-117
Yu, Ge   I-731, I-739, II-256, III-52
Yu, Han   II-455
Yu, Hongfang   II-455
Yu, Philip S.   III-314
Yu, Runlong   III-213
Yu, Xiaoguang   III-425
Yu, Yonghong   II-216
Yuan, Chunyuan   I-615
Yuan, Lin   I-137, III-149
Yuan, Pingpeng   I-250
Yuan, Shuhan   I-395
Yue, Lin   II-289
Yue, Yinliang   I-493
Yun, Hang   II-69

Zang, Tianzi   II-53
Zang, Yalei   II-289
Zeng, Guanxiong   I-353
Zeng, Li   III-430
Zeng, Lingze   I-587
Zeng, Shenglai   II-455
Zettsu, Koji   I-569
Zhai, Yitao   III-331
Zhang, Aoran   II-216
Zhang, Bolei   II-648
Zhang, Cong   II-423
Zhang, Fusang   I-268
Zhang, Han   II-391
Zhang, Hanbing   I-72, I-96, I-476
Zhang, Heng   III-222
Zhang, Jiale   I-552
Zhang, Jianing   I-587
Zhang, Jiujing   I-259
Zhang, Junbo   II-656
Zhang, Kai   I-72, I-96, I-476
Zhang, Lei   III-117

Zhang, Leilei   I-698
Zhang, Li   II-216
Zhang, Lingzi   II-232
Zhang, Luchen   II-523
Zhang, Luhao   II-199, III-351
Zhang, Meihui   I-587
Zhang, Mengfan   III-522
Zhang, Mi   II-69
Zhang, Ming   II-298
Zhang, Mingming   II-540
Zhang, Nevin L.   III-3
Zhang, Peng   I-333, II-359
Zhang, Qianzhen   I-441
Zhang, Ruisheng   II-606
Zhang, Tao   I-38, II-697
Zhang, Tingyi   I-137, III-149
Zhang, Weiyu   I-587
Zhang, Wenkai   III-101
Zhang, Xi   III-230
Zhang, Xianren   I-714
Zhang, Xiaohui   II-281
Zhang, Xiaoli   III-247
Zhang, Xin   I-268
Zhang, Xingyu   II-36
Zhang, Xu   II-166
Zhang, Yan   I-682
Zhang, Yifei   I-250, II-256, III-255
Zhang, Yiming   II-150
Zhang, Yixin   II-315
Zhang, Yuxiang   III-518
Zhang, Zhao   II-439
Zhang, Zhengqi   III-289
Zhang, Zhiqing   I-714
Zhang, Ziwei   III-481
Zhao, Deji   III-181
Zhao, Fanyou   III-522
Zhao, Feng   I-122, I-153
Zhao, Hang   I-72
Zhao, Hui   II-697
Zhao, Jiashu   I-722
Zhao, Lei   I-137, I-191, I-207, III-85, III-149
Zhao, Long   I-459
Zhao, Mengchen   III-389
Zhao, Pengpeng   I-191, III-85
Zhao, Rongqian   III-430
Zhao, Wayne Xin   I-38
Zhao, Weibin   II-216
Zhao, Wendy   III-481
Zhao, Xiang   I-441
Zhao, Yan   III-281

Zhao, Yue   I-682
Zhao, Yuhai   I-459
Zheng, Bo   III-509
Zheng, Gang   III-331
Zheng, Hai-Tao   I-632
Zheng, Shengan   I-341
Zheng, Wei   II-85
Zheng, Yefeng   II-631
Zheng, Yin   I-632
Zheng, Yu   II-656
Zhou, Aoying   I-293, I-309
Zhou, Fang   II-506, II-689
Zhou, Haolin   II-3
Zhou, Jinhua   III-430
Zhou, Jinya   II-272
Zhou, Rui   II-472
Zhou, Shanlin   III-101
Zhou, Wei   I-615
Zhou, Xiangdong   II-375
Zhou, Xin   II-232

Zhou, Zimu   II-606
Zhu, Rui   I-3, III-247
Zhu, Shishun   III-468
Zhu, Shixuan   II-150
Zhu, Wenwu   III-481
Zhu, Xian   I-615
Zhu, Yangyong   III-314
Zhu, Yanmin   II-53
Zhu, Yao   II-697
Zhu, Ying   III-255
Zhu, Yongchun   II-166
Zhu, Zhihua   I-722
Zhuang, Fuzhen   II-166
Zong, Weixian   II-506
Zong, Xiaoning   II-315
Zou, Beiqi   III-85
Zou, Bo   III-425
Zou, Chengming   II-332
Zou, Lixin   I-722

Printed in the United States
by Baker & Taylor Publisher Services